THE
PHOTOGRAPHIC ACTION
OF
IONIZING RADIATIONS

WILEY SERIES ON PHOTOGRAPHIC SCIENCE AND TECHNOLOGY AND THE GRAPHIC ARTS

WALTER CLARK, EDITOR
Kodak Research Laboratories

PHYSICAL AND PHOTOGRAPHIC PRINCIPLES OF MEDICAL
RADIOGRAPHY, *Herman E. Seemann*

OPTICAL AND PHOTOGRAPHIC RECONNAISSANCE SYSTEMS
Niels Jensen

ELECTRO-OPTICAL PHOTOGRAPHY AT LOW
ILLUMINATION LEVELS
Harold V. Soule

AERIAL DISCOVERY MANUAL
Carl H. Strandberg

PRINCIPLES OF COLOR REPRODUCTION
J. A. C. Yule

TELEVISION FILM ENGINEERING
Rodger J. Ross

LIGHT-SENSITIVE SYSTEMS
Chemistry and Application of Nonsilver Halide Photographic Processes
Jaromir Kosar

THE PHOTOGRAPHIC ACTION OF IONIZING RADIATION
R. H. Herz

In Preparation:

PHOTOMICROGRAPHY, Vols. 1 and 2, *Roger P. Loveland*

IN-WATER PHOTOGRAPHY, *L. E. Mertens*

PROPERTIES OF THE PHOTOGRAPHIC IMAGE, *Joseph H. Altman*

PHOTOGRAPHY IN THE VISIBLE AND NEAR VISIBLE
SPECTRUM, *Allan G. Millikan*

THE
PHOTOGRAPHIC ACTION
OF
IONIZING RADIATIONS

in Dosimetry and Medical, Industrial, Neutron,
Auto- and Microradiography

R. H. HERZ

Formerly Research Physicist,
now Consultant to the Research Laboratories,
Kodak Ltd, England

WILEY–INTERSCIENCE
a division of John Wiley & Sons Inc
New York   London   Sydney   Toronto

Composed by The Universities Press, Belfast, N. Ireland.
Printed in the United States of America.

ERRATA

Page 58, paragraph 3, next to the last line: "Oswald" should read "Ostwald."

Page 113, line 8 from top: the word "Figure" should read "page."

Page 135, legend to Figure 4.29, line 2: "light of short duration" should read "light of *long* duration."

Page 279, line 4 from bottom: "inflammable" should be "*non*-inflammable."

Page 290, Figure 8.5: the ordinate should read "10^4" instead of "0^4."

Page 363, line 3 from bottom: the word "of" is to be omitted.

Pages 451 and 452: the diagrams of Figures 10.22 and 10.23 should be transposed.

Page 544, legend: "Olon" should read "Elon."

Page 578, legend to Figure 13.6: "ZnK_1" should read "$ZnK\alpha_1$."

Page 586, line 6 from bottom: the word "by" should be omitted.

Page 615, legend to Figure 14.7: "cm^2/ft" should read "$cm^2/gram$."

Page 629: The entry "Tropical Processing" should have 385 as the page number, not 485.

Herz: RADIATION   Y 471 37430 X

Preface

With the continuing discovery of new physical and chemical applications and improvements in methods in science, medicine, biology, technology and engineering, the use of ionizing radiations is steadily growing. In many of these disciplines the detection and recording of images by photographic means are especially important. There is no doubt that success depends largely on the correct and efficient use of photographic techniques and this in turn depends on the understanding of the photographic action involved.

Every author tries in his preface to justify the publication of his book and I do not attempt to make an exception to this. In my professional experience as a research physicist for over 25 years in the Kodak Research Laboratories and after my retirement, as a consultant to these laboratories, I received so many scientific and technical enquiries that I became aware of the need for a book of this kind. In some cases important experiments in the past were unsuccessful only because of insufficient knowledge of photographic principles and techniques. Furthermore, as far as I know, there does not exist a single compilation of most of the relevant knowledge on the photographic action of ionizing radiations, comparable to those that are available concerning the photographic action of light. The literature on the former subject is scattered over a vast number of publications by individuals in many different journals.

The first chapter of this book deals briefly with some of the 'Physical principles of ionizing radiations', with particular emphasis on absorption processes, because the photographic action naturally depends on the energy absorbed in the photographic layer. This chapter is mainly written for the beginner as an introduction to the subject. A very short chapter follows on 'Ionization measurements and units', as it would be difficult to quote quantitative data without reference to the established units of measurement of ionizing radiations, such as rems, rads and roentgens.

The chapter of the 'Photographic process' is essentially an elementary account of modern views on latent image formation and on the mechanism of development. The chapters on the 'Response of photographic material to photons and particles' form the basis of the book on which many of the techniques in dosimetry, radiography, autoradiography and other subjects depend.

Whilst attempting to write these chapters it was realized that a more constructive way of presenting the material would be to show the reader how and where these photographic principles are applied. This aspect is presented in the various chapters in the second part of this book. One may argue why such apparently unrelated fields as medical and industrial radiography or dosimetry and autoradiography are contained in one volume. The answer to this is that the author hopes to encourage the wider use of methods applied in one field to others where such methods were previously not appreciated fully. It is unavoidable that in a book of this kind some overlap or even repetition occurs in several places. This must happen, if a theoretical account given, for instance, in part I, is repeated for various techniques which are described in part II.

The selection of items for the various applications of radiography presented in part II is based chiefly on the photographic problems and methods involved. In spite of this limitation the reader may consider some of the chapters as technical reviews of the particular fields.

As is to be expected, readers will find certain subjects missing which are related to the main topic of the book. There is hardly any mention of the many finishing methods of radiography, such as duplication, reduction and printing of radiographs; there is also no detailed discussion on artifacts occurring in film handling. The reasons for the omission of these items is that excellent literature on these subjects exists in brochures supplied by the photographic industry, so that there appears to be no need for repetition. Other fields which have not been included are veterinary and painting radiography and xeroradiography. These subjects have not been discussed for the simple reason that the author has not sufficient familiarity with them.

The book is addressed to medical and industrial radiologists, to hospital, research and industrial physicists, to biologists, to radiographers and technicians, in fact, to all those who apply ionizing radiations professionally and who require a deeper insight into the photographic fundamentals. Parts of the book are presented in an elementary manner in order to introduce the beginner to his particular field of application but certain fields are described on a more advanced level which may be profitable to the professional and experienced reader.

Reference is made to many authors who have contributed theoretically and experimentally to the problems discussed in this book. On the other hand, reference to many other excellent publications has had to be omitted, so that the volume of the bibliographic sections remains in a reasonable proportion to the text as a whole.

Omission of these references has no connection whatsoever with their respective values.

I had the great privilege of finding many of my colleagues of the Kodak Research Laboratories in U.S.A. and in England and also other experts prepared to read and comment on various sections of this book. My particular gratitude is due to Mr. M. Cleare and his colleagues, Research Laboratories, Eastman Kodak Company, U.S.A., who read the manuscript as a whole, to Mr. P. McGlinchey, Dr. L. Mullins, Dr. E. Pitts and Mr. E. W. H. Selwyn of Kodak Ltd., England, who devoted a considerable amount of their time to the careful reading of several chapters. Other colleagues also made valuable comments on various sections throughout this book, such as Mr. G. V. Cadogen, Mr. F. Davies, Dr. G. C. Farnell, Miss M. D. Gauntlett, Mr. E. J. Grimwade, Dr. H. Hirsch, Dr. G. Levenson, Mr. S. Marshall, Mr. P. Maynard, Mr. P. G. Powell, Miss M. G. Rumens, Mr. C. J. Sharpe, Dr. E. A. Sutherns and Dr. F. W. Willets.

The chapters on 'Photographic dosimetry' were read by Mr. M. Heard and his colleagues at the Atomic Energy Research Establishment, Harwell and Mr. R. B. Jones of the Radiological Protection Service, Sutton, Surrey. The chapter on 'Neutron radiography' was read by Dr. H. Berger, Argonne National Laboratory, Illinois, U.S.A., that on 'Ionization measurements and units' by Dr. T. E. Burlin of The Polytechnic, London and that on 'Autoradiography' by Dr. S. R. Pelc of the Medical Research Council, Biophysics Research Unit, King's College, London.

To all these ladies and gentlemen, who gave such valuable comments and criticism and especially to Miss I. Hookham, who helped in proof reading, I should like to express my appreciation and my sincere thanks. I am also grateful for the many facilities which were offered to me by the Kodak Research Laboratories and particularly to their Photographic Department who provided several prints for some of the illustrations used in this book.

Wembley, London   R. H. HERZ

Spring 1969

Contents

2   Ionization Measurements and Units

3   The Photographic Process

PART II

6 Photographic Dosimetry

7 Fundamentals of Radiography

8 Medical Radiography

9 The Processing of Radiographs

10 Industrial Radiography

13   Microradiography

14   Appendixes in the text

THE
PHOTOGRAPHIC ACTION
OF
IONIZING RADIATIONS

Part One

1

Physical Principles of Ionizing Radiations

1.1 INTRODUCTION

The emphasis of this book is on the photographic action of ionizing radiations and a full appreciation of this action cannot be attempted without a knowledge of the basic physical properties of the radiations. Hence a brief introduction into this field is given in this first chapter for those who are less familiar with the subject. Special emphasis is given to the mode of absorption of ionizing radiation, as without this no photographic effects could be obtained.

It should be noted that any attempt at presenting the physics of ionizing radiations in such a short space can only be a sketchy presentation of this very wide subject. Furthermore, it is necessarily somewhat biased, as preference is given to those subjects which are likely to be of interest in photographic applications. Reference to more comprehensive literature is therefore given at the end of this chapter.

The treatment is elementary throughout and first deals with the characteristics of electromagnetic and corpuscular radiation. This is followed by a brief presentation of the properties and origin of X-rays, their continuous and characteristic spectra. The various types of absorption processes for photons and particles and their relative contribution to the total absorption are emphasized. The significance of the absorption coefficient, selective absorption and half value layer is stressed, as reference to these terms will be made quite frequently in subsequent chapters. Finally, the attenuation of neutrons and the concept of nuclear cross-section are treated, as far as it appeared necessary, as an introduction to applications of neutrons in dosimetry and radiography.

1.2   ELECTROMAGNETIC AND CORPUSCULAR RADIATION

The energy loosely termed 'radiation' manifests itself in two main forms, wave-like electromagnetic radiation in which no mass is propagated, and particle-like corpuscular radiation which carries mass. Although this wave and particle designation is useful, since it describes certain characteristics in terms which are readily visualized, it has been shown by many experiments and the theoretical interpretation by wave-mechanics that the distinction is not absolute, but is to be regarded as two different aspects of a fundamental property.

A particle in motion can indeed be regarded as a wave whose frequency depends on the momentum, and electromagnetic waves are emitted in discrete packets of energy (quanta or photons) which possess some particle-like attributes. This duality is illustrated by the fact that electrons, neutrons, and other corpuscular radiations show the typical wave-like phenomenon of diffraction when they interact with regular arrays of atoms in a crystal lattice. On the other hand, Compton showed that the laws governing the incoherent scattering of X-rays (electromagnetic wave-like radiation) may be deduced on the assumption that X-ray quanta behave similarly to particles colliding with the electrons present in matter. Nevertheless, for many purposes the adjectives 'wavelike' and 'particle-like' give an adequate description of the main characteristics of the two forms of ionizing radiation, whose effect on photographic materials is the subject of this book.

Radiation is called ionizing when it has the property of either directly or indirectly releasing electrons in matter through which it passes, thereby creating positive and negative charges. Atoms which have lost electrons are called positive ions, and those which have gained electrons negative ions. In every elementary ionization process pairs of ions are formed, and it is this process which induces chemical changes and hence biological effects in living cells. Likewise the operation of many radiation detection devices, including the photographic emulsion, depends on the ionization process.

We shall now briefly describe examples of the two main types of radiation.

1.3   X- AND γ-RAYS

The discovery of X-rays or Roentgen rays was announced in December 1895 by Wilhelm Conrad Röntgen at the University of Würzburg, who was able to demonstrate most of the properties of the radiation within the first two years after his discovery. Amongst these are their penetrating power, their ability to ionize gases, to make certain crystals fluoresce and to blacken photographic plates. Shortly after Röntgen's announcement, Henri Becquerel in France became interested in fluorescent phenomena caused by potassium uranyl sulfate. He was able to show that the radiation emitted by this salt was similar to X-rays, namely that it could penetrate material opaque to light and that it also caused ionization in gases and blackened photographic plates. Following closely on this discovery by Becquerel of natural radioactivity, Pierre and Marie Curie separated the elements polonium and radium from pitchblende. Their work and that of Ernest Rutherford and his colleagues led them to distinguish three different types of radiation emitted by natural radioactive materials, which were termed α-, β-, and γ-rays. The α- and β-rays showed only limited penetrating power, whereas γ-rays were capable of penetrating a considerable thickness of material.

Although historically the radiation produced by means of electrical machines was called X-radiation, and the penetrating radiation from natural radioactive substances was called γ-radiation, there is no essential difference between the properties of these two wavelike electromagnetic radiations. Later work has shown that a convenient definition to adopt is one based on the origin of the radiation. X-Rays are produced by events outside the nucleus of the atom, and γ-rays arise within the nucleus. The frequent supposition that X-rays are of lower energy than γ-rays is not quite correct, particularly since X-rays of extremely high energies can now be produced by special high voltage generators. On the other hand γ-rays of very low energies are emitted from nuclei of certain radioactive isotopes.

We will now turn to the more detailed description of X- and γ-rays and consider α-particles, β-particles, and neutrons later.

X-Rays are produced when fast moving electrons impinge upon matter. The larger part of the kinetic energy of the electrons is then dissipated in the form of heat and the remaining portion produces X-rays. The interaction of high speed electrons with matter, for instance with the metallic anode of an X-ray tube, can be of different kinds and correspondingly we distinguish between continuous and characteristic X-rays.

1.3a   Continuous X-Rays

When electrons travel close to atomic nuclei they are subjected to deflections by the strong Coulomb field surrounding the nuclei. This occurs with the atoms which form the target of an X-ray tube. According to classical electrodynamic theory the change in velocity of an electron in the field of a nucleus causes the emission of a pulse of radiation. X-Rays which are emitted in this type of process are often called 'Bremsstrahlung'. This German word means braking or stopping radiation and refers to the stopping of electrons by the target. It is evident that the electrons lose their energy in a random fashion, that is, a few lose their energy in a single collision whilst others lose their energy gradually by passing deeper into the target material. Hence this process gives rise to a wide range of energies of X-rays, forming a broad continuous wavelength spectrum. It is sometimes referred to as a white spectrum (see Figures 1.1a and b).

It is of great importance to relate the wavelength to the energy of the X-radiation, but here it is only possible to indicate the formal relationship. Those readers who wish to gain a deeper understanding should consult a suitable textbook of physics (see list of references).

The argument runs as follows: suppose the potential difference between the cathode and anode (target) in an X-ray tube is V volts; then a measure of the energy gained by the electron in moving from the cathode to the anode is the product eV, where e is the electronic charge. One electron volt is equal to the energy required to move one electronic charge through a potential difference of one volt. It is customary to use the electron volt as a unit of energy, and we shall frequently mention it throughout this book. As the electron has a charge of 4.80×10^{-10} e.s.u. and 1 volt is 1/300th of 1 electrostatic unit of potential difference, $1 \text{ eV} = (1/300) \times 4.80 \times 10^{-10}$ ergs $= 1.6 \times 10^{-12}$ ergs. For convenience two other units are frequently used, the kilo-electronvolt (keV) and the million-electronvolt (MeV).

Now a fundamental law of quantum-theory states that a quantum of radiation of frequency v has energy hv, where h is Planck's constant. Hence if we assume that the electron loses all its energy and produces one X-ray quantum, we must equate hv to the product eV, thus

$$hv = eV \tag{1.1}$$

and so we obtain a value for the frequency of the radiation

$$v = \frac{eV}{h}$$

Since $v = \dfrac{c}{\lambda}$, where c is the velocity of light and λ the wavelength, we find

$$V = \frac{hc}{e\lambda} \tag{1.2}$$

After inserting into the equation numerical values (see Table 14.1 in Appendix)

$$V = \frac{6.62 \times 10^{-27} \times 3 \times 10^{10}}{1.6 \times 10^{-12} \times \lambda_{\min} \times 10^{-8}} = \frac{12400}{\lambda_{\min}} \tag{1.3}$$

where V is measured in volts and $\lambda_{\min}$ is measured in Ångstrom units (1 Å is equal to 10^{-8} cm). If V is expressed in kilovolts, we have the relation

$$kV = \frac{12.4}{\lambda_{\min}} \tag{1.4}$$

Hence the shortest or the minimum wavelength of the heterogeneous X-ray spectrum corresponds to the maximum potential difference across the X-ray tube. Thus we have obtained the relationship between the kilovoltage and the wavelength in Å.

It should be emphasized, as mentioned earlier, that only a few electrons give up their total energy in a single encounter corresponding to the minimum wavelength and that the majority of the electrons will thus produce X-rays of longer wavelength as indicated by the continuous X-ray spectra in Figures 1.1a and b.

1.3b Characteristic X-Rays

In describing the second origin of X-radiation it will be necessary to assume that the reader has some acquaintance with the concept of the structure of the atom, in particular with the notion that electrons can be regarded as moving about the nucleus in orbits, in each of which the electron has a particular energy. Another mode of expression refers to the electrons as possessing discrete energy levels in the atom. Now characteristic X-rays are produced when fast electrons impinge on material and dislodge electrons from their orbits by collision. If the electron is entirely removed, the atom is ionized. It is possible for an electron from an outer orbit, that is one of higher energy, to move to the vacant orbit resulting from the removal of the electron by the collision, provided this energy level is lower. The balance of energy is emitted as a quantum of electromagnetic radiation, i.e. X-rays of a given wavelength.

Hence the energy of the X-ray quantum $h\nu$ depends on the energy difference between the two levels and this relation is expressed by the equation

$$h\nu = E_n - E_m \qquad (1.5)$$

where E_n and E_m are the respective energy levels of two different atomic orbits. The X-rays thus produced are called characteristic X-rays, because they are characterized by the amount of energy difference between the energy levels of a given atom, i.e. of a given material. The K-electrons from the innermost orbit are the most tightly held electrons. If such an electron is removed and replaced by an L-electron, (that is from the next higher energy level), the difference between the energies of the K and L shells (called K_α radiation) is a fixed amount equal to the energy of the emitted X-ray quantum. Other possibilities are M–K (K_β), M–L (L_α) transitions. As the L shell has three sub-levels, many more transitions are possible, such as L_I, L_II, L_III.

In order to produce K-radiation the electrons hitting the target of an X-ray tube must of course have an energy great enough to remove an electron from the K-orbit. The voltage required is called the excitation voltage and this obviously differs from element to element. The higher the atomic number of the material bombarded by electrons, the greater is the energy required to remove an electron from a K-shell.

Assuming for instance that an X-ray tube target of tungsten is used, an energy of 69.5 keV is required to remove a K-electron from a tungsten atom. By supplying this potential difference across the X-ray tube the characteristic K-radiations in addition to the continuous spectrum will be emitted, whereas the L-radiation is excited by only 12.1 keV.

1.3c   X-Ray Spectra

We have seen that the X-ray spectrum produced by an X-ray tube is composed of continuous radiation of a wide range of wavelengths on which characteristic radiation (line spectra) may be superimposed. Typical X-ray spectra are shown diagrammatically in Figures 1.1a and 1.1b. Figure 1.1a illustrates the relative intensity of radiation as a function of wavelength. The continuous spectra are shown for various kilovoltages. As the peak potential difference across the X-ray tube is increased from 30–50 kV$_\mathrm{p}$, the minimum wavelength decreases and the intensity increases. Furthermore the maximum intensities are shifted towards shorter wavelengths.

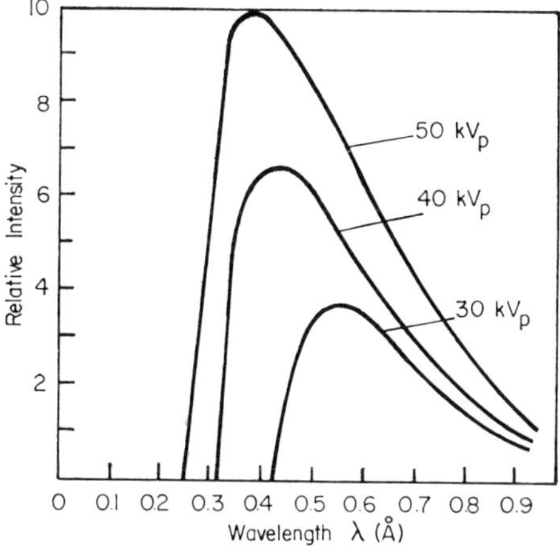

Figure 1.1. (a) Continuous X-ray spectra at 30, 40 and, 50 kV$_p$ (Diagrammatically).

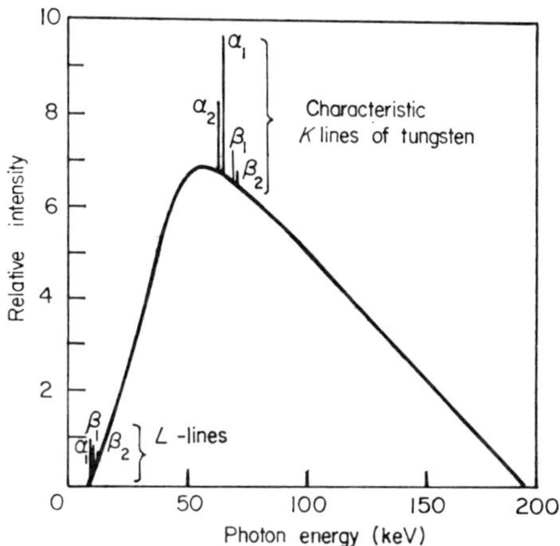

Figure 1.1. (b) Continuous tungsten spectrum generated at 190 kV$_p$. (Diagrammatically) showing also characteristic K- and L-lines.

A theoretical approach to the spectral distribution of Bremsstrahlung has been given by H. A. Kramers (1923) from which it was concluded that the wavelength of maximum emission is about $3/2\lambda_{min}$.

The diagram Figure 1.1b illustrates a spectrum of photon energies corresponding to the respective wavelengths of radiation generated at 190 kV$_p$, i.e. well above the tungsten excitation potential and hence all tungsten characteristic K-lines appear. Only a few of the L-lines are shown in this graph. No characteristic K- or L-lines of tungsten are excited within the range of wavelengths covered by Figure 1.1.(a).

1.3d   keV and kV$_\mathrm{p}$

It is worth mentioning at this stage that the loosely termed specification such as a '100 kV X-ray beam' refers only to the fact that the beam was produced at a maximum kilovoltage (kV peak or kV$_\mathrm{p}$) of 100 kV. This, however, in no way specifies the wavelength distribution of the beam.

On the other hand, if an X-ray beam is specified by '100 keV' this means that it is either a monochromatic beam corresponding to 100 keV, i.e. $\lambda = 0.124$Å, or that it is a beam whose average penetrating power (or wavelength composition) is equal, or is very close, to that of a monochromatic X-ray beam of 100 keV. A method of approaching such a specification experimentally will be given in the section on half value layers.

Another way of specifying very loosely an X-ray or a γ-ray beam is to speak of soft and hard radiation. The term 'soft' refers here to long wavelengths or radiation of low penetrating power or of low kV radiation, as against 'hard' which stands for short wavelengths, high penetrating power, and high kV radiation. The terms soft and hard are only meaningful if used relative to each other.

1.4   ELEMENTS OF RADIOACTIVE DECAY

Before discussing the origin of γ-rays which are produced during radioactive decay, a brief introduction to the basic elements of radioactive disintegration will be useful.

By upsetting the balance of neutrons and protons in atomic nuclei, artificial radioactive isotopes can be produced which have either too many or too few neutrons and thus become unstable. These artificial isotopes* are radioactive in the same way as natural radioactive elements and their stability will be restored by emitting some form of radiation.

The disintegration of radioactive atoms is purely a matter of chance. Although a definite fraction of atoms will disintegrate in unit time, it cannot be predicted which particular atoms will decay. Assuming that, at a given instant, there are N atoms of a particular radioactive species present and that in a short interval of time dt the number of

---

* Isotopes are atoms of elements having the same atomic number but a different number of neutrons, i.e. different mass number.

atoms disintegrating is dN, then the rate of disintegration is represented by dN/dt. Now the number of atoms which disintegrate in a unit interval of time is proportional to the total number of atoms N present, hence

$$-\frac{dN}{dt} = \lambda N \tag{1.6}$$

where λ is a constant called the disintegration or decay constant. The negative sign in the equation indicates that the number of atoms present is decreasing. On integration of the above expression, the result is

$$N = N_0 e^{-\lambda t} \tag{1.7}$$

where N_0 is the number of atoms present at any arbitrary time $t = 0$ and N is the number of atoms remaining after the lapse of the time interval t. The disintegration constant λ may be related to the so called half life $T_{\frac{1}{2}}$ which is the time required for one half of the atoms originally present to disintegrate, thus

$$T_{\frac{1}{2}} = \frac{1}{\lambda} \log_e 2 = \frac{0.693}{\lambda} \tag{1.8}$$

Each radioactive species of natural or artificial origin is characterized by a specific half life and this may be between an extremely small fraction of a second and many millions of years. The rate of particle or photon emission of radioactive nuclei cannot be influenced by temperature, physical environment, or chemical state.

1.4a γ-Rays

A γ-ray photon and an X-ray photon having the same energy are identical radiations, but they differ in their origin, as indicated earlier. γ-Rays are due to transitions between energy levels in radioactive nuclei. In many disintegrations a nucleus is left in an excited state after particle emission and the excess energy is released in the form of γ-rays. These may be emitted as photons of one quantum energy or as a series of several distinct energies (line spectrum), where each is due to a difference of energy levels in the radioactive nucleus. Sometimes the release of γ-rays may be in a cascade where one quantum of a definite energy is released and followed by a second one of a different energy, as seen in the decay scheme of sodium-24 ($^{24}_{11}$Na) Figure 1.2. The subscript 11 is the atomic number and the superscript 24 is the mass number, i.e. the sum of the number of protons and neutrons in the nucleus.

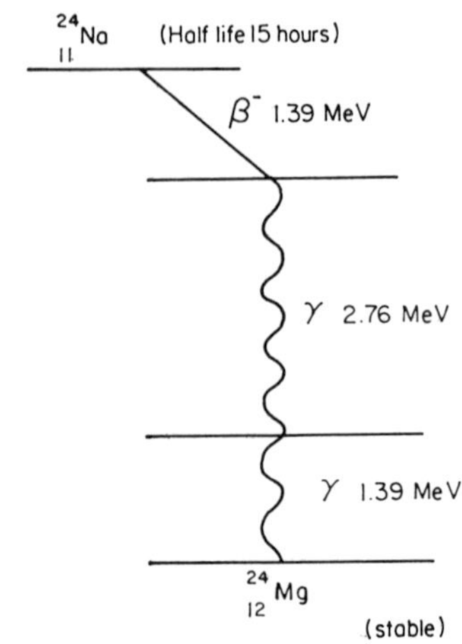

Figure 1.2. Decay scheme of sodium-24.

The decay scheme of sodium-24 is very similar to that of cobalt-60 (see p. 434). In the above scheme a β-particle (electron) of 1.39 MeV is emitted leaving the nucleus in an excited state of $^{24}_{12}$Mg. On returning to the ground state two γ-ray photons are released in cascade, one of 2.76 MeV and the other of 1.39 MeV, as indicated by the wavy lines in Figure 1.2.

γ-Rays emitted by natural and artificial radioactive sources are usually within the range of energies of a few thousand electron volts to a few million electron volts. The absorption of γ-rays which follows the same laws as that of X-rays will be discussed on page 15.

Since photons may be released in other modes of nuclear transformations, two further important processes will be briefly described. These are the 'electron capture' and 'internal conversion' processes.

1.4b   Electron Capture

A nucleus which is unstable due to an excess of protons may capture one of the orbital electrons (e^-) and it is supposed that a proton (p) is converted into a neutron (n) according to

$$p + e^- \rightarrow n + \nu \tag{1.9}$$

where ν is an uncharged particle of very small mass called a neutrino. The product nucleus is often in an excited state and the excess energy may be released as a γ-ray photon. If the capture occurs from a K-shell of an atom, this so called K-capture leaves a gap in the shell which is filled by another electron giving rise to characteristic X-rays.

Electron capture is more likely the greater the positive charge of the nucleus. Its occurrence is more probable if the difference between the energy levels of the parent and daughter nuclei is less than 1.02 MeV, i.e. equivalent to 2 electron masses.

With energy differences exceeding this level another type of nuclear transformation may occur which is characterized by positron (positive electron) emission instead of by electron capture. This effect is less frequent with heavier elements.

1.4c Internal Conversion

In some disintegrations γ-rays emitted from nuclei may knock out their own orbital electrons, i.e. K, L, or M electrons by a process called 'internal conversion'. As a result of this process the whole of the photon energy may be transferred to the electrons according to $E_{ph} - W$, where E_{ph} is the photon energy and W is the respective binding energy of the electrons in the atoms emitting the γ-radiation. The removal of the electrons leave the atoms in an excited state which will then be followed by the emission of X-rays from K, L, and M shells.

The probability that internal conversion takes place is greater for nuclei of high atomic number and increases with decreasing γ-ray energy. Precise spectrographic analyses of the X-ray lines produced in this process have given experimental proof that the internally converted γ-rays are released by transitions occurring in the daughter nuclei rather than in the parent nuclei.

1.5 PARTICLES

A radioactive nucleus may emit an α- or a β-particle to produce a daughter nucleus of a different atomic number. If the daughter nucleus, after particle emission, remains in its lowest state of energy, then the process is completed with the particle emission. As we have seen earlier in many disintegrations the nucleus is left in an excited state after particle emission and the remaining energy is released by γ-ray photons.

1.5a α-Particle Emission

α-Particles are helium nuclei consisting of two protons and two neutrons. α-Particles are thus doubly positively charged and have a mass of four units. Hence an α-particle is denoted by $^{4}_{2}\text{He}$. α-Disintegrations occur chiefly in heavy nuclei. If an α-particle is ejected during radioactive decay, the atomic number of the daughter nucleus is reduced by two units and the number of neutrons is also reduced by two.

A typical α-emitter is radium ($^{226}_{88}$Ra). This decays to radon ($^{222}_{86}$Rn) by the emission of 95.2 per cent α-particles of 4.86 MeV and 4.8 per cent α-particles of 4.68 MeV. As the atom in the latter case has not returned to the ground state, in 4.8 per cent of the events γ-rays of 0.18 MeV (i.e. 4.86 − 4.68 MeV) are emitted.

Typical representatives of natural radioactive sources emitting α-particles are members of the uranium, thorium, and actinium series. A very frequently used α-emitter is polonium-210 emitting α-particles of 5.3 MeV. Representatives of artificial radio-isotopes emitting α-particles are neptunium, plutonium, curium, and other transuranium elements, which are all of high atomic number. Most of the α-particles emitted by radioactive sources have energies of the order of a few to more than 10 MeV.

1.5b   Decay by β-Particles and Positrons

β-Particles ($β^-$) or electrons are negatively charged and their antiparticles ($β^+$), positrons, are positively charged. The mass of an electron and of a positron is 1/1840th of that of the proton (the hydrogen nucleus) and all three have unit charge.

The decay of radioactive nuclei may occur by the emission of negative electrons and positrons. The $β^-$ or $β^+$ emission by a nucleus needs some explanation, as it is generally assumed that an atomic nucleus contains only protons and neutrons and no electrons.

If a negative electron is emitted, it is supposed that a neutron in the nucleus changes into a proton and an electron, whereby the mass of the nucleus remains practically unchanged, but the atomic number increases by one.

On the other hand, if a positron is emitted, it is assumed that a proton changes into a neutron and a positive electron, the atomic number is thus reduced by one.

As is known from the $β^-$ or $β^+$ decay of radioactive nuclei, particles of practically any energy from zero up to a certain maximum energy are emitted. This maximum energy is characteristic for a particular radioisotope. A typical β-ray spectrum of a radioactive isotope is shown in Figure 1.3. On the ordinate we have the frequency, i.e. the number of particles emitted by each energy and on the abscissa the energy of the β-particles. One can see that there are relatively few electrons emitted at the maximum energy.

It was difficult at first to understand why all β-particles are not emitted with the same energy, considering that the energy difference

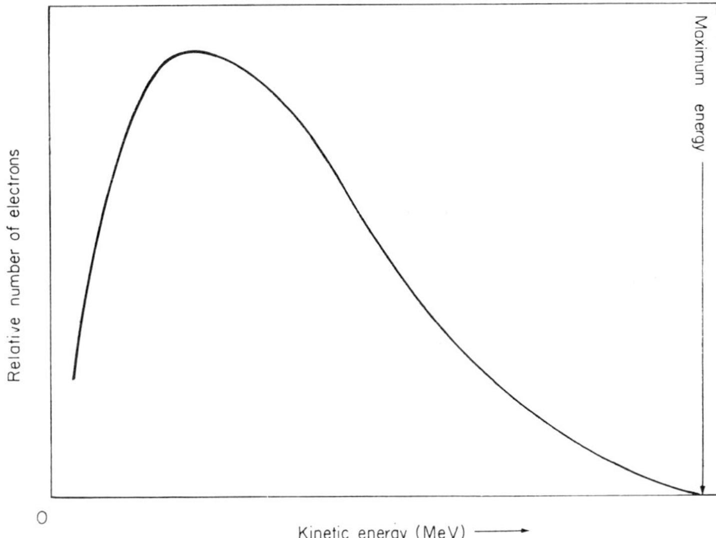

Figure 1.3. Typical β-ray spectrum of a radioactive isotope.

between all the parent and all the daughter nuclei is the same. The explanation for the apparently missing energy was given by the assumption that another particle of no charge and of very small mass shares the missing energy. This particle is the antineutrino ($\bar{\nu}$) in the case of β^- emission and the neutrino (ν) in the case of the positron emission. Thus we can describe the events taking place in β^- and β^+ decay by the following expressions

$$\beta^- \text{ emission: } n \rightarrow p + \beta^- + \bar{\nu} \text{ where } Z^* \text{ increases by } 1$$

$$\beta^+ \text{ emission: } p \rightarrow n + \beta^+ + \nu \text{ where } Z \text{ decreases by } 1$$

1.5c   Neutrons

These are uncharged particles having a mass slightly greater than that of the protons. The atomic nuclei are made up of Z protons and $A - Z$ neutrons, Z being the atomic number and A the mass number. Some of the significant features of neutrons are not only that they made possible the release of energy from atomic nuclei, but that virtually all elements have been found to interact with neutrons.

One of the strongest sources is a nuclear reactor in which neutrons are produced from the fission of ^{235}U or other fissile materials. Weaker sources, more suitable for laboratory purposes, can be obtained by the

* Z = atomic number.

bombardment of certain light elements with α-particles, for example beryllium and boron. A well known neutron source of this kind is a mixture of the α-emitting polonium or radium with beryllium. This leads to a nuclear reaction in which neutrons are emitted and a carbon atom is produced

$$\mathrm{^{9}_{4}Be} + \mathrm{^{4}_{2}He} \rightarrow \mathrm{^{12}_{6}C} + \mathrm{^{1}_{0}}n \tag{1.10}$$

The neutrons obtained from this reaction are not monoenergetic but cover a range of energies. (See Figure 11.3.)

Almost monoenergetic neutrons may be produced by so called photoneutron (γ, n) reactions in which deuterium (i.e. $^{2}\mathrm{H}$) or beryllium (Be) are bombarded with γ-rays. These must have a definite minimum or threshold energy for the reaction to take place. Typical γ-ray emitters used in photoneutron reactions are those mentioned in Table 1.1 below together with the neutron energies obtained.

TABLE 1.1. Photoneutron Sources

Neutron source	γ-Ray emitter	Neutron energy (MeV)
Beryllium	^{124}Sb	0.024
Deuterium	^{72}Ga	0.13
Deuterium	^{24}Na	0.22
Beryllium	^{140}La	0.62
Beryllium	^{24}Na	0.83

Other useful sources are those in which deuterium ($^{2}\mathrm{H}$) or tritium ($^{3}\mathrm{H}$) loaded targets are bombarded by protons or deuterons in an accelerator. Further details of neutron sources suitable for neutron radiography are discussed on page 485–490.

Neutrons are usually classified with reference to their energy or at least with respect to their range of energies. Such a classification, although rather arbitrary, is given in the Table 1.2 below. Other tables given in literature may quote different energies.

TABLE 1.2. Classification of Neutrons with Respect to Energy

Neutrons	Energy range
Cold	< 0.01 eV
Thermal*	$0.01 - 0.3$ eV
Epithermal	$0.3 - 10{,}000$ eV
Fast	$0.01 - 20$ MeV
Relativistic	> 20 MeV

* Thermal neutrons are often specified as having an energy of 0.025 eV.

Cold, thermal and epithermal neutrons are often called 'slow neutrons'. The term 'thermal' is chosen, because thermal neutrons move randomly with a velocity equivalent to that produced during thermal motion of atoms and molecules. Slow neutrons are produced by slowing fast neutrons down by so called moderators. (See section on absorption of neutrons, page 34.)

1.6 ABSORPTION OF PHOTONS AND PARTICLES

1.6a Attenuation of X- and γ-Rays

If X- or γ-rays strike matter, interactions take place with atoms which may lead to a variety of processes depending on the energy of the radiation and the atomic number of the absorber. We shall now describe briefly the chief features of the more important absorption processes.

Photoelectric Absorption

This effect predominates in the low energy range and with material of high atomic number. The absorption of an X-ray quantum in the K-orbit of an atom may cause the removal of a K-electron beyond the periphery of the atom. An electron from a higher orbit will take the place of the ejected electron and simultaneously an X-ray quantum (characteristic X-ray) will be emitted. The energy $h\nu$ of the incident X-ray photon is related to the work W_k required to remove the electron and its kinetic energy $\frac{1}{2}mv^2$ by the equation

$$h\nu = \tfrac{1}{2}mv^2 + W_k \qquad (1.11)$$

The energy W_k is called the binding energy. The electrons ejected from atoms are called 'photoelectrons' and may acquire considerable energies. Whilst they are passing through adjacent atoms they cause further ionizations in them.

Coherent and Incoherent Scattering

Two processes must be distinguished here, the coherent and the incoherent (Compton) scattering.

Coherent Scattering. Incident quanta are scattered by orbital electrons from atoms without changing the energy balance of the atoms involved, i.e. without release of electrons. In this kind of scatter, electromagnetic waves make orbital electrons oscillate which causes them to radiate energy of the same wavelength as the incident radiation. This process of scattering, i.e. change of direction, is more striking for low energy X-rays and elements of high atomic number and plays a predominant part in X-ray diffraction phenomena in crystals. Coherent scattering is also referred to as classical or Thomson scattering.

Incoherent (Compton) Scattering. If the energy of the X-ray quantum is high in comparison with the binding energy of orbital electrons, another process may occur as was first shown by A. H. Compton in 1923. In this case a quantum or photon may collide with a loosely bound orbital electron and may not only suffer a change in direction (scattering), but may also lose part of its energy, i.e. it will become radiation of increased wavelength. The remaining part of its original energy will be transferred to the ejected electron which is called a 'recoil' electron. The energy balance of this type of scattering, which is usually called 'Compton scattering' can be described by the equation

$$h\nu = h\nu' + \tfrac{1}{2}mv^2 \tag{1.12}$$

where $h\nu$ is the energy of the primary photon, $h\nu'$ the energy of the scattered photon and $\tfrac{1}{2}mv^2$ is the kinetic energy transferred to the recoil electron.

Compton treated this process by assuming that the interaction was that of an elastic collision between a photon and a loosely bound, i.e. practically free, electron. From the laws of conservation of energy and momentum, Compton concluded that the change in wavelength $d\lambda$ should be

$$d\lambda = \frac{h}{mc}(1 - \cos\phi)$$

$$= 0.024\,(1 - \cos\phi)\,\text{(in Å)} \tag{1.13}$$

where h is Planck's constant, m is the mass of the electron, c the velocity of light, and ϕ is the angle between the direction of the original and scattered photons. As h/mc is constant, it is seen from the above equation that the change in wavelength depends only on the angle of scatter and not on the energy of the primary photon nor on the scattering material involved. It may be noted that the maximum possible change in wavelength is 0.048 Å, which occurs when $\phi = 180°$. The change in wavelength, as well as the energy of recoil electrons was experimentally verified by several authors in agreement with Compton's theory.

The effect of the scattering process as described above is naturally more complex when it comes to practical considerations of a beam of many millions of photons passing through an absorber, as repeated scattering of the original photons occurs. Furthermore the probability of scatter in various directions has to be considered. It is found that

the probability of scatter in the forward direction, i.e. in the direction
of the primary beam, increases with increasing energy of the primary
quanta. The direction of recoil electrons can be at any angle between
0 and 90° with respect to the direction of the incident beam.

The Compton process is illustrated diagrammatically in Figure 1.4
and it is seen that for any angle of the scattered photon there is a
related angle of the recoil electron, as indicated by corresponding

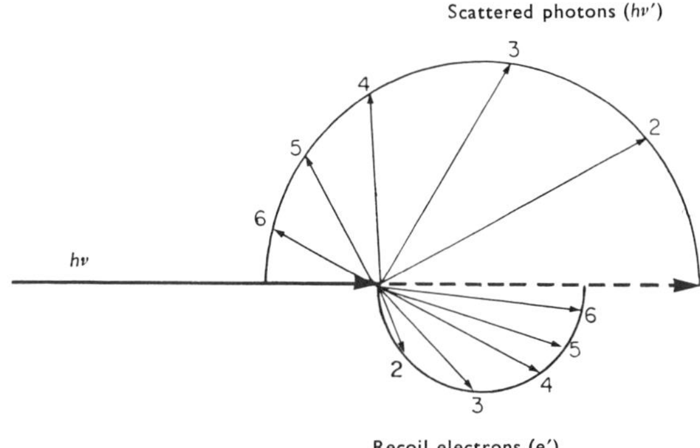

Figure 1.4. Diagram illustrating the Compton effect.

numbers. The scattered quanta and the recoil electrons are shown as
vectors, i.e. each numbered line indicates the relative energy by its
length, and the direction of photons and recoil electrons by the angle
formed between the direction of lines with respect to the direction of
the incident quantum hv.

Pair Formation

This effect occurs only with X- and γ-rays exceeding 1.02 MeV. If
the γ-ray energy is of this level and if the photon passing through
matter is close to a nuclear field, it is possible for the quantum to
disappear and to create an electron-positron pair, which travels in the
direction of the original quantum (see Figure 5.5.). The electron and
positron have each a rest energy of 0.51 MeV, i.e. this energy is
equivalent to their mass according to the well known Einstein
equation $E = m_0 c^2 / \sqrt{1 - v^2/c^2}$. The initial photon energy is thus split
into equal parts. Should the energy of the primary quantum be higher
than the energy required for pair formation to occur, its excess energy
will be shared by the two particles as kinetic energy. Thus the reaction
can be expressed by

$$hv \rightarrow e^+ + e^- + 2E_{kin} \tag{1.14}$$

where e^+ is the positron, e^- the electron and E_{kin} the excess energy shared between the two particles. The probability of pair formation at energies only slightly above 1.02 MeV is very small indeed and the effect only becomes important at higher energies and with material of high atomic number. The conversion of a photon into a positron and an electron is one of the most direct examples showing the equivalence of energy and matter. The exact opposite of this conversion may occur when a positron approaches an electron and they disappear, releasing two quanta of radiation in almost opposite directions. This is referred to as annihilation radiation as it results from the disappearance of the mass of the two electrons. An example of pair formation, as recorded in nuclear track emulsions is shown in Figure 5.5 page 163.

We have thus seen that during the absorption of radiation the photons may be subjected to a variety of different processes whose main features are summarized below.

In the *photoelectric effect* quanta are absorbed and their energy reappears as kinetic energy of electrons and that of characteristic X-rays. In *scattering effects*, quanta are either scattered without loss of energy (coherent scattering) or they are scattered with loss of energy (incoherent Compton scattering), in which the remaining part of the energy is transferred to recoil electrons. In the process of *pair formation*, which is only possible with energies above 1.02 MeV, quanta are converted to matter in the creation of pairs of positrons and electrons.

After having briefly described the more important elementary processes which lead to the absorption of X- and γ-rays we shall now consider the laws of absorption and also the relative contributions of the various processes.

1.6b   The Absorption Coefficient

A beam of X-rays passing through matter is weakened because a fraction of it is absorbed and another fraction is scattered by the material. Both phenomena of scattering and absorption are caused by the elementary processes which have been discussed in the previous section.

When a beam of monochromatic (or monoenergetic) X-rays passes through material of thickness d, it is found that the intensity I transmitted by the absorber is related to the incident intensity I_0 by the relation

$$I = I_0 e^{-\mu d} \tag{1.15}$$

where μ is a constant for the particular wavelength and the atomic composition of the material concerned and d is the thickness of the absorber in cm. Here μ is called the linear absorption (or attenuation) coefficient measured in cm^{-1} and it contains the components arising from photoelectric absorption (τ), coherent scattering (σ_0), Compton scattering (σ) and pair formation (π); hence

$$\mu = \tau + \sigma_0 + \sigma + \pi \tag{1.16}$$

From equation (1.15) it follows that the intensity of the beam decreases exponentially with the thickness of the absorber and that the linear absorption coefficient indicates the fractional decrease in the intensity of the X-ray beam per unit thickness of the absorber.

Apart from the wavelength of the radiation and the atomic number of the absorbing material, the linear absorption coefficient depends also on the physical state of the absorbing material, i.e. on its density. It is therefore more convenient to consider the mass-absorption coefficient μ/ρ, where ρ is the density of the absorber. Thus μ/ρ becomes independent of the physical state of the absorber. Assuming the X-ray beam has a cross-sectional area of 1 cm², $(1 - e^{-\mu})$ is the fractional intensity of radiation absorbed in 1 cm³ and $(1 - e^{-\mu/\rho})$ represents the fraction absorbed from the beam of cross-sectional area 1 cm² in a layer of mass of 1 gram. Hence the mass-absorption coefficient (cm² gram⁻¹) of water is the same regardless of whether the absorber is in the form of ice or steam.

In order to discriminate between the attenuation an X-ray or γ-ray beam suffers in a layer of material, and the kinetic energies of all the charged particles liberated in this layer, a logical distinction is made in medical radiology between

(a) the mass-attenuation coefficient and
(b) the mass-energy transfer coefficient.

(a) The *mass-attenuation coefficient* μ/ρ covers the attenuation to which an X-ray or γ-ray beam is subjected in a layer of material and is defined in analogy to (1.16) by

$$\mu/\rho = \underbrace{\tau/\rho}_{\text{photoelectric}} + \underbrace{\sigma_0/\rho}_{\text{coherent scattering}} + \underbrace{\sigma/\rho}_{\text{Compton scattering}} + \underbrace{\pi/\rho}_{\text{Pair-formation}} \tag{1.17a}$$

(b) The *mass–energy transfer coefficient* μ_κ/ρ however considers only the sum of the kinetic energies of all the charged particles liberated in a layer of material, i.e. all the energy of secondary photon radiation produced is excluded and is thus a more restricted expression than (a). It is defined by

$$\mu_\kappa/\rho = \sigma_a/\rho + \tau_a/\rho + \pi/\rho \tag{1.17b}$$

where σ_a/ρ is the truly absorbed fraction (recoil electron energy) associated with the Compton effect, τ_a/ρ is the photoelectric mass-attenuation coefficient from which the energy due to characteristic radiation is excluded and π/ρ is the pair-formation coefficient. (A further coefficient namely the mass energy absorption coefficient is referred to on page 48).

In the following chapters of this book reference is made only to linear or to mass-absorption coefficients, and unless specifically mentioned, these terms have the same meaning as the linear or mass attenuation coefficients defined above.

With the introduction of the mass absorption coefficient the absorption law can be written in the form

$$I = I_0 e^{(-\mu/\rho)\rho x} \tag{1.18}$$

where the product ρx is considered to be the mass of the absorber material exposed to radiation in a volume of cross-sectional area of 1 cm^2 and a thickness of x cm. The dimensions of ρx are grams/cm^2 or mass per unit area. From such an expression the measurement of I/I_0 permits the value of x to be determined, provided μ is known.

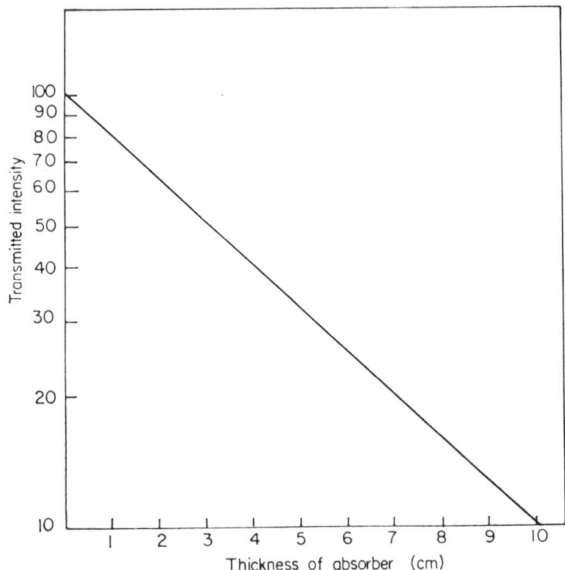

Figure 1.5. Transmission curve of monochromatic X-ray beam using an absorber having a linear absorption coefficient of 0.2 cm^{-1}.

The absorption law is strictly valid only when monochromatic radiation is applied. This can be seen from the following considerations. Let us assume that the linear absorption coefficient μ is 0.2 cm^{-1} and that the intensity of the beam incident on the absorber is 100. Hence the first layer of absorber of 1 cm thickness absorbs 20 per cent, i.e. 80 per cent is transmitted, the next layer of 1 cm thickness absorbs 20 per cent of 80, i.e. 64 per cent is transmitted and the third slab of 1 cm thickness again absorbs 20 per cent, i.e. 51.2 per cent is transmitted and so on. Figure 1.5 shows a graph in which the absorption for further layers is indicated and as this is plotted on semilogarithmic paper a straight line transmission curve is obtained.

However, if we use heterogeneous radiation, as is the case in the majority of practical applications, the absorption conditions as outlined above do not hold. This is because the longer wavelengths of the heterogeneous beam will be more strongly absorbed than the shorter wavelengths. Hence the absorption coefficient changes with each successive filtration until a beam is obtained having a sufficiently narrow wavelength composition, i.e. an approach to a monochromatic beam.

In most practical calculations of absorption the distance of the absorber from the source and the increase of distance presented by the thickness of the absorber has to be considered. Since the intensity of an X- or γ-ray beam decreases inversely with the square of the distance, the absorption formula has to be provided with a correction for the inverse square law of distance and thus reads

$$I = \frac{I_0 e^{-\mu x}\, r^2}{(r + x)^2} \tag{1.19}$$

where r is the distance from the radiation source to the surface of the absorber and x is the thickness of the absorber. Mass-absorption coefficients for various elements and wavelengths are given in the Appendix, page 605.

1.6c Absorption and X-Ray Spectrum

If an X-ray beam is subjected to filtration the softer components of the beam are more readily absorbed than the harder components. The fact, that the longer wavelengths are more preferentially removed than the short ones, has the result that the average wavelength of the filtered beam is shorter than that of the unfiltered beam. The effect of the filtration on the X-ray spectrum is seen diagrammatically in Figure 1.6 which illustrates a continuous X-ray spectrum as emitted by an X-ray tube generated at 100 kV$_\mathrm{p}$. The dashed line curves in the same graph show the resulting spectra after 0.3 and 0.5 mm Cu filtration and these are characterized by a narrowing of the wavelength band due to the effect of increasing filtration of the original spectrum.

1.6d Selective Absorption

If the absorption by a given material is measured as a function of X-ray wavelength, it is found that certain discontinuities occur, as

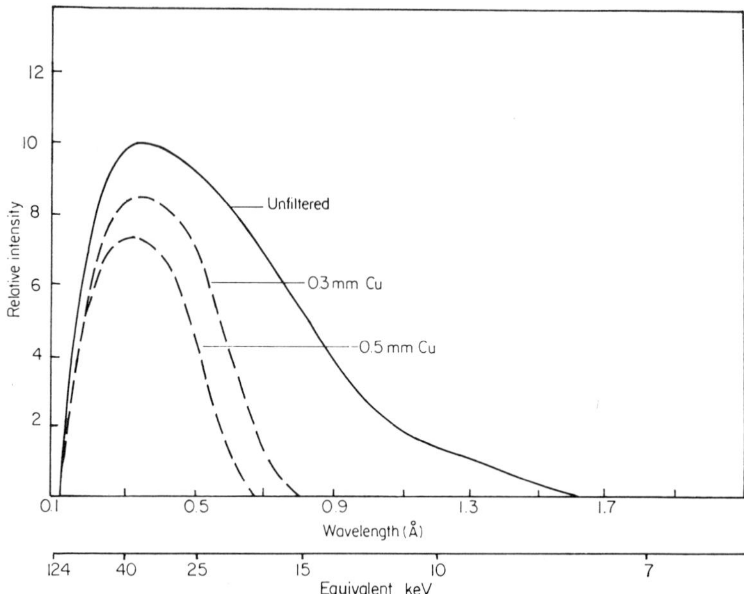

igure 1.6. The effect of filtration on the X-ray spectrum (Max. energy $100 \, \mathrm{kV}_p$) (Diagrammatically)

——————————— unfiltered

– – – – – – – – filtered by 0.3 and 0.5 mm Cu.

illustrated in Figure 1.7. This graph refers to the absorption of X-rays by a layer of silver bromide (the main constituent of photographic emulsion layers) and shows the dependence of the mass-absorption coefficient μ/ρ on the X-ray wavelength in Å. It is seen that μ/ρ increases with increasing wavelength, but that the increase is broken at certain wavelengths, the so called 'selective absorption edges' of silver and bromine. The occurrence of absorption edges can be explained by the theory of X-ray emission. As outlined earlier the formation of the characteristic K-radiation is preceded by the dislodging of a K-electron from its orbit. Although this may be caused by bombardment with electrons, it can also be due to the absorption of X-rays, if the quantum energy of the exciting radiation is greater than the energy required to dislodge the K-electron.

Hence if we follow the decrease of μ/ρ in Figure 1.7 with decreasing wavelength, at 0.918 Å the absorption suddenly increases because the energies of the incident quanta are large enough to dislodge the K-electrons from bromine atoms*. Characteristic K-radiation of bromine will be emitted, as further electrons jump into the places

---

*As silver bromide is an ionic crystal, we have actually to deal with bromine ions and silver ions instead of with atoms.

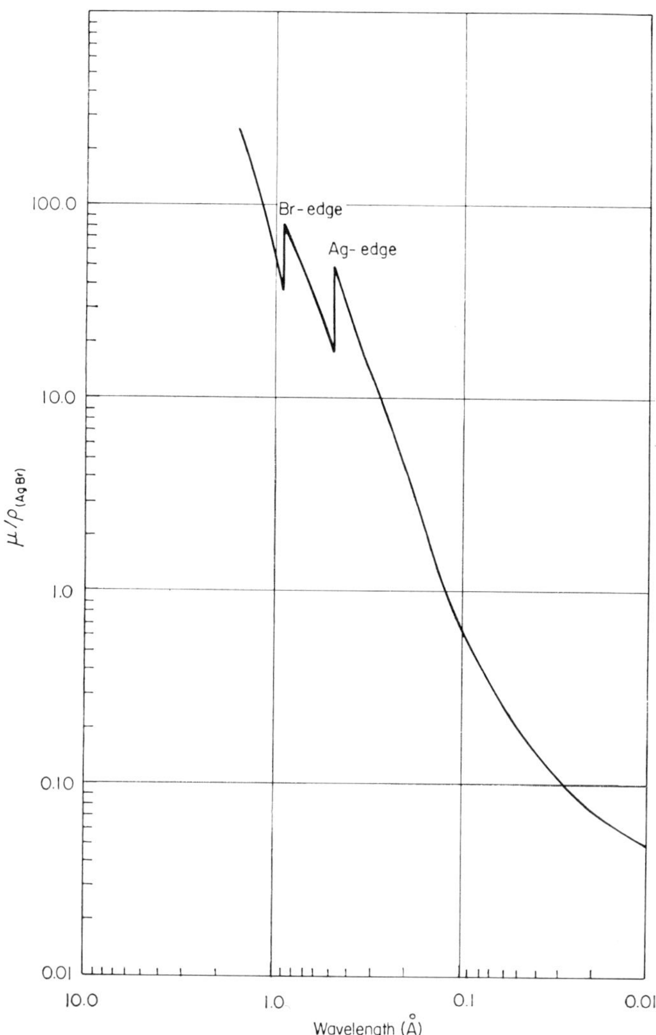

Figure 1.7. Mass-absorption coefficient of AgBr as function of wavelength (Å), showing selective absorption edges of silver and bromine.

of those dislodged. If λ further decreases the absorption drops again until at $\lambda = 0.485$ Å a second absorption edge occurs at a silver atom for a reason similar to that given for the bromine atom.

An X-ray spectrum from a tungsten target taken with a crystal spectrograph is illustrated in Figure 1.8. It reveals the selective absorption taking place at bromine and at silver in the recording medium, the photographic emulsion. Furthermore the *L*-emission lines of tungsten are shown. The *K*-lines are not resolved into their various components owing to lack of dispersion in the short wavelength region.

Selective absorption is of fundamental significance in many observations of the attenuation of X-rays. The role it plays in

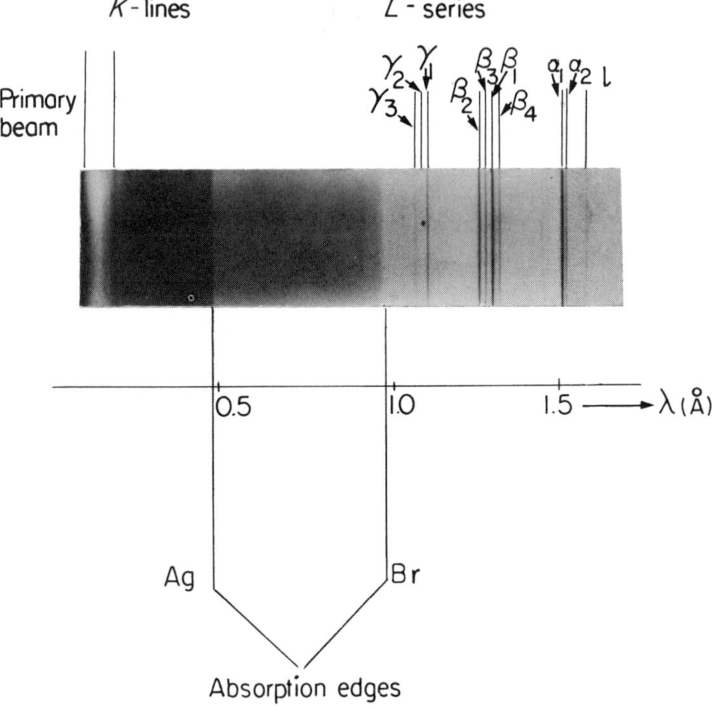

Figure 1.8. X-ray spectrum of tungsten showing Ag and Br absorption edges in photographic emulsion.

processes underlying the spectral response of photographic layers is discussed on pages 124 and 125.

1.6e   The Relative Contribution of Different Absorption Processes

In some considerations of photographic effects and of attenuation problems it is useful to know the relative contribution of different kinds of absorption processes. It is particularly important in the assessment of the photographic response to various photon energies.

For instance a strong contribution of incoherent scattering may modify the wavelength composition of a beam after this has passed through an absorber. This may have the effect that the transmitted radiation is considerably softened compared with the original beam.

In another example one may wish to know whether the scattered radiation transmitted by an absorber is predominantly in the forward direction as for instance with Compton scattering at very high energies, which is of practical importance in γ-ray radiography. On the other hand it may be of interest to know whether the primary beam is hardly influenced by scattering when passing through an absorber, but is chiefly dependent on photoelectric absorption.

The relative contribution of coherent and Compton scattering, photoelectric absorption, pair formation and also the total absorption effects as a function of photon energy are illustrated for various absorbers in Figures 1.9–1.11.

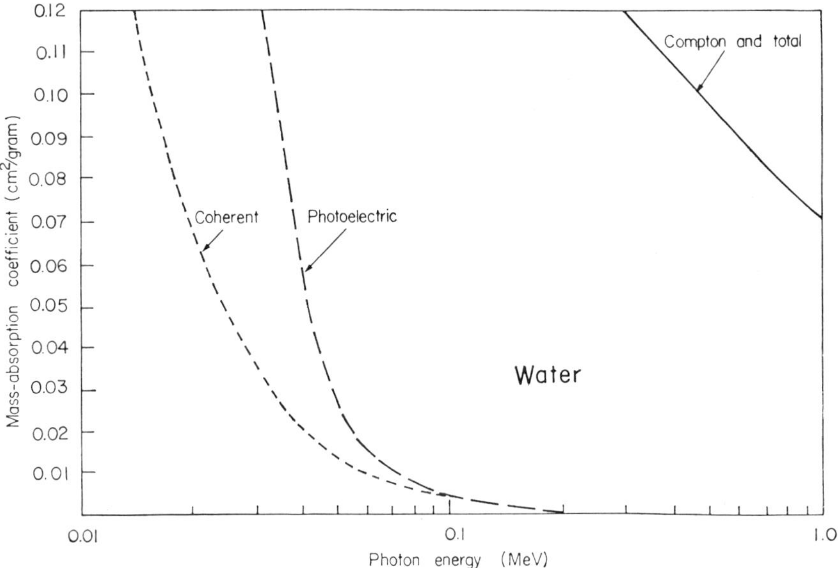

Figure 1.9. Coherent, Compton-, photoelectric and total mass-absorption coefficients for 'water' as function of photon energy. (These curves have been plotted from tables of The National Bureau of Standards, Circular 583, by Gladys White Grodstein.)

In order to demonstrate the striking changes in the relative share of the various absorption processes with photon energy, the mass-absorption coefficients are given in these diagrams for three absorbers which differ widely in their atomic number. When comparing the three

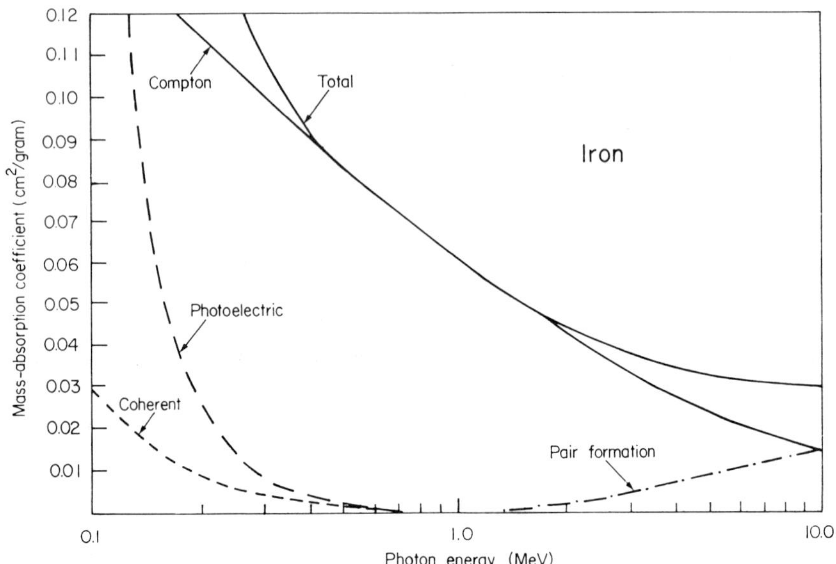

Figure 1.10. Coherent-, Compton-, photoelectric-, pair formation- and total mass-absorption coefficients for 'iron' as function of photon energy. (This diagram is plotted from tables of The National Bureau of Standards, Circular 583 by Gladys White Grodstein.)

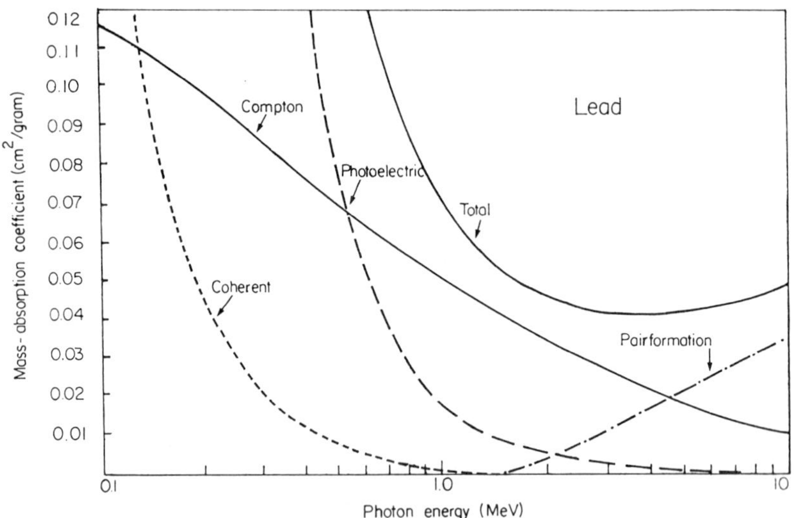

Figure 1.11. Coherent, Compton-, photoelectric-, pair formation- and total mass-absorption coefficients for 'lead' as function of photon energy. (This diagram is plotted from tables of The National Bureau of Standards, Circular 583 by Gladys White Grodstein.)

26

figures it should be noted that the diagram for water as an absorber refers to photon energies from 0.01 to 1 MeV, whereas the graphs for iron and lead refer to photon energies from 0.1 to 10 MeV.

As can be seen from these figures the photoelectric effect is more important at lower photon energies and varies with increasing energy approximately as $1/E^3$, where E is the energy. Coherent scattering becomes quite appreciable with absorbers of high atomic number such as lead. Compton scattering is predominant at medium and at high energies, although it is exceeded by photoelectric absorption at about 500 keV and lower energies, using lead as an absorber. Pair formation plays an increasing role at energies between 5–10 MeV and this is greater the higher the atomic number of the absorber.

1.6f Half Value Layer

Monochromatic radiation is uniquely defined by its wavelength, and the respective mass-absorption coefficients for the given wavelength and absorbing material can easily be found by reference to tables. (See Appendix.) Difficulties occur however in defining the quality of heterogeneous radiation. One approach to this problem is by means of filter analysis, using the so called 'half value layer' (HVL) as a measure of the quality of radiation of mixed wavelengths.

The half value layer is defined as that thickness x_h of a given filter material which reduces the incident radiation flux to half its original value. Thus

$$I/I_0 = 1/2 = e^{-\mu x_h}$$

$$\log 2 = \mu x_h \log e \tag{1.20}$$

$$\text{i.e. } x_h = 0.693/\mu$$

Equation (1.20) shows how the half value layer (HVL) is related to the absorption coefficient μ, but the equation holds strictly only for monochromatic radiation. If one deals with a heterogeneous radiation, use is often made of an equivalent wavelength, which is defined as the wavelength of that monochromatic radiation which shows the same penetrating power, i.e. the same HVL of a given material as the heterogeneous radiation.

The determination of the half-value layer of a heterogeneous beam is rather a 'rough approach' to a specification of the quality of an X-ray beam. This is the more evident, if one considers that X-ray beams having the same HVL may show very wide differences in wavelength distribution. The ambiguity can be reduced by determining consecutive HVL's, as a little consideration will show that due to the loss of softer components the HVL increases with progressive filtration up to a filtration thickness where the X-ray beam becomes almost homogeneous.

In a practical determination of this kind it is convenient to interpose increasing filter thicknesses of the same material, say copper, between the X-ray tube and the detector, such as an ionization chamber. (See p. 43). Half value layer measurements should be carried out with as little interference as possible from modified scattered radiation, as this would otherwise falsify the results. In order to reduce the amount of scattered radiation reaching the detector, a narrow X-ray beam should be used, which is just wide enough to cover the detector adequately. The filters should preferably be placed half-way between the source and the detector, as it has been found that a minimum of scattered radiation reaches the detector under these circumstances.

By plotting the logarithm of the transmitted flux of a heterogeneous X-ray beam against the thickness of the absorber, a transmission curve is obtained, which is illustrated in Figure 1.12. At a certain point of this curve, the softer components of the X-ray beam have been almost eliminated by absorption and only a relatively narrow band of X-ray wavelengths remains. This point in the curve is called the point of homogeneity and is reached by a filtration of 0.9 mm Cu in the example given in Figure 1.12. It represents that region beyond which the HVL remains more or less constant. The HVL may now be determined in the graph by taking differences between two filter thicknesses (within the straight line region) for which the amount of transmitted flux has been halved. For convenience we take as initial radiation flux the value of 2 on the ordinate of Figure 1.12, which is obtained by a filtration of 1 mm Cu and we find that 0.5 mm Cu filtration is required to reduce this radiation flux to half. Hence the half value layer is 0.5 mm Cu, being equivalent to $\mu/\rho = 1.55$ according to equation (1.20), using $\rho_{Cu} = 8.93$. The corresponding wavelength as derived from tables is 0.200 Å or 62 keV. (See Table 14.3 and Figure 14.5.)

It should be emphasized that the accuracy of HVL determination may be influenced by many factors, such as radiation scattering, the diameter of the beam, the position of the ionization chamber, the purity of the filter material and the geometry of the experimental set up.

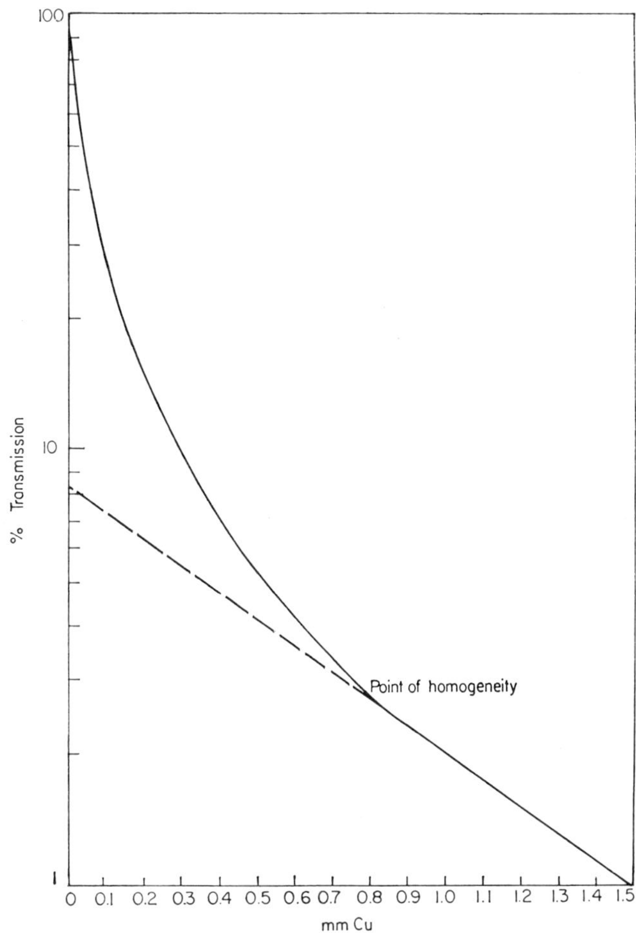

Figure 1.12. Transmission curve of heterogeneous radiation to illustrate half
value layer measurement.

1.7 ATTENUATION OF PARTICLES

1.7.a Interaction of Charged Particles with Matter

If charged particles travel through matter, they lose energy through
their interactions with the atoms which they encounter. These
interactions depend on the charge and energy of the particle and on
the composition of the material through which they pass. The predominant interactions may be elastic scattering from orbital electrons

or inelastic collisions in which orbital electrons are removed (ionization) or radiative processes leading to Bremsstrahlung, when the particles pass close to atomic nuclei. The rate of energy loss per unit length of path (dE/dX) suffered by a charged particle in an absorbing medium can be expressed in a simplified form by the expression

$$ -\frac{dE}{dX} \propto NZ\,\frac{z^2}{v^2} f(v) \tag{1.21} $$

where Z is the atomic number of the absorbing material, z and v are the charge and velocity of the particle respectively, and N is the number of atoms per cm³ in the absorbing material. Most of the energy loss which a particle suffers whilst passing through matter is due to ionization processes. The relation above shows that the rate of loss of energy of a particle per unit length of path is essentially independent of the mass of the particle. It is proportional to the square of the particle charge and inversely proportional to the square of the particle velocity. Hence a charged particle at low velocity will lose more energy per unit length of path than a particle at high velocity. The number of ion pairs produced per unit path length of a charged particle is referred to as the *specific ionization*. The fact that the specific ionization decreases with increasing velocity of a particle is a result of the decreasing time for interaction with atoms.

More elaborate expressions on the energy loss suffered by charged particles have been proposed by B. Rossi and G. Greisen (1941), H. A. Bethe and J. Ashkin (1953), J. H. Smith (1947) and others. A detailed equation which considers the ionization potential of the stopping atoms and the gradual rise of the energy loss at relativistic energies is quoted below

$$ -\frac{dE}{dX} = \frac{4\pi e^4 z^2}{mv^2} \times NZ\left[\log_e \frac{2mv^2}{I} - \log_e (1 - \beta^2) - \beta^2\right] \tag{1.22} $$

where ze is the charge of the particle
  v is the velocity of the particle
  $\beta = v/c$, c is the velocity of light
  N is the number of atoms per unit volume of stopping material
  Z is the atomic number of the stopping atoms
  m is the mass of the electron
and I is the average ionization potential for the stopping atoms.

As in the simplified equation (1.21) the energy loss decreases with the square of the velocity of the particle. At very high velocities however mv^2 approaches mc^2 and the relativistic term in the second part of the equation grows; hence the energy loss passes through a minimum and gradually rises again at very high energies.

The response of photographic material strongly depends on the energy loss a particle suffers in a photographic layer and the reader will find references to this subject on pages 96 and 159.

The rate of energy loss due to ionization of a charged particle traversing matter can be revealed in the form of tracks which a particle leaves in gases, liquids and solid material. The well known methods used for obtaining particle track photographs are, in historical order, the Wilson cloud chamber, the photographic method using nuclear track emulsions, the bubble, and the spark chamber. Particle tracks using nuclear track emulsions are shown in Figures 4.3, 5.3–5.5 and 6.23.

1.7b Linear Energy Transfer (LET)

It is important to realize that radiation effects on biological and on photographic material are influenced essentially by the transfer of energy to the irradiated material rather than by the loss of energy a charged particle suffers whilst passing through matter. We have seen that a charged particle, whilst passing through matter, causes ionizations and excitations along its track. The energy loss produced along the paths of charged particles may be due to elastic or inelastic collisions. In elastic collisions no kinetic energy is transferred to the intrinsic energy of the atoms. For instance the deflection of particles may cause the emission of photons, which leads to a radiative energy loss. Inelastic collisions on the other hand, may cause ionization or excitation of atoms.

The linear energy transfer, abbreviated by LET is defined by

$$\mathrm{LET} = \frac{dE_s}{dX} \tag{1.23}$$

where dE_s is the average energy locally imparted to the medium by a charged particle of specified energy in traversing a distance dX. The LET is thus the amount of energy transferred through inelastic collisions per unit length of track and is usually expressed in kiloelectronvolts per micron.

The LET varies for a given medium with the type of particle and is of a high value when the energy of the particle is low. The specific ionization (SI) of a charged particle which was defined as the number of ion pairs produced per unit length of track is related to the LET by

$$\mathrm{LET} = \mathrm{SI} \times W \tag{1.24}$$

where W is the energy absorption required to create one pair of ions which is found to be of the order of 34 eV per ion pair in air. An example shows that the formation of 120 ion pairs per micron, i.e. 120×34 eV corresponds to 4.08 keV per micron.

1.7c   The Stopping of α-Particles

α-Particles emitted in radioactive disintegrations are easily absorbed even by a sheet of paper or by a few microns of aluminum. This is understandable as at energies of a few MeV the α-particle, which has a charge of two units and a relatively low speed due to its great mass (if compared for instance with a β-particle of similar energy) dissipates much energy by ionization when passing through matter. The total number of ions produced in air by the complete absorption of an α-particle of energy 7.68 MeV (emitted by Ra C′) has been estimated and was found to be 2.2×10^5 ion pairs. From this it was concluded that the average energy loss per ion pair formed is

$$(7.68 \times 10^6)/(2.2 \times 10^5) = 35 \text{ eV}.$$

The range of α-particles in air, i.e. the distance to which they can travel before they lose practically all of their ionizing power is of the order of a few centimeters. Considering that the α-particle of Ra C′ of 7.68 MeV energy produces 2.2×10^5 ion pairs when it is completely absorbed and as its range is about 7 cm in air, it can be concluded that it loses on the average 110000 eV per mm distance in air if the ionization were uniform along the path of the particle. The range mentioned above refers to α-particles emitted by radioactive disintegrations but much higher energy α-particles and thus much greater ranges can be achieved artificially.

The range of α-particles in air can also be expressed by an empirical formula, such as

$$R_\alpha = 0.318\, E^{\frac{3}{2}} \tag{1.25}$$

where R_α is the range in cm of air at atmospheric pressure and 15°C. and E is the energy in MeV. This relation holds well in the energy region between 4–7 MeV.

Instead of quoting the range of an α-particle it is sometimes useful to express this in an alternative form, namely by the equivalent thickness in mg/cm² which is defined by the relationship

Equivalent thickness in mg/cm² = range(cm) × density × 1000.

The equivalent thickness is thus the mass per unit area which is required to absorb the α-particle. Ranges for α-particles referring to various materials are given in the Appendix.

Use is often made of the relative stopping power which is defined as the ratio of the range of α-particles in air to the range in a given material for the same source of radiation. Examples of typical relative stopping powers for Ra C′ α-particles, having an energy of 7.68 MeV, are 1700 for aluminum, 3800 for copper, and 4700 for silver.

1.7d   The Stopping of β-Particles

β-Particles (or electrons) are much more penetrating than α-particles of the same energy. Their specific ionization is therefore very much lower than that of α-particles of the same energy, as the ionizations are distributed over a much longer range. Whereas an α-particle, having a kinetic energy of 3 MeV, has a range of 1.6 cm in air and produces about 5000 ion pairs per mm on the average, an electron of the same energy has a range of about 1300 cm in air and creates only about 5 ion pairs per mm on the average. The reader should be reminded that the respective ion densities of the two particles change along their paths and become considerably greater towards the ends of their passage.

The path length traversed by a *β*-particle is not easy to measure with precision, as owing to its small mass, the particle suffers deflection at collisions with electrons present in the absorber, particularly when its energy is low. Clearly, the path length will not only depend on the energy, but also on the number of encounters (see electron track, Figure 4.3). Using elements of various atomic number as absorbers, the amount of scattering at a given angle is found to be proportional to the square of the atomic number.

Experimentally a measure of the electron range can be obtained as follows. A monoenergetic beam of *β*-particles falls upon a layer of the absorbing material and the number transmitted is measured by some suitable device such as an end-window counter. It is found that the counting rate decreases almost linearly with increasing thickness of absorbing layer, except towards the end of the paths of the particles when the transmission curve tails off owing to multiple scattering. (See Figures 1.13a and b.) The extrapolated range R is found from the point of intersection between the extrapolated straight line of the curve and a line parallel to the abscissa leading through the tail of the curve. With decreasing energy of monoenergetic electrons (<0.2 MeV) the transmission curve deviates more and more from a straight line.

β-particles, as emitted from radioactive isotopes are not monoenergetic but cover a continuous range of energies (see Figure 1.3) and the transmission curves are more complex in character. If the logarithmic counting rate is plotted against the thickness of the absorber an almost linear graph is obtained. (See Figure 1.13b.)

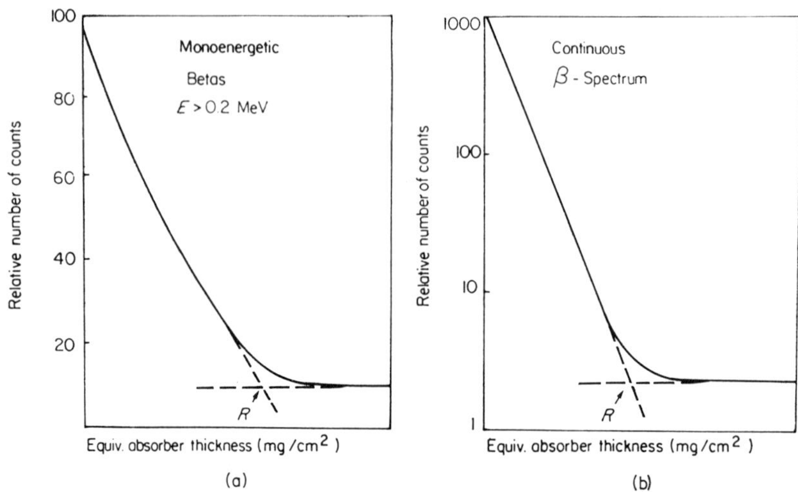

Figure 1.13. β-Ray transmission curves. (a) For monoenergetic β's using a linear graph. (b) For a continuous β-ray spectrum, using a semilogarithmic graph. (R is the extrapolated range in mg/cm².)

The linearity of this curve is accidental and is partly caused by the continuous energy spectrum of the β-particles. Beyond a certain thickness it is often found that the counting rate does not decrease any more but remains practically constant. This phenomenon may in some cases be due to the production of Bremsstrahlung caused by the bombardment of the absorber by electrons.

As for α-particles, β-particle ranges are often quoted in equivalent thicknesses such as in mgm/cm² or in gm/cm². Ranges for β-particles in air and also in nuclear emulsions are given in the Appendix.

A simple empirical formula relating the range of β-particles and their energy is Feather's rule

$$R_e = 0.542E - 0.133 \qquad (1.26)$$

where R_e is the range in gm/cm² and E is the maximum energy in MeV. This rule holds only for $E > 0.6$ MeV. Empirical formulae for other energies are found in the appropriate literature, e.g. Handbook of Radiological Health, see Bibliography.

1.7e Neutron Absorption

Neutrons, being uncharged particles, do not react with orbital electrons of atoms when passing through matter and thus do not directly cause ionization. They may, however, initiate ionization processes indirectly by a variety of interactions with matter.

The interactions of neutrons with matter can be classified chiefly into absorption processes where neutrons are captured by nuclei to form compound nuclei and into scattering processes where neutrons may undergo inelastic and also elastic collisions. In the elastic collisions the kinetic energies of the deflected neutrons can be calculated by the simple laws of Newtonian mechanics from the energies of the incident neutrons and the angles of impact. (See p. 177.) In inelastic scattering collisions (which occur mainly with neutrons having energies above 1 MeV) nuclei may be left in an excited state. This process is more typical for heavier elements.

1.7f Elastic Collisions

Each neutron, when colliding elastically, loses a fraction of the energy it had before the collision took place. This fraction is the greater, the smaller the mass of the nucleus with which the neutron collides, as it will impart a greater amount of energy to a light than to a heavy nucleus. Hence fewer collisions would be required to slow down fast neutrons after impact with light nuclei, e.g. those of hydrogen, than with heavy nuclei. An example may illustrate this:

A fast neutron having a kinetic energy of 1 MeV requires about 18 collisions with hydrogen nuclei to become a thermal neutron, but about 150 collisions would be required with oxygen nuclei to obtain the same degree of thermalization. Materials used in practice for slowing down fast neutrons by means of elastic collisions are, for example graphite, paraffin wax, and other low atomic number materials. Graphite is in widespread use as a moderator in certain types of nuclear reactors.

1.7g Nuclear Cross Sections

The chance of an elastic or inelastic collision occurring between an incident particle and a nucleus is given by the so called nuclear cross-section of the material under consideration. The concept of cross-section can be simply visualized. Suppose we have a layer of matter upon which neutrons are impinging; the dimensions of the nucleus are so small that each one can be thought of as presenting a small 'target area' to the incident radiation. A neutron hitting this 'target' will be involved in a collision, otherwise it will be effectively without interaction. As the geometrical cross-section of a nucleus is of the order of 10^{-24}cm^2, the probability of the number of nuclear processes or reactions occurring is usually expressed in units of 10^{-24}cm^2. The unit of the nuclear cross-section (σ) is called the 'barn'.

Suppose now a parallel beam of I_0 neutrons per cm² all having the same velocity, falls perpendicularly on a layer of thickness dx. Then the volume traversed will be dx per cm² and $N\,dx$ nuclei are present, if there are N nuclei per unit volume. The effective area is however $N\sigma\,dx$, and the number of the incident particles which suffer a collision is therefore $I_0 N\sigma\,dx$. It follows that the number I_1 of neutrons emerging from a layer of thickness x can be computed from

$$I_1 = I_0 e^{-N\sigma x} \tag{1.27}$$

The number of reactions which take place per second in the volume can be calculated from $I_0 - I_1$ which is equal to the number of neutrons lost from the beam. $N\sigma$ is often called the macroscopic cross-section and is usually denoted by Σ. As neutrons may undergo different types of reaction, the total cross section σ_{total} is the sum of a variety of cross-sections, such as those due to scattering, to fission, capture, and to other processes.

Nuclear cross-sections may have values from a small fraction of a barn (millibarn) up to thousands of barns. Although the term cross-section is more frequently used in connection with particles, it is also sometimes applied to the attenuation of photons in the form of an absorption coefficient.

1.7h   Neutron Capture

If a slow neutron approaches a nucleus it does not encounter a potential barrier as a charged particle does. In fact at very close distances, short range attractive forces set in which may lead to the capture of a neutron by the nucleus.

Capture processes follow in general the $1/v$ law (v is the velocity of the neutron). This means that the probability of capture is inversely proportional to $\sqrt{E}$ (E being the energy). It may be noted that the elastic collision cross-section is much greater for biological tissues than the capture cross-section, i.e. fast neutrons will be slowed down thermal energies before they are captured by nuclei.

The capture of a neutron by a nucleus leads to an increase of its ergy by about 7–8 MeV plus the kinetic energy of the neutron. (The latter is of course negligible in the case of thermal neutrons.) Hence the compound nucleus, having captured a neutron, will be in an excited state and attains stability for instance by the emission of a proton, an α-particle or a γ-ray photon. Such reactions are called (n, p), (n, α) or (n, γ) reactions respectively.

One of the most common nuclear processes, in which a slow neutron is captured by a nucleus, is the transmutation of the target element into an isotope of the same element but with a mass number which is one unit higher. Typical examples of (n, γ) reactions of this kind are

$$
\begin{aligned}
{}^{1}_{1}\text{H} + {}^{1}_{0}n &\rightarrow {}^{2}_{1}\text{H} + \gamma \\
{}^{103}_{45}\text{Rh} + {}^{1}_{0}n &\rightarrow {}^{104}_{45}\text{Rh} + \gamma \\
{}^{109}_{47}\text{Ag} + {}^{1}_{0}n &\rightarrow {}^{110}_{47}\text{Ag} + \gamma
\end{aligned}
\qquad (1.28)
$$

The product of the (n, γ) reaction is sometimes radioactive as in the case of ^{104}Rh and ^{110}Ag which emit β-particles. In these reactions the ratio of neutrons to protons in the nucleus has been increased by the capture of a neutron near to the limit of stability. The emission of β's occurs during the replacement of the neutron by a proton. Hence the neutron to proton ratio is returned to the equilibrium value.

Examples of (n, α) and (n, p) reactions are for instance

$$
{}^{10}_{5}\text{B} + {}^{1}_{0}n \rightarrow {}^{7}_{3}\text{Li} + {}^{4}_{2}\text{He}
$$

$$
\qquad (1.29)
$$

$$
{}^{14}_{7}\text{N} + {}^{1}_{0}n \rightarrow {}^{14}_{6}\text{C} + {}^{1}_{1}\text{H}
$$

In the ^{14}N reaction the product ^{14}C is also radioactive and decays by the emission of electrons according to

$$
{}^{14}_{6}\text{C} \rightarrow {}^{14}_{7}\text{N} + e^{-} + E_{\text{max}} \qquad (1.30)
$$

where $E_{\text{max}} = 0.155$ MeV is the maximum energy of the continuous spectrum of electrons emitted. The half life of ^{14}C is 5568 years.

1.7i Resonance Capture of Neutrons

The dependence of the probability of capture cross-sections (σ_a) on neutron energy and on the element concerned is very complex. The cross-section for thermal neutrons is found to be relatively low for lighter elements up to a mass number of about 100 (apart from helium-3, lithium-6, and boron-10) and to have values of fractions of a barn to a few barns. For a considerable number of elements having mass numbers exceeding about 100, (σ_a) values become extremely high at particular energies, so called 'resonance peaks'. Elements showing exceptionally high peaks are for instance cadmium, rhodium, samarium, gadolinium and others. Such peaks occur with neutron energies between a fraction of 1 ev to about 100 eV but a few occur also with higher energies. The typical cadmium resonance peak of 7800 barns at 0.176 eV and other peaks for Au and Rh are shown in Figure 5.11. These high cross-sections for cadmium rhodium, gadolinium, and other elements, are utilized extensively in neutron dosimetry and in neutron radiography (p. 214 and p. 490). With regard to the resonance peak of cadmium the reader is referred to the remarks in the next section.

The fact that the probability of neutron capture may be exceptionally high, as in the resonance peaks, can be interpreted as follows. It is believed that the energy of the compound nucleus, which is formed by the capture of the neutrons, is very close to one of its excited energy levels and if this happens the resonance capture may occur. The reader will find a table of some slow neutron resonance cross-sections in the Appendix.

1.7j Application of Capture Cross-Section

An example of the use of cross-sections is shown in the following calculation. Let us find what fraction of a thermal neutron flux has been lost on passing through 0.1 mm thickness of cadmium foil of 1 cm² area. As mentioned earlier, cadmium has a high capture cross-section for slow neutrons.*

First we have to find the number N of nuclei presented to the flux per unit volume (in cm³). This is derived from

$$N = \frac{N_0 \rho}{A} \tag{1.31}$$

where N_0 is Avogadro's number (6.02×10^{23}), ρ is the density of cadmium (8.65 grms/cm³) and A is the atomic weight of cadmium (112.4). Hence by applying equation (1.31) and using the value of 2700×10^{-24} for the microscopic cross-section (σ_a) of cadmium for thermal neutrons where x is equal to 0.01 cm for the thickness of the foil, we arrive at

$$I_1 = I_0 e^{-N_0 \sigma_a \rho x / A} \tag{1.32}$$

$$= I_0 \exp -\left(\frac{6.02 \times 10^{23} \times 8.65 \times 2700 \times 10^{-24} \times 0.01}{112.4} \right)$$

$$= I_0 e^{-1.24}$$

$$= 0.289$$

Hence 71.1 per cent of the number of incident neutrons are lost by the passage through 0.1 mm of cadmium.

* It should be noted that the cross-section for cadmium used in the following example is 2700 barns instead of 7800 barns as quoted earlier. The reason for this is that natural cadmium shows a cross-section of 2700 barns for 'thermal' neutrons. The 7800 barns refer also to natural cadmium but at a neutron energy of 0.167 eV. On the other hand an isotope of cadmium, namely ^{113}Cd (relative abundance of 12.26 per cent) has a cross-section of 20,000 barns for 'thermal' neutrons. (Thermal neutrons are specified here as having an energy of 0.025 eV.)

BIBLIOGRAPHY

Attix, F. H. and Roesch, W. C. (1968) *Radiation Dosimetry.* Vol. I. Academic Press, London and New York.

Attix, F. H. and Roesch, W. C. (1966) *Radiation Dosimetry.* Vol. II. Academic Press, London and New York.

Compton, A. H., and Allison, S. K. (1953) *X-rays in Theory and Experiment.* Van Nostrand, Princeton.

Evans, R. D. (1955) *The Atomic Nucleus.* McGraw-Hill, New York.

Glasstone S. *Source Book of Atomic Energy.* Van Nostrand, Princeton.

Handbook: *Radiological Health.* U.S. Department of Health Education and Welfare, Public Health Service.

Heitler W. (1944) *Quantum Theory of Radiation.* Oxford University Press, London.

Hine, G. J. and Brownell, G. L. (1956) *Radiation Dosimetry.* Academic Press. New York.

Johns, H. E. (1964) *The Physics of Radiology,* Charles Thomas, Springfield, Ill.

Kaplan, I., *Nuclear Physics,* Addison-Wesley, Reading, Mass.

Lapp, R. E., and Andrews, H. L., *Nuclear Radiation Physics.* Pitman, London.

Meredith, W. J., and Massey, J. B. (1968) *Fundamental Physics of Radiology.* John Wright, Bristol.

Richtmyer, F. K., and Kennard, E. H., *Introduction to Modern Physics,* McGraw-Hill, New York.

Semat, H., *Atomic Physics,* Rinehart, New York.

Shankland, R. S., *Atomic and Nuclear Physics.* Macmillan, New York

Sproull, W. T., *X-rays in Practice,* McGraw-Hill, London-New York.

Tolansky, S., *Introduction to Atomic Physics,* Longmans, London.

REFERENCES

Bethe H. A. and Ashkin J. (1953) *Experimental Nuclear Physics* Vol. 1, Wiley, New York.

Hughes D. J. (1952) *Neutron Cross Sections,* A.E.C.U. 2040, Washington D.C. Office of Technical Services, Dept. of Commerce & Supplements.

National Bureau of Standards, Circular 583, *X-ray Attenuation Coefficients from 10 KeV to 100 MeV,* by Gladys White Grodstein (1957).

Rossi B. and Greisen K., (1941) Cosmic-ray theory, *Rev. Mod. Phys.,* **13,** 249.

Smith J. H. (1947) Theoretical range energy values for protons in air and aluminium. *Phys. Rev.,* (2)**71, 32.**

2

Ionization Measurements and Units

2.1 INTRODUCTION

If photographic applications of ionizing radiation are to be of a quantitative nature, calibration of the radiation-sensitive material is required in terms of units which permit an assessment of the energy dissipated by ionizing radiation. Hence it is natural to discuss the principles of those detection methods which lead to the most direct assessment of ionization in air. These methods were not only historically the first on which the internationally accepted unit of ionization measurement was based, but are still the only generally accepted ones. In this chapter therefore, a brief review is given of the basic principles involved in the ionization measuring method and of the units used in this respect, such as the roentgen, the rad, and the rem.

Although so much has been written on the dangers and hazards of ionizing radiation, it is nevertheless essential in a book of this kind, to reiterate the great importance of this subject. For this reason, a short section at the end of this chapter deals with the maximum permissible radiation quantities for human beings and some protection data referring to the handling of γ-ray sources are presented on page 439. Biological effects caused by ionizing radiations may influence the health of exposed persons and may also influence the individual's descendants (genetic effects); hence familiarity with the quantitative aspects of 'Health Physics' is required to avoid risks involved in radiation exposures. Literature references on these important aspects of protection problems are found at the end of this chapter and information about the literature on 'Health Physics' is given in the bibliography.

2.2 RADIATION DETECTION BY IONIZATION

The detection of ionizing radiation may be carried out by various detection methods such as those based on ionization, calorimetry, chemical effects, photographic effects, fluorescence and scintillation, and solid state physics phenomena. In this chapter only some of the more basic ideas of the air ionization method are briefly discussed, since a review of the other detection methods (apart, of course, from the photographic method) is beyond the scope of this book. The reasons for selecting the ionization method have already been mentioned in the introduction. The ionization in air is not only regarded as the most accurately reproducible method, but also one which, within limits, is independent of the quality (i.e. wavelength composition) of the radiation.

Whenever X-rays or γ-rays pass through air, pairs of ions are formed and these recombine very rapidly after formation, unless they are prevented from doing so by special means. If an electric field is applied to such an ionized gas the ions can be drawn apart. The electric field can be produced by keeping a potential difference between two electrodes which are placed in the volume of air. The positive ions will then move to the negative electrode and the negative ions to the positive electrode. By using first a small potential, and then by gradually increasing it, the current flowing through the circuit increases, owing to the growth of the number of ion pairs which are separated, until a point is reached when most of the ions are prevented from recombination. If the potential is further increased, the current remains practically constant and hence a saturation potential is reached. The value of this depends on the air volume, on the distance between the electrodes, their design, and on the number of ion pairs produced in the volume. If the volume of air is surrounded by walls, preferably of low atomic number material, the device is referred to as a cavity ionization chamber and the ionization current flowing is found to be proportional to the number of X-ray or γ-ray quanta passing through the ionization chamber. Hence a relative measure of the quantity of radiation can be obtained in this manner.

When X-rays strike the walls and electrodes of such a chamber, the radiation is not only attenuated by the walls and electrodes, but also, electrons, are released from the latter, thus modifying the ionization as compared with that obtained in free air. Further, the free path of the electrons released in the air may exceed the linear dimensions of the chamber, especially when hard radiation is used, so that their path length in the air volume is less than their total range. In this situation two approaches are possible, the free air-ionization chamber and the cavity ionization chamber.

2.2a   The 'Free-Air' Ionization Chamber

The first step is to ensure that the walls and electrodes are remote from the volume of air irradiated, so as to allow electrons generated in the air volume to complete their range in air. The so called free-air ionization chamber embodies these precautions and its principle is illustrated in Figure 2.1.

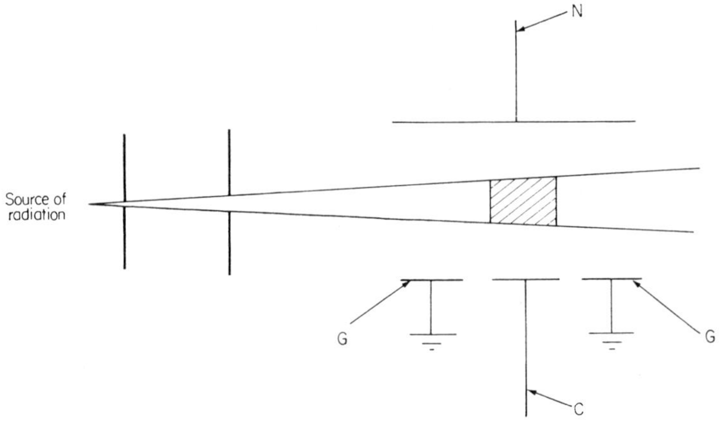

Figure 2.1. Principle of free-air ionization chamber.

A volume of air is defined by means of a collimated X-ray beam and a system of collector plates, but the irradiated volume is not enclosed by walls. The collector system is so designed that all the ion pairs formed by electrons, which are produced within the irradiated volume of air, are collected. The measuring or collecting electrode C is highly insulated from the casing and is connected with an electrometer which collects the charge. The two guard electrodes G are earthed or connected to a casing which itself is earthed. One of the chief purposes of the guard electrodes is to limit the air volume from which ions are collected by the electrode C. The opposite electrode N carries a collection potential and all four electrodes are arranged parallel to the direction of the X-ray beam. The electrostatic forces on the ions in the volume of air concerned all act in a direction perpendicular to the electrodes. The current collected is proportional to the number of ions produced by X-rays generated in the shaded volume.

It is evident that not all the ions produced within the shaded volume arise from the air within this volume, as some originate from the irradiated air in front and beyond the shaded volume. Others are lost from the shaded volume. Generally, it is assumed that the number of ions gained and lost in this way compensate each other to form an equilibrium. In order to determine the absolute number of ions produced per unit volume, several correction factors have to be introduced concerning field distortion, temperature and pressure of air, X-ray absorption in air, and others.

The free-air chamber which is sometimes referred to as the standard free-air ionization chamber just described is of rather large dimensions and therefore inconvenient and unsuitable for transport. It is however used as the parent instrument in the national physical laboratories of the various countries for the calibration of other types of ionization chamber.

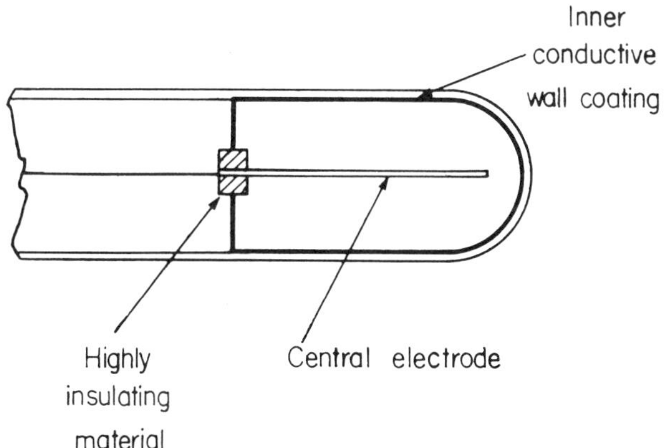

Figure 2.2. Thimble ionization chamber.

2.2b Cavity Ionization Chamber

In this second approach an air volume which is often quite small (for instance 1 cm^3) is completely surrounded by solid material, consisting of the collecting electrode, the high tension electrode, and the insulator. One variety of cavity ionization chamber (sometimes referred to as a thimble chamber) is illustrated in Figure 2.2.

In contrast to the large free-air ionization chamber, where only a small part of the volume of the chamber is irradiated by X- or γ-rays, in the cavity chamber the whole volume of air is irradiated. In this chamber there is an optimum wall thickness to establish electronic equilibrium (see p. 107), depending on the quality of the radiation applied and on the atomic number of the wall material used. The latter should preferably be of low atomic number, e.g. plastic, so that the absorption of X- or γ-rays and the corresponding electron emission are as close as possible to that of air. With these precautions the measurements are nearly independent of the wavelength of the radiation, at least within a fairly large range of photon energies. For the above reasons the inner electrode and the conductive coating of the inner wall should also be made of low atomic number material.

A typical circuit for measuring ionization currents by means of a thimble chamber is shown in Figure 2.3.

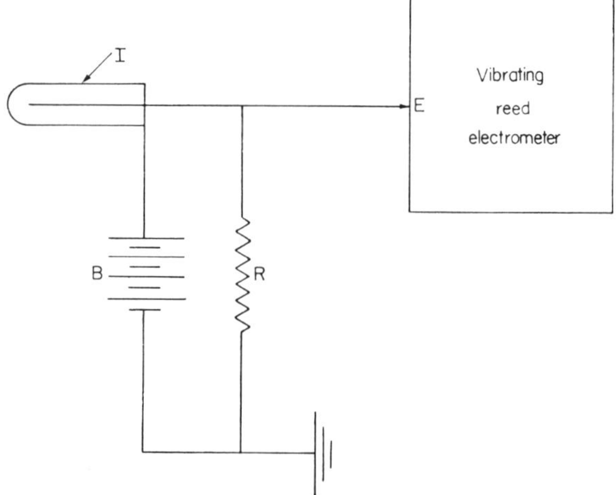

Figure 2.3. Typical circuit for a thimble ionization chamber.

I is the thimble chamber, B a battery supplying saturation potential, R a high resistance and E an electrometer. If X-rays strike the cavity chamber, the air in the chamber becomes conducting and the current supplied by the potential of the battery flows through R to earth. Hence the potential drop caused across the resistance R is measured, for instance, by a vibrating reed electrometer. Similarly if the resistance is replaced by a condenser the charge will be accumulated by it and the total ionization can be measured. Hence the available ionization measuring instruments are either rate meters or integrating meters and some of the more elaborate devices are designed for both purposes. One variety of the integrating instruments is built as a pocket ionization chamber which can be worn by a person likely to be subjected to radiation hazards.

2.3 THE I.C.R.U.

After having briefly reviewed the basic principles on which ionization measurements in air are carried out, we shall now turn to the derivation of units on which such measurements are based. Before doing so, brief mention should be made of the organization which is mainly responsible for the standardization of these units. The International Commission on Radiological Units (I.C.R.U.) is the organization which deals with the development of internationally acceptable recommendations regarding;

1. Quantities and units of radiation.

2. Procedures suitable for the measurement and application of these quantities in clinical radiology and radiobiology.

3. Physical data needed in the application of these procedures, the use of which tends to ensure uniformity in reporting.

The same Commission also considers radiation quantities and units in the field of radiation protection in close cooperation with the International Commission on Radiological Protection. (I.C.R.P.)

The reports of the I.C.R.U. have been published in a series of most useful handbooks by the National Bureau of Standards, Washington D.C., and the reader will find a reference to the titles of some of the more recent handbooks at the end of this chapter. The content of the following sections is chiefly based on the Handbooks No. 84 and No. 85, as it is of the utmost importance that international definitions and units should be adhered to as closely as possible to avoid confusion and to facilitate the interchangeability of experimental results in scientific work.

As this book deals with various types of ionizing radiation, it is opportune to give an extract of the I.C.R.U.'s distinctions between various types of radiation.

1. Directly ionizing particles are charged particles (electrons, protons, α-particles etc.) having sufficient kinetic energy to produce ionization by collision.

2. Indirectly ionizing particles are uncharged particles (neutrons, photons etc.) which can liberate directly ionizing particles or can initiate a nuclear transformation.

3. Ionizing radiation is any radiation consisting of directly or indirectly ionizing particles or a mixture of both.

2.4   THE BRAGG–GRAY PRINCIPLE AND THE ABSORBED DOSE (RAD)

The concept of a quantity of ionizing radiation arises from the need to relate the radiation incident at a point to a given physical, chemical or biological effect. Such effects are not related to the radiation flux at a point in the medium but rather to the radiation energy deposited in the medium at this point. This leads to the concept of a radiation dose as distinct from a radiation intensity.

The fundamental principle which describes the energy absorbed per unit mass of irradiated material in terms of the ionization per unit volume was laid down in its initial stage by W. H. Bragg (1912) and was expressed independently in its most fundamental form by L. H. Gray (1936, 1937). In the Bragg-Gray principle it is assumed that a small gas filled cavity is introduced in the irradiated medium under consideration. This cavity is considered to be so small that the flow of ionizing particles through it is undisturbed, i.e. that they have the same energy spectrum and number in the gas as in the medium. If the effective mass stopping power for these particles is s_m for the medium and s_g for the gas and their ratio s_m/s_g is S_M then the absorbed dose D expressed in ergs per gram is, according to the Bragg-Gray principle,

$$D = J_g W S_M \qquad (2.1)$$

where J_g is the number of ion pairs formed per gram of gas and W is the average energy in ergs expended by the ionizing particles crossing the cavity per ion pair formed. *The unit of absorbed dose is called a 'rad' and 1 rad is equivalent to 100 ergs per gram of any material.* This means that

$$D = 0.01 J_g W S_M \text{ rads.} \qquad (2.2)$$

The absorbed dose rate is dose per unit time and a particular absorbed dose rate is any quotient of the rad by a suitable unit of time such as rad/minute, rad/hour, rad/week etc.

2.5   THE EXPOSURE (ROENTGEN)

Although the absorbed dose (in rads) is undoubtedly the most satisfactory physical unit of radiation dose, its measurement is accompanied by considerable technical complications. Historically the rad was preceded by the definition of the roentgen unit (r) which was regarded until recently as the so called exposure dose. The I.C.R.U. however in its latest report wants to retain the term 'dose' for one quantity only and this is for the absorbed dose. Whilst the term roentgen retains basically its previous definition it is now regarded as an 'exposure' unit only and is abbreviated by a capital R instead of the small r.

The definition of the roentgen unit, which was first internationally accepted in 1928, was extended in 1937 to include γ-radiation. Its wording is:

'The roentgen shall be that quantity of X- or γ-radiation such that the associated corpuscular emission per 0.001293 gram of air produces, in air, ions carrying one electrostatic unit of electricity of either sign'. It should be noted that 0.001293 gram of air is equivalent to the mass of 1 cc of dry air at 0°C and 760 mm mercury pressure.

The energy required to produce one ion pair is 33.73 eV and the liberation of 1 e.s.u. per cc of air means the production of 2.082×10^9 electronic charges,* being equal to the number of ion pairs and 1 eV = 1.6×10^{-12} ergs. Hence

$$1 \text{ roentgen} = \frac{2.082 \times 10^9 \times 33.73 \times 1.6 \times 10^{-12}}{0.001293}$$

$$= 86.9 \text{ ergs/gram in air} \tag{2.3}$$

In the I.C.R.U. Report 10a (Handbook 84 (1962)) the attempt is made to define the various fundamental quantities within a common pattern and complex quantities are therefore defined in terms of the simpler quantities of which they are composed.

Hence 'exposure' (X) is defined as $\Delta Q/\Delta m$, where ΔQ is the sum of the electric charges on all the ions of one sign produced in air, when all the electrons (negatrons and positrons) liberated by photons in a volume element of air, whose mass is Δm, are completely stopped in air. Δ is a symbol indicating an averaging procedure. The special unit of exposure is the roentgen (R) and

$$1R = 2.58 \times 10^{-4} \text{ coulomb/kg} \tag{2.4}$$

This unit is numerically identical with the old one defined as 1 e.s.u. of charge per 0.001293 gram of air. Exposure rate is any quotient of the roentgen by a suitable unit of time, such as R/sec and R/minute, R/hour etc.

---

* 1 e.s.u. = $1/(4.80 \times 10^{-10})$ = 2.082×10^9 electronic charges. $(4.80 \times 10^{-10}$ electrostatic units being the charge on 1 electron.)

2.6   RELATION BETWEEN RAD IN AIR AND ROENTGEN

The absorbed dose at a point in air which is surrounded by a thickness of air equal to or greater than the range of secondary electrons and which is uniformly exposed to 1 roentgen of X- or γ-radiation is equal to 86.9 ergs per gram. Since the rad is equal to an absorbed dose of 100 ergs per gram, it will be seen that one roentgen is equivalent to 0.869 rad in air.

Thus if an ionization chamber or other measuring instrument that has been calibrated in roentgens against a standard free-air ionization chamber (under specified conditions of electronic equilibrium, see Handbook N.B.S. No. 85) records an exposure of X roentgens, then the absorbed dose $D_{(air)}$ in rad is given by

$$D_{(air)} = 0.87X \tag{2.5}$$

2.7   THE MEASUREMENT OF THE ABSORBED DOSE IN MATERIAL OTHER THAN AIR

If the absorbed dose in a medium other than air is to be measured indirectly by an ionization chamber which is calibrated in roentgens, the chamber should ideally be so small that the errors due to the insertion of the chamber into this medium should be negligible. If the roentgen unit has been measured in this way (and the specified conditions of electronic equilibrium can be assumed to hold) then the relationship between the dose absorbed in the medium and the number of roentgens (X) is given by

$$D_{med} = 0.87 \frac{(\mu_{en}/\rho)_{medium}}{(\mu_{en}/\rho)_{air}} X \text{ rads} \tag{2.6}$$

where $(\mu_{en}/\rho)_{med}$ and $(\mu_{en}/\rho)_{air}$ are the mass energy absorption coefficients for the medium and for air respectively and X is the number of roentgens. It should be noted that the mass energy absorption coefficient (μ_{en}/ρ) is derived from the total mass-absorption coefficient $\mu/\rho = (\tau/\rho) + (\sigma_0/\rho) + (\sigma/\rho) + (\pi/\rho)$ by excluding from it those components which do not present a real loss of energy. These components are those due to Compton scattered photons, characteristic radiation, Bremsstrahlung and annihilation radiation (see page 19). Hence μ_{en}/ρ indicates that fraction of the incident energy which being converted into energy of moving electrons is finally responsible for photographic, fluorescent and biological effects.

The ratio

$$\left(\frac{\mu_{en}}{\rho}\right)_{med} \bigg/ \left(\frac{\mu_{en}}{\rho}\right)_{air}$$

(see equation 2.6) is between 0.88–0.96 for water and muscle at energies of 0.01, 0.1 and 1 MeV; the ratios for heavier media are however quite considerable, and are for bone 3.58, 1.47 and 0.927 respectively for the three energies quoted above. Further values for other materials are given in Handbook N.B.S. No. 85.

2.8 THE DOSE EQUIVALENT (REM)

As the same dose produced by different types of radiation does not necessarily produce the same biological effect, it became customary in recent years to refer to the relative biological effectiveness (RBE) for the measurements of different types of radiation. The unit of 'Rem' was introduced for this purpose and until recently it was defined by

$$\text{Dose}_{(rem)} = \text{Dose}_{(rad)} \times \text{RBE} \tag{2.7}$$

Many arguments were put forward against the use of the RBE, as it was realized that the effects of two different types of radiation cannot always be uniquely distinguished by one factor, if high precision is required. The following example will illustrate this.

The same absorbed dose might be delivered to a biological medium by two different types of radiation, but if the specific ionization (i.e. the average number of ions per micron path) of the particles and their local distribution differ, the biological effect might differ as well. The use of the RBE is now restricted to the field of radiobiology only and other terms have been introduced instead for purposes of radiation protection. One of these is the Quality Factor (QF) which is related to the Average Specific Ionization (ASI), i.e. the average number of ions produced per micron of the medium concerned. For most practical purposes the following QF values can be used for the various types of radiation in use. (See Table 2.1.)

TABLE 2.1. Quality Factors

Type of radiation	QF
X-rays, γ-rays, β-particles	
of energies > 30 keV	1
of energies < 30 keV	1.7
Protons up to 10 MeV	5–10
α-particles from radioactive decay	10
Heavy recoil nuclei	20
Neutrons (dependent on energy)	2.5–10.5
Thermal Neutrons	3

There are other modifying factors which may influence the biological effect, such as the distribution factor which expresses the modification of the biological effect due to non-uniform distribution of internally deposited isotopes.

The product of the absorbed dose (rad), the Quality Factor (QF), and any other necessary modifying factor, such as the Distribution Factor (DF), is called the dose equivalent $DE_{(rem)}$, which may be written

$$DE_{(rem)} = Dose_{(rad)}QF \times DF \qquad (2.8)$$

2.9   THE CURIE

A unit of great practical importance in the use of radioactive substances, whether these are of natural or artificial origin is the 'curie', which is defined as the activity of a radioactive nuclide such that 3.7×10^{10} disintegrations per second occur in it. The term activity stands for the rate of transformation of a nuclide. The abbreviation of the "Curie" is Ci.

2.10   MAXIMUM PERMISSIBLE DOSE-RATES

The maximum permissible dose-rate is the recommended upper limit of dose (in rems) which may be received during a specified period (one hour, one week, one calendar quarter, one year etc.) by a person exposed to ionizing radiation. So far as is known, a normal person receiving the following dose rates will suffer no harmful effects. The value of 2.5 millirems per hour for occupational workers, given in the last column of Table 2.2, is based on a 40 hours working week.

TABLE 2.2. Maximum Permissible Dose Rates

Exposed group	Maximum permissible dose in rems for various periods (Whole body exposure)		
	per year	per quarter year	per week	millirems per hour
Occupational workers	5	3*	0.1	2.5
Members of the Public**	0.5			

* and ** see the following explanatory notes.

Explanatory notes: * An apparent discrepancy appears in the table above, namely that a maximum of 3 rems is permitted per calendar quarter, but only 5 rems per year. This means that a person may receive a relatively high dose of 3 rems within a quarter of a year, but should not exceed a dose greater than 5 rems per year. In other words the discrepancy refers only to the time distribution of the dose.

Occupational workers are persons who are employed in work involving the storage, manipulation, operation, use or installation of sealed sources or apparatus generating ionizing radiations.

** Members of the public are divided into special groups, such as

(a) Adults who work in the vicinity of controlled areas but who are not themselves employed in work causing exposure to radiation.

(b) Persons who enter controlled areas occasionally in the course of their duties, but are not regarded as occupational workers.

(c) Members of the public living in the neighborhood of controlled areas.

(d) The population at large.

It should be noted that the I.C.R.P. have made recommendations for each of the various exposure categories, but these are not discussed in detail here.

Maximum Permissible dose in Relation to Age

The maximum permissible dose to gonads, the blood forming organs and the lenses of the eyes for persons at any age over 18 years shall be governed by the relation

$$D = 5(N - 18) \tag{2.9}$$

where D is the dose in rems and N is the age in years.

It should be emphasized that the preceding notes on protection are very incomplete, since the dose rates mentioned above refer only to whole body exposures. Larger dose rates are permissible e.g. for extremities and skin. For data on these and other specialized conditions of personnel protection the vast available literature on this subject may be consulted. (See Bibliography at end of this chapter.) Many countries have issued their own recommendations and regulations for the protection in research work, in hospitals and in industry. Furthermore the reader may refer to the literature on the maximum permissible concentrations of radionuclides in drinking water and in air.

BIBLIOGRAPHY

Abbatt, J. D., Lakey, J. R. A., and Mathias, D. J. (1961) *Protection Against Radiation*, Cassel, London.

Attix, F. H. and Roesch, W. C. (Eds.) (1968) *Radiation Dosimetry.* I Academic Press, London and New York.

Attix, F. H. and Roesch, W. C. (1966) *Radiation Dosimetry.* II Academic Press, London and New York.

Barnes, D. E., and Taylor, D., *Radiation Hazards and Protection*. George Newnes, London.

Hine, G. J. (1967). *Instrumentation in Nuclear Medicine*. Academic Press, New York.

Ionising Radiations: Precautions for Industrial Users, Ministry of Labour, Her Majesty's Stationery Office, London. New Series No. 13.

Morgan, K. Z., and Turner, J. E. (1967). *Principles of radiation protection, An Introduction to Health Physics*. Wiley, New York.

Recommendations of the International Commission on Radiological Protection. (1966), I.C.R.P. Publication 9. Pergamon, London.

Recommendations of the International Commission on Radiological Protection (1964) I.C.R.P. Publication 4. *Protection Against Electromagnetic Radiation above 3 MeV and Electrons, Neutrons and Protons*. Pergamon, London.

Rees, D. J. (1967) *Health Physics*, Butterworths. London.

Code of Practice for the Protection of Persons Against Ionizing Radiations Arising from Medical and Dental Use. (1964). Ministry of Education and Science and Ministry of Health, Her Majesty's Stationery Office, London.

Code of Practice for the Protection of Persons exposed to Ionizing Radiations in Research and Teaching. (1964) Ministry of Labour, Her Majesty's Stationery Office, London.

REFERENCES

Bragg, W. H. (1912). *Studies in Radioactivity*, Macmillan, London.

Gray, L. H. (1936). An ionization method for the absolute measurement of gamma-ray energy. *Proc. Roy. Soc. A*. (London) 156, 578.

Gray, L. H. (1949). The experimental determination by ionization methods of the rate of emission of beta- and gamma-ray energy by radioactive substances. *Brit. J. Radiol.*, 22, 677.

National Bureau of Standards, Washington. *Radiation Quantities and Units*. I.C.R.U. Report 10a Handbook 84.

National Bureau of Standards, Washington, *Physical Aspects of Irradiation*. I.C.R.U. Report 10b, Handbook 85.

National Bureau of Standards, Washington. *Radioactivity*. I.C.R.U. Report 10c, Handbook 86.

National Bureau of Standards, Washington, *Radiobiological Dosimetry*. I.C.R.U. Report 10e, Handbook 88.

National Bureau of Standards, Washington. *Methods of Evaluating Radiological Equipment and Materials*. I.C.R.U. Report 10f, Handbook 89.

Spiers, F. W. (1956). Radiation units and theory of ionization dosimetry. In G. J. Hine and G. L. Brownell (Eds.), *Radiation Dosimetry*, Academic Press, New York.

3

The Photographic Process

3.1 INTRODUCTION

In this chapter a general review is given of the fundamentals of the photographic process using silver halides. The presentation must necessarily be limited to the photographic principles involved in the field of ionizing radiation. Familiarity with these fundamentals is useful for an understanding of the response of photographic material exposed to these radiations.

The unique position of silver bromide or generally of silver halides as the medium used in radiation sensitive layers lies in the possibility that the primary effect caused by the action of radiation can be amplified by a factor of about 10^9. This means that the few atoms of silver, formed in a crystal of the sensitive layer on exposure to radiation, can be multiplied by means of the developing process by a factor of about one thousand million times. A further outstanding property of the photographic layer is that it is capable of integrating the effects of weak intensities of radiation over long periods of time so that a permanent record of this effect of integration is formed.

In the following we shall deal first with the photographic emulsion and its preparation, then with the formation of the latent image, and finally with the basic principles and mechanisms of processing.

3.2 THE PHOTOGRAPHIC EMULSION

The radiation sensitive layer, usually called the photographic emulsion, is a product of very delicate chemical and physical operations. Essentially a photographic emulsion consists of silver bromide

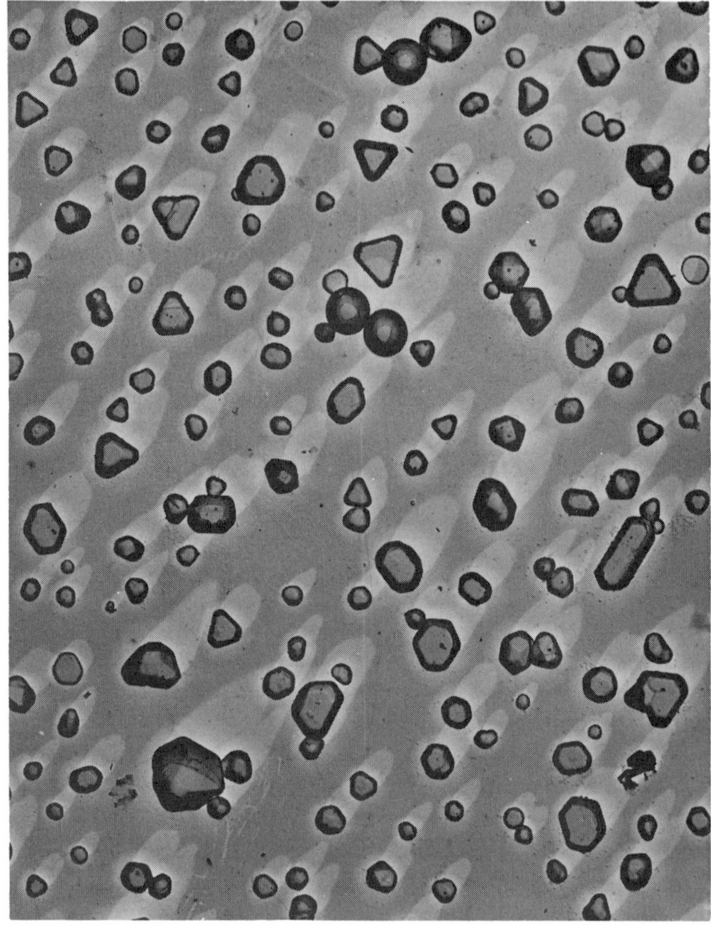

Figure 3.1. Shadowed carbon replica of typical photographic emulsion grains. Electronmicrograph ($\times 15000$) *by courtesy of* R. B. Flint, Research Laboratories, Kodak Ltd., Harrow, England.

or silver chloride crystals or grains embedded in gelatin. Figure 3.1 shows typical grains as they appear under electron-microscopic inspection. The term 'emulsion' for both liquid and dried coating is not accurately chosen as a description for the physical–chemical condition of a sensitive layer, but it will be used throughout this book as this term has been generally accepted in the literature. The silver halide which is predominantly used in high speed emulsions is *'silver bromide'* with or without small additions of iodide. Some of the slower photographic emulsions contain silver chloride and others are made up of a mixture of silver bromide and silver chloride. In the finished product the emulsion is coated on a suitable flexible support, usually made of either cellulose triacetate, cellulose mixed esters or of polyester. The support is called the film base.

Gelatin

The most suitable medium in which silver halide grains may be embedded has been found to be gelatin, as it satisfies a number of requirements.

1. It keeps the grains well dispersed and prevents clumping and sedimentation of the grains.

2. It protects the unexposed grains from reduction by a developer.

3. It allows the processing solutions to diffuse easily and permits access of the reagents to the grains.

4. It interacts chemically with silver halides and has an important sensitizing action, which is caused by minute impurities present in the gelatin.

5. In general it does not produce undesirable effects on the photographic behavior of the silver halide grains, such as fogging, loss of sensitivity etc.

Gelatin being a natural product can exhibit great variations in physical as well as in chemical properties. For photographic use careful selection is necessary and gelatin should be chosen or blended for the purpose of satisfying all the above mentioned requirements and making it possible to produce emulsions with higher sensitivity to ionizing radiation and light than can be obtained with any other dispersing medium. The correct choice of gelatin for emulsion making is of fundamental importance.

Silver Halides

Silver bromide and silver chloride grains are ionic microcrystals and possess the sodium chloride, i.e. the face centered cubic structure, as seen in Figure 3.2. In a silver halide crystal positively charged silver ions are in alternate positions with negatively charged halide ions. Hence the crystal lattice consists of two interpenetrating lattices of silver and halide ions, as seen in Figure 3.2 for the silver bromide lattice. The distances between two nearest ions is 2.88×10^{-8} cm in the silver bromide lattice and 2.77×10^{-8} cm in the silver chloride lattice.

In many emulsions the sensitivity is increased by adding small amounts of potassium iodide during the course of precipitation of silver bromide. As a result, a certain percentage of the bromide ions in the

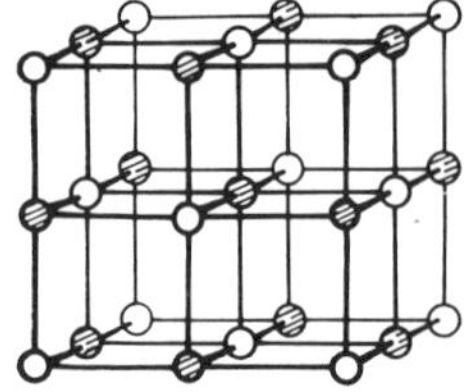

Figure 3.2. Face centered cubic structure of silver-bromide. ● Silver ions, ○ Bromide ions.

crystal is then replaced by iodide, the iodide forms a solid solution with the silver bromide, but the crystal structure of the silver bromo-iodide grains remains face centered cubic.

The minute silver bromide grains may vary considerably in their external appearance, when viewed under the microscope, depending on the conditions of precipitation. In certain types of emulsion triangular and hexagonal tabular shaped crystals are formed, in others the grains are rounded off and some are of intermediate shapes. Certain emulsions reveal cubically and octahedrally shaped grains. The size of the grains varies over a very wide range from a few microns (1 micron = 1 thousandth of a millimeter) to 0.1 micron and less. Most of the faster X-ray type emulsions have grain diameters from 0.2–3 microns, i.e. 0.0002–0.003 millimeters and most emulsions may have a wide distribution of grain size. This distribution in a given emulsion is often presented as a grain size distribution curve, typical shapes of which are illustrated in Figure 3.3. An even more significant characteristic is the distribution of the projected areas of the grains, as this represents the total area exposed to radiation quanta. There are about 10^9–10^{12} grains per cm² in an emulsion of the type used for X-rays.

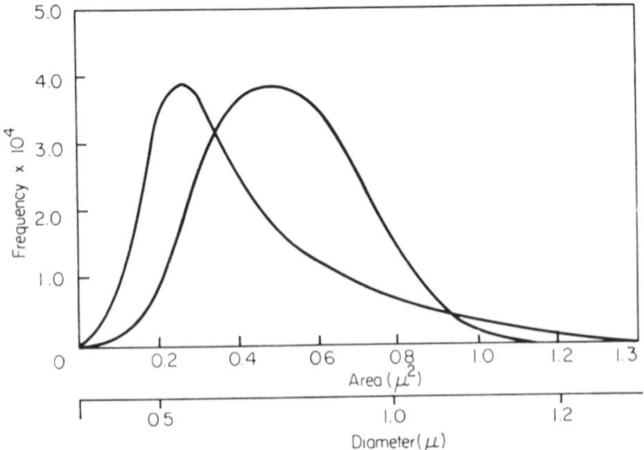

Figure 3.3. Typical grain size distribution curves.

3.3 THE PREPARATION OF AN EMULSION

The making of a photographic emulsion is a complicated and extremely critical operation, which is carried out partly in complete darkness or in subdued light of wavelengths to which the emulsion is least sensitive. A high degree of purity is required for the ingredients and solvents to be used in emulsion making, and great care must be taken that no contamination occurs in the equipment in which the operations take place. The variety of existing types of commercial emulsion is enormous and their manufacture differs not only between the types of emulsion but also between the various manufacturers. It would thus be beyond the scope of this book to touch upon any details of the operations and only some of the bare principles will be outlined. The interested reader is however referred to the relevant literature at the end of this chapter.

The majority of emulsions as they are used in work with ionizing radiations is of the so called 'negative emulsion' type. In the preparation of negative emulsions basically 5 operations may be considered.

1. The precipitation and physical ripening stage.
2. The removal of excess soluble salts from the emulsion. (Washing).
3. The second digestion or after-ripening during which the final sensitivity is reached.
4. Final additions before coating.
5. Coating.

3.3a The Precipitation and Physical Ripening Stage

A solution of silver nitrate is added to a diluted solution of gelatin containing an alkali halide, such as potassium bromide which will lead to a reaction in which silver bromide is precipitated according to

$$AgNO_3 + KBr \rightarrow AgBr + KNO_3$$

The alkali halide is usually present in excess over that required for complete reaction with the silver. The presence of the gelatin prevents coalescence of the grains and serves as a protective colloid, thus preventing the coagulation of the precipitate. In many cases an ammoniacal solution of silver nitrate is used, and emulsions prepared in this way are referred to as ammoniacal, whilst the former are called neutral emulsions. The ammonia, being a silver halide solvent, has a strong influence on the growth of the crystals and on their size distribution in the physical ripening stage. As mentioned before, potassium iodide is usually added to fast negative emulsions, so that the final iodide content in the lattice of the crystals amounts to a few mol per cent with respect to the total halide. The potassium iodide may be added to the potassium bromide solution before or during the precipitation.

There exist a great variety of conditions at which a precipitation procedure may be carried out. The silver nitrate solution may for instance be added at various rates of addition to the halide-gelatin solution; this is called a single jet precipitation. The silver nitrate and bromide may also be added simultaneously to the gelatin solution, in which case one speaks of a double jet method, or the halide solution may be added to the silver gelatin solution. In order to mention a few more variables a coprecipitation of silver iodide and bromide may also be carried out, the gelatin may be used in different concentrations, the temperature and agitation may be varied and all these parameters naturally influence the final product. Hence with the selection of a particular precipitation technique, the final dispersion and grain size, distribution of grain size and sensitivity can be controlled to a wide degree.

The precipitate is now subjected to a physical ripening stage. In this stage, which may partly overlap the precipitation stage, the final size and shape of the grains will be reached. The temperature of the solutions during the precipitation ranges between 50–70°Centigrade. The grain size and its distribution are largely influenced by the temperature of the solutions and by the rate of addition of silver nitrate. In general the slower the rate of precipitation the larger will be the grain size. This increase of grain size may occur in different ways either by the larger particles growing at the expense of the smaller ones, owing to differences in rate of solution (Oswald ripening) or by coalescence of particles which may then undergo recrystallization.

3.3b   The Removal of Excess Soluble Salts from the Emulsion. (Washing).

When the precipitation and the first ripening stage are completed, the emulsion still contains potassium nitrate and excess potassium bromide and in the case of an ammoniacal emulsion, ammonia. Some of these products may crystallize out during the drying process, i.e. after coating the emulsion on film base or glass, and may therefore influence the physical and photographic properties of the emulsion. Hence the emulsion has to be washed in order to remove these products. The washing process is facilitated by adding further gelatin and by allowing to set to a jelly. This is then diced into cubes or shredded into fragments, called 'noodles' by pressing the jelly through perforated plates. Washing is then continued to completion by monitoring the silver ion concentration or electrical conductivity of the wash water.

3.3c The Second Digestion or After-Ripening

After washing, the emulsion is melted and more gelatin is added. The emulsion is kept at a temperature of about 50°C during the after-ripening (or chemical ripening) period which may last 1–2 hours. The main purpose of the after-ripening is to attain the desired sensitivity of the emulsion to radiation with a minimum of fog. Before the second digestion the sensitivity is extremely low.

It was shown by S. E. Sheppard (1925) and by S. E. Sheppard and J. H. Hudson (1930) and other coworkers that the sensitivity of an emulsion was dependent on traces of certain sulfur compounds which are present in gelatin. They called these chemical sensitizers. They postulated that these compounds lead to the formation of sensitivity specks in the grains and that they consist of silver sulfide. Apart from sulfur sensitization there are other important chemical sensitizing methods, which can be applied to increase sensitivity, such as the so called 'reduction' and 'metal salt' sensitization. Reduction sensitization derives its name from the incorporation of reducing agents into emulsions. These may be organic or inorganic compounds, such as stannous salts for example. The effect of sulfur and reduction sensitization is additive in many cases. Well known metal salt sensitizers are gold compounds, such as auric chloride or potassium aurous thiocyanate, which can be added to an emulsion at the beginning of the after-ripening period. Of some interest for certain X-ray emulsions are bivalent lead or cadmium ions which may be coprecipitated with silver halides. These have the tendency to increase the X-ray and γ-ray sensitivity of iodobromide emulsions whilst they depress the light sensitivity of these emulsions.

It is thus assumed that sensitivity specks are formed and grow during the after-ripening. The rate of ripening as measured by the rate of rise of sensitivity depends on a number of factors such as sensitizer concentration, temperature, pAg, pH etc. The rise in sensitivity is accompanied by an increase of fog, which is due to grains becoming spontaneously developable without being exposed to radiation. The second digestion has to be stopped when optimum sensitivity is reached and before the proportion of fog grains becomes appreciable.

Optical dye sensitizers are chiefly of interest with respect to changes in the spectral sensitivity of photographic emulsions. Since silver bromide emulsions possess a natural sensitivity to X-, γ-rays, electrons and other charged particles, as well as to ultraviolet and blue light, optical sensitization is of less interest for emulsions exposed to ionizing radiations or blue light, as used in conjunction with intensifying screens, when exposed to X-rays.

3.3d   Final Additions Before Coating

Before an emulsion is coated, additions have to be made which give the emulsion its final physical and photographic properties. Some of the additions are hardening agents to impart mechanical strength in the coated layer, preservatives which prevent decomposition of gelatin due to bacterial attack, antifogging agents, antistatic agents, surface active agents which aid the coating procedure, and others. Since water evaporates during the drying stage, the concentration of the emulsion ingredients changes and thus a control of the bromide and hydrogen ion concentration in the coated layer becomes necessary from the point of view of keeping quality and fog increase.

3.3e   Coating

The emulsion is finally coated on some suitable film base, glass, or paper in the case of photographic papers. The film base is treated with a subbing solution in order to establish proper adhesion between the hydrophilic emulsion and the hydrophobic cellulose base. Film base for X-ray film is generally of the order of 200 microns in thickness. The ratio of silver halide to dry gelatin in the final coating of most commercial negative emulsions is of the order of $1:1$ by weight and the coating weight of silver bromide is of the order of $1-2$ mg/cm^2 for some types of X-ray emulsions and is less in most other negative emulsions. In X-ray emulsions the thickness of the dried coating is between 10–20 microns, depending on the type of film and may reach thicknesses of 3–7 times that when swollen by processing solutions. The silver halide to gelatin ratio in nuclear emulsions may be $5:1$ by weight, so that in this type of emulsion the grains are very much more closely packed than in ordinary emulsions. After the coating of emulsion for X-ray film (usually on both sides of the base), a supercoat, of gelatin for instance, is given on top of the emulsion layers to prevent damage due to abrasion when handled in the darkroom. The thickness of the supercoat varies between 0.5–1 micron in most commercial films.

The coating procedure is a very delicate process requiring most elaborate machinery in order to achieve a high degree of uniformity. It is usually carried out on large rolls of film base in an absolutely dust free atmosphere. The coatings are then dried in so called drying alleys under exactly controlled conditions of temperature and humidity. Finally films are cut into the appropriate sizes, checked for any defects and packed in the desired manner depending on the purpose of the particular material.

3.4 SENSITIZED MATERIALS

X-Ray Films. As has been mentioned in the previous section, X-ray films are usually coated on both sides of the film base in order to make use of the penetrating power of X-rays and so to allow the simultaneous response of both emulsion coatings (see also p. 279). There are mainly two types of X-ray film commercially available (a) films used for direct X-rays (no-screen films) and (b) films for use in conjunction with so called salt intensifying screens. These latter are cardboard or plastic foils coated with substances which fluoresce with a bluish light when excited by X-rays. The screen film, which is chiefly blue light sensitive is then sandwiched between a pair of such intensifying screens for exposure. (See pages 284 and 297.) Screen type films are available in several types, differing mainly in speed and contrast. A variety of no-screen films are available differing in sensitivity and average grain size. In general the higher the sensitivity the larger the average grain size. In special applications of medical radiography, such as mass miniature and image amplifier radiography, single coated films of different spectral sensitivities are used. The various types of X-ray film in different applications of medical and industrial radiography, and also in photographic dosimetry, will be described in more detail in the relevant chapters.

Other Sensitized Material. Apart from X-ray films, there exist a great variety of usually single coated sensitized materials, e.g. for the detection of nuclear particles, such as nuclear emulsion plates, coarse grained and fine grained stripping and non-stripping plates for autoradiography and rather insensitive, but extremely fine grained emulsions for microradiography and other uses. The applications of these materials are discussed in the appropriate chapters in the second part of this book.

Photographic emulsions can be made sensitive to a wide range of radiations of the electromagnetic spectrum and to corpuscular radiations as well. In the latter case the paths of individual particles can also be detected as a row of grains (particle tracks), when viewed under the microscope. Even indirectly ionizing radiations, such as slow and fast neutrons can be detected photographically by means of slow neutron converters releasing γ-rays and electrons to which the photographic emulsion is sensitive, or, in the case of fast neutrons, by means of collision processes with hydrogen nuclei in the gelatin of the emulsion (proton tracks.) (See pp. 177 and 238.)

3.5   THE FORMATION OF THE LATENT IMAGE

Whether the photographic emulsion is exposed to light, to X-rays or to charged particles, some change must occur in the silver halide grains which renders only the adequately exposed grains developable, the remainder being left undevelopable, except for the small fraction developable as fog. The understanding of the intricate mechanism of these minute changes in the grains has occupied the minds of many scientists since the beginning of photography with silver halides. The change which causes the grains to be rendered developable on exposure is considered to be the formation of the latent image. The latent image itself, which is the product of the change, is so minute that it cannot be detected either by physical or chemical means apart from development which destroys the latent image. Before giving a brief exposition of our present ideas on the mechanism of latent image formation, it is useful to examine first the experimental facts which led to our modern concept.

3.5a   The Substance of the Latent Image

Although neither physical nor chemical methods permit the direct detection of the individual latent image, it is practically certain that it consists of silver. Previous controversies concerning the substance of the latent image have ceased, as it is almost generally accepted now that it is composed of an aggregate of a few silver atoms. Several experiments will be described which give some indirect and even direct indication that the substance of the latent image is silver.

If a photographic emulsion is very heavily over-exposed to light or to X-rays (print-out exposure), and is then subjected to quantitative chemical analysis, it is found that metallic silver has been formed during the exposure which was not present before exposure. Such print-out effects require 10^6–10^9 times the exposure necessary for the formation of a latent image. Although no direct conclusion can be drawn from this experiment as to the nature of the latent image, there are no very plausible reasons why the substance formed during the extremely high exposure should be different from that in the

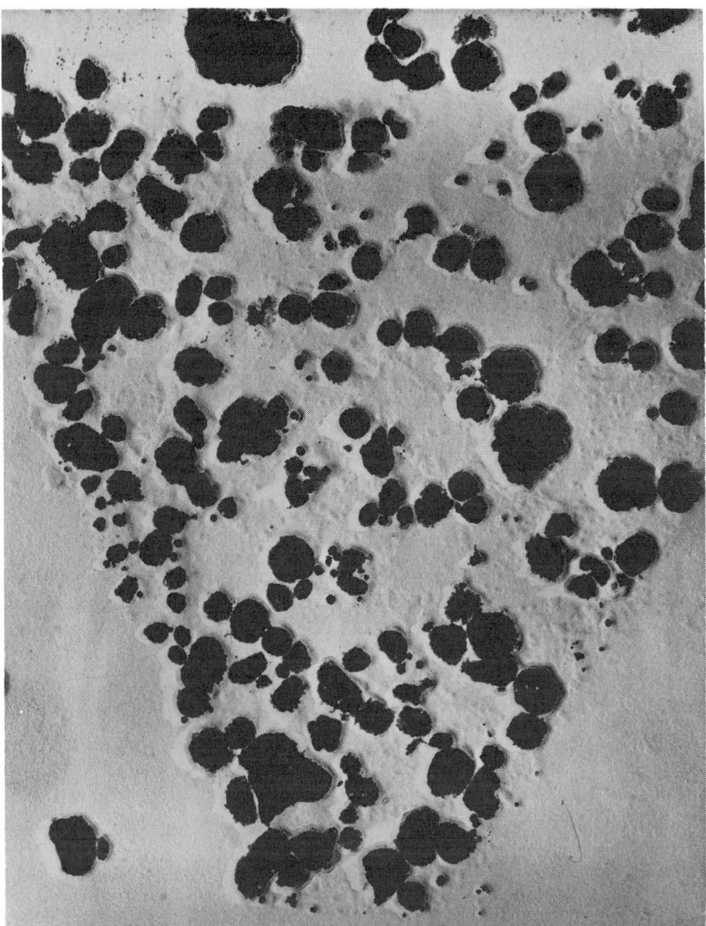

Figure 3.4. Print-out effect in a tabular silver bromide grain. Electronmicrograph
($\times$ 70000). *By courtesy of* Dr. G. C. Farnell and R. B. Flint, Research Laboratories, Kodak Ltd., Harrow, England.

latent image region. The print-out effect within a region of a single
tabular silver bromide grain, being due to exposure without development, is shown in the electron micrograph of Figure 3.4.

There are other indications that the latent image may consist of
silver. P. P. Koch and H. Vogler (1925) and many others have
identified the existence of silver in X-ray diffraction experiments on
strongly overexposed, but unprocessed silver bromide emulsions.
Similar tests have been carried out by means of electron diffraction.

A further indirect demonstration is provided by the fact that the latent image is destroyed by strong oxidizing agents, such as chromic acid or potassium persulfate.

Using an extremely sensitive method, direct evidence within the latent image region of exposure has been given by H. Tellez Placensia (1959, 1960). He applied a neutron activation technique after X-ray exposure and thus identified silver. This technique has apparently not yet been successful with light exposures.

It may be mentioned that the directly observable silver formed on exposure to radiation is referred to as 'photolytic silver' and is accompanied by the formation of free halogen. The halogen may be removed by certain halogen acceptors such as sodium nitrite, so preventing recombination of the halogen with silver.

3.5b   The Size of the Latent Image

The number of quanta necessary to render a grain developable has been the object of many investigations. This number gives some guide as to the number of silver atoms which have to be aggregated to one another or to some pre-existing speck in order to render a grain developable. The more recent investigations (G. C. Farnell & J. B. Chanter (1961), A. Marriage (1961)) show that a latent image can be formed in the fastest available grains of commercial emulsions by exposure of only 4–10 quanta of light, thus indicating that latent image specks may consist of aggregates of 4–10 silver atoms. In an X-ray exposure however it has been shown that on the average 1000 silver atoms are formed per X-ray quantum absorbed in a grain. (See p. 66.) From this it follows that an X-ray exposure involves an unavoidable waste of energy.

3.5c   Basic Experimental Results

In the following sections the more important experimental facts will be presented on which are based our present views on the mechanism of the formation of the latent image. Although most of our knowledge is derived from experiments with light, there is no reason to believe that there is any fundamental difference in latent image formation by ionizing radiation. However, quantitative differences exist which are due to the difference in energies of X-ray and light quanta.

Sensitivity Specks.   As previously mentioned on page 59 Sheppard and his coworkers (1925) demonstrated that the sensitivity of a photographic emulsion is dependent on the presence of certain

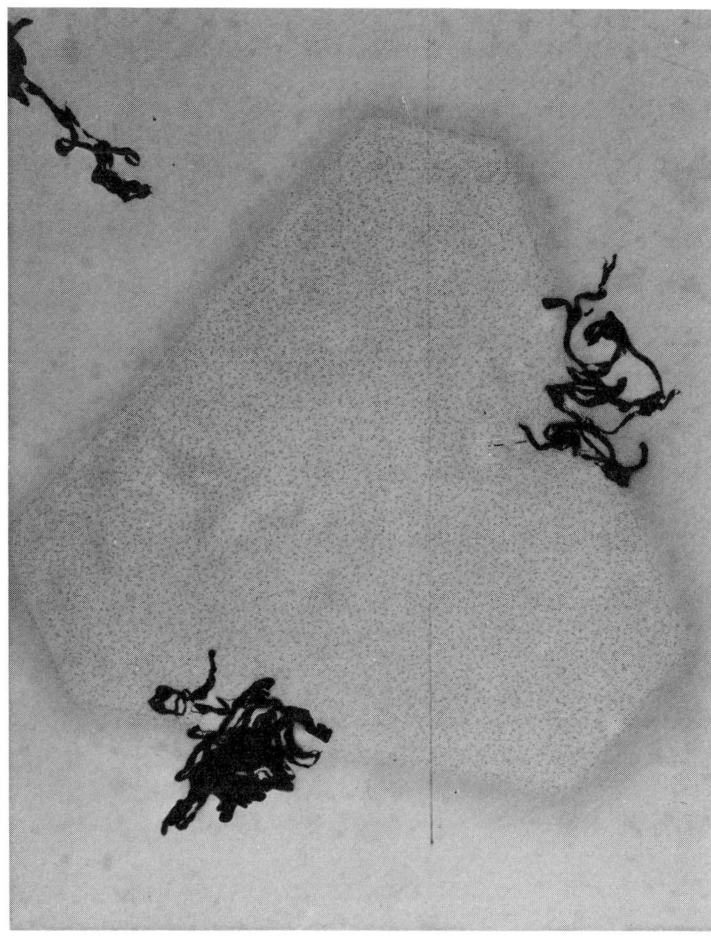

Figure 3.5. Gelatinate shell of a tabular silver bromide grain with retained development centers. Electronmicrograph ($\times 40000$). *By courtesy of* Dr. G. C. Farnell, R. B. Flint and B. Chanter, Research Laboratories, Kodak Ltd., England. *J. Phot. Sci.* **13,** 25 (1965)

impurities in the gelatin. They suggested that these impurities form specks of silver sulfide and that these act as nucleating centers for the formation of the latent image. The sites at which latent image centers are formed can be made visible microscopically by partial development and these so called development centers may in some cases coincide with sensitivity specks. An electron-microscope image of development centers is shown in Figure 3.5.

Quantum Yield of Photolysis. J. Eggert and W. Noddack (1921, 1925) determined whether Einstein's law of quantum equivalence is applicable to latent image formation, i.e. whether each quantum ($h\nu$) of light absorbed per grain leads to the release of one silver atom. Thus

$$AgBr + h\nu = Ag + Br$$

or

$$Br^- + h\nu = Br + e^-$$

$$Ag^+ + e^- = Ag$$

Eggert and Noddack found that the number of silver atoms formed in the photolysis of silver bromide and silver chloride emulsions is equal to the number of light quanta absorbed, i.e. the quantum yield is

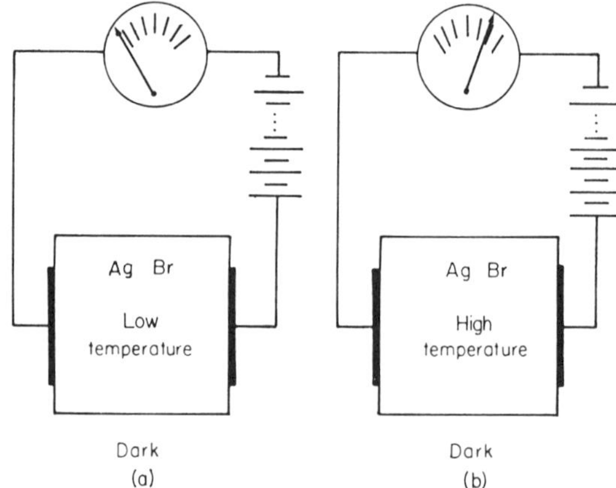

Figure 3.6. Ionic conductivity.

unity. The wavelength of the light used was close to 4000 Å. J. Eggert and W. Noddack (1927, 1928) also examined the effect of X-rays (of wavelength 0.45 Å) and found, as indicated before, that about 1000 silver atoms were formed for each X-ray quantum absorbed. This is not surprising, since about 10^4 times the energy of a light quantum is dissipated in the absorption of an X-ray quantum. The fact that only 1000 instead of the expected 10,000 silver atoms are released is probably due to recombination effects. P. Günther and W. Tittel (1933) found that the number of silver atoms liberated per X-ray quantum absorbed was almost proportional to the quantum energies involved within the region of 0.245–1.54 Å.

Ionic Conductivity. Important experiments on the ionic movements in silver halides were carried out by R. W. Pohl (1937), W. Lehfeldt (1935), E. Koch and C. Wagner (1937), W. Jost (1938) and many others. If a potential is applied across a large silver halide crystal in the dark (see Figure 3.6) at a low temperature, the crystal behaves as an insulator, i.e. no current flows between the electrodes. On raising the temperature, a conductivity is observed, which increases rapidly with temperature. Several workers have shown that the current represents the motion of positive silver ions through the crystal. This movement is explained by the assumption that some silver ions are in 'interstitial' positions in the crystal lattice (Frenkel defects) and that silver ion vacancies are present as shown in the two dimensional lattice of silver bromide in Figure 3.7. Silver ions can move from one interstitial position to another, while other silver ions can move from their normal lattice sites into vacant ones.

Although one would expect that both positive silver ions and negative bromide ions may occupy interstitial positions, only the

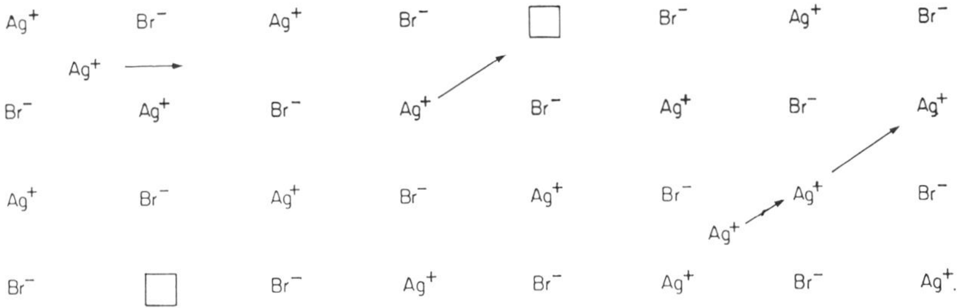

Figure 3.7. Model to illustrate electrolytic conductivity in silver bromide. (*After* J. H. Webb.)

former do, as the halide ions are too large relative to the lattice spacing to move into such positions.

The ionic conductivity rises with temperature because more silver ions and vacancies become available and their mobilities increase. The activation energy, i.e. the energy required to move a silver ion from its normal position into an interstitial one in the lattice, is found to be 1.27 eV, while the activation energy for migration of a silver ion from one interstitial position to another is only 0.15 eV.

Photoconductivity. In the investigations described above the silver bromide crystal was assumed to be in the dark. If the crystal is exposed to light (see Figure 3.8) a further conductivity can be observed which was thoroughly studied by R. Hilsch and R. W. Pohl (1930) K. Hecht (1932), W. Lehfeldt (1935), and their coworkers. The relationship between voltage and current was studied for constant illumination and it was found that the voltage-current curve 1 (see Figure 3.9), rises first linearly with the voltage and then gradually reaches saturation at higher voltages. The current is due to the move-

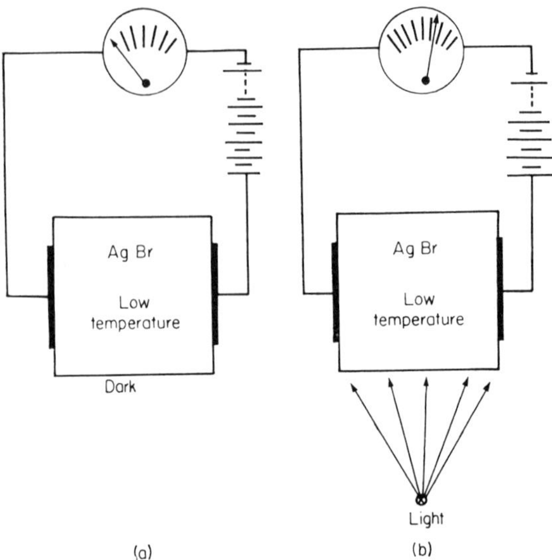

Figure 3.8. Photoconductivity.

ment of electrons liberated by illumination from the bromide ions. They travel through the crystal under the influence of the electric field. At low voltages across the crystal, some of the electrons appear to be caught before they reach the positive electrode and thus a smaller current flows than in silver halides at higher voltages, when they reach the positive electrode. The photoconductivity is not strongly dependent on temperature.

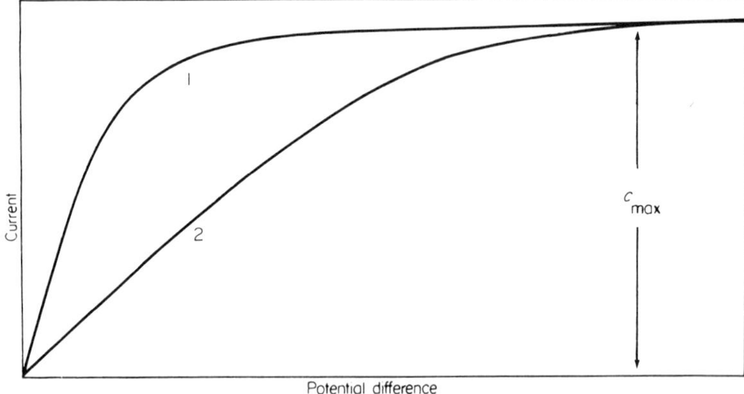

Figure 3.9. Current-voltage curves for photo conductance in (1) Pure Silver bromide and in (2) Silver bromide containing colloidal silver.

If the crystal contains impurities, such as colloidal particles of silver for example, higher electric fields are required to obtain the same photocurrent than are required for pure silver bromide, as shown by curve 2 of Figure 3.9. A plausible interpretation of this effect is that the colloidal silver specks act as traps for the electrons, thus reducing the average distance of travel through the crystal. It has further been established that, in silver bromide, a close relationship exists between the variations of photoconductivity and the absorption of light with wavelength.

Conduction Band and Electron Trapping. The phenomenon of photoconductivity just described can be interpreted by reference to

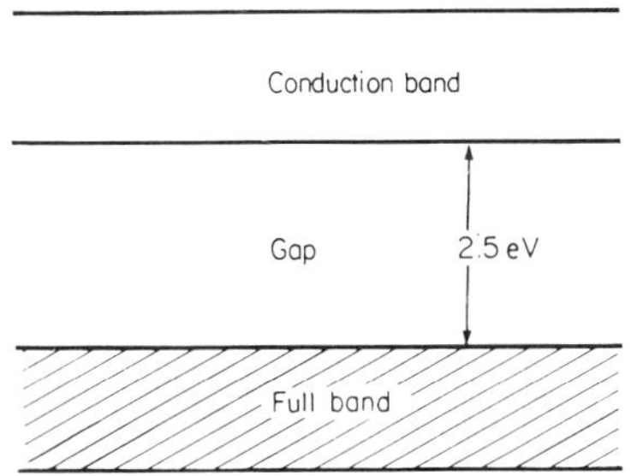

Figure 3.10. Energy levels in silver bromide.

the electron energy band model of silver bromide. In a crystal the discrete energy levels of atoms and ions become smeared out to bands of allowed energy levels through which electrons can freely move and forbidden zones through which no motion is possible. In the silver bromide crystal the valence electrons associated with the bromide ions completely fill the lower energy band. The upper band, which is associated with the silver ions, is empty, and separated from the lower band by a relatively wide gap of about 2.5 eV (see Figure 3.10). Hence an electron liberated from a bromide ion in the valence band due to the absorption of a quantum of radiation exceeding 2.5 eV can move freely through the conduction band of the crystal. The conduction levels of impurities in silver-bromide, e.g. silver or silver sulfide, are below those of silver bromide. This means that an electron which passes from silver bromide into silver or silver sulfide will be held for a period ('trapped') until lifted back into the conduction band by receipt of thermal energy.

Positive Hole Movement. Little attention has been paid so far to the change occurring in the silver bromide crystal after an electron has been removed from the bromide ion by absorption of a quantum. By the act of liberation of an electron the bromide ion becomes in effect a bromine atom. The latter will be converted to a bromide ion again, if an electron jumps from a neighboring bromide ion to the bromine atom. The previous bromine atom presents a positive charge in the crystal lattice and is therefore called a 'positive hole'. Its mobility is due to a replacement mechanism carried out by the transfer of electrons as indicated before. This mechanism is illustrated in Figure 3.11. The electron transfer requires relatively low activation energies. It is essential for the formation of free silver atoms that positive holes are removed or trapped in some way, as otherwise recombination may occur between bromine atoms and electrons, leading again to the formation of bromide ions, or between bromine atoms and silver to reform silver bromide.

Step 1	*Step 2*	*Step 3*
$Br^-Ag^+Br^-Ag^+Br^-$	$Br^-Ag^+Br^-Ag^+Br^-$	$Br^-Ag^+Br^-Ag^+Br^-$
$Ag^+Br^pAg^+Br^-Ag^+$	$Ag^+Br^-Ag^+Br^-Ag^+$	$Ag^+Br^-Ag^+Br^pAg^+$
$Br^-Ag^+Br^-Ag^+Br^-$	$Br^-Ag^+Br^pAg^+Br^-$	$Br^-Ag^+Br^-Ag^+Br^-$

Figure 3.11. Displacement mechanism of positive hole movement.

3.5d   The Gurney-Mott Mechanism of Latent Image Formation

The knowledge presented in the previous sections led Gurney and Mott (1938) to propose a mechanism of latent image formation, which still forms the basis of our present views. Two successive steps in the formation of a latent image are postulated which are indicated in Figure 3.12, 1–4. The triangles represent silver bromide grains and in Figure 3.12(1) the small circles indicate interstitial silver ions; the shaded half circle on the outside of the grain indicates an external silver sulfide speck (or a sensitivity speck). In the first step (Figure 3.12(2)) an X-ray quantum (or a charged particle) is absorbed in the grain which releases a high energy electron, whose track is shown by the wavy line. This creates an avalanche of low energy electrons as shown by the short streaks arising from the wavy line through the grain. The electrons liberated move through the conduction band of the crystal until they are trapped by a sensitivity speck which thus becomes negatively charged. In a second step (Figure 3.12(3)) interstitial silver ions, which have positive charges are attracted towards the negatively charged specks, so neutralizing the charge with the formation of metallic silver. This is indicated by the blackened surface speck in Figure 3.12(4).

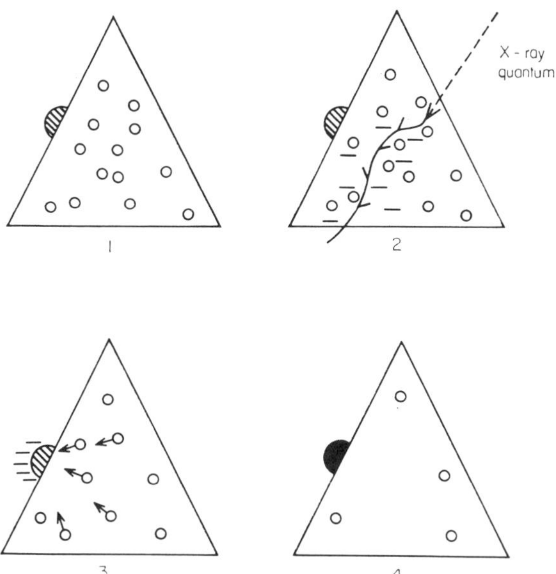

Figure 3.12. Mechanism of latent image formation according to Gurney and Mott. (1) Grain showing interstitial silver ions ($\bigcirc$) and a surface sensitivity speck. (2) Absorption of an X-ray quantum with the simultaneous release of an electron from a bromide ion. (3) The speck is charged negatively and positive silver ions move towards the speck. (4) Neutralization of the charge and formation of metallic silver.

The basic concepts of the Gurney-Mott theory are still widely accepted and much experimental evidence for their soundness has accumulated in recent years. However, important criticisms of some of the features of the theory have been raised since its announcement. Before discussing these, the results of some recent experiments on 'pulsed exposures' will be described, as these have provided strong support for the basic concepts.

3.5e Pulsed Exposures

One of the most striking physical experiments carried out on emulsion grains was by means of what are called 'pulsed exposures'. These were carried out by J. H. Webb (1955), J. F. Hamilton, F. A. Hamm and L. E. Brady (1956), and J. F. Hamilton and L. E. Brady (1959). In these exposures emulsion grains were subjected simultaneously to brief electric field pulses in combination with light flashes. An electron micrograph of an emulsion grain which has been subjected to a heavy light exposure without an electric field is shown in Figure 3.13. The grain itself is not visible in this picture, because it has been fixed and only the impression it left in the gelatin is seen. The black dots shown in this photograph are photolytically formed silver particles which have been formed during exposure and their distribution is random. The dark region around the outline of the grain is due to bromine liberated during exposure.

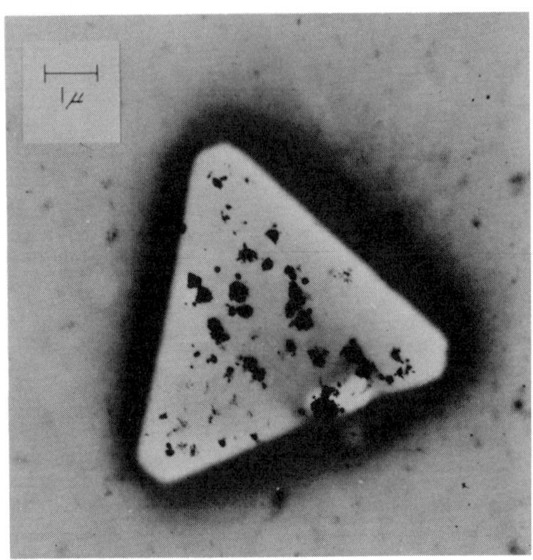

Figure 3.13. Random distribution of photolytic silver due to light exposure without electric field. (Dark edge due to liberation of halogen.) (*By courtesy of* J. F. Hamilton and L. E. Brady, Eastman Kodak Research Laboratories, Rochester, U.S.A., and *J. Appl. Phys.* **30,** 1902 (1959).)

Figure 3.14 shows a grain which also has had a heavy exposure to light, but in a pulsed electric field. The silver particles all appear on one side of the grain and the bromine on the opposite side. The direction of the electric field is indicated in the upper part of the picture and it will be seen that the bromine atoms (positive holes) which were left after the electrons were released have moved towards the negative electrode (black region), whereas the electrons have moved towards the positive electrode to form silver particles.

These experiments have demonstrated the electric movement of the positive holes (see p. 70) and have given evidence of the liberation of bromine near the negative electrode and of the formation of silver

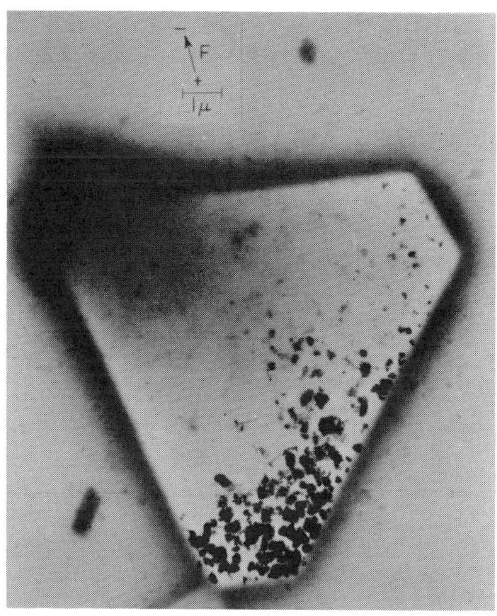

Figure 3.14. Displacement of photolytic silver and bromine due to the electric field, the silver forms near the positive electrode, the bromine near the negative. (*By courtesy of* J. F. Hamilton and L. E. Brady, Eastman Kodak Research Laboratories, Rochester, U.S.A., and *J. Appl. Phys.* **30,** 1902 (1959).)

near the positive electrode. E. Klein and R. Matejec (1959, 1961), using steady electric fields, demonstrated the validity of the ionic motion which is predicted by the Gurney-Mott mechanism.

3.5f Structural Imperfections

Another important contribution to the understanding of the mechanism of latent image formation was made by Mitchell and his coworkers. They showed in photomicrographs of larger silver bromide crystals that silver deposits photolytically at mosaic boundaries and at structural imperfections. These investigations gave evidence of the importance of internal dislocations in the crystal structure of silver bromide grains with respect to the location of photolytically formed silver deposits. J. M. Hedges and J. W. Mitchell (1953); T. Evans and J. W. Mitchell (1954) (see Figure 3.15).

Structural imperfections have also been revealed in microcrystals of commercially available emulsions using various types of X-ray diffraction methods. (C. R. Berry, 1956; R. H. Herz, E. Pitts and G. C. Terry, 1964.)

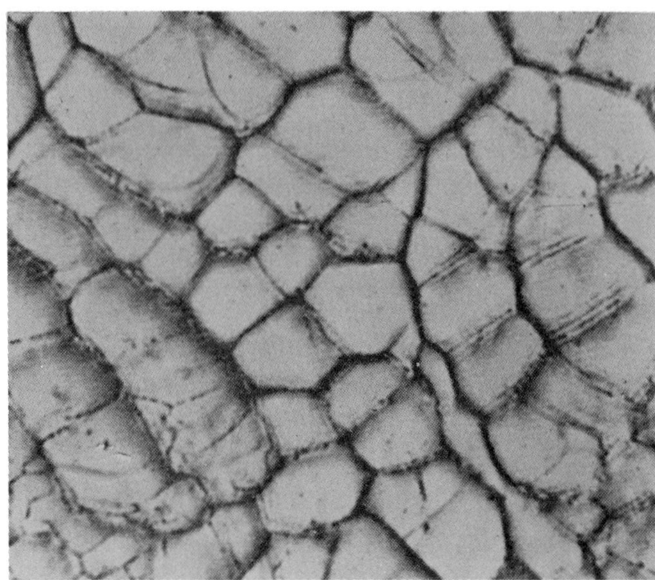

Figure 3.15. The substructure of a large single silver bromide crystal made visible by the separation of particles of silver along the dislocation lines of the sub-boundaries ($\times$ 850). (*After* J. W. Mitchell *J. Phot. Sci.* **6,** 57 (1958).)

3.5g Criticism of the Gurney-Mott Model

As mentioned earlier several aspects of the Gurney-Mott theory have been criticized in recent years. Although the controversy is still in progress a brief review of the criticism will be given, which can be attributed primarily to J. W. Mitchell (1955, 1957a, 1957b, 1958) and his coworkers. The major arguments relate to the following problems.

Electron Traps

What is the detailed mechanism of the trapping of electrons, if no sulfide specks are available, as for instance in chemically unsensitized emulsions, in which latent images are also formed?

Mitchell suggests that an efficient trap would be formed by an interstitial silver ion close to a silver ion kink site, i.e. a structural imperfection of the crystal lattice.

Availability of Interstitial Silver Ions

Is there a sufficient number of interstitial silver ions available in the small emulsion grains to act as postulated by Gurney and Mott? From extrapolation of bulk data Mitchell concludes that there are about 10^{12} interstitial silver ions per cm³ of silver bromide and hence there may not be one ion in a grain having a diameter of a fraction of a micron. To overcome the difficulty Mitchell proposed a mechanism by which interstitial silver ions are created by the trapping of positive holes. However, from conductivity measurements on emulsion grains J. F. Hamilton and L. E. Brady (1959) have shown that the silver ion supply is quite adequate and that there is no necessity for Mitchell's interstitial silver ion generating mechanism.

Fate of Positive Holes

What is the fate of the positive holes and why do they not recombine with electrons?

In the interior of a face centered crystal any ion is neutral and surrounded by six ions. A bromide ion at a kink site however, being surrounded by only 3 ions, has a negative charge. According to Mitchell a hole is trapped at a bromide ion kink site and its negative charge is converted into a positive one. According to Hamilton's view the trapped hole has the same charge as a positive silver ion at a kink site. Hence the electron is no more likely to be captured by the hole than by any silver ion site or groups of silver atoms.

3.5h   Mitchell's Hypothesis.

By the absorption of a quantum of light as in the Gurney-Mott model, an electron is released and a positive hole is created. The positive hole is trapped, e.g. by a surface halide ion at a kink site or at a sensitivity speck, and this occurs before conduction electrons are trapped. On trapping of the hole an interstitial silver ion is formed and adsorbed at a dislocation, i.e. at a structural imperfection (e.g. kink site) of the crystal lattice. This silver ion traps the electron. The first silver atom thus formed is called a 'latent pre-image'. On repetition of this process a further silver atom is formed and the speck consisting of two silver atoms is called a 'latent sub-image.' By the creation of a third silver atom the 'latent image' is formed. Mitchell (1957a) considers that the latent image is positively charged and thus, by attracting electrons and repelling positive holes, prevents the recombination between an electron and a positive hole. (J. W. Mitchell, 1957c, 1963.)

3.6   THE BASIC PRINCIPLES OF PROCESSING

In order to render the effect of exposure, i.e. the latent image, properly visible, a photographic emulsion layer has to be treated first with a developer solution and secondly with a fixing solution. In the former the exposed silver halide grains are reduced to metallic silver grains, in the latter the unexposed silver halide grains are dissolved and removed from the layer. In the following pages the basic principles of the 'processing' of photographic emulsions will be discussed. We shall deal with the composition of processing solutions, the function of their various ingredients and with the mechanism of development.

3.6a   Developers

Development is carried out by bringing the exposed photographic emulsion layer into direct contact with a developer solution. The essential constituents of a developer solution are

1. The developing agent
2. The accelerator
3. The preservative
4. The restrainer
5. The solvent and diluent.

1. *The developing agent.*   The purpose of the developing agents is, as already mentioned, to convert the exposed silver halide to metallic silver. The most important developing agents are organic compounds of which several hundreds are known, but of these only relatively few are of practical significance. The majority of useful developing agents are benzenoid compounds. X-Ray emulsions are generally developed in developer solutions containing hydroquinone with either 'Metol' or 'Phenidone' (1 phenyl 3 pyrazolidone). 'Metol' (methyl-p-aminophenol sulfate) is also known under certain trade names, such as Elon, Veritol, Photol and others. Metol is a developing agent of high activity which produces high emulsion speed, but images of low contrast. Hydroquinone (p-dihydroxybenzene) is a developing agent of low activity giving a low emulsion speed, but images of high contrast. The two agents Metol and hydroquinone combined give a higher activity than each separately, a phenomenon which is known as super-additivity. They are both used in X-ray developers and by changing the ratio of Metol to hydroquinone, developers of a wide range of characteristics can be obtained. 'Phenidone' is a developing agent with properties similar to Metol when mixed with hydroquinone. The quantity of Phenidone required to give the same photographic effect as Metol is only about 1/10th of the latter. It is therefore particularly useful for concentrated liquid developers. The storage life of Phenidone is limited in alkaline solutions, and under tropical conditions its storage life is shorter than that of Metol, but it will keep satisfactorily in moderate climates.

The developing agents, e.g. hydroquinone together with Metol or Phenidone, convert the exposed silver halide to metallic silver.

2. *The accelerator* is an alkali which is essential for the function of the developing agent. It consists usually of sodium or potassium carbonate or hydroxide. Its main purpose is to adjust the pH, i.e. the hydrogen ion concentration of the solution, to the optimum value. Generally, the more alkaline the solution the higher the activity of the developer, but there are practical limitations because of the increase in grain size and rate of oxidation, softening effect upon gelatin, and other effects.

3. *The preservative* prevents the developing agent from being too rapidly oxidized. The addition of the preservative is necessary mainly for two further reasons, because some of the oxidation products have a retarding effect on the developing action and because they cause staining of the emulsion. The most satisfactory preservatives are potassium and sodium sulfite. They are used in fairly large quantities in X-ray developers.

4. *The restrainer*, usually potassium or sodium bromide, is added chiefly for the purpose of ensuring that only the exposed silver halide grains are reduced to silver, thus preventing the formation of fog. More recently organic restrainers have been introduced also, such as 'benzotriazole' which is used together with potassium bromide.

With increasing use of the developer more bromide will be accumulated and this causes the activity of the developer to be reduced. The accumulation of bromide in the solution is caused by its release from the exposed silver bromide.

5. *The solvent and diluent.* Water is used to dissolve the constituents of the developer. Excessive amounts of iron, copper and tin in the mains water supply can cause fogging and rapid oxidation of the developer. Fluoride salts which are added to many water supplies generally do not cause harmful effects owing to the low concentration at which they are used.

One of the most important requirements in the processing of films, be it in film monitoring procedures or in radiography, is to obtain reproducible results. This can be achieved by keeping the activity of the developer solution constant, by maintaining constant temperature and agitation of the solution and by using the same developing time. In order to keep the activity at the same level over the working lifetime of the developer solution, a *replenisher* solution is required.

3.6b   Exhaustion and Replenishment

As more films are developed the activity of the developer solution decreases because increasing amounts of exposed silver halide have been converted into metallic silver. Furthermore, oxidation of the solution occurs together with an increase of the soluble amount of bromide (restrainer), leading to further retardation. Apart from the exhaustion of the developer a certain volume of the solution is physically carried over to the next processing bath, thus gradually reducing the original volume of the bath.

A replenisher is thus used (a) to maintain the initial activity of the developer and (b) to replace the solution which is carried away with each developed film. The replenisher contains greater quantities of developing agents than the developer, it has an increased content of alkali and is free from any restrainer (e.g. potassium bromide.) Certain methods of topping up the developer by replenishing solutions are recommended in practice, in order to restore the activity of the developer and to keep its level in the tank. How this is carried out will be shown in more detail on page 381.

3.6c   Rinsing

After development the photographic emulsion should be rinsed, preferably in an acid solution, before it is placed in the fixing bath. This is done in order to stop development quickly by neutralizing the alkaline developer remaining in the emulsion layer, thus avoiding contamination of the acid fixing solution by the alkaline developer. Without rinsing the continued action of the developer taken over into the fixing solution may cause stain in the emulsion layer.

3.6d   Fixing Bath

After development and rinsing the undeveloped silver halide crystals remaining in the emulsion must be removed. The halide removed must also be converted into soluble compounds without damaging the silver in the emulsion. The operation which serves this purpose is called 'fixation'. There are not many substances which function successfully as fixing agents and the most important ones are ammonium and sodium thiosulfate. The latter often called 'hypo' is the most common fixing salt. Ammonium thiosulfate, which is used in most concentrated liquid fixers, is more rapid in its action than hypo and the time for an emulsion to clear in it is considerably shorter than it would be in hypo. Ammonium thiosulfate also permits about twice the number of films of the same total area to be fixed before fixing time becomes excessive. The use of ammonium thiosulfate suffers however from the disadvantage of having a slight bleaching action on the silver image, if left too long in the fixing bath. This is a drawback to be considered in fixing monitoring films from which the X-ray dose is derived, whereas in radiography its effect is practically negligible.

Apart from ammonium or sodium thiosulfate, fixing baths usually contain an acid, such as acetic acid which neutralizes the alkali in the developer carried over in the emulsion and a preservative such as sodium sulfite which prevents the decomposition of the thiosulfate by the acid in the fixing bath. The sodium sulfite also prevents the formation of colored oxidation products. A further requirement is usually a hardening agent such as potassium alum, the addition of which contributes to the hardening of the gelatin of the emulsion, thus preventing softening during the following washing procedure. The hardened film is mechanically stronger than the unhardened film and dries more quickly.

3.6e   Chemical Action of Fixing Bath

Fixing agents are substances which easily form soluble complex ions with silver ions. Basically it can be said that the first reaction of thiosulfate ions with silver bromide leads to the formation of argento-thiosulfate ions, which are strongly adsorbed to silver halide, H. Baines (1955), according to

$$AgBr + S_2O_3{}^{2-} \rightarrow AgS_2O_3{}^- + Br^-$$

The existence of further silver thiosulfate complex ions in solution is well established such as $Ag(S_2O_3)_2{}^{3-}$ and $Ag(S_2O_3)_3{}^{5-}$.

3.6f   Washing

During washing, thiosulfate ions and soluble silver thiosulfate ions are removed from the fixed photographic material.

The practical purpose of washing after processing an emulsion coating is to make the silver images permanent. If washing is insufficient the color of the silver image may change during the storage of the coated emulsions, particularly at high humidity and temperature. Washing may be carried out either by frequent changes of the washing water or by continuous flow of fresh water, details of which will be discussed on pp. 236 and 383

3.6g   Drying

This may be carried out by natural evaporation, preferably after immersing the films in a water bath containing a wetting agent, as this minimizes drying marks and reduces drying time. In most modern radiographic departments special hot air drying cabinets are used.

3.7   THE MECHANISM OF DEVELOPMENT

The fundamental reaction occurring during development is the reduction of silver ions to silver atoms. This reaction may be presented in the form

$$Ag^+ + e^- = Ag$$

A theory of the mechanism of development should be capable of explaining the distinction a developer makes between the exposed and the unexposed grains. It is found that the sufficiently exposed grains are reduced to silver in a considerably shorter time than the insufficiently exposed or unexposed grains. It is thus the latent image which is responsible for the preferential reduction of the exposed grains. A theory should therefore indicate in which way the latent image can accelerate the reduction process.

3.7a   Physical Development

It is useful to refer first to the so called 'physical development' which, although of limited direct use, plays an important part in transfer processes (see p. 367) and in research on photographic problems. In physical development the silver which is to form the image is incorporated into the developer solution. Reduction of silver metal from the solution will take place slowly but the presence of the latent image particles which are metallic in character serves to catalyze a more rapid reduction. This is deposited as in silver plating upon the latent image nuclei. The silver which is added to the physical developer may originate from the silver halide grains of the emulsion or it may be added quite separately to the solution. It is worth noting that if the exposed emulsion is fixed carefully in an alkaline hypo solution, the latent image material will survive the fixation and it can be physically developed later.

The mechanism of the physical development is thus explained by a catalytic reduction of silver ions. The silver ions from the solution are adsorbed to the latent image silver nuclei and are reduced to silver atoms. The electrons required for this process are supplied by the reducing agent of the developer.

The knowledge of the effect revealed by 'physical development', namely that silver ions adsorbed by silver nuclei can be reduced in the presence of a reducing agent, contributed much to the present understanding of the mechanism of development.

Many theories on the action of developers have been suggested in the past and some controversy still exists about the final solution of the problem. According to T. H. James and G. C. Higgins (1960) and T. H. James (1966) it can be assumed that mainly two different mechanisms are operating simultaneously. These are 'Direct or chemical development' and 'Solution physical development'.

3.7b Direct or Chemical Development

In this process it is assumed that the developer donates electrons to the silver nuclei of the exposed grains. These nuclei then act as tiny electrodes at which silver ions become reduced. This process proceeds until eventually the whole grain is reduced to metallic silver.

The precise mechanism by which a latent image produces a discrimination in development between the exposed and unexposed grains is still a matter of debate. Several mechanisms have been proposed and each appears to offer advantages according to the particular aspect of development under discussion. By whatever mechanism, the essential step in development is the transfer of an electron from the developing agent to a silver ion from the silver halide grain. Both silver ion and developing agent adsorb to the silver latent image and hence the latent image affords a site where the two reactants can meet temporarily with sufficient closeness for electron transfer to take place. Without such a site the probability of a favorable interaction between silver ions and developer molecules is greatly reduced.

3.7c Silver Filaments

If the developed silver image is examined under the electron microscope, it is found to consist of interwoven silver threads or filaments, see Figure 3.16. Various arguments have been put forward to explain the formation of filaments. In one of them (W. Gurney and N. Mott (1938) W. F. Berg (1943); N. F. Mott (1948)) it is believed that the filaments are formed by the interstitial silver ions becoming reduced at and deposited upon the silver halide side of the latent image center, which is the electrode in the cell set up between developer and silver halide. The accumulation of silver at the silver halide side of the latent image results in the formation of an elongated crystalline fibre of silver.

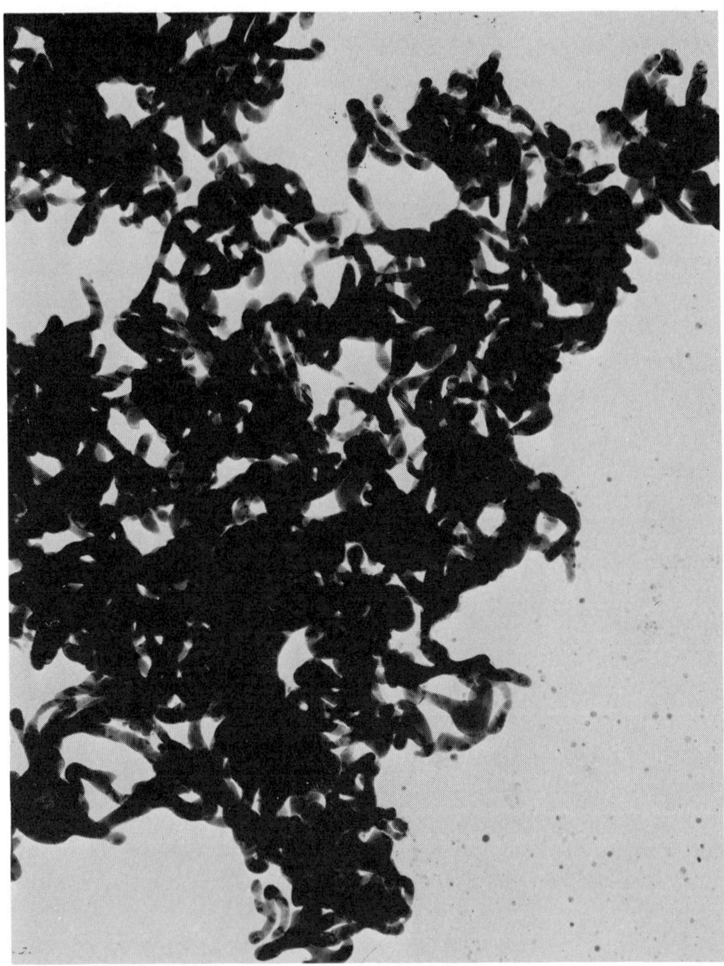

Figure 3.16. Part of a developed grain showing filamentary silver. Electron-micrograph ($\times$ 100000). (*By courtesy of* Dr. G. C. Farnell and R. B. Flint, Research Laboratories, Kodak Ltd., Harrow, England.)

3.7d   Solution Physical Development

Due to the slight solvent action of the developer, silver ions from the grains pass into solution. These silver ions move towards latent image nuclei or to silver formed by direct development, where they are reduced to silver. Thus the process is similar to that of physical development with the exception that the silver ions are not supplied by the developer but by the silver halide crystal itself. The relative contributions of solution physical development and direct chemical development may differ for different types of developers and may even alter during the development process.

3.7e   The Formation of Fog

Any reduction taking place in emulsion grains which is not due to latent image formation can be regarded as 'fog'. This may be caused by special conditions occurring in the grains or in the developer or in both. Emulsion fog can arise from a catalytic action of certain impurities of specks of silver or silver sulfide or other substances which may lead to spontaneous developability just like a latent image. Similarly, certain impurities in the developer solution may act as fogging agents. Amongst the many conditions which may lead to the formation of fog is the oxidation by air of certain developing agents whilst an emulsion is being developed. (Aerial fog).

Fog is obviously undesirable and increases with the storage of film materials. The formation of fog is accelerated by storing materials under unfavorable atmospheric conditions, such as high humidity and high temperature.

Summarizing we may say that the mechanisms of development are essentially

1. The latent image center increases the probability of the interaction between developing agent and silver ion.

2. The silver ions can be reduced directly at the interface between the latent image and the silver halide. (Direct or chemical development.)

3. Due to the solvent action of the developer, silver ions from the grains pass into the solution and are then reduced at a silver surface. (Solution physical development.)

The mechanism of the development process is much more involved than indicated in this highly simplified treatment and many other effects which have not been mentioned play an important part. For a discussion on this the reader is referred to the original literature and C. E. K. Mees and T. H. James (1966).

3.8   CONCLUSIONS

Vast scientific efforts into the study of the fundamentals of photographic processes have led to an improved understanding of the mechanisms involved in emulsion making, in the formation of the latent image, and processing. Hence emulsion making is no longer an art, as it was considered in the past, but has developed to a technology which is based on sound scientific principles which explain most of the stages of production and also permit their control.

Although many of the ideas explaining the mechanisms of the formation of the latent image and those of processing are still in flux, it appears that the basic approach is in the right direction. It is not surprising that a field like that of the photographic process, in which such minute physical, chemical, and organic reactions occur, still presents us with a wealth of puzzles, which may only gradually be solved by further research and possibly by the development of new research tools.

BIBLIOGRAPHY

Berg, W. F. (1963). Latent Image Theories. In A. Hautot, (Ed.) *Photographic Theory*, Liège Summer School, Focal Press Limited, London, New York.
Berg, W. F. (1943). Latent Image Formation. *Trans. Faraday Soc.*, **39,** 115.
Glafkides, P. (1957). *Chimie Photographique*. P. Montel, Paris.
Hautot, A. (1962). *Photographic Theory*. Liège Summer School. Focal Press, London, and New York.
Hay, A. and Rohr, M. (1932). *Handbuch d. wiss Photographie*, Vol. 5, Chapter II, Springer, Vienna.
James, T. H. and Higgins, G. C. (1960). *Fundamentals of Photographic Theory*. Wiley, New York.
Klein, E. and Matejec, R. (1961). *Offene Probleme in der Theorie der Photographischen Elementarprozesse*. Mitt. Forschungslaborat. der Agfa, Leverkusen-München, Vol. III. Springer-Verlag, Berlin.
Mees, C. E. K., and James, T. H. (1966). *The Theory of the Photographic Process*, 3rd Edition. Macmillan, New York, Collier-Macmillan, London.
Mitchell, J. W. (1957). Photographic Sensitivity. *Report. Progr. Phys.* **XX,** 433.
Mott, N. F., and Gurney, R. W., *Electronic Processes in Ionic Crystals*. Clarendon Press, Oxford.
Mott, N. F. (1948). Notes on Latent Image Theory, *Phot. J.*, **88B,** 119.
Neblette, C. B., *Photography, Its Principles and Practice*. Chapman and Hall, London.
Zelikman, V. L., and Levi, S. M., *Making and Coating Photographic Emulsions*. Focal Press, London, New York.

REFERENCES

Baines, H. (1955). Fixation—The chemistry of the hypo bath. *J. Phot. Sci.* **3,** 175.
Berg, W. F. (1963). Latent Image Theories. In A. Hautot (Ed.), *Photographic Theory*, Liège, Summer School. Focal Press Limited, London, New York.

Berry, C. R. (1956). Convergent beam X-ray analysis of mosaic structure in poly-crystals. *J. Appl. Phys.*, **27**, 6, 636.

Eggert, J., and Noddack, W. (1921) Über die Prüfung des Photochemischen Aquivalentgesetzes. *Sitzungsber, Preuss. Akad. Wiss.*, 631.

Eggert, J., and Noddack, W. (1925). Zur Prüfung des Quanten Äquivalentgesetzes an einigen Halogen-silber-emulsionen. *Z. Phys.*, **31**, 922.

Eggert, J., and Noddack, W. (1927, 1928). Uber die Quantenausbeute bei der Wirkung von Röntgenstrahlen auf Silberbromid. I and II. *Z. Phys.*, **43**, 222. *Z. Phys.*, **51**, 796.

Evans, T. and Mitchell, J. W. (1954). Defects in crystalline solids (*Report of Conference Bristol* 1954). The Physical Society, London.

Farnell, G. C., and Chanter, J. B. (1961). The quantum sensitivity of photographic emulsion grains. *J. Phot. Sci.*, **9**, 2, 73.

Günther, P., and Tittel, W. (1933). Die Bildung von Silber in der Photographischen Schicht unter dem Einfluss von Röntgenstrahlen. *Z. Electrochem.*, **39**, 646.

Gurney, R. W., and Mott, N. F. (1938). The theory of the photolysis of silver bromide and the photographic latent image. *Proc. Roy. Soc.* (London) Ser. A., **164A**, 151.

Hamilton, J. F., and Brady, L. E. (1959). Electrical measurements on photographic emulsion grains. I. Dark conductivity. II. Photoelectronic carriers. *J. Appl. Phys.*, **30**, 12, 1893.

Hamilton, J. F., Hamm, F. A., and Brady, L. E. (1956). Motion of electrons and holes in photographic emulsion grains. *J. Appl. Phys.*, **27**, 8, 874.

Hecht, K. (1932). Zum Mechanismus des lichtelektrischen Primärstromes in isolierenden Kristallen. *Zt. Phys.*, **77**, 235.

Hedges, J. M., and Mitchell, J. W. (1953). Some experiments on photographic sensitivity. *Phil. Mag.*, **7**, 44, 357.

Herz, R. H., Pitts, E., and Terry, G. C. (1964). A comparison of X-ray diffraction methods in the study of imperfections. *Acta Cryst.*, **17**, 265.

Hilsch, R., and Pohl, R. W. (1930). Zur Photochemie der Alkali- und Silberhalogenid Kristalle. *Zt. Phys.*, **64**, 606.

James, T. H., The mechanism of development. (1966). In C. E. K. Mees and T. H. James (Eds.), *The Theory of the Photographic Process*, Chap. 15. 3rd Ed. Macmillan, New York; Collier-Macmillan, London.

Jost, W. (1938). The energies of disorder in ionic crystals. *Trans. Faraday Soc.*, **34**, 860.

Klein, E., and Matejec, R. (1959). Formation and growth of silver centers on silver halide microcrystals. *Zt. Elektrochem.* **63**, 8, 833.

Koch, P. P., and Vogler, H. (1925). Uber die Ausscheidung von Silber aus Silber-Halogeniden durch Intensive Belichtung. *Ann. Phys.*, **4**, 77, 495.

Koch, E., and Wagner, C. (1937). Mechanismus der Ionenleitung in festen Salzen auf Grund von Fehlordnungsvorstellungen. I. *Zt. Phys. Chem.*, **38B**, 295.

Lehfeldt, W. (1935). Zur Elektronenleitung in Silber-und Thallium Halogenid-Kristallen. Nachr. Ges. Wiss. Göttingen. *Math. Phys. Kl.*, **1**, 171.

Marriage, A. (1961). How many quanta? *J. Phot. Sci.*, **9**, 2, 93.

Mitchell, J. W. (1955). The photographic process. In W. E. Garner (Ed.), *Chemistry of the Solid State*, Ch. 13. Butterworths, London.

Mitchell, J. W. (1957a). The nature of photographic sensitivity. 11th Renwick Memorial Lecture, *J. Phot. Sci.* **5**, 49.

Mitchell, J. W. (1957b). The nature and formation of the photographic latent image. *Phil. Mag.*, **2**, 21, 1149.

Mitchell, J. W. (1957c). Photographic sensitivity. *Report Prog. in Phys.* **20**, 433.

Mitchell, J. W. (1958). Photographic sensitivity. *J. Phot. Sci.*, **6**, 57.

Mitchell, J. W. (1963). *Photographic sensitivity.* In S. Fujisawa (Ed.), (Tokyo Symposium 1962) Vol. 3. Maruzen, Tokyo.

Mott, N. F. (1948). Notes on latent image theory, *Phot. J.*, **88B**, 119.

Pohl, R. W. (1937). Electron conductivity and photochemical processes in alkali halide crystals. *Proc. Phys. Soc.*, **49**, 3.

Sheppard, S. E. (1925). Photographic gelatin. *Phot. J.*, **65**, 380.

Sheppard, S. E. and Hudson, J. H. (1930). Determination of labile sulfur in gelatin and proteins. *Ind. Eng. Chem. Anal. Ed.*, **2**, 73.

Tellez Plasencia, H. (1959). Emploi des Radio-Isotopes pour l'Analyse de l'Argent Primaire ou Développé dans les Emulsions Photographiques. III. Mesure de l'Argent Photolytique de l'Image Latente. *Sci. Ind. Phot.*, (2) **30**, 385.

Tellez Plasencia, H. (1960). Sur les Quantitées d'Argent Photolytique libérées à la Surface et dans la Masse des Microcristaux de Bromure d'Argent. *Sci. Ind. Phot.* (2) **31**, 465.

Webb, J. H. (1955). Ultrashort light and voltage pulses applied to silver halide crystals by turbine-driven mirror and spark-gap switch. *J. Appl. Phys.*, **26**, 1309.

4

The Direct Photographic Response to X- and γ-rays.

4.1 INTRODUCTION

In this chapter we shall deal with the direct response of photographic emulsions to X- and γ-rays. By 'direct' response we mean the direct effects of these radiations on emulsion layers, i.e. without the aid of fluorescent screens or metal foils which are used chiefly for the intensification of the photographic effect. For a description of the action of screens and their use in conjunction with films the reader is referred to the chapters on medical, industrial and neutron radiography.

In some of the discussions of the previous chapter we have described the phenomena occurring in individual silver halide grains; in this chapter we shall be concerned with the effects of the emulsion layer as a whole. In the first part the essential methods of measuring the response of photographic material, such as the optical density and its relation to the number of grains will be discussed. Before dealing with the response, some of the fundamental concepts of the photographic action of X- and γ-rays have to be introduced. We shall then deal with the basic methods of relating the exposure and the density, which are summarized by the term 'sensitometry'. This involves in turn a discussion of the density-exposure curve and of the characteristic curve. Later sections of this chapter will be devoted to discussions on spectral sensitivity, the effect of the rate of exposure, the distribution of the latent image, latent image fading, graininess, directional effects, the influence of temperature during exposure, and other effects.

4.2   OPTICAL DENSITY AND ITS MEASUREMENT

4.2a   Density

In any quantitative application of photographic methods, the degree of blackening caused by the silver deposit of exposed and processed films or plates serves as a measure of the response of the photographic material to a given quantity and quality of radiation. The light transparency of a given blackened area of a film depends on the light absorbing and scattering properties of the deposit, i.e. on the area, volume, distribution, and number of silver grains per unit area. The greater the population of developed grains per unit area in the film, the greater will be the quantity of light absorbed and the darker the area appears to the eye. The blackening of a photographic emulsion is usually called 'optical density' or in short 'density'. The term density (D) which was originally defined by F. Hurter and V. Driffield (1890) is the logarithm of the ratio of the radiant light flux (P_O) incident on the sample, to the radiant light flux (P_T) transmitted by the sample, i.e.

$$D = \log_{10} \frac{P_O}{P_T} = \log_{10} \frac{\text{incident radiant flux}}{\text{transmitted radiant flux}} \tag{4.1}$$

where P may, for instance, be measured in ergs per second. The ratio P_O/P_T is known as the opacity (O), the reciprocal ratio, namely P_T/P_O, is called transmittance (T). The density is thus the logarithm of the opacity, i.e. $D = \log O = \log 1/T$. (See for instance British Standard BS1384:1947.)

The original argument which led Hurter and Driffield to their definition of density was that they found it directly proportional to the mass of silver per unit area. Later investigations have shown that this linear relationship does not hold precisely for a given emulsion chiefly because of the considerable spread of grain sizes in most photographic emulsions.

There are several advantages in using the logarithm of the ratio instead of the ratio. Firstly, in light exposures a great part of the useful range of densities is proportional to the logarithm of the exposure (i.e. the product of intensity and time) and secondly, the numerical values of a series of densities can simply be added together to give the total density. Hence two areas of film, each having a density say of 0.5 would, if super-imposed, measure a density of 1.

Suppose the incident radiant flux (P_O) on a film is 1000 units and the transmitted flux (P_T) is 100 units, i.e. 1/10th of the incident flux, the resulting density D is then

$$\log_{10} \frac{P_O}{P_T} = \log_{10} \frac{1000}{100} = \log_{10} 1000 - \log_{10} 100 = 3 - 2 = 1 \quad (4.2)$$

Hence if the transmitted radiant flux is 1/10th of the incident flux, then $D = 1$, if it is 1/100th, or 1/1000th of the incident flux, then $D = 2$ or $D = 3$ respectively.

If photographic material having paper as a base is used, density has to be measured by reflection instead of by transmission. The definition of the reflection density is

$$D_r = \log \frac{1}{r} \quad (4.3)$$

where r is the ratio of the light reflected by the base to that reflected by the image.

4.2b   The Measurement of Density

If light is impinging upon an exposed and processed photographic layer, it is partly absorbed, partly reflected, and partly scattered by the layer. The numerical density value of a given silver deposit depends on the geometry of the optical arrangement used in measuring P_0 and P_T. Thus the measurement of density can be carried out in different ways. Depending on the optical arrangement one distinguishes three different basic types of density

(a) diffuse density
(b) specular density
(c) double diffuse density

If the light flux is incident normally on the sample and all the flux transmitted is measured, as seen in Figure 4.1(a), a value of the 'diffuse density' is obtained. If only a fraction of the transmitted flux is measured, namely its normal component, as in Figure 4.1(b), one speaks of 'specular density'. If the incident flux is completely diffuse, as in Figure 4.1(c) and all the transmitted flux is collected and measured, we speak of 'double diffuse' density.

As the ratio P_O/P_T for a given P_O is higher the smaller the transmitted flux, so the specular density of a given sample leads to a higher value than that of the diffuse density. The ratio of specular to diffuse density is known as the Callier coefficient or Callier Q factor (see p. 141) which increases with grain size.

Generally, use is made of the diffuse density and all the densities referred to in this book are diffuse densities, unless another type of density is specifically mentioned. The type of density evaluation carried

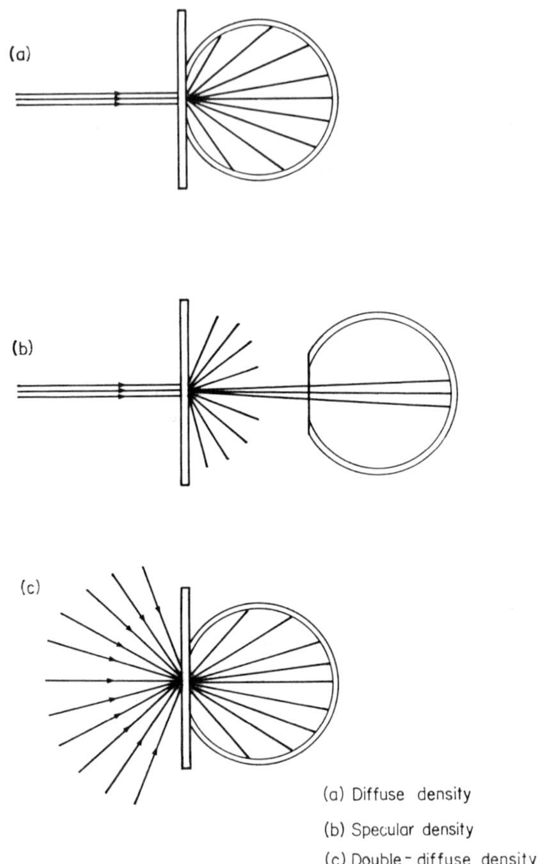

Figure 4.1. Optical arrangement used for measuring density.

out in practice is however, of less importance than the reproducibility
of the measurement. It is advantageous to make use of the method which
is most frequently applied, as it permits comparison of measurements
with those of other workers.

If two samples of different color are to be compared (e.g. different
color of bases or stain in the emulsion) the spectral absorption of the
incident beam will differ for the two measurements. Hence care has to
be taken if comparison tests are to be made, as the change of color of
the samples may invalidate the accuracy of the result.

4.2c Densitometers

The instruments used to measure optical density are called 'densitometers'. Broadly speaking, two different varieties of densitometers may

be distinguished, the visual type based on matching transmitted light fluxes and the physical type based on photoelectric measurement.

Visual Type Densitometer. A representative of the visual type of densitometer is that designed by J. G. Capstaff and R. A. Purdy (1927). In this a light source furnishes two light beams. One passes through a movable circular photographic density wedge through an opal glass on to the film to be measured; the other beam is reflected by a system of mirrors to provide a comparison beam. The observer looks through an eyepiece and sees a circular spot illuminated by the opal glass. This spot is surrounded by a comparison field illuminated from the mirrors.

In order to measure a density, the photographic wedge, having been calibrated precisely in terms of density, is rotated until the brightness of the two fields seen in the eyepiece appear equal to the eye. Using this type of instrument, reproducibility of density readings down to a density of about 0.02 can be achieved, if the instrument is used by the same operator. Greater variations may occur if the results of two different observers are being compared. One of the advantages of this type of visual densitometer is that any variations of the light intensity of the illuminating lamp will affect both viewing fields equally and thus does not influence the density reading.

If no densitometer is available and one wants to check whether a diffuse density of approximately 1 has been achieved in a certain area of a processed film, this can easily be done in the following way. One places the blackened film on a piece of printed white paper. If the film is inspected in a well illuminated room (or illuminated for instance by a 50 watt lamp at a distance of about 2 feet), then the printed letters can just be seen through the film in those areas where the density is approximately unity. If the printed letters are easily seen, then the density is below 1 and if they cannot be detected, the respective density is above 1. Although this is by no means an accurate method of measurement, it is a very practical and useful technique, if no densitometer is available.

Photoelectric Types. Owing to the great advances in the development of electronic methods, photoelectric densitometers are now more widely used. In one of the simplest forms of design a light beam traverses an opal glass, the sample, and then impinges on a barrier layer cell, the photocurrent of which can be read on a microammeter. The scale of the measuring instrument may be calibrated in terms of density. The readings of this rather elementary device may be influenced by fluctuations of the intensity of the primary light source and may require frequent adjustments, unless the constancy of the light output can be guaranteed by some means, (R. H. Herz, 1938; M. H. Sweet, 1950, 1959). Slightly more elaborate densitometers are provided with null or balancing systems in which any changes of the light flux incident on the sample will be automatically compensated for by the light flux transmitted by the sample. In another design the null or balancing principle may be achieved by the use of two photocells and a split beam from one light source, (C. Tuttle 1936).

Some of the more expensive and refined instruments may have not only the merit of greater accuracy and reproducibility, but also permit the results to be obtained with greater speed, even if used by unskilled operators. In these instruments the photoelectric devices may consist of barrier layer cells, gas or vacuum photocells, or photomultipliers provided with amplifier circuits. Advances of stabilization by negative feedback make such instruments independent of mains' fluctuation. Logarithmic amplifiers and servo mechanisms also make them convenient to use. Automatically operating devices which permit the plotting of densities at high speed rates have been found particularly useful in large scale film monitoring services, (see for instance p. 250). The extremely sensitive photomultiplier tubes of modern design combined with the use of high intensity light sources also permit the measurement of very high densities up to about $D = 6$, whereas the majority of instruments measure only up to about $D = 3$ with fair accuracy.

In some experiments it is essential to measure with precision small densities within the range of $D = 0$ to 0.5. A densitometer specially designed for this purpose is described by A. S. Cross (1955).

Very complex types of densitometer are now commercially available, such as recording densitometers, microdensitometers and recording microdensitometers; for details of these the reader is referred to the publications of the electrical, optical and photographic industries. The basic principles of densitometry are described in standards such as those issued by the American Standard Association and the British Standards Institution. (See list of references.)

4.2d   Covering power

It has been shown how a quantitative evaluation of the developed image can be carried out, utilizing the light absorbing properties of the silver deposit. This method of measuring the optical density is the one most commonly adopted for the evaluation of the photographic image.

In order to consider the chemical reactions involved in the photographic processes, it appears logical to determine the mass of silver deposited per unit area. Such measurements were carried out by means of gravimetric and volumetric methods. The ratio between the mass of silver per unit area and the optical density was originally called the photometric constant, but is now referred to as the photometric equivalent (P), as it was shown later that P is far from being a constant. P is expressed by

$$P = \frac{\text{grams of Ag}/100 \text{ cm}^2}{\text{optical density}} = \frac{1}{\text{covering power}} \tag{4.4}$$

$1/P$ is called the covering power. The importance of the covering power has been investigated more recently as new analytical and rapid methods became available for the determination of the mass of silver per unit area. In one of these the mass of silver is measured in a commercially available X-ray spectrometer and in another by converting the silver image into a silver iodide image using radioactive iodine-131, (A. E. Ballard, G. W. W. Stevens and C. W. Zuehlke 1952).

Investigations using the above mentioned methods showed that the relation between mass of silver per unit area and optical density is greatly influenced by the type of photographic material used, by the method of processing and also by the mode of drying the emulsion. (T. H. James and L. J. Fortmiller 1961 and, C. E. K. Mees and T. H. James 1966).

4.2e. Relation Between Density and Number of Grains

A quantitative evaluation of the developed image may not only be carried out by means of the optical density or by the determination of the mass of silver per unit area, but also in certain cases, by counting the number of developed grains per unit area. Such measurements are of practical importance in autoradiographic applications (see p. 556) and also in certain problems of film monitoring. It is of significance in those cases where the sensitivity of the photographic material, or the available amount of radiation, is too low to obtain records resulting in measurable optical densities.

The relationship between optical density, the mean developed grain area and the number of developed grains per unit area was first derived theoretically by P. G. Nutting (1913) (see p. 150). This relation is expressed by

$$D = \bar{a}n \log_{10} e = 0.4343\bar{a}n \tag{4.5}$$

where $\bar{a}$ is the average projected area of a developed grain and n the number of developed grains per unit area (cm^2). After a thorough check on the practical validity of this expression, G. C. Farnell and L. R. Solman (1963) introduced a slight modification to take account (a) of the increase in absorption as a result of light scatter within the layer and (b) of the fact that the mean projected area of the fully developed grains in general exceeds that of the undeveloped grains. The nearest approach of Nutting's formula to satisfying the conditions of most modern emulsions, which have been developed in an energetic developer is found to be

$$D = \bar{a}nk \log_{10} e \tag{4.6}$$

where k is a constant $= 2.6$.

This equation is useful in assessing the approximate density to be expected if the number of quanta absorbed per unit area in an emulsion and the average size of the developed grains (projected area) are known. (See pp. 148–152 and 528.)

Considering that the mass of silver per unit area is equal to the product of the number of grains (n) per unit area, their volume (v) and their specific density (ρ), i.e. $nv\rho$, the covering power ($1/P$) can be expressed by

$$\frac{1}{P} = \frac{D}{nv\rho} = \frac{0.4343\bar{a}nk}{nv\rho} = \frac{0.4343\bar{a}k}{v\rho} \tag{4.7}$$

Thus, because the expression on the extreme right is proportional to the inverse of the average thickness, the covering power is inversely proportional to the average thickness. This is what might be expected. If a layer of grains of given volume is compressed between two glass plates the thickness decreases but the density increases because the grains are flattened out to larger areas.

With emulsions having a wide range of grain sizes the covering power was found to increase with increasing quantity of radiation absorbed. This can be explained on the basis that in the initial stages of exposure, the larger grains have a greater probability of being hit than the smaller ones, simply because of their larger volume. Hence the average size of developed grains being hit should decrease with increasing exposure, thus leading to an increase of covering power.

It may be mentioned that most of the more highly sensitive X-ray emulsions are not directly suitable for developed grain count measurements, unless a certain volume of the emulsion is diluted and spread out. Methods of grain counting are more easily applied to nuclear and autoradiographic emulsions. (See p. 556.)

4.3   FUNDAMENTAL ASPECTS OF THE PHOTOGRAPHIC ACTION OF X- AND γ-RAYS

After having discussed the basic methods with which the effect of radiation on photographic materials can be assessed quantitatively, it seems appropriate to consider first some fundamental aspects of the photographic action of X- and γ-rays. The knowledge of these will undoubtedly assist the understanding of the more detailed response of emulsions when exposed to ionizing photons.

4.3a The Release of Electrons by Photon Absorption

If X- or γ-rays pass through a coated emulsion layer a certain fraction of the number of quanta incident will be absorbed, some will be scattered and the greatest number will be transmitted. In this absorption process ionizations occur whereby electrons are released from the silver halide grains. Electrons originate not only from the grains but also from gelatin, from the base and from the surroundings of the layer. It should be emphasized that the formation of latent images in the grains is not caused directly by X-ray photons or quanta absorbed but indirectly by the electrons released.

The X-ray absorption in emulsion occurs primarily by three mechanisms as referred to on pages 15–18. These are photoelectric absorption, Compton scattering and pair formation. The latter, which theoretically does not occur below 1.02 MeV, only becomes of practical importance with X- or γ-rays of a few million volts energy and is associated with electron-positron pairs. The electrons released in any of these processes may acquire very high kinetic energies and are thus capable of travelling through many grains and rendering them developable (S. R. Pelc 1945; R. H. Herz 1949a). Their range in the emulsion depends, in the important region of photoelectric absorption, on the energy of the photon absorbed minus the energy required to release an electron from *K-L* or other shells of the ions or atoms concerned. The *K*-fluorescent radiation emitted in this process may give rise to a further photoelectric absorption in the same or in another grain (H. Tellez-Plasencia 1950, 1953; J. R. Greening 1951).

4.3b The Rate of Energy Dissipation of Electrons

The response of the photographic material is largely governed by the behavior of these electrons whilst travelling through the emulsion, i.e. by their rate of energy dissipation or, if expressed in another way, by their rate of transfer of energy to the grains. It is therefore useful to discuss first their mode of energy dissipation (see also pp. 29 and 159).

The ionizing power of a charged particle of high kinetic energy, such as an electron, is inversely proportional to the square of its velocity (see p. 30). In Figure 4.2 the average rate of loss of energy (E) of an electron per unit length of path (x), i.e. dE/dx in an X-ray emulsion, in keV/micron length is shown on the ordinate and the kinetic energy, (in keV) of electrons is plotted on the abscissa axis. It is seen that

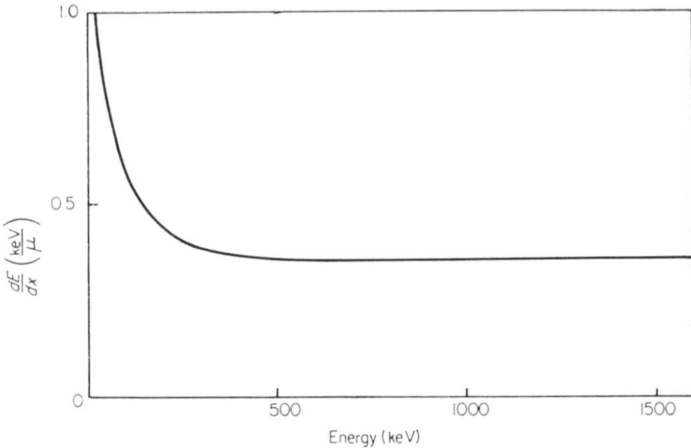

Figure 4.2. Average rate of energy loss of electrons per micron in X-ray emulsion as function of energy (keV) (calculated).

dE/dx decreases first rapidly with increasing energy of electrons to about 200–300 keV (0.2–0.3 MeV), then the curve flattens and reaches a minimum of energy loss (near 0.6 MeV) or, as it is often called, the region of minimum ionization (see also p. 159). Hence if an electron is released from a grain by the absorption of a high energy X-ray or γ-ray quantum, it travels first with a high initial kinetic energy but will gradually slow down, mainly through its interactions with the ions of silver and bromine, until it reaches quite a small kinetic energy and finally stops or comes to rest. The graph in Figure 4.2 shows that the rate of energy loss of an electron, i.e. the rate of linear transfer of energy to grains along its path, is considerably higher at low than at high kinetic energies. Thus an electron is capable of highest ionization per length of path when it slows down, i.e. towards the end of its path through the emulsion. Some of the electrons will obviously be scattered out of the emulsion and thus are lost to the photographic process unless they are scattered back into the emulsion from surrounding material (see also page 107 or electronic equilibrium).

The individual paths of electrons cannot be recognized in an X-ray emulsion by microscopic inspection; this is chiefly because the grains are too far from each other and are of irregular size. In a nuclear particle emulsion however, in which the grains are very closely packed and have very similar size, the paths or tracks can easily be recognized by microscopic inspection. A photomicrograph of such an electron track is shown in Figure 4.3. This electron had an initial kinetic energy of about $300\,\mathrm{keV}$ and it is seen that the number of grains rendered developable per unit length of path, i.e. its grain density along the track, is greatest towards its end when the particle has slowed down (left side in Figure 4.3). The highest grain density is due to the maximum number of ionizations the particle has caused per unit length of path. The number of ion pairs per unit length of distance along the track is also known as specific ionization. It is further seen in Figure 4.3 that an electron is easily scattered in the emulsion due to its small mass. This scattering is due to electrostic interactions with orbital electrons which it encounters whilst travelling through the emulsion layer. The scattering is increased with reduction in speed of the

Figure 4.3. Photomicrograph of an electron track as seen in a nuclear track emulsion.

particle. From about 1 MeV towards higher energies the path of the electron is almost straight.

The optical density in an X-ray image is thus formed by the superimposition of electron tracks, in spite of the fact that their paths are not recognizable in an X-ray emulsion for the reasons mentioned above. The concept of electron tracks naturally breaks down when the kinetic energy of an electron is so low that it cannot pass through more than one grain on the average. This occurs with low energy X-rays generated at approximately 30 keV, using a fast coarse grained X-ray emulsion. The effect naturally depends on the average stopping power of grains and gelatin for electrons and is thus dependent on grain size.

4.3c   Quantum Efficiency

We may conclude from the facts mentioned above, that one X-ray quantum absorbed in a grain leads to the developability of at least one grain and that, with higher X-ray energies, several grains may be rendered developable per X-ray quantum absorbed.

The number of grains rendered developable per X-ray quantum

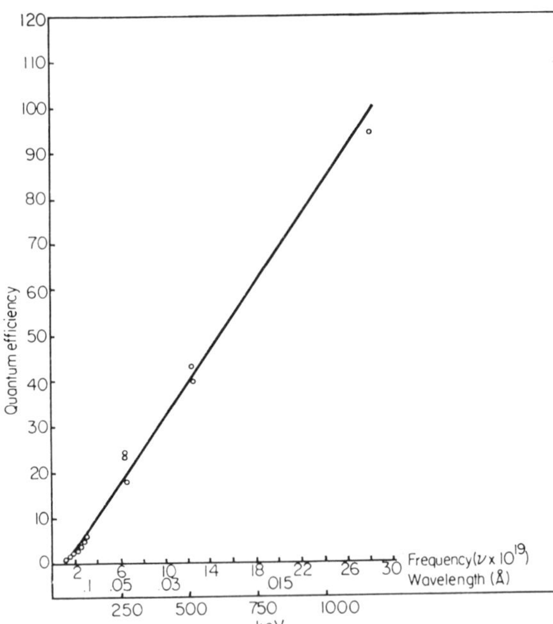

Figure 4.4. Quantum efficiency as a function of X-ray energy (keV), wavelength (Å) and frequency ($\nu \times 10^{19}$). (After D. Bromley and R. H. Herz (1950). *Proc. Phys. Soc. B.* LXIII, 90 (Institute of Phys. & Physical Soc. London).)

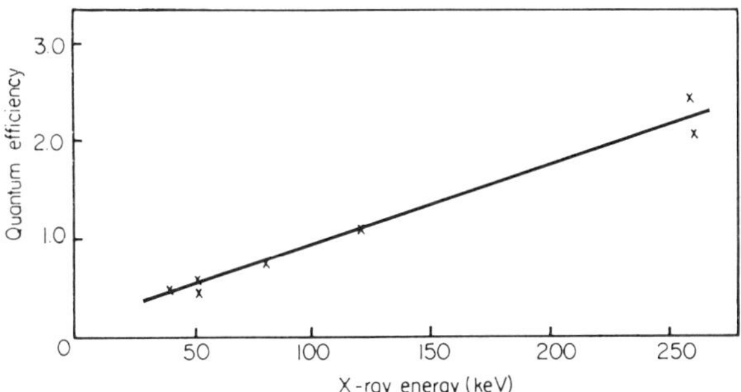

Figure 4.5. Quantum efficiency as a function of X-ray energy (keV) of a slow chlorobromide emulsion. (After D. Bromley and R. H. Herz (1950). *Proc. Phys. Soc. B.* LXIII, 90 (Inst. of Phys. & Phys. Soc. London).)

absorbed is referred to as the 'quantum efficiency' and this has an important effect on the behavior of photographic emulsions when exposed to X- or γ-rays, as will be pointed out in later sections of this chapter (see p. 105).

In view of the low energy of light quanta, it is not surprising that it was found that several light quanta are required to render a grain developable, even in a very sensitive emulsion. Comparing the few volts of energy of a light quantum with the high voltage energy of an X-ray quantum (say 30,000 or 100,000 volts) a one quantum hit developability can be expected. Experiments first carried out by J. Eggert and W. Noddack (1927, 1928) have shown that about 1000 silver atoms (see p. 66) were formed on the average in a grain by the absorption of one X-ray photon and that this led to developability of a grain. (See also R. Glocker 1927a, 1927b.)

Experiments carried out by S. R. Pelc (1945), D. Bromley and R. H. Herz (1950), H. Tellez-Plasencia (1948) and J. R. Greening (1951) revealed that with further increase of energy several grains can be made developable by one quantum absorbed. In the graph Figure 4.4 the quantum efficiency is plotted against the energy (in keV) of the X-ray photons concerned. It is seen that in this case as many as 100 silver grains were produced due to the effect of one quantum of cobalt-60 γ-rays (approximately equivalent to 0.01 Å in wavelength). The graph shown in Figure 4.4 refers to a fast X-ray emulsion. However the quantum efficiency may fall below unity using emulsions of low sensitivity and radiation of long wavelength or of low energy transfer (e.g. as with γ-rays at minimum ionizing power). It is seen from Figure 4.5 that the quantum-efficiency drops below unity for an emulsion of rather low sensitivity at long wavelengths. The explanation for this behavior is simply that the transfer of energy to a grain by a single hit is insufficient to render a grain developable. For this slow chlorobromide emulsion a quantum efficiency of less than unity was found even when using X-rays near 100 kV$_p$ (see Figure 4.5).

It should be mentioned that the experimental evaluation of quantum efficiency is less accurate within the region of high energy radiation, because many electrons will be scattered out and others scattered into the emulsion layer from its environment. Any assessment of quantum-efficiency is thus ambiguous unless electronic equilibrium is reached, i.e. just as many electrons are reflected back into the emulsion as are lost by scattering out of the emulsion (see p. 107).

4.4 X- AND γ-RAY SENSITOMETRY

Broadly speaking, the response of a given photographic material to direct X- or γ-rays depends on the number and the energy of the quanta absorbed per unit area of an emulsion, and on the efficiency of the conversion of the absorbed radiation energy into developable silver halide. The absorption effects are accompanied by photon and electron scattering in the photographic layer and by secondary radiation effects, all of which are also influenced by the type and thickness of the material surrounding the layer during exposure. The conversion of the absorbed radiation energy into developed silver is dependent on the efficiency of the developer solution, on its temperature, on the development time and the degree of agitation of the film in the solution, on the time elapsing between exposure and development, on the chemical fog produced, and on several other effects.

The combined result of all these factors can be evaluated by a method called X-ray or γ-ray sensitometry, i.e. the study of the quantitative relationship between exposure and optical density.

4.4a Principles of Sensitometry

In sensitometry with light, exposure is defined as the product of radiation intensity and time. In X-ray and γ-ray work exposure is usually expressed by the appropriate exposure unit, the roentgen (R) (see p. 47). The rad, being an absorbed dose, also may be used instead of the roentgen (see p. 46). In this case it must be known whether the rad is defined as an absorbed dose in air or in another medium. It is important to realize that a rad in 'air' has a constant relationship to the roentgen which is independent of the energy of radiation, at least above 20 keV, namely 1 rad$_{(air)}$ = 0.869 R. For any medium other than air this relationship depends on the energy of radiation (see p. 48).

For an absolute assessment of the response of a given material to radiation in terms of roentgens, the use of an ionization chamber (see p. 43) or other type of dosimeter is required, unless a radioactive source of known activity is used.

X-Ray or γ-ray sensitometry may be carried out basically by two methods, time and intensity scale exposures, the choice of which depends on the convenience of application. The rate of exposure, in general, does not influence the result using direct X-ray exposures (see reciprocity rule, p. 130).

Time Scale Exposure.   In this case successive exposures are given on adjacent areas of a photographic layer by using a constant exposure rate at increasing times of exposure. An example may illustrate the procedure. For a given exposure rate of say 0.02 R per second, one may expose a film 1/10th, 2/10th, 4/10th, 8/10th . . . seconds in order to give adjacent areas exposures of, for instance, 0.002, 0.004, 0.008, 0.016 . . . R. The exposure rate and the increment have been chosen arbitrarily to give an example only and any other values may be selected to suit individual conditions.

A time scale exposure can be carried out by using a sensitometric device. This may consist of a sheet of lead having a slit in the center. A film in its cassette or in its paper wrapping may be shifted, e.g. by mechanical means, below the slit for successive exposures, as shown in Figure 4.6.

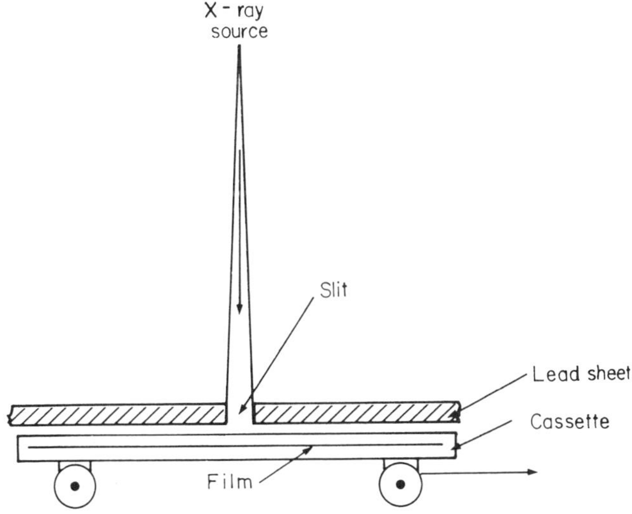

Figure 4.6. X-ray sensitometer for time scale exposure.

It is important that the lead thickness is so chosen that the film does not display any measurable density (after processing) below the area protected by lead. If absolute exposures are not required, the use of an ionization chamber dosimeter is not necessary, but instead relative exposures can be achieved, using an accurate and reliable time switch. (See check on timing by spinning top, p. 331.) The exposures may be expressed as multiples of some arbitrarily chosen quantity, the actual value of which need not be known, for example in terms of mA secs.

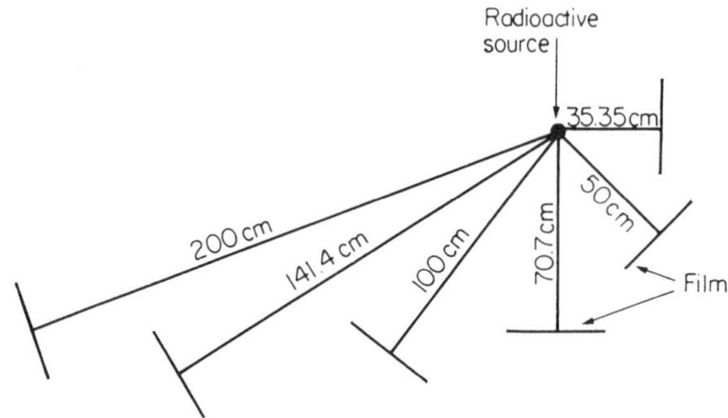

Figure 4.7. Arrangement for intensity-scale exposure according to inverse square law of distances. (Distances are given in cm and exposure values are increasing by a factor of 2 from 200 cm towards short distances.)

Intensity Scale Exposure.   In this method so named due to its similarity to that used in light sensitometry, increasing exposures are taken (usually on various film strips) at a given time. The following example may illustrate the intensity scale exposure. Assume that a radioactive source emitting hard γ-rays is used, several strips of the same sheet or type of film may be placed at various distances from the source in such a way that the films nearer to the source do not act as absorbers for those more remote from the source. The relative positions of the film strips with respect to the source may be arranged so that in each position the exposure is double that of the next one. The positions are thus chosen according to the inverse square law (see p. 21) and one exposure time only is to be given to all the films simultaneously (see Figure 4.7.)

As mentioned before, this method is particularly useful when applying a radioactive isotope source emitting hard γ-rays, as it would be difficult or impossible to make use of a time scale exposure. This is because extremely great thicknesses of lead would be required to protect adjacent areas of a sheet of film from unwanted exposures, owing to the great penetration of γ-rays emitted by certain radioactive isotopes, such as for instance, cobalt-60.

4.4b Stepwedge Exposures

A further method of studying the photographic response is to interpose a stepwedge made of aluminum, copper, plastic or other material between the source of radiation and the film. The results of exposures carried out in this way differ basically from those presented by a density-exposure curve. The main difference is that each step of the wedge alters the quality of the radiation transmitted. Hence the degree of preferential absorption of the long wavelengths differs from step to step, with the result that the transmitted radiation from each successive step is of a different quality. The change of quality might be only slight, depending on the quality of the incident radiation in relation to the change in thickness and material of the step wedge. Furthermore, the radiation flux transmitted by each step is not easily determined experimentally, as it is also influenced by scatter from adjacent steps; hence it is hardly possible to plot density against exposure expressed in quantitative terms. On the other hand it must be admitted that a stepwedge exposure is a closer approach to a practical exposure in radiography as objects are interposed between radiation source and film. Thus if a comparison of films without the aim of a quantitative analysis is required, a stepwedge exposure might be useful.

Stepwedges specially designed to provide a geometric increment of X-ray flux on the film, i.e. a constant log exposure interval per step for a given range of kV_p, have been described by H. E. Seemann and B. Roth (1960, 1962) and G. M. Corney and H. E. Seemann (1947). The wedge is thickest in the middle and tapers symmetrically to the thinnest steps at both ends which gives the operator a check on uniformity of development and on the incident X-ray flux. The physical dimensions of the wedge are chosen so that the amount of scattering will be constant using the recommended conditions of exposure. It was found that the sources of error due to changes of quality of radiation by each step were small enough to be ignored for practical radiographic work.

4.4c Density as Function of Exposure

After the sensitometric exposure has been carried out on one or on a series of film strips, these are then processed according to a standard procedure. If the response to radiation is to be studied on several types of film, it is preferable to process these simultaneously, in order to avoid errors which might otherwise occur during successive processing.

Fog Value.   It is necessary to process an unexposed film of the same batch as the exposed film simultaneously with each type of film used in the sensitometric exposure. This permits the determination of the background fog (i.e. the density which is unrelated to exposure) for which the measured densities due to exposure have to be corrected. The measured and corrected densities of the exposed film are referred to as 'net densities, i.e. density minus fog.

Shape of the Density-Exposure Curve

The relationship between net density and exposure is then obtained by plotting them on linear scales. The exposure may be expressed in terms of roentgens or in terms of exposure time, if the radiation flux can be taken as being constant. In Figure 4.8 we see a typical density-exposure curve in which the exposure is expressed in roentgens. In the initial part of the curve it is seen that the net density increases proportionally with exposure in roentgens. This is to be expected because

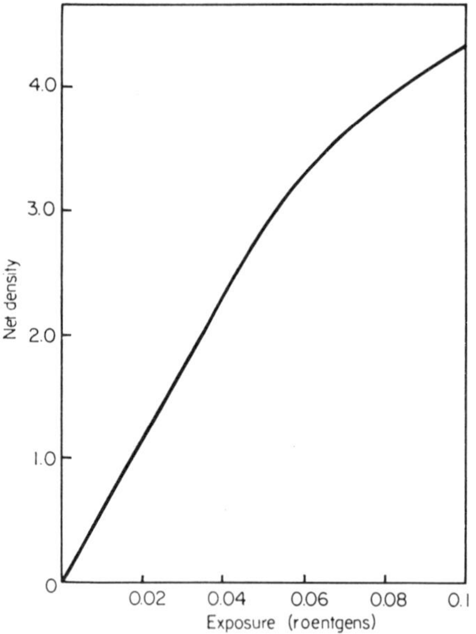

Figure 4.8. Relationship between density and exposure (in roentgens) for X-rays generated at 80 kV$_p$ and filtered by 0.3 mm Cu).

the number of grains made developable by absorption of one quantum is statistically constant for all quanta during exposure. Thus the number of developed grains is proportional to exposure. The number of developed grains can also be taken as being proportional to density. This proportionality holds as long as the population of unaffected grains is large in comparison to the number of affected grains. Thus as the exposure is increased, the probability of absorbed quanta affecting grains already hit by previous absorptions increases. The relation between density and exposure then ceases to be one of proportionality and the curve tends to turn into a flat maximum at which all grains have been made developable.

Many modern X-ray emulsions exhibit proportionality up to a density of 2 and some even up to 3–4. The length of the straight line depends largely on the grain size and its distribution, on the intrinsic sensitivity of the grains, on the quality of the radiation applied (in extreme cases) and on the development procedure. The length of the straight line can sometimes be increased within limits by prolonging development time. In general it can be said that the length of the density–exposure curve for which this proportionality holds is greater, the larger the grain population per unit volume of the emulsion coating, i.e. the smaller the grain size, assuming that the quantum efficiency is unity or higher.

If the intrinsic sensitivity of grains decreases beyond a certain limit at which one quantum hit does not lead to developability of a grain, the proportionality ceases to occur. This is often found in low sensitive emulsions and is either revealed by a very short straight line density-exposure curve or even by a bend in the initial region of the curve, and it is also found with low energy X-rays and more sensitive emulsions exposed to radiation in the region of 10–60 Ångströms, i.e. approximately equivalent to 1.2–0.2 keV. (F. R. Hirsh (1938, 1942), Z. Szilagyi and E. Blackman (1966) and P. A. Atkinson (1966).) Similar effects using slow emulsions may also occur in the region of hard γ-rays where many of the electrons released are close to minimum ionization (see p. 96). Hence these will cause a low energy transfer to the grains and multiple hits per grain may be required.

The proportionality is often regarded as an extremely useful property of emulsions, as it facilitates scientific investigations in many applications of X-rays, particularly in X-ray crystallography, in dosimetry and in quantitative autoradiography (see p. 553). The advantage is that the relative intensities or exposures to radiations can be read directly from the densities within the region of proportionality.

Comparison Between Light and X-Ray Exposure

If the density-exposure relationship for X-rays is compared with that revealed by light exposures, as shown in Figure 4.9, it will be seen, as mentioned before, that there exists a toe in the curve (called the induction period) which is generally absent for X-ray exposures. One can assume that within the region of the foot of the curve on the average about 4–10 light quanta are required for a fast emulsion to

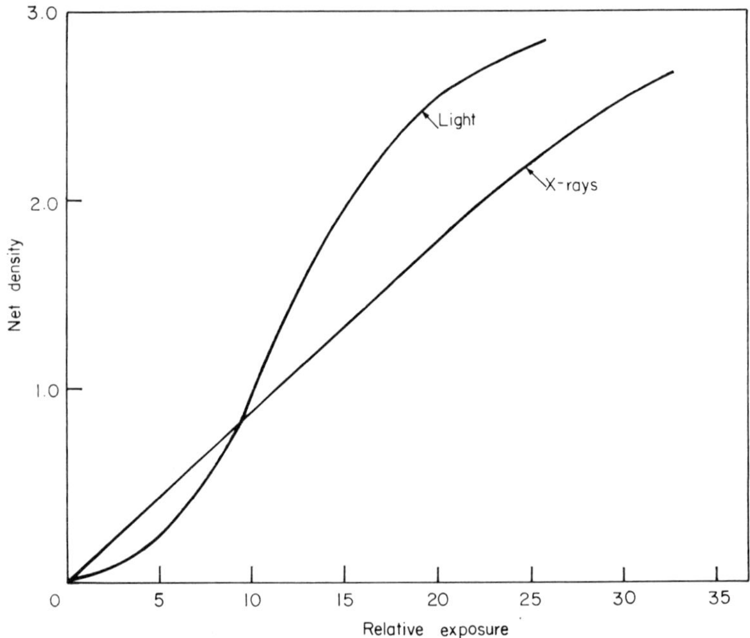

Figure 4.9. Comparison between density-exposure curves for light and X-rays.

render a grain developable. No such threshold value exists for X-ray exposures, apart from those made at very low X-ray energies and taken with slow emulsions.

Speed or Sensitivity

If the density-exposure curves of two emulsions having different properties are to be compared, as seen in Figure 4.10, it is found that a lower exposure is required for film I than for film II, in order to obtain the same net density. Hence film I is said to have a higher sensitivity or higher speed than film II at a given density. It may also be said that film I is faster than film II. The term speed can be defined as the exposure required to obtain a certain density. Hence it may be

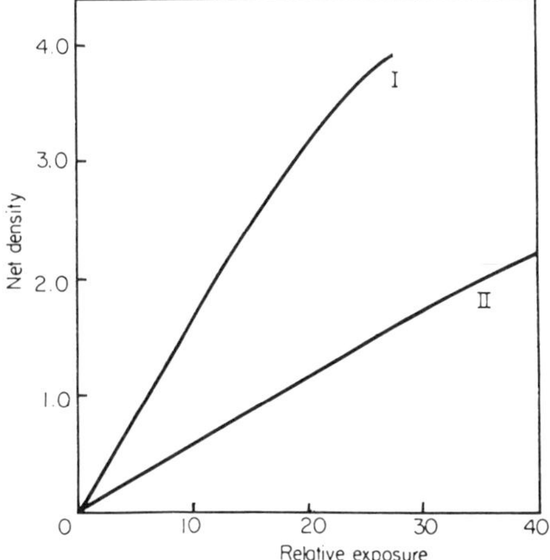

Figure 4.10. Density-exposure curves of two films having different speed.

expressed by D/R, where R is the exposure in roentgens and D the density. Thus the slope dD/dR of the density exposure curves is a measure of the speed of the emulsion. Generally *speed* or *sensitivity* are referred to as the *reciprocal of the exposure*, in this case as the reciprocal of the number of roentgens required to obtain a given density. Preferably one chooses the density within the straight line region, as the speed is then constant throughout this region. Practical D/R values for modern no-screen X-ray emulsions are given on pages 309 and 401.

It was found that the speed of an X-ray emulsion is largely influenced by the mass of the grains and by the coating weight. J. Eggert (1931) established that the sensitivity of X-ray emulsions increases approximately with $m^{2/3}$, where m is the mass of silver bromide. The increase in speed with mass is not surprising, as one would expect that the chance of hitting grains during the random absorption process of X-ray quanta should depend on the projective area of the grains. Furthermore the absorption is obviously enhanced by increasing coating weight of silver bromide, leading also to an increase in emulsion speed.

Electronic Equilibrium

Some of the results achieved by sensitometry of X- or γ-rays may be rather ambiguous because a certain fraction of the electrons released during exposure are scattered out of the emulsion and are thus lost in an assessment of emulsion speed. This effect may be compensated for, partly or fully, by embedding the X-ray film in a material the thickness of which is of the order of the range of the electrons. Hence an 'electronic equilibrium' can be achieved when for each electron lost from the emulsion one of approximately the same energy is scattered into the emulsion. Up to about 250 keV the wrapping material of films is usually sufficient to obtain equilibrium. Using higher quantum energies Perspex (Lucite) sheets may be placed in contact with both sides of a double coated X-ray film and the necessary thicknesses required to obtain electronic equilibrium may be determined experimentally.

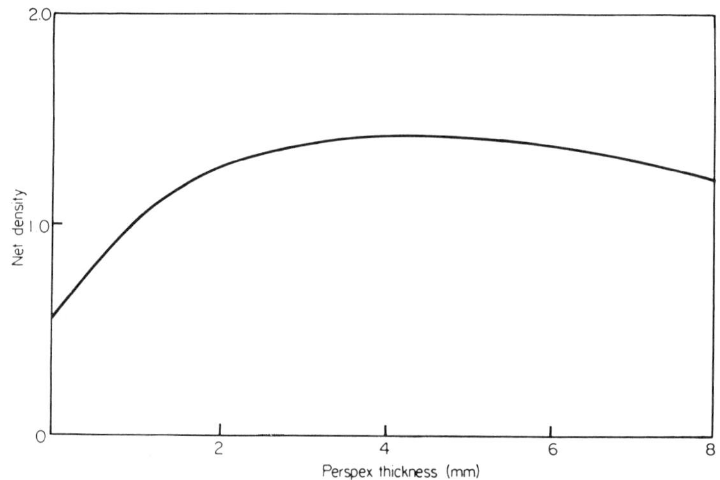

Figure 4.11. Electronic equilibrium reached at 3 mm Perspex (Lucite) thickness on each side of double coated X-ray film. (cobalt-60 γ-ray exposure).

In Figure 4.11 the increase of density as a function of the thickness of Perspex sheets between which a double coated X-ray film was sandwiched, is shown for γ-rays of cobalt-60. The flat density maximum was obtained with about 3 mm of Perspex, whence the equilibrium is attained. The curve then gradually drops to lower densities with further increase of Perspex thickness owing to the fact that the attenuation of γ-rays exceeds that of the contribution of electrons.

The electronic equilibrium effect also depends on the diameter of the γ-ray beam, as can be seen from the graph in Figure 4.12. This shows the relationship between the ratio of the relative response of a film sandwiched between 3 mm Perspex sheets to that of the same film without Perspex, as a function of the diameter of a γ-ray beam emitted from a cobalt-60 source. In both cases the films were exposed in paper envelopes. It is seen from Figure 4.12 according to unpublished work

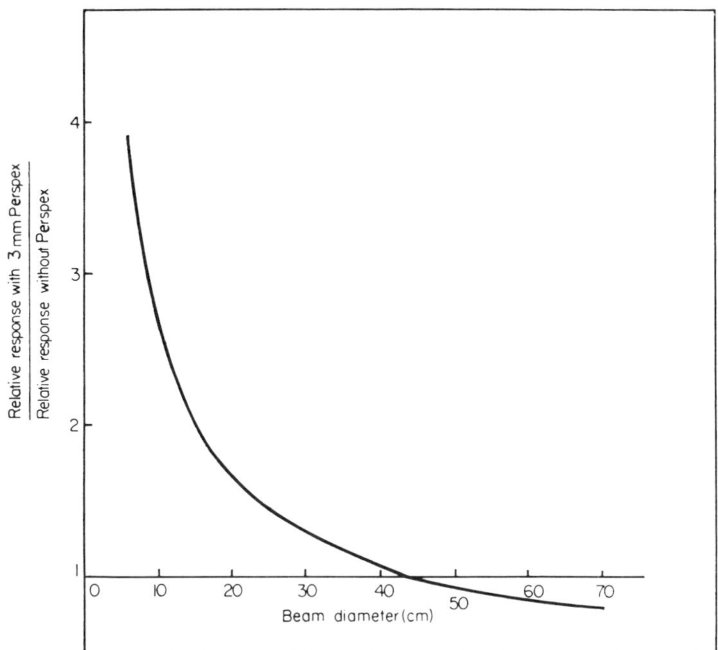

Figure 4.12.
Relative response with 3 mm Perspex (Lucite)
————————————————————————————
Relative response without Perspex (Lucite)
as a function of γ-ray beam diameter.

by D. Frost and R. H. Herz that the ratio of the response with and without Perspex increases appreciably towards smaller beam diameters.

This effect can be interpreted as follows. With increasing beam diameter the number of electrons generated in air becomes so great that it compensates for the electrons scattered out of the non-Perspex film. Hence there exists a certain diameter (45 cm in this particular experiment) where the response with and without Perspex is the same. With further increase of the diameter beyond 45 cm, the ratio of the response becomes smaller than unity, i.e. the contribution of electrons from outside the film exceeds that lost from the film using no Perspex. If a metal cassette is used the effect is very much reduced.

4.4d The Characteristic Curve

Although it has been found convenient in many laboratories to plot the density-exposure curve for calibration purposes, as shown in the previous sections, it is more usual to present this relationship by plotting density (or net density) against the logarithm of exposure. In light exposures this presentation is often preferred, because a considerable portion of the curve reveals a straight line relation between density and log exposure in the most important range of image densities. In X- and γ-ray exposures such a proportionality between density and log exposure is rarely found.

The graph obtained by plotting density against the logarithm of exposure is called the 'characteristic curve',* because the presentation is characteristic of the properties of the particular type of emulsion, the type of radiation to which it was exposed, and the processing conditions to which it was subjected. (Obviously the density-exposure curve is also characteristic of these factors). Figure 4.13 illustrates a

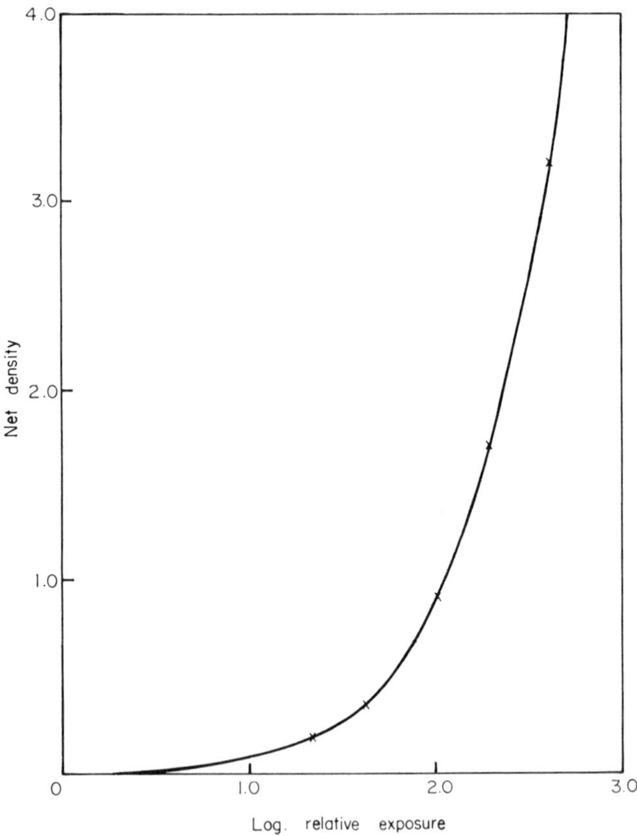

Figure 4.13. Characteristic curve of fast no-screen X-ray film exposed to direct X-rays, i.e. without intensifying screens.

* The characteristic curve is also often called the H and D curve, after Hurter and Driffield who first used it in 1890.

typical characteristic curve of a no-screen X-ray film exposed to direct X-rays. The ordinate shows the net density and the abscissa the log relative exposure (E). The exposure may also be expressed in absolute terms, for instance in log. roentgens.

Most modern no-screen X-ray films are so thickly coated on both sides of the base, that the ordinate of the characteristic curve of such films would reach a density of about 12 or even more, if the exposure were continued beyond the value shown in the graph of Figure 4.13. As this would be of no practical advantage, most characteristic curves of modern no-screen X-ray films are only drawn up to a density of about 3–4.

If however a characteristic curve of a film is plotted, which is actually designed for intensifying screen exposures (see p. 303), one can easily follow the characteristic curve up to the maximum density. This type of film in general does not exceed densities beyond about 3.5–5. However the characteristic curve of such a screen X-ray film exposed without screens is shown in Figure 4.14 in order to illustrate the typical shape of a full characteristic curve based on a direct X-ray exposure. As seen from Figure 4.14 the curve consists essentially of 4 main parts.

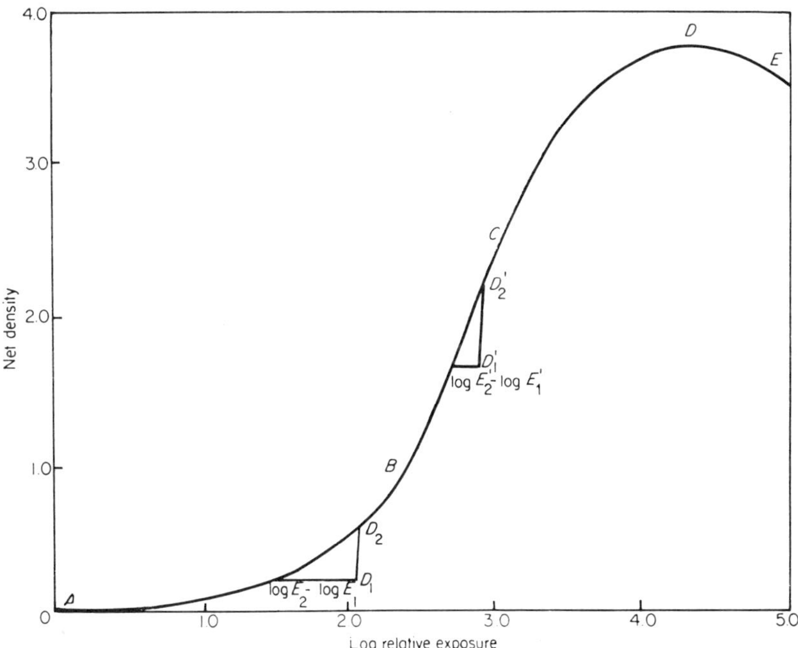

Figure 4.14. Characteristic curve of a screen-type X-ray film exposed to direct X-rays without screens.

1. A curved region (A-B) which is often called the toe region or the foot of the characteristic curve.

2. An almost linear part (B-C).

3. A shoulder (C-D), where D indicates the maximum density D_{max}. At this density it is assumed that practically all the available silver bromide grains of the emulsion have been reduced to metallic silver.

4. The region from D-E, which is called the 'reversal' or 'solarization'. In this region the density gradually decreases again with further increase of the exposure(see p. 116).

Gradient

For any given small difference of log exposure ($\Delta \log E$) in the characteristic curve there is a corresponding small difference of density (ΔD) and the ratio $\Delta D/\Delta \log E$ is called the gradient G. Characteristic curves of emulsions which have been exposed to light usually show a straight line region between B-C, as has been mentioned before. Hence the gradient G within this straight line region, i.e. the slope is constant and is referred to as the 'gamma'. This is equal to the tangent of the angle α formed between the straight line and the $\log E$ axis. Hence

$$\text{gamma} = \frac{D_2 - D_1}{\log E_2 - \log E_1} = \tan \alpha \qquad (4.8)$$

This gamma is not of importance in direct X-ray or γ-ray sensitometry, as the characteristic curve of X-ray and γ-ray exposures does not in general reveal a straight line region but the slope increases gradually towards higher densities.

The gradient or the slope which can be determined at any point of the characteristic curve by the tangent at that point is taken as a measure of the inherent contrast of the emulsion, which should not be confused with the objective contrast defined as the difference between the densities $D_2 - D_1$, for instance, of two adjacent areas of an image (see p. 274) If the gradient of the characteristic curve exceeds unity, the ratios of the radiation intensities falling on the photographic material are recorded in an exaggerated manner, i.e. *the film acts as a 'contrast amplifier'*.

It is seen in Figure 4.14 that initially, within the region ($A - B$), the gradient ($\Delta D/\Delta \log E$) and hence the inherent contrast is only slowly increasing. This means that in radiographic exposures, where contrast is of fundamental importance, the essential region of an image should not be taken at very low densities where contrast is insufficient. X-ray and γ-ray exposures reveal a steady increase of slope and thus of 'inherent contrast' from $B - C$ until the region of saturation of the curve is reached. Typical modern X-ray films designed for no-screen exposure show saturation only at extremely high densities, as indicated earlier. Hence the gradients of some of the no-screen X-ray films may reach values of up to 6 or more at densities of about 3 (see Figure 403) and are of the order of 3.5 at a density of 1.5 depending on the type of film.

Speed

It has been mentioned earlier in connection with the density-exposure curve that the speed (or sensitivity) of a film to direct X-rays is revealed by the slope of the density-exposure curve (see Figure 4.10). Using a characteristic curve however, the speed is determined from the relative position of the characteristic curve with respect to the log exposure axis, as will be shown in greater detail in the next section.

The speed which was defined earlier as the reciprocal of the exposure to obtain a given density may be based on relative or on absolute exposure (e.g. roentgen) values. In the former case we speak of relative, in the latter case of absolute speed. There is no internationally accepted definition of speed for X-ray films, but the following standards ASA PH 2.8.1964, ASA PH 2.9.1964 and ASA PH 2.10.1965 give recommendations for speed measurements in X-ray and γ-ray work.

Evaluation of Gradient and Speed

An example may illustrate the determination of gradient and speed of two different emulsions, the characteristic curves of which are shown in Figure 4.15(a) and (b). The two sets of curves of Figures (a) and (b) are identical.

Instead of determining the gradient at many points of the characteristic curve, it is convenient to describe the contrast properties of a film by one single gradient value. This can be done by determining an average gradient between two arbitrarily chosen densities which are of importance in the film record under consideration. The net densities chosen in the following example of Figure 4.15(a) are between 0.5 and 2.5. The Table 4.1 shows the corresponding difference of log exposure values at the selected densities for film I and II of

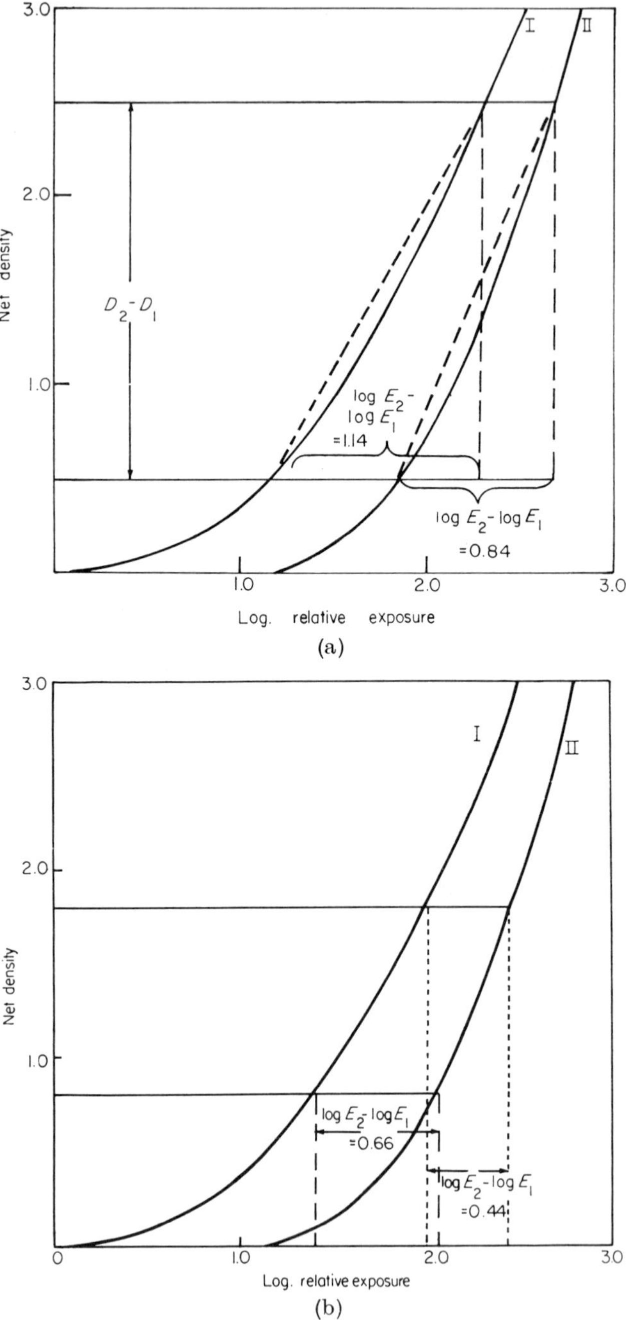

Figure 4.15. (a) Characteristic curves of two different types of no-screen X-ray films. Determination of average gradients. (See dashed lines.) (b) Characteristic curves of two different types of no-screen X-ray films (same used as in Figure 4.15. (a)) Determination of speeds at $D = 1.8$ and at $D = 0.8$ above fog.

114

Figure 4.15(a). In the third column of Table 4.1 the average gradients are quoted which have been calculated, as seen from the Table 4.1.

It is seen that the average gradient of film II is 1.4 times greater than that of film I.

The determination of the relative speed from the characteristic curves of the emulsions I and II of Figure 4.15(b) can be carried out in the following way.

TABLE 4.1. Average Gradients for Films I and
II Between Selected Densities of 2.5 and 0.5
(see Figure 4.15(a))

Film	$\log E_2 - \log E_1$	Average gradient between D_2 and D_1 expressed by $\dfrac{D_2 - D_1}{\log E_2 - \log E_1}$
I	1.14	$2/1.14 = 1.75$
II	0.84	$2/0.84 = 2.38$

As the absolute values of exposure are not given in Figure 4.15(b) the relative speeds may be determined by assigning an arbitrary speed value of unity, for instance, to film II. It is seen in Figure 4.15(b) that the difference of log exposure between films II and I at a density of 0.8 is 0.66 and at a density of 1.8 it is 0.44. The corresponding antilogarithms are 4.57 and 2.75 respectively. Hence we conclude that the speed of film I is 4.57 times that of film II at a density of 0.8 and 2.75 times that of film II at a density of 1.8. One may also say that film I is 4.57 times or 2.75 times faster than film II at the densities 0.8 and 1.8 respectively. The fact that the relative speeds differ at the two selected densities is because the characteristic curves are not parallel.

Apart from the quantitative data derived in the preceding paragraph, qualitative information can be derived from mere inspection of the characteristic curves. For instance, it is evident from Figures 4.15(a) and (b) that film I is faster than film II, as for any given log exposure film I reveals a higher density than film II. Furthermore the steeper increase of the curve representing film II reveals a higher average contrast than the flatter curve corresponding to film I.

A rather different example is shown in Figure 4.16 in which the characteristic curves of two X-ray films cross each other at point P, thus indicating that films I and II have equal speed at P. In the lower

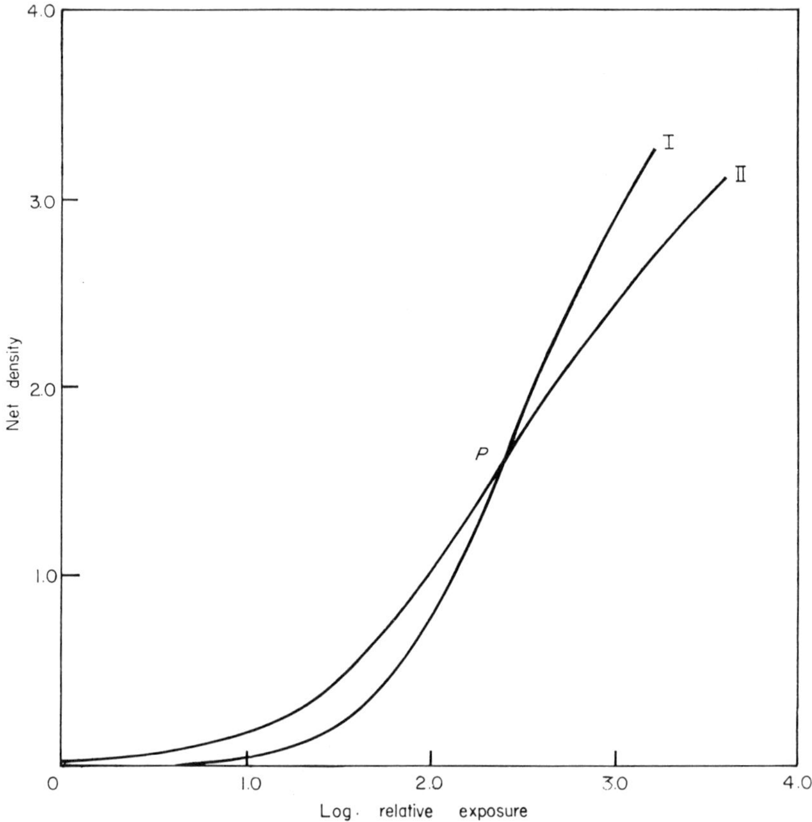

Figure 4.16. Characteristic curves of two X-ray films having equal speed at point *P* and different gradients.

density region (below *P*) film II is faster than film I, whereas in the higher density region (above *P*) film II is slower than film I. From the respective slopes of the curves, conclusions can be made as to the relative inherent contrast exhibited by the two films in various regions of density.

Solarization

If an emulsion is exposed beyond the region where the maximum density occurs, a decrease of density (see Figure 4.14 region *D–E*) will be observed. The region beyond the maximum density towards still higher exposures, in which the curve shows the tendency to drop towards lower densities is known as the region of reversal or solarization. This term is so called because the effect was first noticed as a result of exposures of photographic material to the sun's radiation. Exposures in the region of reversal tend to exhibit a partially or a completely positive image in contradiction to the negative image obtained within the ascending part of the characteristic curve. A discussion of the reversal effect and of its possible mechanism will be given on page 136.

Film Latitude

The term 'film latitude' is used to indicate the difference of log exposure in the characteristic curve which corresponds to a useful range of densities in the radiograph. Taking films I and II in Figure 4.15(a) as an example, for which the useful range of densities may be between $D = 0.5$ to $D = 2.5$, the corresponding difference in $\log E$ is 1.14 for film I and 0.84 for film II. Hence the range of exposure which can be revealed in film I is almost twice that of film II, as found from the antilogarithms of the differences. It is seen that the latitude of a film depends on its gradient or contrast. The higher the contrast, the less is the film latitude. The term latitude is not always used in the same sense in X-ray sensitometry as it may cover a variety of other meanings, such as the range of exposures for a given development time which may lead to an acceptable radiograph, or the range of development times for a given exposure which may lead to an acceptable radiograph. In the latter applications the proper terms to be used are 'exposure latitude' and 'development latitude' respectively.

In the field of radiation monitoring where uniform density areas rather than image formation are required, the meaning of film latitude is interpreted as follows. A film of great latitude is capable of recording a wider range of doses than a film of small latitude, but a given difference of dose will correspond to a greater difference in density in a film of small latitude than in a film of great latitude.

Some Fundamental Aspects of the Characteristic Curve

It has been shown in the previous sections on the density-exposure curve that the initial part of this curve reveals, in general, proportionality between density and exposure and that the length of the straight line differs for different types of emulsion. Within these limits it follows that

$$D = kE \qquad (4.9)$$

where D is the density, k a constant and E the exposure. Taking the natural logarithm of both sides

$$\log_e D = \log_e k + \log_e E \qquad (4.10)$$
$$\log_e D = \log_e k + (\log_e 10) \log_{10} E \qquad (4.11)$$

Then on differentiation

$$\frac{1}{D} \times \frac{dD}{d(\log_{10} E)} = \log_e 10 \tag{4.12}$$

thus

$$G = \frac{dD}{d(\log_{10} E)} = D \log_e 10 = 2.3D \tag{4.13}$$

This means that the slope of the toe region of the characteristic curve depends on D and not explicitly on E. All possible characteristic curves

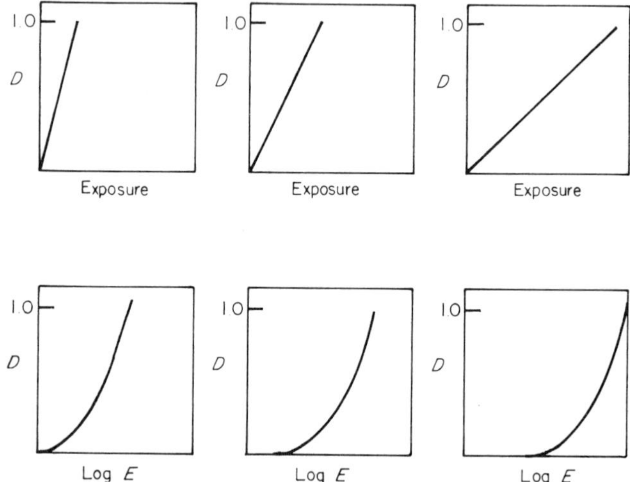

Figure 4.17. Density-exposure curves (top) and corresponding characteristic curves (bottom) of the initial relationship between density and exposure and log exposure respectively. (Schematically.)

of this kind have therefore the same shape for all densities for which $D = kE$, namely $G = 2.3D$, where G is the gradient $dD/d \log_{10} E$, but may have any position relative to the $\log E$ axis, as shown in Figure 4.17.

The general shape of the characteristic curve for X-rays can be derived theoretically, if certain simplifying assumptions are made, such as

1. That each grain is rendered developable by a single hit, i.e. absorption by one quantum.

2. That all the grains present the same projective area to a perpendicular incident radiation.

It can then be shown (M. Blau and K. Altenburger, 1923; L. Silberstein and A. P. H. Trivelli, 1930; J. H. Webb, 1939, 1941) that

$$N = N_0(1 - e^{-an}) \tag{4.14}$$

where N_0 = total number of grains per unit area

N = number of grains rendered developable

n = number of photons falling on unit area

and a = projective area of grains.

Assuming that the total number of grains per unit area N_0 is replaced by the maximum density D_m and the number of grains rendered developable per unit area N be replaced by density D, then

$$\frac{D}{D_{max}} = (1 - e^{-an}) \tag{4.15}$$

For a small value of an, i.e. for a small exposure

$$\frac{D}{D_{max}} = an \tag{4.16}$$

hence the density is proportional to exposure, as found experimentally. On plotting equation (4.15) the typical shape of a characteristic curve applied to X-ray exposures is revealed. In Figure 4.18 the excellent fit between the theoretical density-exposure curve based on the equation (4.15) and an experimental curve is illustrated. If the grains of an emulsion are not sensitive enough to be rendered developable by a

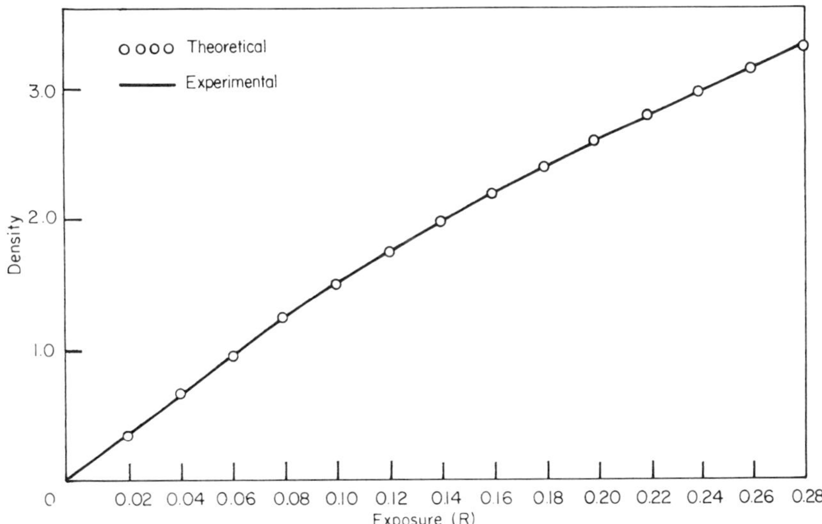

Figure 4.18. Density-exposure curve.

one-hit process, but require several hits $r = 2, 3, 4$, etc. See Figure 4.19 and if $y = an$, the characteristic curve can be expressed by

$$\frac{D}{D_{max}} = 1 - e^{-y}\left(1 + y + \frac{y^2}{2!} + \cdots \frac{y^{r-1}}{(r-1)!}\right) \qquad (4.17)$$

For example, this leads for $r = 2$ to

$$\frac{D}{D_{max}} = 1 - e^{-y}(1 + y) \qquad (4.18)$$

The resulting characteristic curves, in which the fraction of grains rendered developable are plotted against the logarithm of exposure, are shown in Figure 4.19. One can see how the shape of the characteristic curve changes with the number of quantum hits (r) required

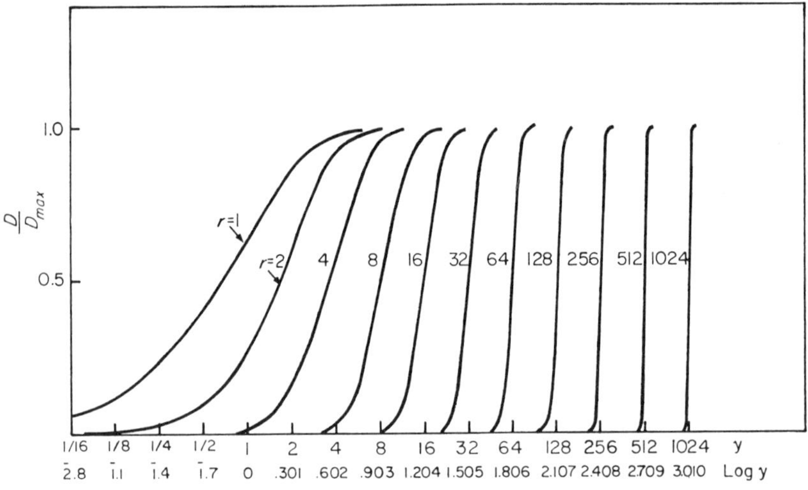

r = mean number of absorbed quanta per grain.

Figure 4.19. Characteristic curves for increasing number of quantum hits (*After* J. Webb (1939).)

for rendering a grain developable. Obviously the curve corresponding to $r = 1$ is the curve typical of X- and γ-ray exposures. It is to be expected and has also been experimentally verified that the density-exposure curves for r exceeding unity exhibit a toe.

In recent years the theory of the characteristic curve has been extended to take account of the sensitivity spread of various grain size classes. (H. Frieser and E. Klein, 1961; G. C. Farnell, 1966.)

Influence of the Quality of Radiation on the Characteristic Curve

The shape of the characteristic curve is practically independent of the quality of radiation, unless extremely soft radiation or very slow emulsions are used. However it has been found occasionally by some workers that a comparison of the shape of characteristic curves of commercially available X-ray films obtained with X- and with γ-rays did not reveal a perfect match in the region of saturation densities. This is probably caused by the fact that fewer grains in a γ-ray exposure were rendered developable on a single hit basis than with X-rays. It was found further that prolonged development tends to reduce the slight differences in shape. Characteristic curves originating from exposures due to different radiation qualities are shifted on the log exposure axis as the speed changes with the quality of radiation. This will be discussed in more detail in section 4.5.

Effect of Development Conditions on the Characteristic Curve

It has been mentioned that the shape of the characteristic curve is not only typical for the type of emulsion and the radiation to which it was exposed, but also for the development conditions to which it was subjected. The development conditions i.e. development time, temperature and agitation for a given developing solution, and obviously the type and concentration of the developer solution have a marked influence on the speed and inherent contrast of a certain film material. Figure 9.1 shows a typical family of characteristic curves of X-ray screen film, developed at constant temperature of 68°F for different times of development. It is seen that the speed and slope rise with increasing times of development, and that the fog value also gradually increases with development time. Hence the optimum time of development for a given film is derived from the maximum slope and the highest speed, combined with a tolerable fog value. Similar curves as shown in Figure 9.1 result from the use of increasing temperatures of the developer solution keeping the developing time constant.

4.5 QUANTITATIVE CONSIDERATIONS ON SPECTRAL SENSITIVITY

4.5a. Speed as a Function of the Quality of Radiation

The dependence of sensitivity of photographic emulsions on the quality of X-rays plays an important part in many applications, and its influence is a particular problem in the field of photographic dosimetry. On the other hand, the ionization chamber used as a dosimeter is independent of the radiation quality at least over a wide although limited range of wavelengths.

It is one of the characteristic properties of photographic emulsions that they are wavelength dependent, i.e. that their response to a given exposure (for instance in roentgens) changes with the quantum energy of the incident radiation. Some workers in the field of photographic dosimetry consider this property as an advantage, as it may provide an indication of the quality of radiation to which a film was exposed. The majority of workers, however, consider the energy dependence as a considerable drawback (see Chapter 6, p. 192). In the past many efforts have been made either to influence the wavelength dependency in emulsion manufacture or to devise suitable methods of compensating this disturbing effect. These attempts will be dealt with in more detail in Chapter 6.

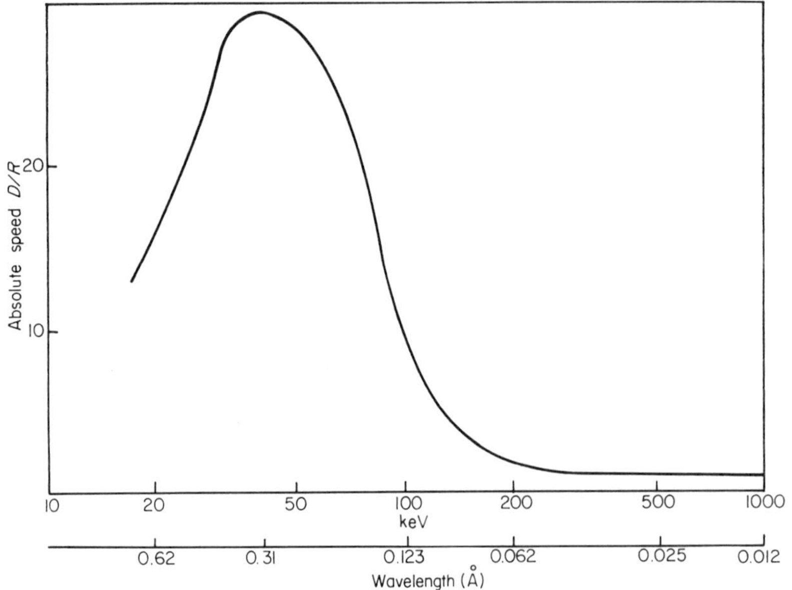

Figure 4.20. Absolute speed D/R as a function of energy of radiation (keV) and wavelength (Å) for a medium fast no-screen X-ray film.

In the following sections the basic facts of energy or wavelength dependence and the causes will be discussed.

The spectral sensitivity of a given X-ray film can be presented by plotting the speed (D/R) against the energy (in keV) of the radiation. (See Figure 4.20.) Apart from the energy, the corresponding wavelengths in Å are also given in the graph. It can be seen that a peak sensitivity exists at about 40 keV. Towards higher and lower quantum energies the speed decreases and reaches a fairly flat part in the region of very high energies.

The ratio of the peak to the minimum sensitivity in the high energy region changes with the type of emulsion and developing conditions, but is usually between 20–50. This ratio also depends strongly on the heterogeneity of the radiations used, as can be seen from Figure 4.21 due to R. B. Wilsey (1951). The experimental results shown in this figure have been obtained in one case using heavily filtered radiations and in the other case with only weakly filtered radiations. Using the latter the trough is flattened due to the use of the heterogeneous radiation applied at each kilovoltage. This is expected to have a smoothing effect on the curve. More heavily filtered radiation, i.e.

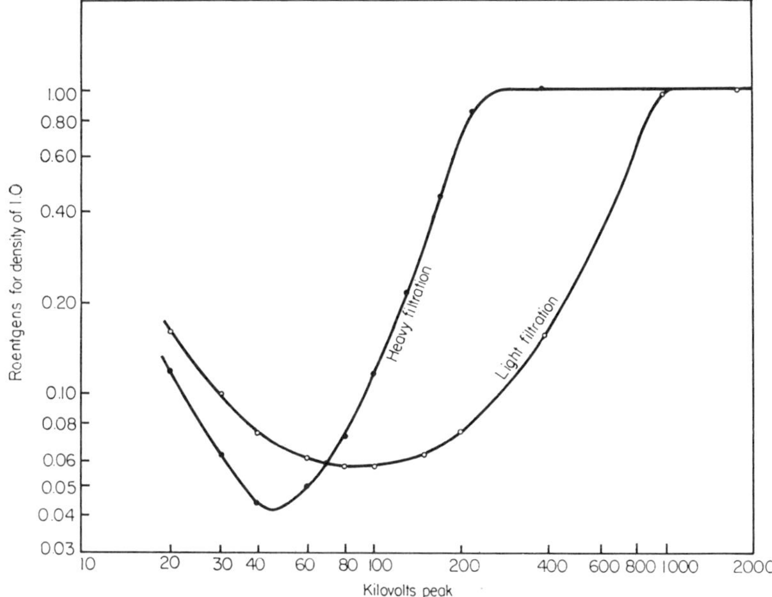

Figure 4.21. Number of roentgens required for a density of unity as a function of X-ray tube voltage for X-rays used with light and with heavy filtration on Kodak Blue Brand Medical X-ray film. (*After* R. B. Wilsey, *Radiology* **56**, 2, 230 (1951).)

more narrow wavelength bands, have been used in obtaining the other curve which has a more pointed minimum.

4.5b. Energy Absorption and Speed

An explanation for the wavelength dependency of the photographic response, as shown in Figures 4.20, and 4.21, must be based essentially on the relative absorption of X-rays in the silver bromide emulsion to that in air. The reason for introducing the air absorption is that the speed D/R refers to roentgen units and that the definition of this is based on the absorption of X-rays in 1 cc of air. Hence one would

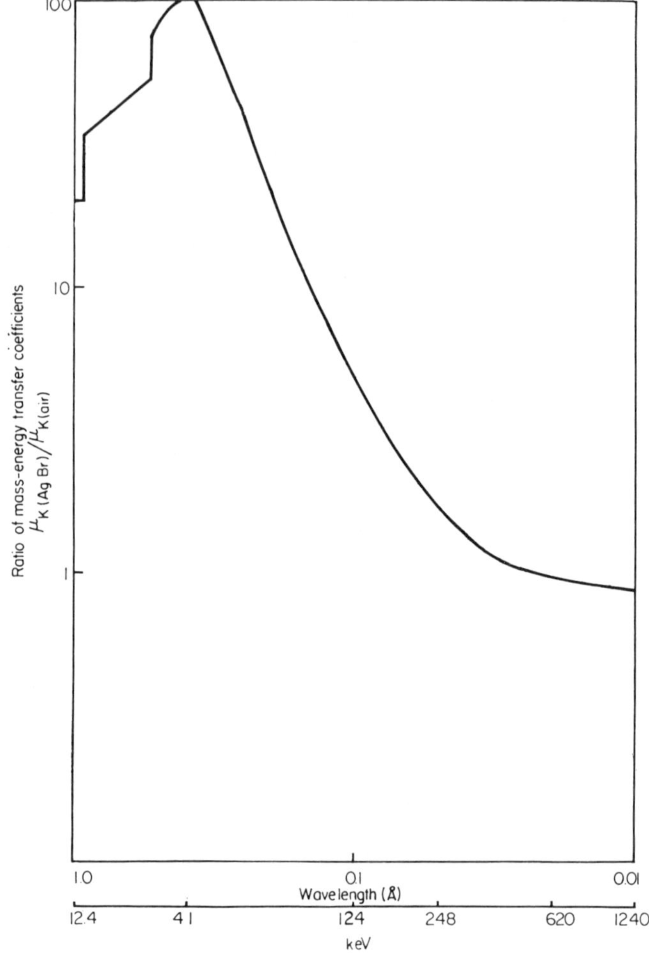

Figure 4.22. Ratio of mass-energy transfer coefficients of silver bromide over that of air as a function of keV and wavelength of X-rays.

expect that the change of speed with wavelength should be proportional, in the first approximation, to the ratio of the absorption coefficient of silver bromide (neglecting temporarily the absorption in gelatin) to that of air within the range of X-ray wavelengths concerned. This ratio is plotted as a function of quantum energy in Figure 4.22 in terms of mass enegy transfer coefficients.

By comparing the experimental curves, Figure 4.20 showing D/R against keV (and wavelength) and Figure 4.22 showing $\mu_{K(AgBr)}/\mu_{K(air)}$

against keV (and wavelength), it is found that the general trend of the two curves is similar, although there are two distinct differences.

There is the sudden change in the curve of Figure 4.22 at low energies compared with the smooth peak of Figure 4.20. The sudden change is due to the selective absorption edges of silver and bromine which are smoothed out in Figure 4.20 because of the use of less narrow wavelength bands in this figure. Furthermore the ratio of the maximum to the minimum mass-energy transfer coefficients $\mu_{\kappa(AgBr)}/\mu_{\kappa(air)}$ is found to be about 110, which is considerable higher than the corresponding ratio of maximum to minimum speed found in X-ray emulsions for the wavelengths of 0.3 and 0.01 Å (i.e. 40 keV and 1240 keV respectively.) This speed ratio is between 20–50 in commercially available emulsions, as mentioned before. The reason for this discrepancy is due to the contribution, neglected so far, of electrons released by quantum absorption in the gelatin, film base and the surrounding (e.g. wrapping material), which plays an important role in the region of high quantum energies where photoelectric absorption is practically negligible. This contribution is associated with Compton recoil electrons, mainly from the gelatin, and tends to diminish the numerical discrepancy mentioned above. However, the problem is more complex as other corrections have to be introduced. These refer to the energy loss due to electrons created in silver bromide, but whose paths are not wholly confined to the emulsion. Other corrections should be made with respect to the loss of electron energy due to absorption of electrons by gelatin in the emulsion, and others to the energy gained by silver bromide due to electrons originating from surrounding material. The latter effect is very considerable at high quantum energies.

The effects are even more complex if one considers the directional distribution of photo- and recoil electrons, their respective ranges in the emulsion layer and their varying specific ionizing power along their paths. All these factors, the production of Auger-electrons*, and several minor effects, should be considered for a precise account of the energy imparted to silver bromide. Many workers have contributed experimentally and theoretically to this problem (R. Glocker, 1927a, 1927b; A. Charlesby, 1940; M. Sobolew, 1945; S. R. Pelc, 1945; D. Bromley and R. H. Herz, 1950; J. R. Greening, 1951; H. Tellez-Plasencia in about 15 papers from 1948–1959; H. Hoerlin and V.

---

* If a K_α photon is generated by the drop of an electron from an L- into a K-shell, it frequently acts photoelectrically on another L or M electron of the same atom and ejects it. This latter electron is called an Auger electron.

Hicks, 1947; H. Hoerlin, 1949, 1951; W. Mauderli, 1957; K. Becker, 1960, 1961, 1962 and others).

Probably the most detailed approaches to this field of work were made by Greening and Tellez Plasencia to whose original publications the reader is referred for further information. The complex treatment has been summarized in a simplified form by Greening whose presentation is given in the following notation.

$$\frac{\text{Ergs/gram of AgBr}}{\text{Ergs/gram of air}} = \frac{[(\tau + \sigma_a)/\rho]_{\text{AgBr}}}{[\tau + \sigma_a/\rho]_{\text{air}}} \times F_1 \times F_2 \times F_3 \quad (4.19)$$

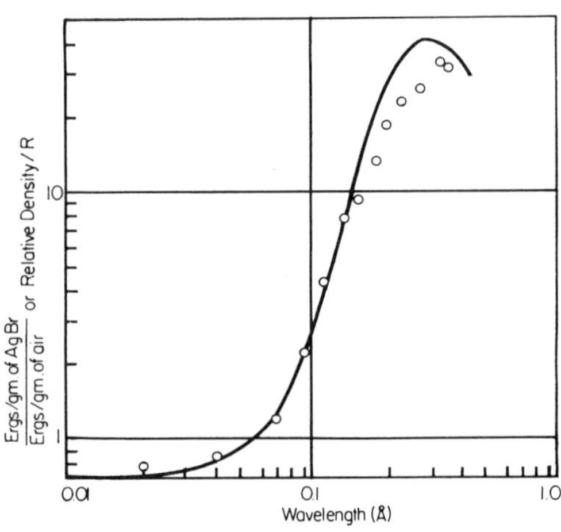

Figure 4.23. Theoretical wavelength variation of the ratio of the energy absorption per gram of silver bromide to the energy absorption per gram of air (full line) compared with experimental values of relative density per Roentgen (points) for Ilford Line film. (*After* J. R. Greening, *Proc. Phys. Soc.* London B **64, 977** (1951) Inst. of Physics and Phys. Soc. London.)

whereby τ is the photoelectric absorption coefficient, σ_a is the true absorption coefficient associated with Compton scattering and F_1, F_2, and F_3 are correction factors for energy lost by fluorescence (characteristic radiation), for electron energy lost from the emulsion and for energy lost to the gelatin respectively. Greening has found reasonable agreement between theory and experimental results, although he has only checked this on one commercial type of emulsion, as seen from Figure 4.23. In this graph the ratios of ergs per gram of AgBr to ergs per gram of air and the relative density per roentgen (D/R) for Ilford Line film are plotted against the wavelength of radiation. From the fair agreement between the full line and the dotted curve, it can be concluded that *the optical density is practically proportional to the energy absorbed* by the emulsion within the range of wavelengths studied by Greening.

The spectral response of photographic emulsions to X-rays is usually presented in relation to roentgen units, because the roentgen is the most practical measure of the exposure to X- and γ-rays. It should be realized that the shape of the spectral response curve (see Figure 4.20) is due to the use of the roentgen unit.

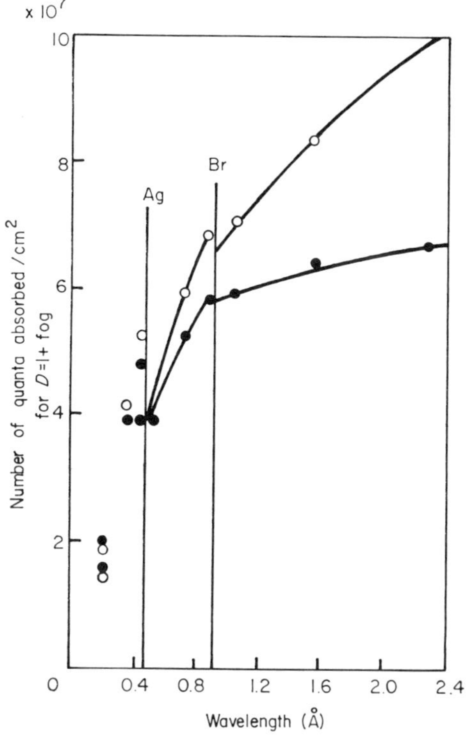

Figure 4.24. Spectral sensitivity of two different types of no-screen X-ray film. (Number of quanta per cm² absorbed for $D = 1$ + fog) as a function of wavelength, using monochromatic radiation. (*After* H. Seemann, (1950). *Rev. Sci. Instr.*, **21**, 314 (1950).)

In the investigations carried out by H. Hoerlin (1949) and by H. Seemann (1950) the roentgen unit has been replaced by more appropriate physical units, such as quanta per cm² and ergs per cm² absorbed in the emulsion. The results of H. Seemann which are shown in Figure 4.24 present the number of absorbed quanta per cm² for a density of $D = 1$ as a function of X-ray wavelength. Seemann's experiments were carried out with monochromatic radiation, using two different types of X-ray emulsion. The number of quanta per cm² required to obtain a density of unity decreases with shorter wavelengths as seen in Figure 4.24, thus indicating that the high energy quanta are more effective than the low energy quanta, but not in a manner which is proportional to their energy. This means that more energy is wasted in the absorption process of the higher energy quanta in spite of the fact that several grains are rendered developable by one quantum hit, as shown on page 98.

H. Hoerlin (1949) has carried out similar experiments, but he used filtered heterogeneous radiation from 1 Å to 0.01 Å. He showed that the absorbed energy in ergs/cm^2 as a function of wavelength for two different types of X-ray emulsion, decreases steadily with decreasing wavelength without revealing a peak in the curves (see also G. M. Corney 1966).

4.6   THE INFLUENCE OF DOSE-RATE

An important and advantageous rule controlling the photographic behavior of X-ray and γ-ray exposures is the Bunsen–Roscoe law. This states that *the amount of product in a photochemical reaction depends on the product of radiation intensity and time and is independent of the rate at which the exposure is delivered.* The validity of this rule for the photographic recording of ionizing radiations means that it is immaterial whether a large dose is delivered in a short time or a small dose in a long time, as long as the product of dose and time and the quality of radiation are kept constant (H. Kroencke, 1914; R. Glocker, 1921; F. Luft, 1933; G. E. Bell, 1936; and others). Without the validity of this so called reciprocity rule one may argue that photographic dosimetry would hardly be possible. This is because in dosimetric practice it is uncertain whether an operator receives a certain dose in a few seconds or in a few days. Hence if the photographic emulsion were rate-dependent, the dose-rate to which a person was subjected would have to be known in each instance.

4.6a.   Reciprocity Failure for Light Exposures.

In contradiction to the validity of the Bunsen–Roscoe rule for X- and γ-rays and for charged particles, light exposures do not obey it and reciprocity failure is revealed. A typical reciprocity failure curve based on the behavior of an X-ray emulsion exposed to light is shown in Figure 4.25. The curve illustrates the log It values (I = intensity, t = time) i.e. the log exposure values required for a constant net density of 0.2 as a function of log I. It is seen that an optimum value of log I is obtained at the minimum of the curve and further that more exposure is required at low and at high intensities to obtain a constant density. The region towards the left of the minimum of the curve is referred to as the region of low intensity failure and that towards the right as the region of high intensity failure. The general behavior as illustrated in Figure 4.25 is typical for many emulsions when exposed to light. At very short light exposure times, smaller than about 10^{-5} seconds, reciprocity failure ceases to occur. (W. F. Berg, 1940.)

It was suggested by Schwarzschild (1900) that the photographic reciprocity failure, as it occurs in light exposures, can be expressed by $It^p = $ constant for a given density, where I is the intensity, t is the exposure time, and p is a constant for a given photographic material.

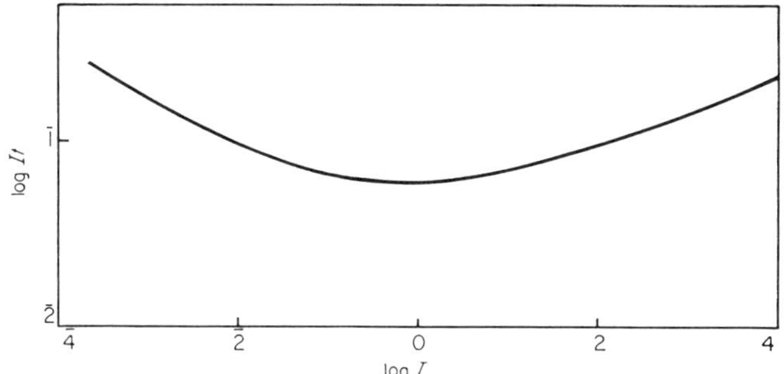

Figure 4.25. Reciprocity failure. Log It versus log I for light exposure at $D = 0.2$ above fog.

p is found to be of the order of 0.8. Further investigations however have shown that the Schwarzschild rule which governs the reciprocity failure, holds only for a limited range of exposure times and intensities and is not only dependent on the type of emulsion but also on the wavelength of light and on the temperature during exposure. (C. E. K. Mees and T. H. James, 1966.)

4.6b Interpretation of Reciprocity Rule and its Failure

Low Intensity Failure

It is believed that the low intensity (long time) reciprocity failure is caused by an instability of the latent image in its early stages of growth. Since low intensity failure is much reduced at low temperature, it is concluded that the regression of the latent image is connected with thermal disintegration, whereby an electron is ejected and a silver ion diffused away. In a low intensity exposure the time intervals between consecutive arrivals of electrons are relatively long, so as to allow disintegration to occur.

High Intensity Failure

Several interpretations have been given for the mechanism of the high intensity failure for light exposures. From Gurney and Mott's hypothesis (see p. 70) it is concluded that the failure is due to sluggishness of the migration of interstitial silver ions. In a high intensity exposure light quanta follow each other so rapidly that the motion of silver ions towards the electron traps cannot keep pace. Hence the formation of silver atoms at the required rate is reduced. Furthermore the chance increases that electrons are repelled by the negatively charged specks (which have not been neutralized by silver ions), thus the rate of electron trapping is diminished. The fact that no reciprocity failure is observed in light exposures below about 10^{-5} seconds can be explained by the assumption that all electrons have been ejected before the first step in the formation of a latent image has occurred.

4.6c.   Reciprocity Rule for X- and γ-Rays

The fact that the reciprocity rule is obeyed in X-ray and γ-ray exposures as well as in exposures with charged particles, can be explained without any reference to the mechanism of latent image formation. The condition that each X- or γ-ray quantum absorbed in a grain renders this developable is not dependent on the rate at which this occurs. Hence there is no reason why the reciprocity rule should not be obeyed.

Although no further explanation for the validity of the reciprocity rule with ionizing radiations is required, it should be mentioned, that any X- or γ-ray exposure must be regarded as a high intensity exposure since it is expected that a multitude (or avalanche) of electrons is released within a grain by a single quantum absorption, so that the interval between successive releases of electrons can be regarded as being much shorter than 10^{-5} seconds. The time of transit of an electron released by absorption of an X-ray quantum in a grain is thought to be of the order of 10^{-12} to 10^{-14} seconds. This may be taken as a further reason why the reciprocity rule should hold.

4.6d.   Exceptions to Reciprocity Rule

Exceptions to the reciprocity rule for X-rays do occur if the intrinsic sensitivity of a grain is insufficient for a one hit process, i.e. if the quantum-energy is below a certain limit with respect to the sensitivity of the grains. As some of the recoil electrons emitted in γ-ray exposures are of very high energy (i.e. close to minimum ionizing power see p. 96) reciprocity failure may occur in slow emulsions. This has been shown by N. J. P. Chassende-Baroz (1961) whose results are illustrated in Figure 4.26.

An exception of a different kind has been observed by M. Ehrlich (1956) in the region of high densities, using a medium fast X-ray emulsion, as illustrated in Figure 4.27. The exposures were carried out

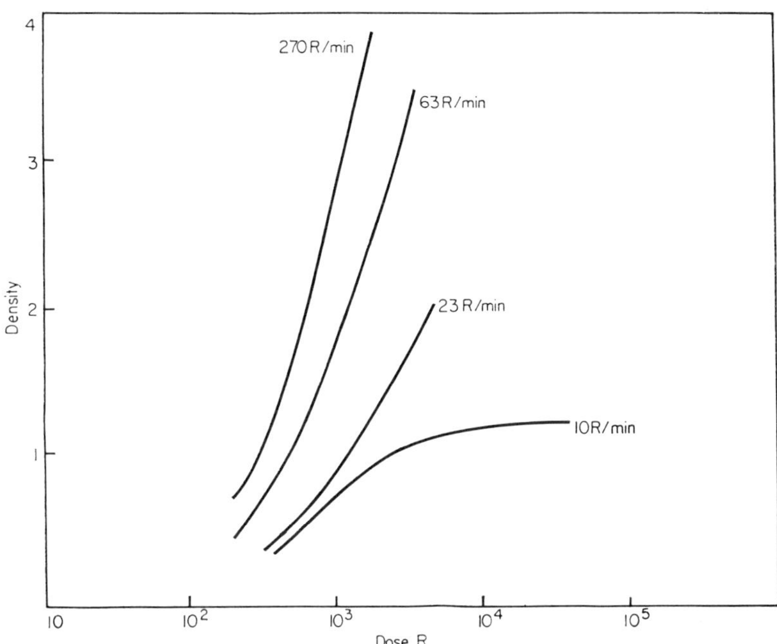

Figure 4.26. Characteristic curves of a special emulsion for different dose rates. (*After* N. J. P. Chassende-Baroz, Selected Topics in Radiation Dosimetry IAEA Wien, (1961) p. 285.)

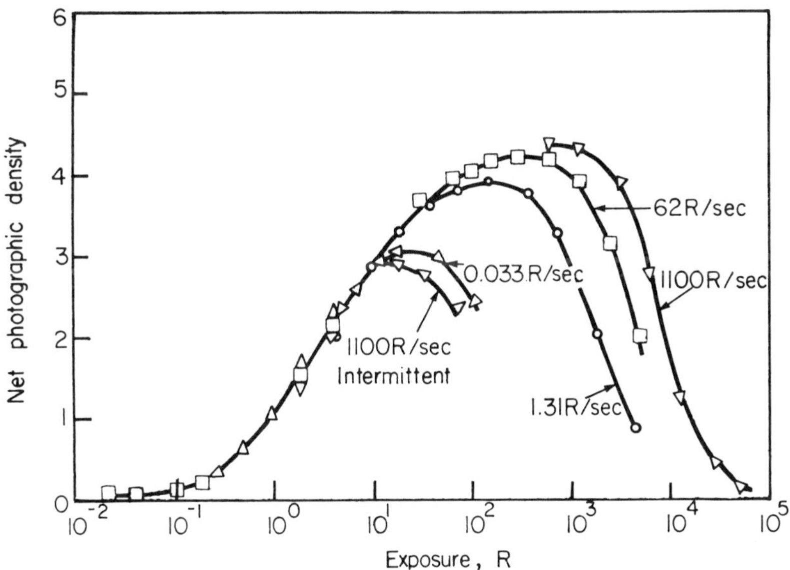

Figure 4.27. Characteristic curves for X-ray exposures at various dose rates. (*After* M. Ehrlich, *J. Opt. Soc. Am.* **46,** 801 (1956).)

with 50 kV X-rays using dose rates from 0.033 R/sec to 1100 R/sec. It is seen that the maximum density increases from low to high dose rates and that the region of solarization varies with dose-rate. If the exposure is given by intermittent radiation not only a reverse behavior to that shown above is observed, but also lower maximum densities are reached than with continuous exposures. The mechanism on which these results are based appears to be rather complex and no attempt will be made here to explain it.

4.6e.   Intermittency Effect

If a number of exposures using X- or γ-radiation is given at intervals on the same area of a film, the total density will be the sum of the densities obtained from each individual exposure. (C. A. Schleussner and W. Kaempfert, 1924). Whether the time intervals are of short or long duration, provided that no latent image fading (see p. 144) occurs between the exposures, the total photographic effect will not be influenced. M. Ehrlich and W. L. McLaughlin (1961) showed that even for subsequent exposures to ^{60}Co γ-rays of different exposure rates, the additive photographic effect was not influenced at low and medium densities, but it showed variations in the region of solarization. The possibility of simply summing up a series of exposures arithmetically is obviously a great advantage generally and in particular in the field of photographic dosimetry.

From the fact that the reciprocity rule is essentially valid for direct X-ray exposures (apart from the exceptions mentioned above), it is to be expected that the frequency of the intermittent exposure should not influence the photographic result if compared with a continuous exposure, assuming that the total exposures given in both cases are equal. This expectation should be valid, as long as the one quantum hit event is fulfilled. The reader may be reminded in this context that the emission of X-rays from the majority of X-ray units is not continuous, it being intermittent and having the same frequency as the mains supply. This should have no effect on the results of 'direct' X-ray exposures.

In general, as the reciprocity rule fails with light exposures, so does the intermittency effect, because the two effects are closely related (J. Webb, 1933). This means that the sum of the interrupted exposures leads to a different photographic density than that obtained with a continuous exposure even if both total exposures are equal. The failure with light exposures occurs within certain critical frequency limits which are related to the high and low intensity reciprocity failure characteristics of the particular emulsion.

4.7 THE DISTRIBUTION OF THE LATENT IMAGE

It has been shown that the number of silver atoms formed during photolysis is equal to the number of light quanta absorbed in silver halide grains. This is referred to as the law of quantum equivalence. Investigations with X-ray exposures have shown that the number of silver atoms formed is roughly proportional to the energy of the absorbed quanta within the range of 0.245 to 1.54 Å (see p. 66). Since the X-ray quantum energy is so much greater than that of a light quantum, many more silver atoms are produced in one grain by one X-ray quantum than by one light quantum absorbed. Hence one may contemplate whether the distribution of the photolytic silver formed in grains by X-rays differs from that by light. Further it has been established that the initiation of development not only depends on the number of latent images present, but also on their position within grains.

A considerable amount of work has been done in studying the distribution of latent image silver in grains for various types of exposure. For instance developer solutions, having little or no solvent action on silver halides, mainly react at the latent image silver specks situated at the surface of the grains. Many workers have however given evidence that latent image silver can also exist in the interior of silver halide grains. These are referred to as internal latent images. W. F. Berg, A. Marriage and G. W. W. Stevens (1941a, 1941b) and G. W. W. Stevens (1942) proposed a convenient and efficient technique by which the surface latent image can be distinguished from the internal latent image. As already mentioned a surface developer containing little or no solvent action should essentially affect only the surface latent image. On the other hand, if an exposed emulsion layer is first bathed in a solution containing an oxidizing agent, such as chromic acid, potassium ferricyanide, acid-dichromate and others, the surface latent image can be completely destroyed before the development of the internal image. Using a solvent developer the effect of the internal latent image can be revealed.

For example, low sulphite Metol hydroquinone-carbonate or Phenidone ascorbic acid solutions have been recommended as surface developers, whereas for the internal image an energetic Metol hydroquinone developer(such as an X-ray developer) to which thiosulfate is added, may be used. The relative amounts of surface and internal latent image that will be observed, depend greatly on the composition and acting time of the solutions applied and, in the case of the internal image, on the degree of etching.

Although the interpretation of the results achieved by these methods is often difficult and ambiguous, several definite effects have been established. One of the outstanding effects revealed in the study of latent image distribution is the fact that it depends on the rate of exposure. It has been found that, in general, using a low intensity light and a long exposure time the latent image is situated preferentially on the surface of grains. Under these conditions sensitization of emulsions is most effective. Using a high intensity and short exposure time a considerable amount of internal image is formed as well. The latter result agrees well with the observation that X-ray exposures, which are regarded as high intensity exposures (see p. 130), in general also reveal a high contribution of internal latent image, as is illustrated in Figure 4.28. In this diagram is is seen that the contributions of internal and surface latent image are approximately equal. A further interesting contribution to this subject has been made by D. A. Nepela and H. F. Nitka (1960). These authors have shown that the ratio of surface to internal latent image increases with the energy of the radiation, from $65\ kV_p$ to cobalt-60 γ-rays.* Hence the proportion of the internal image in γ-ray exposures was found to be considerably smaller than that for low kV X-rays and was found to be very close to that of a low intensity light exposure (see Figure 4.29). This observation can be explained on the basis that the linear energy

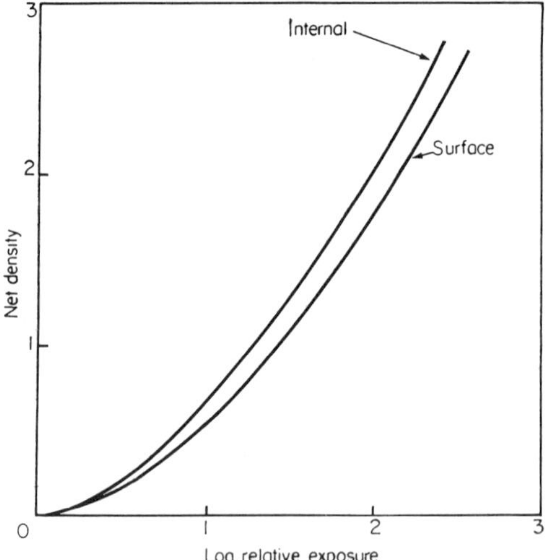

Figure 4.28. Characteristic curves showing the relative contribution of internal and surface latent image for X-ray exposure on a no-screen X-ray film.

* Equivalent to about 1,200 keV.

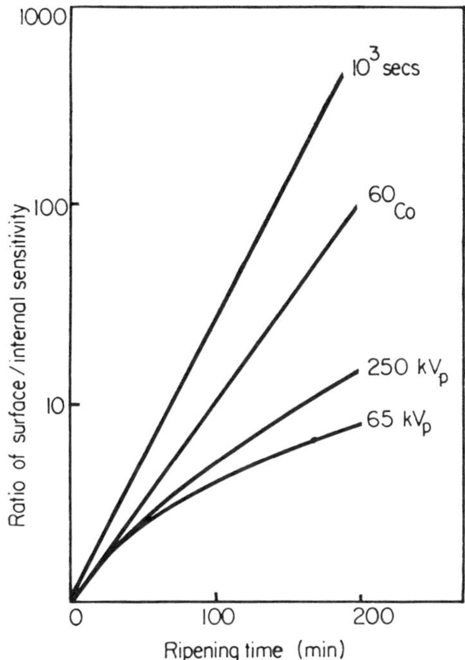

Figure 4.29. Ratio of surface to internal sensitivity as a function of ripening time for X-rays of various energies, cobalt-60 γ-rays and light of short duration. (*After* D. A. Nepela and H. F. Nitka, *Phot. Sci. and Eng.* **4**, 12 (1960).)

transfer to grains using high energy γ-rays such as those emitted by a cobalt-60 source is much smaller and more dispersed than that with 65 kV$_\mathrm{p}$ X-rays. It should be remembered that approximately 1 MeV energy corresponds to minimum ionizing electrons.

Nepela and Nitka have also given evidence that chemical ripening of an emulsion has considerably less effect on X-ray than on γ-ray exposed emulsions. It was further found that the chemical ripening effect on light exposed films was similar to that on γ-ray exposed films.

Considering the two aspects described above, it can be concluded that the photographic action of high energy γ-rays is closely related to that of light in several respects. Similar, although less systematic and unpublished experiments carried out by the author confirm the observations by Nepela and Nitka.

Since latent image distribution is dependent on the rate of exposure, it is not surprising that different reciprocity failure effects are revealed with light exposures depending on whether surface or internal latent images are involved. The latent image distribution is of further importance in the study of other latent image effects, such as solarization, Villard and Sabattier effects. (See following pages and P. J. Hillson and E. A. Sutherns (1966).)

4.8   REVERSAL EFFECTS

4.8a.   Solarization

As pointed out on page 116 the density of a photographic emulsion reaches a limit and then decreases with increasing exposure. With further increase of exposure the density may rise again and may decrease a second time. The effect of the first decrease of density is called 'reversal' or 'solarization' and the second increase of density is known as the re-reversal.

The reversal effect is known to occur with all kinds of radiation, but it has been observed that in general it occurs more readily with X-rays than with light. (R. B. Wilsey and H. A. Pritchard, 1926; A. May, 1943). This agrees well with the observation that solarization with light of high intensity takes place at a lower exposure than with light at low intensity. It has been shown that an X-ray exposure must always be considered as a high intensity exposure, hence the apparent agreement. The experimental study of solarization with X-rays is usually difficult because very heavy exposures are needed and thus very intense X-ray sources are required. Hence it is not surprising that most investigations have been carried out with light and relatively little work has been done with solarization due to X- or γ-rays.

Solarization differs widely between various types of emulsion, and depends not only on the rate of exposure but also on the type of development. With underdeveloped emulsions the solarization is small, but increases with development time.

It has been found that solarization with light exposures is connected with the distribution of the latent image within grains. (W. F. Berg, A. Marriage and G. W. W. Stevens, 1941.) If an emulsion layer is developed in a silver halide solvent free developer, its action is confined to the surface latent images only. By treating an emulsion with a developer containing a mild solvent for silver halide, internal latent images, if present, are made accessible to the developer solution, and it was found that in this case solarization was reduced. This reversal effect is also diminished in the presence of halogen acceptors, such as sodium nitrite. Both these effects indicate that solarization is connected with rebromination of silver. It is believed that in the case of heavy exposures the amount of bromine released is too great to be entirely removed by the gelatin of the emulsion and may attack the surface latent image by rebromination. The fact that the reversal for X-rays usually occurs at a smaller exposure than for light may be connected with the greater dispersion of the photolytic silver produced in X-ray exposures.

The most widely accepted mechanism to explain solarization is the rebromination hypothesis. According to this, photolytic bromine released when an internal image is formed, destroys the surface latent image. This mechanism is also supported by the action of halogen acceptors to suppress solarization. Further evidence in favor of this mechanism was given by G. C. Farnell, J. B. Chanter and F. S. Judd (1964) by their observation of the formation of print out silver in the interior of the emulsion grains, which was accompanied by the disappearance of print out silver on the surface.

The literature on the reversal effect and on the proposed mechanisms of solarization is very extensive. Some unknown factors may play a part in solarization and it is difficult to isolate fundamental facts which are common to all solarization phenomena. (P. J. Hillson and E. A. Sutherns, 1966.)

4.8b.   Villard Effect

If an X-ray exposure is given to a photographic material within the ascending part of the characteristic curve of a photographic emulsion a negative image is obtained. It was observed however by P. Villard (1899) that the latent images formed by X-rays could be partly or completely destroyed by a subsequent exposure to moderate light intensity, thus leading to a positive or partly positive image. This phenomenon is called the Villard effect. A similar effect can be achieved if the first exposure is given by charged particles followed by light of high intensity. It was further observed that the Villard effect could be more easily achieved when using emulsions which readily reveal solarization phenomena. Depending on the intensity and duration of the first and the subsequent exposures, various degrees of the Villard effect have been observed. (H. Arens, 1955, 1959); H. Tellez-Plasencia, 1960.) Under certain conditions the first X-ray exposure produced a sensitization of the emulsion for the second exposure. This effect has been called the Becquerel effect, although the original observation by E. Becquerel in 1841 was actually different from the one just described.

Investigations by Tellez-Plasencia have shown that the Villard effect is produced largely by surface latent images and the Becquerel effect by internal latent images, but no unambiguous explanation of the mechanism of both effects has been given so far.

4.8c.   Sabattier Effect

Another reversal effect which is easily distinguished from the Villard and Becquerel effect is that discovered by M. Sabattier (1860). Whereas the previous phenomena are observed by second exposures before any development took place, the Sabattier effect occurs if a film has been first exposed, then developed and washed, but not fixed before a second uniform exposure is given. Then the film is subjected to further development leading to a partially or completely positive image. It is immaterial whether the first exposure is due to light or X-rays.

With partial reversal the density first decreases with increasing first exposure, passes through a minimum and then increases again. The effect has been observed in medical and in industrial radiography, when light has fallen accidentally on an exposed and partially developed X-ray film, leading usually to a partially positive image.

Several other reversal and desensitizing effects, such as Albert, Clayden, Herschel, low intensity desensitization, internal image desensitization effects are well known in photographic literature but these are mainly concerned with light exposures and a detailed description of these is given in C. E. K. Mees and T. H. James (1966).

Experiments related and dealing with more complex effects are referred to under the title of 'latensification' on page 227.

4.9   GRAININESS AND GRANULARITY

4.9a.   Graininess

The visual appearance of the irregularity of optical density of a uniformly exposed region of a photographic emulsion coating is called 'graininess'. According to L. A. Jones and N. Deisch (1920) graininess is defined as 'the sensation or impression of non-uniformity in a photographic deposit produced upon the consciousness of the observer when such a deposit is viewed.'

In 1938 the author observed that areas of an X-ray film uniformly exposed to X-rays generated at 200 kV_p appeared to be more grainy than those exposed to X-rays generated at 40 kV_p. This is illustrated in Figure 4.30 in photomicrographs taken from X-ray films exposed to 200 and 40 kV_p It should be mentioned that the two films were exposed to the same density level.

In experiments carried out by J. Eggert and E. Schopper (1939) and by R. H. Herz (1949b) the cause of this effect was investigated. It was proposed that the grainy appearance is connected with the formation of grain clumps at the ends of electron tracks (see p. 97). The higher the quantum energy, the higher is the kinetic energy of electrons

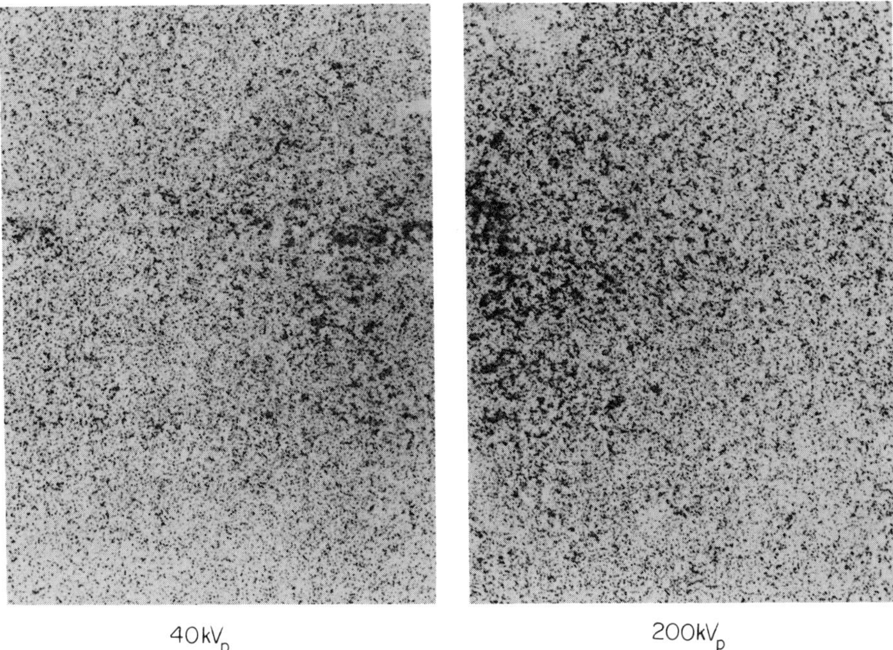

Figure 4.30. Photomicrographs of no-screen X-ray film exposed to X-rays generated at 40 kV$_p$ (left) and 200 kV$_p$ (right). Both photographic records have the same density level but show different graininess.

released; hence greater lengths of tracks (i.e. rows of grains) will be formed in the emulsion layer. Towards the ends of these tracks the specific ionization increases and with it the grain density, thus causing clumping of developed grains (see Figure 4.3). With further increase of quantum energy towards 1 MeV γ-rays, the appearance of graininess decreases again. This can be explained by the fact that the majority of electrons released lies within the region of minimum ionization where the transfer of energy to the grains is more dispersed, and fewer of the ends of tracks are within the emulsion layer.

The effect of the increase of graininess is apparently a similar manifestation of the randomness of quantum absorption as that observed when exposures with salt-intensifying screens are used, which leads to the phenomenon of quantum mottle (see pp. 335 and 342).

The increase of graininess with higher energy of X-rays can be observed with the naked eye and can be very disturbing in radiographs taken in the region of higher quantum energies as for instance in industrial radiography (see pp. 412-413).

4.9b.   Granularity

Apart from the visual appearance of graininess, the effect may be subjected to physical measurements in which case the property measured is referred to as 'granularity.' This latter term has been adopted as an expression for physical measurements of the statistical fluctuations of density over the area of a photographic emulsion. The basis of the granularity measurement used in such investigations is the scanning of a sample of emulsion by a small spot of light (diameter of the order of 0.08 mm) and the resulting irregular fluctuations of the transmitted

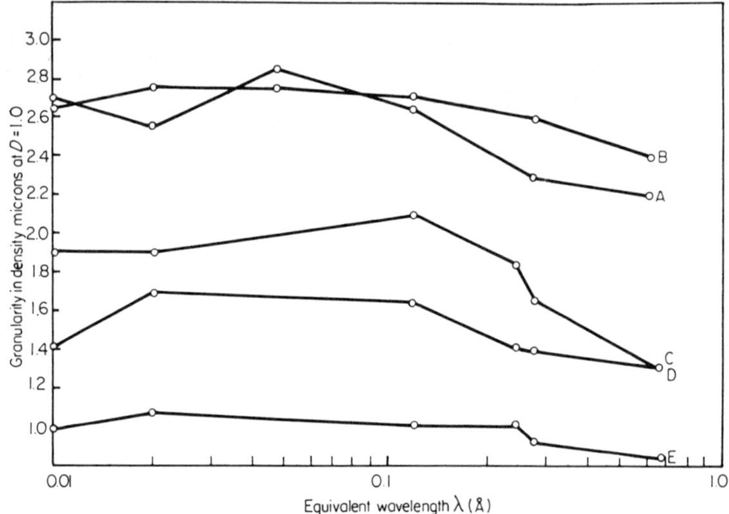

Figure 4.31. Relationship between Selwyn granularity and X-ray wavelength (Å) for five different types of X-ray films. (Measured at a density of 1.0 above fog.) (*After* R. H. Herz, *Phot. J. Section B.* **89B**, 89 (1949).)

light, measured by a photocell are taken as a measure of granularity. (E. W. H. Selwyn, 1935, 1939). The actual fluctuations of density depend upon the size of the scanning spot, increasing as the scanning spot diameter becomes smaller. It is convenient to quote the granularity (G) as the root mean square fluctuations of density (σ_a) multiplied by the square root of the area (a) of the scanning spot in square microns, hence $G = \sigma_a \sqrt{a}$. The term granularity is expressed in units of density microns according to Selwyn and is usually called the 'Selwyn' granularity.

A physical measurement of the increase of granularity with increasing X-ray quantum energy was carried out by R. H. Herz (1949b) for 5 different emulsions, using a Selwyn granularity-meter and the results are illustrated in Figure 4.31.

In this figure the Selwyn granularity is plotted on the ordinate and the wavelengths of the radiation on the abscissa axis. The results refer to a density of unity above fog. The letters A–E denote various types of X-ray emulsion. A and B are coarse grained no-screen X-ray films, C and D refer to finer grained X-ray emulsions and E is a very fine grained X-ray film. It is seen that the granularity increases with shorter wavelengths for all the emulsions and decreases again towards very short wavelengths such as those of ^{60}Co γ-rays, except in the case of emulsion A. It has further been observed that the granularity increases inversely with grain separation, as has been shown by exposing water swollen emulsions and emulsions of decreasing ratios of silver to gelatin by weight. (R. H. Herz, 1949b.)

4.9c. Callier Coefficient

Apart from the measurement of granularity carried out on a granularity-meter based on the Selwyn definition, other methods of measurement of graininess have been suggested by J. Eggert and A. Küster (1938). These authors make use of the ratio of specular density to diffuse density (D''/D^*) which is called the Callier coefficient (Q). (see also p. 89). Eggert and Küster pointed out that the value of Q is correlated with the effect of graininess. It has however been found that Q is also influenced by other emulsion properties, such as the gradient of the emulsion. For instance it has been shown that Q initially rises rapidly with increasing diffuse density, then reaches a maximum and subsequently decreases at higher diffuse densities. This increase of Q is markedly influenced by the gradient and rises sharply with the latter. Although it has been demonstrated that Q is related to the average diameter of the developed grains, it can only be taken as a measure of graininess if the latter is determined solely by the grain diameter. This is also shown by the fact, that using direct X-ray exposures for which the graininess increases with increasing quantum energy, the Callier coefficient remained constant (E. H. Belcher and G. Spiegler, 1953).

4.10 DIRECTIONAL EFFECTS

In all the previous sections of this chapter it has been assumed that the radiation was incident normally on the emulsion layer. In many cases, for instance in the practice of photographic dosimetry, an X-ray or γ-ray beam may fall obliquely on an emulsion. Hence it is necessary to study the response of an emulsion as a function of the angle at which the radiation is incident and how this effect may be influenced by the quality of the radiation applied.

Assuming that X-rays are incident on a film at an angle α, then the quantity of radiation I_1 absorbed per unit area of a photographic emulsion is

$$I_1 = I_0 \cos \alpha (1 - e^{-\mu d/\cos \alpha}) \tag{4.20}$$

where I_0 is the intensity of the X-ray beam, μ is the linear absorption coefficient and d is the thickness of the emulsion layer. For a small value of $(\mu d/\cos \alpha)$ the intensity of radiation absorbed by the emulsion is $I_0 \mu d$; hence the absorption is almost independent of the angle of incidence. If $\mu d/\cos \alpha$ is large i.e. when using soft radiation or

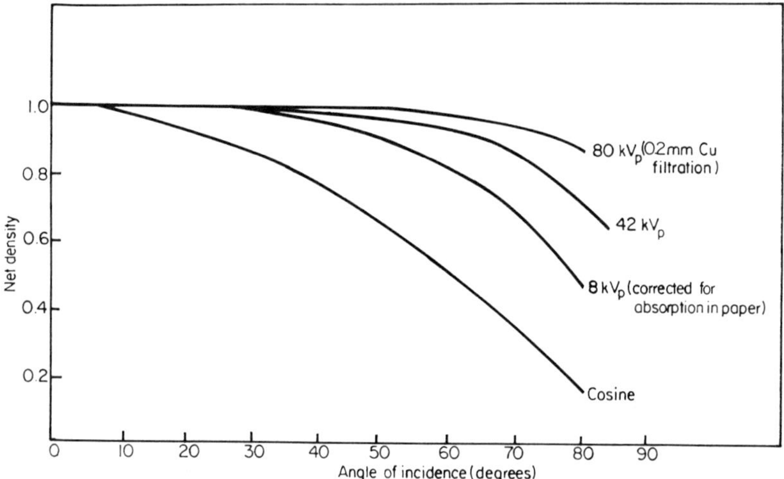

Figure 4.32. Change of density as a function of the angle of the incident X-ray beam for radiations generated at 8, 42 and 80 kV_p. The calculated response using the cosine of the angle is also shown.

large angles of incidence with respect to the normal beam, the energy removed from the X-ray beam tends to be greater. It is to be expected that the drop in film response with increasing angle will be greater the softer the radiation, and will be more marked if the film material is wrapped by paper or if the X-rays have to pass the front sheet of a cassette or filter material before they hit the emulsion. Much experimental work has been carried out on this subject by J. R. Greening (1951), D. A. Palmer (1954) M. Ehrlich (1954), R. B. Wilsey and others (1956), M. J. Heard, J. E. Cook and P. D. Holt (1960), K. Becker (1960, 1962) and others. A typical example of the decrease of density with increasing angle of incidence with respect to the normal beam is shown in Figure 4.32 due to D. A. Palmer (1954) (private communication). The figure illustrates the effect of radiation generated at three different kilovoltages namely $80 \, kV_p$ with $0.2 \, mm \, Cu$ filtration, $42 \, kV_p$, and $8 \, kV_p$ both unfiltered. Only the curve for $8 \, kV_p$ was corrected for the absorption of the paper wrapping. The lowest curve is that due to the cosine law. At 80 kV a slight drop of density is observed only at angles beyond 50° with respect to the normal beam. The loss of density using 80 kV at an angle of 80° is only about 10 per cent. As the effective thickness of the emulsion layer traversed increases with increasing angle, the drop in density is much more striking with soft radiation such as that generated by only 8 kV. The marked effect of filters placed close to the film (as in photographic dosimetry) when using oblique beams is discussed on page 203.

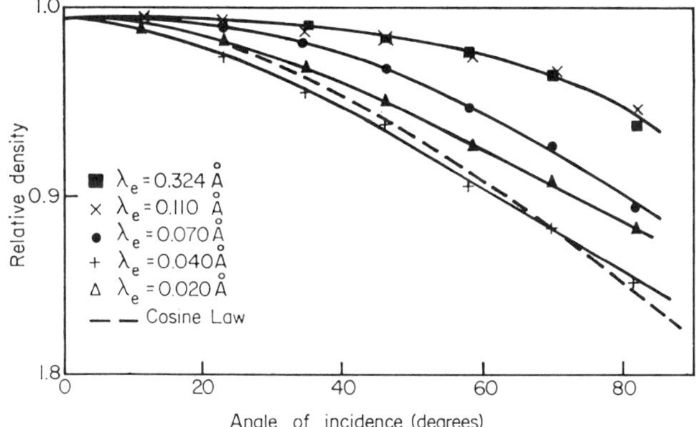

Figure 4.33. Variation of density with angle of incidence for radiations of decreasing wavelength. (*After* J. R. Greening, Proc. Phys. Soc. London B. 64, 977 (1951).)

The results of J. R. Greening (1951) are of interest as he established that the deficiency of density with angle of obliqueness of the incident beam increases towards shorter wavelengths and then decreases again for still harder radiation. His results are shown in Figure 4.33 and his interpretation of this effect is given below. At long wavelengths (0.324 and 0.110 Å) the dependency of density on the angle of incidence is relatively small. With shorter wavelengths an obliquity effect of different origin is to be expected as most of the density is produced by electrons originating from light material surrounding the film. These electrons are mainly recoil electrons and their general direction of motion is on the average the same as that of the X-rays producing them (see Figure 1.4). If these electrons are completely absorbed in the emulsion, the energy absorption will be proportional to $\cos \alpha$. Hence with further decrease of wavelength (0.070 Å) see Figure 4.33, the density as a function of angle becomes even smaller and follows approximately the cosine law at a wavelength of 0.040 Å. Using a still shorter wavelength (0.020 Å) the ranges of these recoil-electrons are considerably greater than the emulsion thickness. Hence they lose only a small fraction of their energy in the emulsion and less change of density with angle of incidence is to be expected. Thus the angular dependence becomes similar to that of medium wavelength X-rays, as revealed by the graph of Figure 4.33.

4.11   FADING OF THE LATENT IMAGE

If an exposure has been made on a photographic material and processing is postponed for a relatively long time, the optical density obtained may be smaller than if processing had followed the exposure immediately. This effect, which is called 'latent image fading', is caused by an instability of the latent image and may first appear to be of a similar nature to that leading to reciprocity failure. There are however marked differences between the two phenomena. The instability occurring in reciprocity failure is that of a not fully grown latent image, whereas that of latent image fading is an instability in a fully grown latent image and no direct correlation between the two effects has been found so far.

Experiments have shown that the degree of fading depends on the relative humidity of the atmosphere in which an emulsion is stored between exposure and processing. The higher the humidity the greater is the tendency for fading. Fading is also influenced by temperature chiefly because relative humidity is temperature dependent. Furthermore it has been shown that latent image fading can be reduced by storing photographic emulsions in an oxygen free atmosphere during the time between exposure and processing. In general it can be said that the tendency for fading is greater in slow fine grained emulsions than in fast coarse grained ones and that the effect is also influenced by the quality of radiation applied. Generally the higher the specific ionization of the electrons passing through the emulsion, the less will be the chance for fading (see Figure 4.35).

The influence of relative humidity on latent image fading is shown in Figure 4.34 due to M. J. Heard, J. E. Cook and P. D. Holt (1960). In this figure the relative response of an X-ray film is shown as a function of the time of storage in weeks and it is seen that the response at 30 per cent relative humidity is unaltered over the whole period of 12 weeks, whereas it is halved at 8 weeks, if kept at 60 per cent RH and it is reduced to 25 per cent of the original response at 80 per cent RH after 4 weeks storage time.

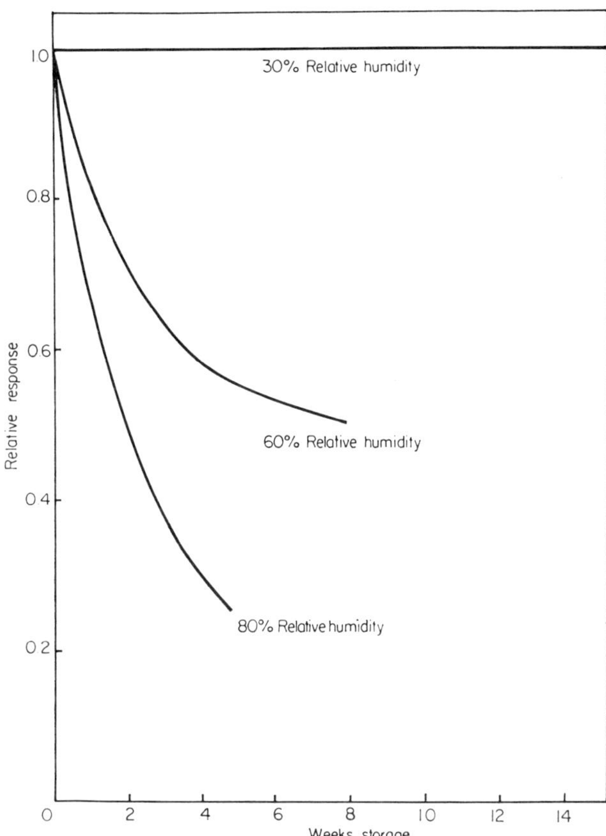

Figure 4.34. Latent image fading as a function of humidity at a temperature of 30°C for Ilford PMI film with exposures in the range 0.5 to 2.0 R. (*After* M. J. Heard, J. E. Cook and P. D. Holt, AERE-Report 3300 (1960).)

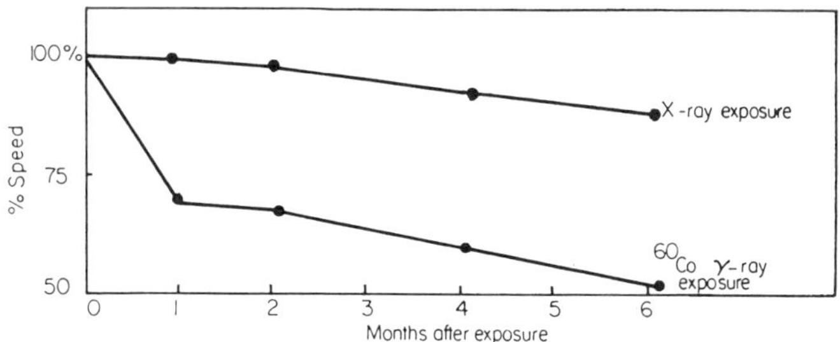

Figure 4.35. Latent image fading in slow X-ray film. Percentage speed as a function of the time elapsed in months between exposure and processing, using 80 kV$_p$ X-rays and ^{60}Co γ-rays.

145

The effect of the quality of radiation on latent image fading is illustrated in Figure 4.35 derived from some unpublished experiments due to the author. An X-ray film was exposed sensitometrically to X- and γ-rays, and the film strips were processed immediately after exposure and also after periods of one, three and six months. The curves in Figure 4.35 show the speed loss as a function of storage time between exposure and development and it is seen that the loss for γ-rays is considerably greater than that obtained for X-rays. This may be explained by the fact that the specific ionization of the electrons released in the X-ray exposures is greater, on the average, than that of the electrons released by γ-rays. The latter are near minimum ioniza-

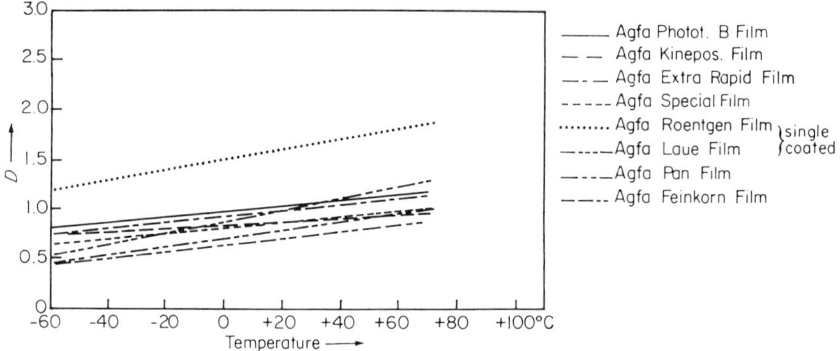

Figure 4.36. Dependence of optical density on temperature (C) during exposure for a constant flux of X-rays for each type of Film. (*After* J. Eggert and F. Luft. Veröffentlichungen des Wiss. Zentral-Lab. der Phot. Abteilg. Agfa **II, 9** (1931).)

tion, and may cause a regression of the latent image in the interval between exposure and processing more easily than those released by X-rays.

Latent image fading may be particularly troublesome in photographic neutron dosimetry and generally in the recording of particle tracks, as this is usually done in fairly fine grained slow emulsions. The delayed processing is inherent in this technique. The reader will find more details on pages 245 and 558.

4.12   THE INFLUENCE OF TEMPERATURE ON SENSITIVITY

A certain amount, although not quite enough, effort has been devoted to the study of the sensitivity of photographic emulsions as a function of temperature whilst exposed to X-rays or γ-rays. The dependence of density on temperature within the range from $-60°$ to $+70°C$ was investigated on various types of emulsion by J. Eggert and F. Luft (1931) and their results are shown in Figure 4.36. The films were exposed to Cu K-radiations ($\lambda = 1.54$ and 1.39 Å) and to heterogeneous radiation at 50 kV$_\text{p}$ equivalent to a mean wavelength of $\lambda = 0.45$ Å. No essential difference was found between the results obtained using the various radiation qualities. Only a relatively small increase of density of the order of $D = 0.6$ was obtained by increasing the temperature from $-60°$ to $+70°$C. This corresponds roughly to an increase of $D = 0.1$ for an average change of $18°$C.

W. F. Berg and K. Mendelssohn (1938) found that the X-ray sensitivity at the temperature of liquid helium was relatively less reduced than the light sensitivity, using the same photographic material. J. Reekie (1939) investigated the change of X-ray sensitivity with temperature from $5°$K to $300°$K and he observed that at low temperature a very slow rise of sensitivity is produced with increasing temperature up

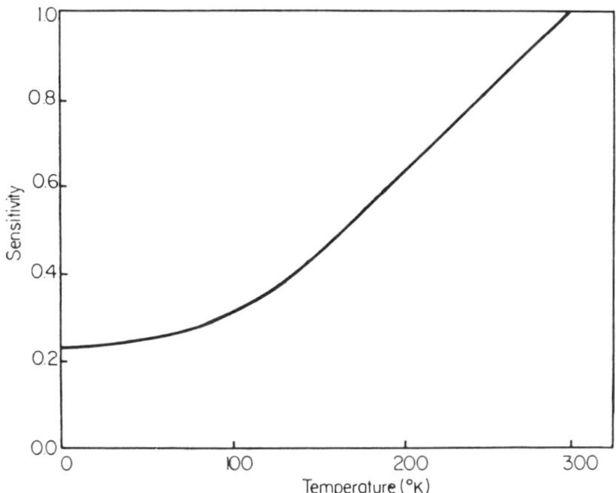

Figure 4.37. X-Ray sensitivity as function of temperature (°K). (*After* J. Reekie, *Proc. Phys. Soc. London* **51**, 683 (1939). *Inst. Phys. & Phys. Soc. London.*)

to about $70°$K and then an almost proportional increase with the higher temperature, as shown in Figure 4.37.

The decrease of sensitivity with lower temperature is believed to be due mainly to the reduced mobility of silver ions which are necessary to form a latent image. (W. Lehfeldt, 1935.) This was particularly supported by the light exposure experiments carried out by J. H. Webb and C. H. Evans (1938) which are in good agreement with theoretical expectations according to the Gurney-Mott mechanism of latent image formation (see p. 70). Webb and Evans showed that, if an emulsion is exposed to light flashes at $-186°$C, with increasing interruptions between the flashes and warmed up to room temperature after each flash, a greater photographic effect is achieved than if the emulsion is not warmed up during the intervals between the flashes, apart from warming up after the final flash. This result indicates that the warming up between the flashes permits the silver ions to combine with the photoelectrons to form latent images whereas they are prevented from doing so if no warming up between the flashes has occurred.

A further contribution to the loss of sensitivity at lower temperatures is thought to be caused by the light fluorescence effect which silver halides exhibit during exposure at low temperature. (G. C. Farnell, P. C. Burton and R. Hallama, 1950.) As this fluorescence was found to already occur at medium temperatures with X-ray exposures, it might be speculated that it is the reason for the relatively smaller total drop of sensitivity with X-ray exposures than with those of light exposures. This argument is based on the assumption that the fluorescence effect might have a gradually diminishing effect on X-ray sensitivity almost from room temperature downwards, with the result that the total drop of sensitivity is smaller if compared with that of light exposures. The fact that Eggert and Luft (1931) have shown that α- particle exposures are independent of temperature within the range from $-60°$ to $+70°C$, agrees well with the argument stated above, as light fluorescence of silver halides bombarded by α-particles can even be observed visually at room temperature using for instance a polonium source. (R. H. Herz, 1954.) Although this argument may play a part in the effect, it is realized that the mechanism is much more complex than that presented above.

4.13   COMPUTATION OF THE PHOTOGRAPHIC RESPONSE

It is sometimes useful to predict approximately the photographic response to X- or γ-rays before an unusual experiment is to be carried out. This might save not only time, but also valuable apparatus.

In order to calculate the response, three relationships have to be known and these are

1. The relation between the photon flux/cm² per roentgen and photon energy.

2. The relation between the mass-absorption coefficients of silver bromide and the energy of radiation.

3. The relation between the optical density and the number and the average projective area of developed grains. (Nutting's formula, which has been mentioned on page 93.)

In the following the items 1 and 3 will be derived and item 2 is shown in Figure 1.7. Later an example will be given which shows how such calculations may be carried out in a practical case.

1. The Relationship Between the Number of Photons per cm² per roentgen and Photon Energy

This relationship was originally introduced by W. V. Mayneord, (1950).

We have first to derive the energy flux/cm² per roentgen and from this the number of photons/cm² per roentgen. The energy absorbed per gram of air can be calculated by using the mass-energy absorption coefficient $(\mu_{en}/\rho)_{air}$, (see Figure 14.7), which is closely related to the mass-absorption coefficient μ/ρ from which however the fraction of energy which is not really absorbed is excluded.

The energy truly absorbed per gram of air is

$$E_{abs} = E \times \left(\frac{\mu_{en}}{\rho}\right)_{air} \text{ ergs per gram} \qquad (4.21)$$

In Chapter 2 (Equation 2.3) it has been shown that 1 roentgen is equivalent to 86.9 ergs/gram in air. Hence

$$E_{abs} = 86.9 \text{ R ergs/gram} \qquad (4.22)$$

Thus from equations (4.21) and (4.22) it follows that

$$E(\mu_{en}/\rho)_{air} = 86.9 \, R$$

or

$$\frac{E}{R} = \frac{86.9}{(\mu_{en}/\rho)_{air}} \text{ ergs/cm}^2 \text{ R} \qquad (4.23)$$

In order to arrive at the photon flux per cm² roentgen we have to find the number of photons N in the incident flux. If each photon has energy W MeV ($\equiv W1.6 \times 10^{-6}$ ergs)

$$E = 1.6 \, NW \times 10^{-6}$$

and

$$\frac{N}{R} = \frac{86.9}{(\mu_{en}/\rho)_{air} W \, 1.6 \times 10^{-6}} = \frac{54.31 \times 10^6}{(\mu_{en}/\rho)_{air} \, W} \text{ photons/cm}^2 \text{ R} \qquad (4.24)$$

The relationship between the number of photons per cm² per roentgen and photon energy is plotted in Figure 4.38, using the known values of $(\mu_{en}/\rho)_{air}$ for the various energies, as given in the appendix Figure 14.7 page 615.

2. Mass-Absorption Coefficients of Silver Bromide

The mass-absorption coefficients of silver bromide are found in Figure 1.7 from the data of the mass-absorption coefficients of silver

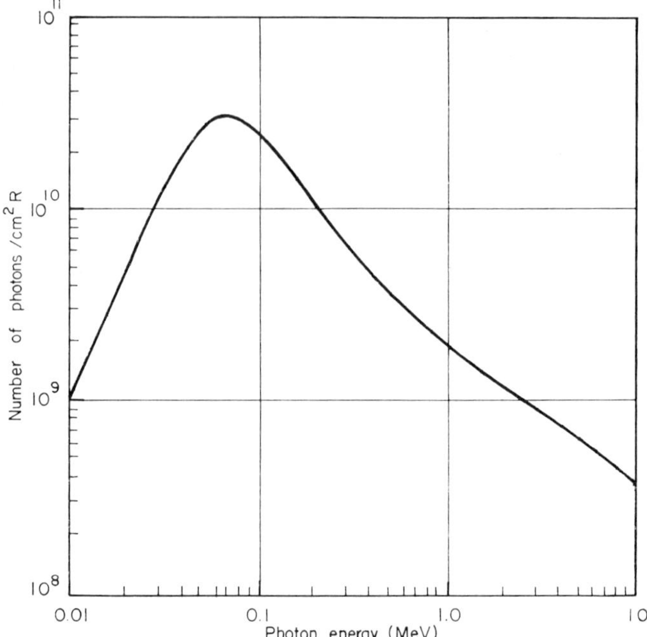

Figure 4.38. Number of photons per cm² per roentgen as a function of photon-energy in MeV.

and bromine given in Table 14.3 in the appendix, by using the expression

$$\left(\frac{\mu}{\rho}\right)_{AgBr} = \frac{P_1}{P_1 + P_2} \times \frac{\mu_1}{\rho_1} + \frac{P_2}{P_1 + P_2} \times \frac{\mu_2}{\rho_2} \qquad (4.25)$$

where P_1 and P_2 are the molecular weights, ρ_1 and ρ_2 the physical densities of silver and bromine respectively.

3. Relationship Between Optical Density and Number of Developed Grains per unit area. (Nutting's Formula, see p. 93)

Let us imagine a photographic emulsion built up of n layers, each containing one developed grain per unit area. The transmission of light by one grain layer will be $(1 - a)$, if a is the fraction of the incident light intensity absorbed by one grain. Adding another layer with one grain at random position within this layer, the transmission will be $(1 - a)(1 - a)$. If n layers are superimposed the transmission T is

$$T = (1 - a)^n \qquad (4.26)$$

Now by definition

$$D = -\log_{10} T = -0.4343 \log_e T \tag{4.27}$$

$$D = -0.4343\, n \log_e (1 - a) \tag{4.28}$$

Since a is of the order of 10^{-8} cm², we may expand the logarithm and to a good approximation obtain

$$D = 0.4343\, na \tag{4.29}$$

A multiplying factor of 2.6 was introduced by Farnell and Solman (1963) for the reasons mentioned on page 94, so that Nutting's formula reads

$$D = 2.6 \times 0.4343\, na \tag{4.30}$$

Example of Trend of Calculation

In the following an example is given in which the photographic response is computed using the relationships just shown.

A research worker is engaged in an experiment in which an X-ray beam equivalent to 48 keV is incident on a fast no-screen X-ray film and the exposure at the surface of the film is 0.5×10^{-2} roentgens per hour. He wants to know whether he will obtain a measurable density in one hour's exposure, but he does not know the relationship between density and exposure in roentgens for this film.

Number of photons/cm² Incident.   From Figure 4.38 we find that a flux of 2.8×10^{10} photons is required per roentgen at 48 keV. Hence an exposure of $2.8 \times 10^{10} \times 0.5 \times 10^{-2} = 1.4 \times 10^{8}$ photons per cm² would be given to the film within one hour.

Percentage Absorption.   The coating weight of silver bromide (both coating layers together) of the particular film in use, neglecting the weight of gelatin, has been found to be 4.5×10^{-3} grams per cm². The mass absorption coefficient of silver bromide (see Figure 1.7) for $\lambda_{(Å)} = 0.26$ is 8.3. ($\lambda_{(Å)} = 12.4/48 = 0.26$) see Equation (1.4) Chapter 1. Hence the percentage absorption can be found from

$$I_1 = I_0 e^{-8.3 \times 4.5 \times 10^{-3}} = I_0\, e^{-0.037} = I_0\, 0.963.$$

Thus 3.7 per cent of the incident radiation is absorbed by the film.

Optical Density.   Assuming that 1 quantum absorbed in the emulsion renders on the average 3 silver bromide grains developable using 48 keV X-rays, the number of grains affected is the product of the percentage absorption, the number of incident quanta per cm², and the quantum efficiency, which is $0.037 \times 1.4 \times 10^8 \times 3 = 1.55 \times 10^7$. The average grain size of the emulsion was found to be 1.5×10^{-8} cm². Thus using the Nutting's formula, we have

$$D = 0.4343 \times 1.5 \times 10^{-8} \times 1.55 \times 10^7 \times 2.6$$

$$= 0.26$$

The actual density obtained using the conditions quoted above was 0.18. The result shows that the density obtained in 1 hour is too low for the recording of average images. It should be mentioned that the results do not always turn out so close to the experimental findings, as this depends very much on the accuracy of the data involved in the calculation.

Similar calculations can be carried out from the relationship between roentgen per hour per millicurie at 1 cm and photon energy, using radioisotope sources (see Figure 14.6).

REFERENCES

Arens, H., (1955). Über den Villard Effekt. *Z. Wiss. Phot.* **50,** 392.

Arens, H., (1959). Über den Villard Effekt, III. *Z. Wiss. Phot.* **53,** 157.

American Standard Association, (1964). *Method for the Sensitometry of Medical X-ray Films.* A.S.A. PH **2, 9.**

American Standard Association, (1964). *Method for the Sensitometry of Industrial X-ray Films for Energies up to 3 Million Electron Volts.* A.S.A. PH **2,** 8.

American Standard Association, (1965). *Method for Evaluating Films for Monitoring X-ray and Gamma rays having Energies up to 3 Million Electron Volts.* PH **2,** 19.

American Standards Association, (1959). *Diffuse Transmission Density* (ISO R5). PH **2,** 19.

Atkinson, P. A., (1966). *The Investigations of Several X-ray Films in the 2 to 67Å region.* M.Sc. Thesis, University of Leicester.

Ballard, A. E., Stevens, G. W. W., and Zuehlke, C. W., (1952). The Determination of silver in photographic images using ^{131}I, radioactive iodide. PSA Journal *Phot. Sci. Tech.,* **18B,** 27.

Becker, K., (1960). Probleme and Ergebnisse der Filmdosimetrie Ionisierender Strahlen. *Phot. Korr.,* **96,** 83, 99, 115.

Becker, K., (1961). Beitrag Zur Filmdosimetrie Energiereicher Quantenstrahlung. I. Theoretische Grundlagen. *Fortschr. Geb. Röntgenstr. Nuklearmed.,* **95,** 694.

Becker, K., (1962). Filmdosimetrie, Springer-Verlag. Berlin-Göttingen-Heidelberg.

Belcher, E. H. and Spiegler, G. (1953). The density, granularity and silver content of photographic film exposed to ionizing radiation. *J. Phot. Sci.* **50**.1.86.

Bell, G. E., (1936). The photographic action of radium gamma rays. *Brit. J. Radiol.*, **9**, 578.

Berg, W. F., (1940). Reciprocity failure of photographic materials at short exposure times. *Proc. Roy. Soc.*, **174A**, 559.

Berg, W. F., Marriage, A., and Stevens, G. W. W., (1941a). Latent image distribution. *Phot. J.*, **81**, 413.

Berg, W. F., Marriage, A., and Stevens, G. W. W., (1941b). Latent image distribution. *J. Opt. Soc. Am.*, **31**, 385.

Berg, W. F., and Mendelssohn, K., (1938). Photography sensitivity and the reciprocity law at low temperatures. *Proc. Roy. Soc.*, **168A**, 168.

Blau, M., and Altenburger, K., (1923). Über Einige Wirkungen von Strahlen. II, *Z. Phys.*, **12**, 315.

British Standards Institution. (1947). *Measurements of Photographic Transmission Density.* B.S. 1384.

Bromley, D., and Herz, R. H., (1950). Quantum efficiency in photographic X-ray exposures. *Proc. Phys. Soc.*, (London) **B.63**, 90.

Capstaff, J. G., and Purdy, R. A., (1927). A compact motion picture densitometer. *Trans. S.M.P.E.*, **31**, 667.

Charlesby, A., (1940). The action of electrons and X-rays on photographic emulsions. *Proc. Phys. Soc.*, (London), **52**, 657.

Chassende-Baroz, N. J. P., (1961). Utilisation des émulsions photographiques pour la measure de forte doses—influence du débit de la source de rayonnement. *Proc. Symposium on selected topics in radiation dosimetry.* Vienna.

Corney, G. M., (1966). Photographic effects of X-rays and gamma rays. In C. E. K. Mees and T. H. James, (Eds.), *The Theory of the Photographic Process.* MacMillan, New York.

Corney, G. M., and Seemann, H. E. (1947). A logarithmic step tablet for X-rays. *Non Destructive Testing*, **6**, 27.

Cross, A. S., (1955). An instrument for the accurate measurement of low optical densities. *J. Sci. Instr.* **32**, 59.

Eggert, J., (1931). Die Empfindlichkeit Photographischer Emulsionen für Röntgenstrahlen in Abhängigkeit von der Korngrösse. *Veröff. Wiss. Zentral Lab. Agfa*, Vol. **II**, 1.

Eggert, J., and Küster, A., (1938). Grain size determination and other applications of the Callier effect. *J. Soc. Mot. Pict. Eng.*, **30**, 181.

Eggert, J., and Luft, F., (1931). Die Temperaturabhängigkeit des Photographischen Prozesses. *Veröff. Wiss. Zentral. Lab. Agfa.*, **II.**

Eggert, J., and Noddack, W., (1927, 1928). Über die Quantumausbeute bei der Wirkung von Röntgenstrahlen auf Silberbromid. I. *Z. Physik* **43**, 222; II. *Z. Physik* **51**, 796.

Eggert, J., and Schopper, E., (1938, 1939). Körnigkeit Photographischer Schichten bei Bestrahlung mit energiereichen Quanten. *Z. Wiss. Phot.*, **37**, 221, 1938. *Agfa Wiss. Veröff.*, **6**, 159 1959.

Ehrlich, M., (1954). Photographic dosimetry of X- and gamma rays. *Nat. B. St.* (U.S.), Handbook 57.

Ehrlich, M., (1956). Part II. Validity for high intensity exposures in the negative region. *J. Opt. Soc. Am.*, **46**, 801.

Ehrlich, M., and McLaughlin, W. L., (1956). Reciprocity law for X-rays Part. I. Validity for high intensity exposures in the negative region. *J. Opt. Soc. Am.* **46**, 797.

Ehrlich, M., and McLaughlin, W. L., (1961). Photographic response to successive exposures of different types of radiation. *J. Opt. Soc. Am.*, **51**, 1172.

Farnell, G. C., (1966). The relationship between density and exposure. In C. E. K. Mees and T. H. James, (Eds.), *The Theory of the Photographic Process*. MacMillan, New York.

Farnell, G. C., Burton, P. C., and Hallama, R., (1950). The fluorescence of silver halides at low temperatures. Part I. The pure halides. *Phil. Mag.*, **7**, 157. Part II. Mixed crystals of silver halides. *Phil. Mag.*, **7**, 545.

Farnell, G. C., Chanter, J. B., and Judd, F. S., (1964). Photolytic effects to theories of solarization. *J. Phot. Sci.*, **12**, 1.

Farnell, G. C., and Solman, L. R., (1963). The relationship between covering power at saturation density and undeveloped grain size. *J. Phot. Sci.*, **11**, 347.

Frieser, H., and Klein, E., (1961). Die Theorie der Schwärzungskurve und die Berechnung der Empfindlichkeitsverteilung. *Mitt. Forsch. Lab. Agfa*, Vol. III. Springer-Verlag. Berlin.

Glocker, R., (1921). Das Photographische Schwärzungsgesetz, *Phys. Zt.*, **22**, 345.

Glocker, R., (1927a). Über das Grundgesetz der physikalischen Wirkungen von Röntgenstrahlen verschiedener Wellenlänge, *Z. Physik*, **43**, 827.

Glocker, R., (1927b). Über den Energieumsatz bei einigen Wirkungen von Röntgenstrahlen, *Z. Physik*, **40**, 479.

Greening, J. R., (1951). The photographic action of X-rays. *Proc. Phys. Soc.*, (*London*), **B64**, 977.

Heard, M. J., Cook, J. E., and Holt, P. D., (1960). Photographic emulsion dosimetry and the A.E.R.E. film dosimeter, *A.E.R.E.*, R.3300.

Herz, R. H., (1938). Ein besonders einfaches lichtelektrisches Photometer, *Zt. Wiss. Phot.* **37**, 5, 107.

Herz, R. H., (1949a). The recording of electron tracks in photographic emulsions. *Phys. Rev.*, **75**, 478.

Herz, R. H., (1949b). Granularity measurements of X-ray emulsions exposed to X-rays with increasing quantum energy. *Phot. J.*, **89B**, 89.

Herz, R. H., (1954). Sekundäre Strahlungseffekte bei der photographischen Wirkung von Alpha-Strahlen. *Strahlentherapie.* **943**, 455.

Hillson, P. J., and Sutherns, E. A., (1966). Disposition of the latent image as determined by variation of exposure and development techniques. In C. E. K. Mees and T. H. James, (Eds.), *The Theory of the Photographic Process*, Chap. 6. MacMillan, New York.

Hirsh, F. R., (1938, 1942). The blackening of photographic plates by long wavelength X-rays. I. *J. Opt. Soc. Am.*, **28**, 463. II. *J. Opt. Soc. Am.*, **32**, 219.

Hoerlin, H., (1949). The photographic action of X-rays in the 1.3 to 0.01 Å range. *J. Opt. Soc. Am.*, **39**, 891.

Hoerlin, H., (1951). Die Quantenausbeute der photographischen Wirkung der Rontgenstrahlen als Funktion der Wellenlänge und der Sensibilisierung. *Z. f. Naturforsch*, 6a. 344.

Hoerlin, H., and Hicks, V., (1947). Quantitative relations between the photographic response of X-ray films and the quality of radiation. *Nondestructive Testing.* **6**, 25.

Hurter, F. and Driffield, V. C., (1890). Photochemical investigations and a new method of determination of the sensitiveness of the photographic plates. *J. Soc. Chem. Ind.*, **9**, 455.

James, T. H., and Fortmiller, L. J., (1961). Dependence of covering power and spectral absorption of developed silver on temperature and composition of the developer. *Phot. Sci. Eng.*, **5**, 297.

Jones, L. A., and Deisch, N., (1920). The measurement of graininess in photographic deposits. *J. Frank. Inst.*, **190**, 657.

Kroencke, H., (1914). Über die Messung der Intensität and Härte der Röntgenstrahlen. *Ann. Phys.*, **43**, 687.

Lehfeldt, W., (1935). Zur Elektronenleitung in Silver-und Thalliumhalogenidkristallen. *Nachr. Ges. Wiss. Göttingen, Math. -Phys. Kl.*, **1**, 171.

Luft, F., (1933). Der Schwarzschild-Effekt bei Röntgenaufnahmen *Agfa Veröff.* **3**, 245.

Mauderli, W., (1957). Dosimetrie von Röntgen-und Gammastrahlen mittels photographischer Filme. I. Physikalische Grundlagen. *Fortschr. Geb. Röntgenstrahl. Nuklearmed*, **86**, 634.

May, A., (1943). Solarization with X-rays and visible light, *J. Opt. Soc. Am.*, **33**, 81.

Mayneord, W. V., (1950) Some applications of nuclear physics to medicine. *Brit. J. Rad. Supplement 2. London.*

Mees, C. E. K., and James, T. H., (1966). *The Theory of the Photographic Process.* MacMillan, New York. p. 419/20.

Nepela, D. A., and Nitka, H. F., (1960). Effect of chemical ripening on the photographic response to light, X- and gamma radiation, *Phot. Sci. Eng.*, **4**, 12.

Nutting, P. G., (1913). On the absorption of light in heterogenous media. *Phil. Mag.*, **26**, 423.

Palmer, D. A., (1954). Private Communication.

Pelc, S. R., (1945). The photographic action of X-rays, *Proc. Phys. Soc.*, (*London*). **57**, 523.

Reekie, J., (1939). The sensitivity of photographic films to X-radiation at very low temperatures. *Proc. Phys. Soc. (London).* **51**, 683.

Sabattier, M., (1860). *Bull. Soc. Franc. Phot.* **6**, 306.

Schleussner, C. A., and Kaempfert, K., (1924). Über die Wirkung intermittierender Röntgenstrahlung auf die photographische Platte. *Fortschr. Geb. Röntgenstrahl.*, **32**, 593.

156 *The Photographic Action of Ionizing Radiation*

Schwarzschild, K., (1900). On the deviations from the law of reciprocity for bromide of silver gelatine. *Astrophys. J.*, **11**, 89.

Seemann, H., (1950). Spectral sensitivity of two commercial X-ray films between 0.2 and 2.5 Å., *Rev. Sci. Instr.*, **21**, 314.

Seemann, H. E., and Roth, B., (1960). New stepped wedges for radiography. *Acta Radiol.*, **53**, 215.

Seemann, H. E., and Roth, B., (1962). A new stepped wedge design for industrial radiography. *Nondestructive Testing*, **20**, 37.

Selwyn, E. W. H., (1935). A theory of graininess. *Phot. J.*, **75**, 571.

Selwyn, E. W. H., (1939). Experiments on the nature of graininess. *Phot. J.*, **79**, 513.

Silberstein, L., (1922). Quantum theory of photographic exposures I. *Phil. Mag.*, **44**, 257.

Silberstein, L., (1941). On the number of quanta required for the developability of a silver halide grain. *J. Opt. Soc. Am.*, **31**, 343.

Silberstein, L., and Trivelli, A. P. H., (1930). The quantum theory of X-ray exposures on photographic emulsions. *Phil. Mag.* **7**, 9, 787.

Sobolew, M., (1945). Schwärzungsgesetze Photographischer Platten für Röntgenstrahlen, *Zh. Eksperim. i. Teor. Fiz.*, **15**, 431.

Stevens, G. W. W., (1942). A technique for examining latent image distribution. *Phot. J.*, **82**, 42.

Sweet, M. H., (1950). An improved photomultiplier tube color densitometer. *J. Soc. Mot. Pict. Eng.*, **54**, 35.

Sweet, M. H., (1959). Direct-reading zero-to-six range densitometer incorporating digital output. *Phot. Sci. Eng.*, **3**, 97.

Szilagyi Z., and Blackman, E., (1966). Behaviour of some photographic materials exposed to radiation in the 8 angstrom region. *Phot. Sci. Eng.*, **10**, 2.

Tellez-Plasencia, H., (1948). Contribution à l'étude de la théorie quantique de l'effet photographique des rayons X. I. Energie utilisable. Courbes Caracteristiques. *Sci. Ind. Phot.*, **19**, 45.

Tellez-Plascencia, H., (1950). Réabsorption du rayonnement secondaire de fluorescence dans une émulsion photographique impressionnée par les rayons X. *Sci. Ind. Phot.*, **21**, 361.

Tellez-Plasencia, H., (1953). Etudes sur le noircissement photographique produit par les rayons X I., *Sci. Ind. Phot.*, **24**, 1.

Tellez-Plasencia, H., (1954, 1955, 1957). Etudes sur le noircissement photographique produit par les rayons X. II Parcours des électrons, rendement de l'énergie utilisable. *Sci. Ind. Phot.*, **25**, 41. (1954) III Energie dépensée et énergie utilisée. *Sci. Ind. Phot.*, **25**, 225, 1954. IV Influence de la lumination. Les seuils quantiques apparent et réel. *Sci. Ind. Phot.*, **25**, 425, 1954. V Traces d'électrons dans les émulsions radiographiques ordinaires. *Sci. Ind. Phot.* **26**, 312, 1955. VI Influence de la solarisation sur les courbes caractéristiques des images totales, internes et externes. *Sci. Ind. Phot.*, **28**, 144, 1957. VII Effets de l'hétérogénéité des grains de l'émulsion, la courbe de noircissement, fonction du volume des grains et de l'énergie des photons. *Sci. Ind. Phot.*, **28**, 235, 1957.

Tellez-Plasencia, H., (1960). Introduction à la gammagraphie I. Lois Générales Effets d'absorption directe des photons gamma incidents. *Sci. Ind. Phot.*, **2,** 31, 49.

Tuttle, C., (1936). A recording physical densitometer. *J. Opt. Soc. Am.*, **26,** 282.

Villard, P., (1899). Les Rayons X et la Photographie. *Soc. d'encouragement pour l'industrie nationale.* **4,** 518, Paris.

Webb, J. H., (1933). Relationship between reciprocity law failure and the intermittency effect in photographic exposure. *J. Opt. Soc. Am.*, **23,** 157.

Webb, J. H., (1939). Graphical analysis of photographic exposure and a new theoretical formulation of the H and D curve. *J. Opt. Soc. Am.*, **29,** 314.

Webb, J. H., (1941). Number of quanta required to form the photographic latent image as determined from mathematical analysis of the H and D curve. *J. Opt. Soc. Am,* **31,** 348, 559.

Webb, J. H., and Evans, C. H., (1938). An experimental study of latent image formation by means of interrupted and Herschel exposures at low temperature. *J. Opt. Soc. Am.*, **28,** 249.

Wilsey, R. B., (1951). The use of photographic films for monitoring stray X-rays and gamma rays. *Radiology,* **56,** 2, 229.

Wilsey, R. B., and Pritchard, H. A., (1926). A comparison of X-ray and white light exposures in photographic sensitometry. *J. Opt. Soc. Am.*, **12,** 661.

Wilsey, R. B., Strangways, D. H., and Corney, G. M., (1956). Experiments in the photographic monitoring of stray X-rays. *Radiology,* **66,** 408.

5

The Photographic Response to Particles

5.1   INTRODUCTION

As the photographic response to *X-rays* is mainly caused by electrons released in the emulsion, the basic principles of the response to the 'direct exposure' of electrons should essentially be the same. (In *γ-ray* exposures however, many of the electrons striking an emulsion originate from surrounding media.) Any changes to be expected in the response of an emulsion exposed to an electron beam as against that produced by a long or medium wavelength X-ray beam, must partly be of a geometric nature, as the electrons in the latter case are released chiefly within the emulsion whereas in the former case they strike the emulsion from outside. Greater differences are involved if particles other than electrons, such as α-particles, protons or neutrons, are incident on photographic material.

Apart from the recording of electrons, we are chiefly concerned in this chapter with the action of neutrons and protons, as these are of interest in photographic dosimetry and the former also in neutron radiography. The large field of recording other nuclear particles such as the various types of mesons, hyperons and the many other so called elementary particles will not be considered in this book.

The action of charged particles on the photographic emulsion may lead either to the formation of individual tracks (see p. 97) revealing the paths of the particles under microscopic inspection, or to measurable optical densities depending on the exposure given. As the optical density can be regarded as the result of the superimposition of many individual tracks, the study of the behavior of individual particles should logically precede that of the optical density.

5.2   PARTICLE TRACKS

5.2a   On the Formation of Tracks

The average rate of energy loss per unit length of path of an electron in an X-ray emulsion as a function of its energy has been illustrated in Figure 4.2 p. 96. The general shape of this curve is typical for any singly charged particle as seen from Figure 5.1 in which the rate of

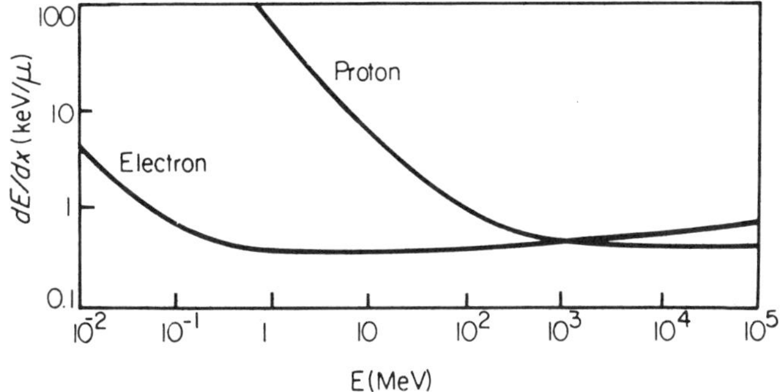

Figure 5.1. Energy loss dE/dX (keV/μ) as a function of the energy of electrons and protons in a nuclear track emulsion.

energy loss dE/dX (keV/μ) for electrons and protons in a given nuclear emulsion is shown as a function of energy in MeV. This graph is based on the simplified equation (1.21) Chapter 1 page 30, which is modified by a more elaborate formula in which the average ionization potential of stopping atoms and a relativistic correction have been considered (see equation (1.22) page 30). The graph shows that the rate of energy loss decreases with increasing energy to a minimum (called minimum ionization) and then gradually rises again with further increase of the energy of the particle. The separation between the energy loss curves of the proton and electron with respect to the abscissa axis, in Figure 5.1, is in proportion to their respective masses. The proton rest mass is 1840 times that of the electron.

The average energy loss in keV/μ for electrons is slightly lower in Figure 4.2 (which refers to X-ray films) than that indicated in the above Figure 5.1. This is caused by the much greater gelatin content of an X-ray emulsion if compared with that of a nuclear emulsion. In the latter the energy loss (see Figure 5.1) is about 500 eV/μ in the region of minimum ionization. The energy required for the liberation of an electron from silver bromide is found to be about 5.8 eV (K. A. Yamakawa, 1951). Hence the average number of electrons produced in a 0.4 micron diameter grain, when an electron at minimum ionizing power passes through it, would be $500 \times 0.4/5.8 = 35$ electrons. This means that on the average this number of electrons (or number of ionizations) should be adequate to form a developable latent image if the grain is capable of detecting electrons at minimum ionization.

An example of the determination of the rate of energy loss of another particle, for instance of an α-particle, assuming that the energy loss of the proton is known, is given below. The formula holds for the case that both the α- and the proton particle have the same velocity, and is

$$\frac{dE}{dX_\alpha} = \left(\frac{z_\alpha}{z_H}\right)^2 \left(\frac{dE}{dX}\right)_H \tag{5.1}$$

where z_α is the charge of the α-particle and z_H that of the proton. Thus the energy loss of an α-particle having the *same velocity* as a proton is 4 times that of the proton.

Similarly, the ranges of α-particles, i.e. their path lengths in the emulsion, are related to those of protons, assuming again that both have the *same velocity*, by

$$R_\alpha = \left(\frac{z_H}{z_\alpha}\right)^2 \left(\frac{m_\alpha}{m_H}\right) R_H \tag{5.2}$$

where R_α and R_H are the ranges, and m_α and m_H the masses of the α- and proton respectively. This can be illustrated by a practical example. An α-particle of a polonium source, having an energy of 5.3 MeV, has a range of about 22 microns in a highly concentrated nuclear emulsion (see Figure 14.2). A proton having the same velocity as the α-particle has a quarter of its energy, namely about 1.3 MeV. As can be seen from the range curve (Figure 14.4 in the Appendix) a proton of 1.3 MeV energy also has a range of 22 microns in the samy type of emulsion.

5.2b   Characteristics of Particle Tracks

In Figure 5.2 a diagrammatic representation is given of a section through a nuclear emulsion which illustrates the characteristics of various particle tracks as seen by microscopic inspection. The white circles indicate unaffected grains, the black circles developed grains. It is seen that the grains are very closely packed, as is typical for a nuclear particle emulsion in which there is relatively little gelatin if compared to that of an ordinary emulsion, e.g. an X-ray film.

If for example (see Figure 5.2) an *α-particle* travels through such an emulsion, it renders each grain developable along its path, and a straight

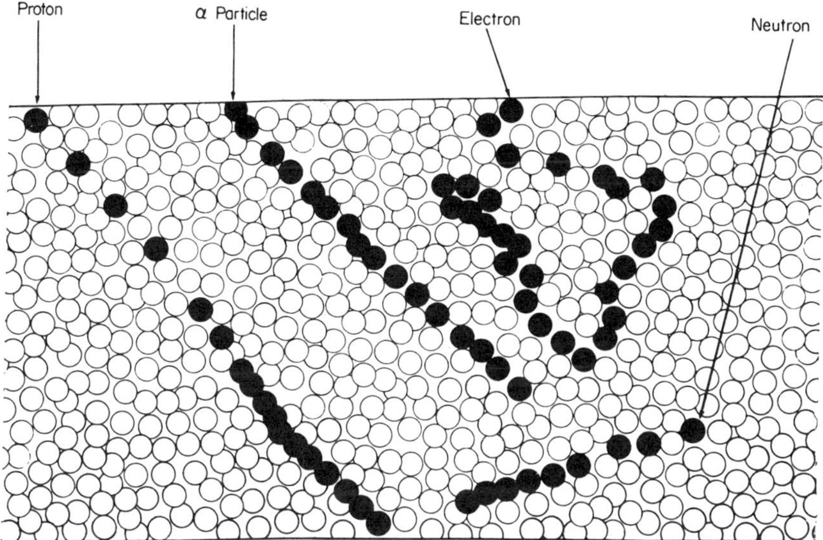

Figure 5.2. Characteristics of atomic particle tracks (Diagrammatically).
○ : Unexposed grains
● : Exposed grains

line row or track of silver grains becomes visible under the microscope after processing. This is because the α-particle possesses a high ionizing power (see p. 32) which is partly due to its double charge and partly due to the fact that it has for a given energy, a relatively low velocity on account of its great mass.

A proton on the other hand, having the same energy as the α-particle, has one unit charge only and travels much faster because of its smaller mass compared with that of the α-particle. Hence it ionizes less at the beginning of its passage through the emulsion, but its specific ionizing power increases when the proton is slowing down. The grain density, i.e. the number of grains rendered developable per unit length of path is therefore greater towards the end of the proton track.

A neutron, having no charge, causes no direct ionization. However, if a neutron (see Figure 6.23) collides with a nucleus of hydrogen (due to the gelatin in the emulsion, film base, and wrapping material) it may impart some or the whole of its energy to the hydrogen nucleus, hence a proton track is produced in the emulsion layer. The counting of proton tracks produced in a nuclear particle emulsion can therefore be used as a method of photographic neutron dosimetry (see p. 236). Thermal, i.e. slow neutrons, are photographically effective chiefly by indirect effects due to nuclear reactions, as will be discussed on pages 174 and 490.

An electron, with its small mass and its relatively high velocity even at comparatively low energies, will not ionize very strongly unless it slows down, when it acquires high specific ionizing power. The energy of an electron for minimum ionization lies near 0.6 MeV. Because of its small mass the electron is easily scattered within the emulsion, due to interactions with atoms which it encounters. A photomicrograph of an electron track in a nuclear particle emulsion is illustrated in Figure 4.3 page 97. Other particle tracks are shown in Figures 5.3–5.5. Figure 5.3

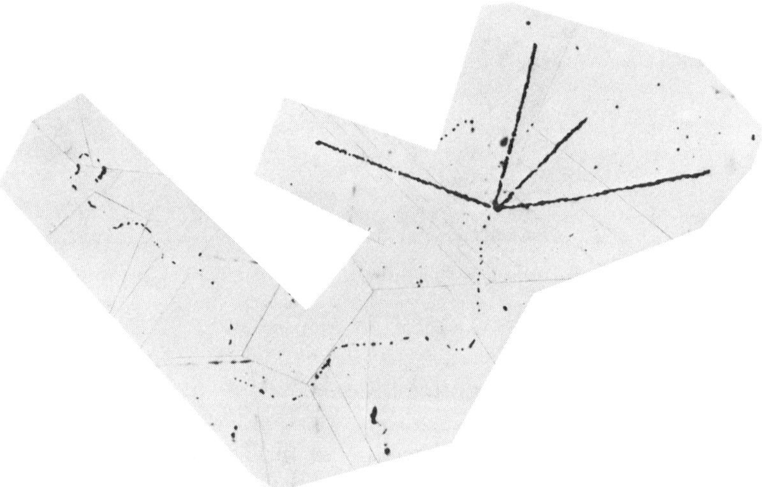

Figure 5.3. Disintegration of a radiothorium atom in a nuclear track emulsion loaded with thorium nitrate. The photomicrograph shows four α-tracks and one electron track emerging from the same origin. (*By courtesy of* Research Laboratories of Kodak Limited, Harrow, England.)

illustrates the disintegration of a radiothorium atom into four α-particles and one electron. This photomicrograph shows the dense grain population of α-tracks which is due to the high ionizing power of the α-particles. The picture also reveals the typical scattering to which an electron is subjected and the relatively low grain density of the electron track. This photomicrograph is a mosaic of twenty-four individual photomicrographs each taken whilst the grains are in focus under the microscope, as the tracks lie in different depths of the emulsion layer.

Figure 5.4 reveals the advantage of the high concentration (right hand image) of silver bromide in a nuclear emulsion having a narrow grain size distribution, as against an emulsion of low concentration,

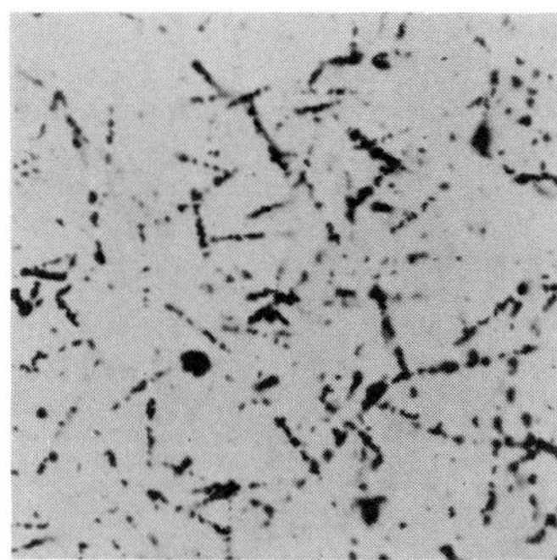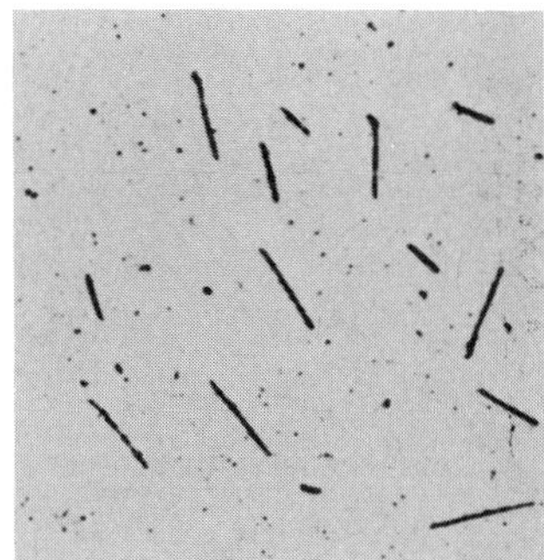

Figure 5.4. Tracks of α-particles in a medium fast X-ray film (left) and in a
nuclear track emulsion (right).

having a wide grain size distribution (left image). The separation
between silver grains in the tracks (on the left) is due to the higher
gelatin content of this emulsion.

Figure 5.5 illustrates the event of pair formation (see p. 17)
whereby a positron and an electron are emitted during the annihilation
of a high energy γ-ray quantum. The tracks of both particles, having
minimum ionizing power, should show the reader the typically low
grain density combined with small particle scatter (A. C. Coates and
R. H. Herz, 1949). A photomicrograph of proton tracks is shown in
Figure 6.23, page 238.

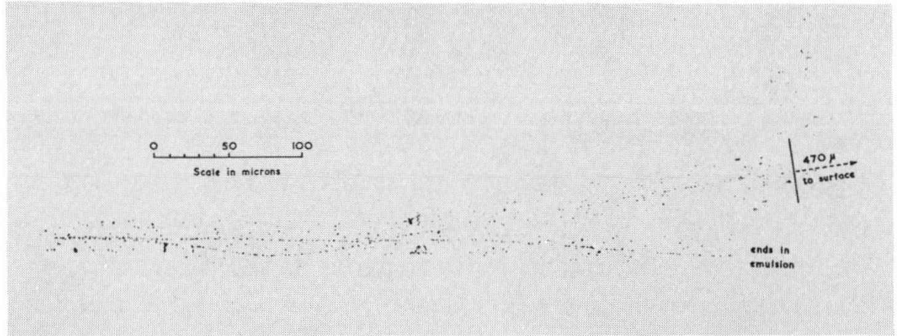

Figure 5.5. Pair formation in nuclear track emulsion. One of the two electron
tracks is due to a positron, the other one to an electron, both at minimum
ionizing power. (Produced by exposure to 25 MeV X-rays from a synchroton.)
(*After* A. C. Coates and R. H. Herz, *Phil. Mag.* **XL 7.** 1088 (1949).)

The photographic recording of particle tracks is one of the important tools in nuclear physics and in cosmic ray studies. As mentioned before, a more detailed treatment of this subject is beyond the scope of this book and the interested reader is referred to the excellent literature which is mentioned in the list of references at the end of this chapter. P. Demers, 1958; (C. F. Powell, P. H. Fowler, and D. H. Perkins, 1959; W. H. Barkas, 1963).

5.3   OPTICAL DENSITY DUE TO ELECTRONS

The photographic effect of fast moving electrons is the result of a series of interactions between the electrons and the atoms composing the emulsion. Some of these reactions lead to inelastic collisions, whereby the electrons ionize some of the atoms through which they pass and may thus render silver bromide grains developable; other atoms may be transferred to a state of excitation. Some of the electrons may approach nuclear fields whereby X-ray Bremsstrahlung (see p. 4) is produced. The passage of electrons may also lead to elastic collisions between the particles and the orbital electrons of atoms with which they interact, thus leading to deflections or scattering. The electrons released in the events of ionization may cause further secondary ionizations and also further Bremsstrahlung processes. Part of the electron energy is utilized for the production of δ-rays, which are electrons of low energies.

5.3a.   Effects on Single and Double Coated Film

Assuming a stream of electrons impinges perpendicularly on an emulsion layer, then its photographic response depends predominantly on the number of electrons per unit area of cross-section of the beam and on the energy of the electrons. Although in some practical applications of electron beams the particles originate from a continuous energy spectrum, as for instance from radioisotopes, it is instructive to discuss first the response to monoenergetic electrons, for instance as used in an electron microscope.

In Figure 5.6 we see a section through a double coated X-ray film and the paths of electrons are indicated in relation to the section. It is shown that with low kinetic energies up to about 0.1 MeV an electron penetrates increasing depths of the first emulsion layer, where it suffers a considerable amount of scatter. This first emulsion layer reaches about maximum sensitivity just below 0.1 MeV. Beyond this energy the highest specific ionization of the particle towards the end of its path occurs in the film base, where it is obviously lost for the photographic response. Hence it is expected that the sensitivity of the

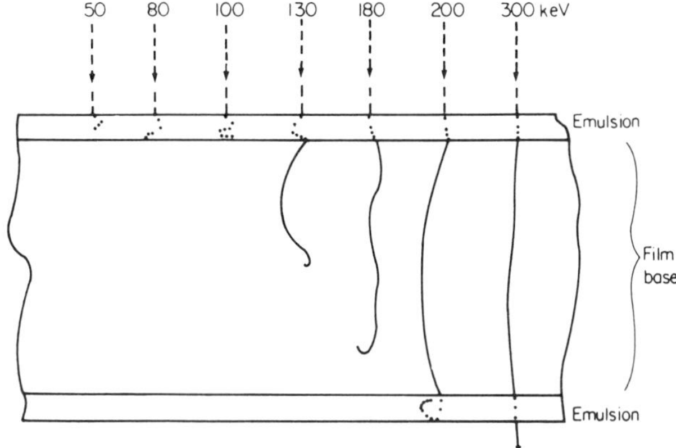

Figure 5.6. Electrons of increasing energies passing through a double coated
X-ray film.

film will decrease. With further increase of energy (up to about 200 keV),
the region of maximum specific ionization reaches the second emulsion
layer (see Figure 5.6). Hence a further increase of sensitivity would be
expected. Using still higher energies, the maximum specific ionization
is expected to lie outside the double coated X-ray film, resulting in a
decrease in sensitivity which then should remain fairly constant with
still higher energies. Naturally the detailed effects depend entirely on
the coating thicknesses of the emulsion concerned.

This pictorial presentation of the events taking place with electrons
of increasing energy was checked experimentally by R. A. Dudley
(1951a, 1951b) using an electron spectrograph. Dudley's results are
shown in Figure 5.7 which illustrates the sensitivity (in terms of rads
for a density of 0.3) against the energy of electrons at perpendicular
incidence for a single coated and for a double coated no-screen X-ray
film. The crossed curve corresponds to the single coated, the circled
curve to the double coated film. The other curve in this graph which
refers to diffuse incidence will be discussed later. It is seen that the
sensitivity rises with electron energy and reaches practically the same
value for the first peak at about 0.1 MeV for the single and double
coated film. Beyond the peak the speed of the *single coated* film decreases with further increase of electron energy as the maximum
specific ionization of the electrons occurs chiefly in the film base and is
thus not utilized by the emulsion. Towards still higher energies the
single film sensitivity remains practically constant. A similar decrease
of sensitivity beyond 0.1 MeV is first shown by the *double coated film*,
but the sensitivity increases again at about 0.2 MeV, when the second

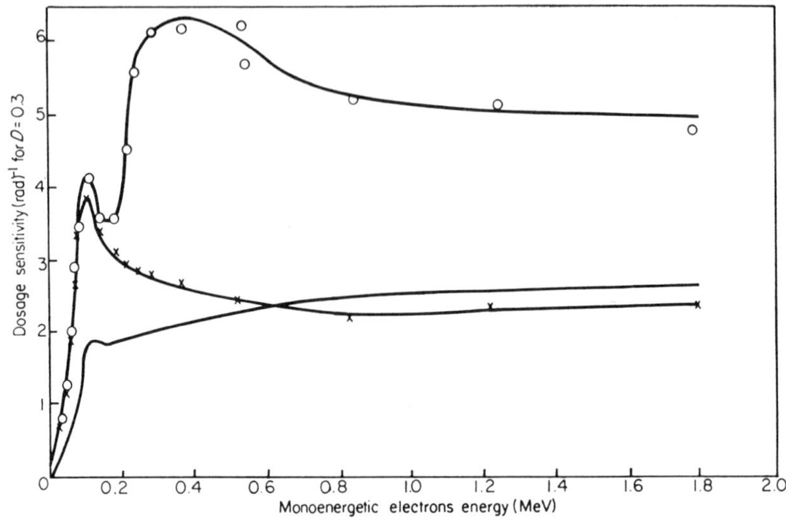

Figure 5.7. Dosage sensitivity for a density of 0.3 of Eastman Kodak X-ray film as a function of the energy of electrons (*After* R. A. Dudley, Photographic film dosimetry in G. J. Hine and G. L. Brownell, *Radiation Dosimetry*, Academic Press, New York, 1956).

film coating is reached by the electrons. The maximum sensitivity of the double coated film then becomes almost twice that of the single coated film, but finally decreases slightly, when the ends of the electron paths lie outside the emulsion.

It is thus seen that the theoretical expectation has been proved experimentally to be correct.

5.3b   Optical Density and Energy Absorption

R. Glocker (1960) examined the relationship between the energy of electrons absorbed and the optical density using Dudley's experiments, (Figure 5.7). From his calculations he concluded that the optical density was proportional to the energy of electrons absorbed within the range from 0.1–2 MeV. Glocker expected this to happen on theoretical grounds.

5.3c   Characteristic Curves of Electron Exposures

Detailed theoretical and experimental studies on the photographic effects of electrons by H. Frieser and E. Klein (1958), and H. Frieser,

E. Klein, and E. Zeitler (1961) have shown that the equation

$$D = D_{max}(1 - e^{-ka})$$

(5.3)

where k is a constant and a is the number of incident electrons, holds for the characteristic curves produced by exposures to electrons, assuming one electron hit per developable grain. (See also equation (4.15) as given for X-ray exposures, p. 119). Hence the initial region of the density-exposure curve is linear, as it is with X-rays. The shape of the

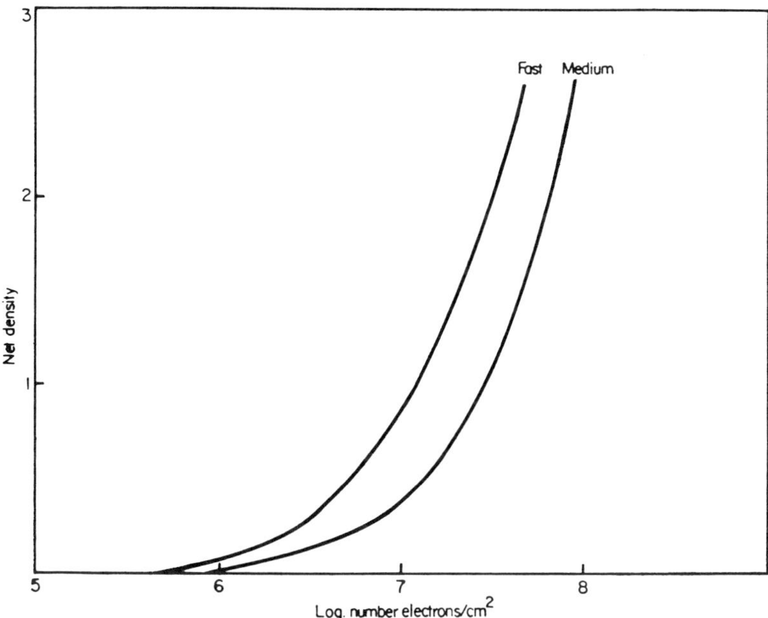

Figure 5.8. Characteristic curves of a fast and a medium fast no-screen X-ray film exposed to electrons from radioactive carbon (^{14}C). Max. energy: 0.155 MeV.

characteristic curve for the same type of single coated emulsion is independent of the energy of electrons applied, as long as the range of the electrons is not smaller than the emulsion thickness. For the reasons mentioned in the previous section, the shape of the characteristic curves of double coated X-ray films as a function of energy must depend on the penetration of the electrons through the two layers at various energies.

Typical characteristic curves of a fast and a medium fast X-ray film exposed to electrons from a carbon-14 source are shown in Figure 5.8. The electron emission from this source covers a wide range of energies having a maximum at 0.15 MeV. It is seen that the fast emulsion requires about 10^7 electrons to obtain a density of unity above fog. The relative sensitivity between different emulsions exposed to electrons was found to be identical with that found for X-ray exposures.

5.3d   Angular and Diffuse Incidence of Electrons

Measurements on the photographic sensitivity as a function of the angle of incidence of monoenergetic electrons have been carried out by R. A. Dudley (1956). In Figure 5.9 the ratio of the sensitivity at various angles of incidence to that at vertical incidence (S_{angle}/S_0) is plotted

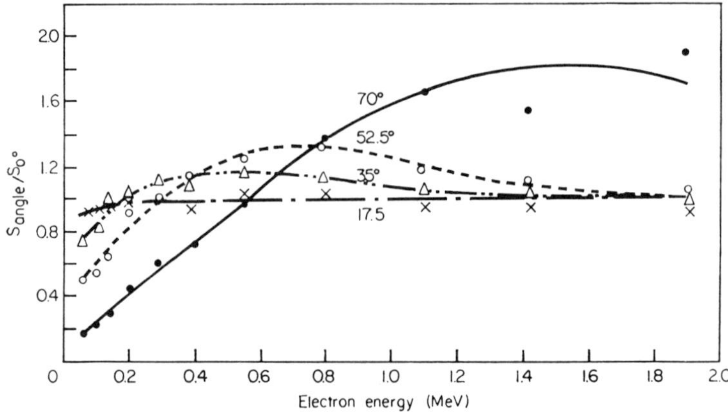

Figure 5.9. Dosage sensitivity of single Eastman Kodak no-screen emulsion as a function of monoenergetic electron energy for various angles of incidence relative to that for perpendicular incidence (0°). (*After* R. A. Dudley, Photographic film dosimetry in G. J. Hine and G. L. Brownell, *Radiation Dosimetry* Academic Press, New York (1956).)

against the energy in MeV, using a no-screen X-ray film. According to Dudley's interpretation, at low energies the path lengths of electrons are chiefly confined to the emulsion at all angles of incidence, hence the sensitivity varies approximately proportionally to the cosine of the angle. At medium energies the sensitivity increases with increasing angle of incidence, but at high energies, when absorption and scatter are less influential, the sensitivity becomes less dependent on the angle of incidence.

In most practical cases electrons are so much scattered within the emulsion particularly at the lower energies that one deals with a diffuse incident beam rather than with a parallel one. Using diffused incidence the sensitivity was computed by Dudley by appropriate weighting of the results shown in Figure 5.9. The results for diffuse incidence are illustrated in Figure 5.7 by the full line curve.

5.3e   The Effects of Absorbers on Continuous β-ray Spectra

Much of the work done with electrons is carried out with radioisotopes and photographic recording is frequently concerned with autoradiographic and dosimetric applications. β-radiation emitted by radioisotopes covers a wide spectrum of energies and it is of importance to realize that the shape of the spectral emission curve of a radioisotope is influenced relatively little by the use of absorbing filters, unless this is

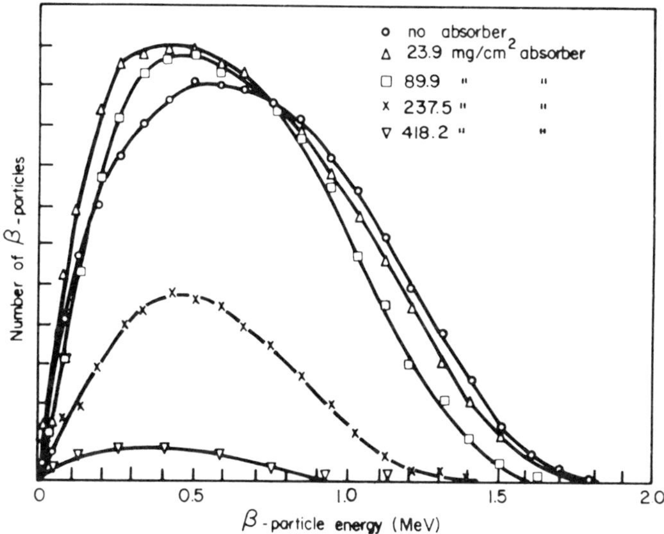

Figure 5.10. Effect of the Absorption by Perspex in mg/cm² on the energy spectrum of perpendicularly incident β-rays of radioactive phosphorus (^{32}P). (*After* G. L. Brownell in G. J. Hine and G. L. Brownell (Eds.), *Radiation Dosimetry*, Academic Press, New York (1956).)

done to the extreme. This has been clearly shown by G. L. Brownell (1952), using ^{32}P β-rays and Perspex (Lucite) absorbers of various thicknesses expressed in mg per cm² (see Figure 5.10). On the ordinate of this graph is plotted the number of β-particles and on the abscissa axis the β-energy and the curves refer to different absorbers in mg/cm². It is seen from this figure that the maximum energy of 1.7 MeV electrons is reduced only by a small extent with increasing absorber thickness. The fact that the spectral energy distribution does not alter materially with increasing filtration facilitates the interpretation of photographic recordings, as long as the same isotope is in use.

Unless exposures can be carried out in the dark, i.e. without absorber between source and photographic layer, the attenuation by wrapping materials may be considerable. It is found that the half thickness in terms of mg/cm² for continuous β-ray spectra emitted by radioactive sulfur-35 (maximum energy 0.17 MeV) and that of thallium-204 (maximum energy 0.78 MeV) are 3 and 29 mg/cm² respectively. (R. A. Dudley, 1954).

5.3f   Reciprocity Rule for Electrons

As is to be expected with electrons exhibiting a one grain hit developability, the reciprocity rule is valid. (N. Colombie, 1959; R. C. Ray and G. W. W. Stevens, 1953). This validity appears to be dependent on the same parameters as in X-ray exposures. If the intrinsic grain sensitivity or the transfer of energy is too low, or if both effects are present together, then reciprocity failure is to be expected. Failure has been observed by several authors. (N. G. Shushkin and I. A. Kovner, 1948; I. A. Fomina, 1959 and others.)

5.3g   Low Energy Electron Exposures

Most X-ray films are provided with a supercoat, for instance of gelatin of the order of $\frac{1}{2}$–2 microns thickness. Using electrons of very low energy and thus of correspondingly low penetrating power these are either not able to pass through the supercoat layer, or suffer such a considerable absorption that their photographic effect is very much reduced. This has been experienced particularly in exposures using tritium (³H) as an electron emitter. The maximum energy of ³H electrons is 18 keV and the mean energy is about 6 keV. The range of these electrons is on the average about 0.9 microns in silverbromide, with the result that practically only one grain layer is affected by ³H electrons. The remaining layers may respond to the higher energy electrons and to X-ray Bremsstrahlung produced by the bombardment of electrons. Because of this extremely short range tritium electron exposures are preferably carried out with 'non supercoated' fast X-ray films, which permit a speed increase to about 10 times that obtained with the same type of supercoated X-ray films. Unfortunately such emulsions are not commercially available because of their restricted use and because of the abrasion hazard connected with them. Non supercoated emulsions of low gelatin content in which the grains are practically sticking out of the gelatin on the surface of the emulsion layer are available from a few manufacturers. These are, for instance, the Kodak-Pathé SC-5 and SC-7 plates, the Ilford Q plates, the

Eastman Kodak SWR plates and the Eastman Kodak Special Film Type 101–01. Although these plates are strongly subject to abrasion hazards and very delicate to handle, they are extremely sensitive to slow electrons. Exposure times of the order of 1/40th that used for fast double coated X-ray films (with supercoat) have been achieved with ³H electrons by the author with some of the plates mentioned above. (H. Frieser and E. Ranz, 1962.)

It is possible to record electrons of very low energies down to about 4–5 eV photographically according to V. V. Kryachko (1960).

5.4 EXPOSURE TO NEUTRONS

General. Neutrons, being uncharged particles, do not cause ionizations directly but may do so indirectly by imparting their kinetic energy to nuclei in collision processes, or by releasing ionizing radiations when undergoing nuclear reactions. Hence photographic effects arise from elastic and inelastic collisions. In elastic collisions with atomic nuclei of the photographic emulsion it is mainly the lightest nuclei, i.e. those of hydrogen of the gelatin or base, to which neutrons of higher energy impart their kinetic energy, wholly or partly, depending on the angle of impact. In inelastic collisions neutrons may be captured, i.e. taken up by nuclei, for instance of atoms or ions composing the emulsion in which case nuclear reactions will occur. Essentially thermal neutrons are subject to nuclear capture events.

Thermal Neutrons. The direct effect of thermal neutrons on photographic emulsions is relatively small and inefficient, as will be shown quantitatively in the next section. It is chiefly caused by the capture of some of the neutrons in silver, nitrogen and bromide nuclei of an emulsion. This effect can however be intensified by means of so called converter screens containing certain natural isotopes having high cross-sections for thermal neutrons. On the impact of thermal neutrons the screens which are placed in contact with an emulsion layer emit α-, β- or (and) γ-rays, and thus expose the emulsion. Similar methods can be used with low energy epithermal neutrons.

Fast Neutrons. These also have little direct effect on photographic emulsions, as will be shown in the next section, but their action can be recorded in nuclear particle type plates in which elastic collisions occur between fast neutrons and hydrogen nuclei of the emulsion and its surrounding, thus producing knock-on protons. These leave rows of developed grains, i.e. proton tracks in the emulsion as mentioned on p. 161.

5.5   THE DIRECT AND INEFFICIENT RESPONSE
TO THERMAL AND FAST NEUTRONS

Before referring to conventional methods of neutron recording, as have been indicated in the preceding section, it is interesting to see how X-ray emulsions respond to slow and fast neutrons, when neither screens are used nor tracks can be detected. M. Ehrlich (1960) has studied the direct response of several commercially available X-ray films to 3 MeV monoenergetic and also to thermal neutrons, and has calculated the response to be expected for the atomic composition of these emulsions, allowing for their thickness and average grain size. Considering the uncertainties in such a theoretical approach and the fact that various simplified assumptions had to be made, the agreement between the experimental response and the theoretical results was found to be quite satisfactory for both fast and thermal neutrons. The number of fast and thermal neutrons required per cm² of emulsion, to obtain a density of 0.1 on three different types of film, is shown in the table below.

TABLE 5.1. Number of Fast and Thermal Neutrons per cm² Required to Obtain a Net Density of 0.1. (After M. Ehrlich)

Dupont X-ray film type	(Experimental data)	
	Number of fast neutrons/cm² (3 MeV monoenergetic)	Number of thermal neutrons/cm²
502	4.9×10^9	7.0×10^8
510	9.8×10^9	2.0×10^{10}
606	6.8×10^{10}	4.5×10^{10}

Considering the high flux of both types of neutrons required to obtain a density of 0.1, one may conclude that the sensitivity of these films is very low indeed. This becomes more evident if one compares the above data with the average number of electrons required, namely 10^7, to obtain a density of unity for a similar type of emulsion.

The photographic response to thermal neutrons is mainly due to the capture of these in silver and bromine nuclei. The nuclear activation products and their emission which is responsible for the photographic effect, will now be discussed in more detail. In Table 5.2 p. 173, the natural isotopes of silver and bromine are given on the left of the table together with their relative abundances. The next column shows the cross-sections in barns. The 4th and 5th columns contain the activation

TABLE 5.2. The Activation Products of Ag and Br after Thermal Neutron Capture

Natural isotopes	% abundance	Cross-section (barns)	Activation products	Half lives	Principal emission energies (MeV)	
					γ	β
^{107}Ag	51.4	45	^{108}Ag	2.3 min	0.62	1.77
^{109}Ag	48.6	113	^{110}Ag	24 sec	0.66	2.87
		3.2	^{110m}Ag	253 days	0.89	0.085
					0.66	
^{79}Br	50.57	2.9	^{80m}Br	4.6 hrs.	0.049	—
		8.5	^{80}Br	18 min.	0.6	2.0
						1.38
^{81}Br	49.43	3.3	^{82}Br	35.9 hrs.	0.7	0.46
					0.55	

products with their corresponding half-lives and the 6th and 7th columns the type of radiation and their maximum energies.

In Table 5.2 the activation products ^{110}Ag and ^{80}Br are quoted as having two different half-lives. This is because of the occurrence of isomers. An isomer is one of several nuclides having the same number of neutrons and protons but is capable of existing for a measurable time in different quantum states with different energies and radioactive properties.

The activation products are β- and γ-emitters of complex decay schemes with half lives from a few seconds to several hundred days. The emission consists of radiations of low and high energies, as seen in the last column of the table. Apart from the thermal neutron capture reactions mentioned, there are also (n_{th}, p) reactions taking place in carbon and in nitrogen. The latter ^{14}N(n_{th}, p) ^{14}C has an even lower nuclear cross-section than those mentioned in the table and has not been considered in Ehrlich's experiments. Thus the low efficiency of emulsions for thermal neutrons is caused by the low nuclear cross-sections of silver and bromine.

5.6 CORRELATION BETWEEN DIRECT PHOTOGRAPHIC EFFECTS OF NEUTRONS AND PRACTICAL REQUIREMENTS

Thermal Neutrons. Photographic records with thermal neutrons are of practical interest in dosimetry and in neutron radiography. Considering the former, it may be useful at this stage to assess the direct photographic effect of thermal neutrons, for instance, in relation to the maximum permissible dose rate for thermal neutrons of 0.1 rem per week. In Table 6.2 page 239 we find that this dose corresponds to 9.6×10^7 neutrons per cm^2 per week. On the other hand we found in Table 5.1 that about 7×10^8 thermal neutrons per cm^2 were required to obtain a density of only 0.1 in fast X-ray emulsions. Hence we may conclude that a direct thermal neutron exposure is of the order of 10–100 times too low for dosimetric purposes.

Fast Neutrons.   Similarly it has been shown in Table 5.1 that the photographic response to fast neutrons is equally inefficient. As mentioned before it is mainly due to elastic interactions with hydrogen and to a smaller extent with carbon of the emulsion and base. The low direct sensitivity of emulsions to fast neutrons becomes apparent by the following consideration. E. Tochilin, B. W. Shumway and G. D. Kohler (1956) have calculated the optical density attainable for a dose of 0.1 rem of fast neutrons in an X-ray film, assuming it were only made up of proton tracks and they found that this density is only 1/10th–1/100th of that required for a dose of 0.1 rad of cobalt-60 γ-rays.

It may seem surprising at first that the proton track sensitivity is so low compared with that of γ-rays of cobalt-60. This is particularly so if one considers that the absorption cross-section for fast neutrons and γ-rays in silver bromide is of the same order of magnitude. M. Ehrlich (1963) argues that a 1 MeV proton suffers about 100 times the energy loss per unit length of path than a 1 MeV electron released by γ-rays (see Figure 5.1). Since the developability of a grain requires only the energy of that supplied by a fast electron, the high specific ionizing power of the proton is rather wasted. Hence the efficiency of producing developable latent images is correspondingly lower for the proton than for the electron.

5.7   THE EFFICIENT DETECTION OF THERMAL NEUTRONS

As already indicated, an efficient response to thermal neutrons can be achieved by the use of so called converter screens having high thermal neutron cross-sections. These act as intensifiers when applied in contact with X-ray and other film materials. The screens convert the absorbed neutron energy into radiations to which the photographic material is sensitive. The choice of the material for the converters depends (a) on their respective cross-sections for thermal neutrons, (b) on the type and quality of radiation during and after activation and (c) on the half life of the decaying isotope. The radiation emitted may be α-, β- or (and) γ-rays. Details of the use of these isotopes in dosimetry and in neutron radiography are given in the specialized chapters on these subjects (Chapters 6 and 11). It may be mentioned that whilst the converters are used in conjunction with the film during thermal neutron exposure, in some cases of neutron radiography it is advantageous to bring them in contact with the film after activation (see p. 490).

Some of the most efficient convertor foils recommended in practice contain isotopes such as lithium-6, boron-10, rhodium-103, cadmium-113, indium-115, gadolinium-155 and 157, gold-197 and others. For

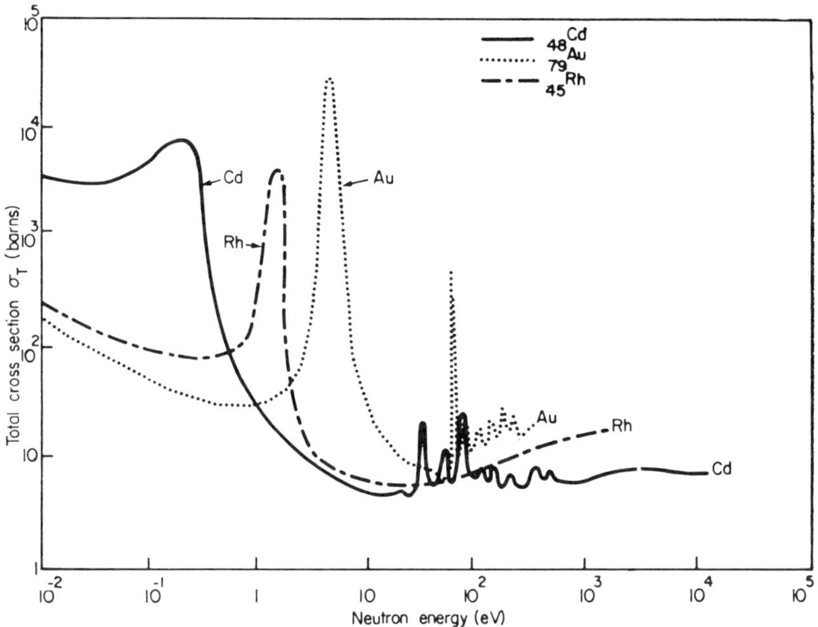

Figure 5.11. Total neutron cross-section σ_T (in barns) as a function of neutron energy (eV) for cadmium, rhodium, and gold.

epithermal neutrons tungsten-186, lanthanum-139, maganese-55 have also been found useful. In Table 11.2 details of the percentage abundance of some of the isotopes involved, nuclear cross-sections and respective half-lives are quoted, and the optimum foil thicknesses recommended are found in Table 11.3, page 492.

5.7a Energy Dependence of Some Neutron Cross-Sections

The energy dependence of the probability of neutron capture for various converter materials is shown in Figure 5.11. This figure illustrates the relationship between the total neutron cross-section (σ_T) for various elements and neutron energy, i.e. from slow to fast neutrons.

It is seen that high cross-sections apply only to a very limited range of neutron energies and that the cross-section in general becomes extremely low in the region of medium energy neutrons.

5.7b.   Loading of Emulsions

A further increase in sensitivity of emulsions exposed to thermal neutrons can be obtained by loading the emulsions with lithium or boron compounds (E. W. Titterton, 1949a, 1949b, and E. W. Titterton and M. E. Hall, 1950). These authors made use of the following reactions

and
$$^{10}\mathrm{B}(n_{\mathrm{th}},\ \alpha)^{7}\mathrm{Li}$$

$$^{6}\mathrm{Li}(n_{\mathrm{th}},\ \alpha)^{3}\mathrm{H}$$

in Ilford C2 Nuclear Research plates containing 16 mg/cc lithium and 23 mg/cc boron respectively in which particle tracks are produced. They then compared the effects obtained on these plates with an unloaded Ilford C2 plate. In the Ilford C2 plate used in their tests 4.3 per cent of the atoms were nitrogen. The relative yields obtained are given in the Table 5.3 below. The $^{14}\mathrm{N}(n_{\mathrm{th}},\ p)^{14}\mathrm{C}$ reaction mentioned

TABLE 5.3.   Relative Track Densities and Track Lengths in Loaded Emulsions

Reaction	Loading of element in grams/cc	Relative track density	Track length in microns
$^{14}\mathrm{N}\,(n_{\mathrm{th}},\ p)^{14}\mathrm{C}$	0.067	1	6
$^{6}\mathrm{Li}\,(n_{\mathrm{th}},\ \alpha)^{3}\mathrm{H}$	0.016	18.4	42
$^{10}\mathrm{B}\,(n_{\mathrm{th}},\ \alpha)^{7}\mathrm{Li}$	0.023	183	8

first in the table above is, as has been shown before, only one of the reactions occurring in the direct exposure to neutrons, and the loading 0.067 grams/cc refers to the nitrogen content of the emulsion.

The lengths of tracks using the lithium reaction are considerably greater than those occurring in the other reactions and are about 7 microns for the α-track and 35 microns for the triton ($^{3}\mathrm{H}$) track, i.e. a total of 42 microns. (The triton is the nucleus of tritium ($^{3}\mathrm{H}$).) By using the pure isotopes instead of the natural ones, the yields can be increased still further, as the relative abundances of $^{6}\mathrm{Li}$ and $^{10}\mathrm{B}$ in natural lithium and boron are only 7.5 and 18.5 per cent respectively. Hence the track density should be greater by a factor of 13 for pure $^{6}\mathrm{Li}$ and of 5.5 for pure $^{10}\mathrm{B}$.

5.7c   Characteristic Curves for Thermal Neutron Exposures

The shape of the characteristic curves for thermal neutron exposure naturally depends on the type of converter used, i.e. on the type of radiation emitted by a particular isotope, and not on the neutrons themselves. The directional dependence of the response can be practically ignored, as the emission of α-, β-, and γ-rays from converters is isotropic, i.e. diffuse radiation.

5.8 THE RESPONSE TO FAST NEUTRONS (PROTON TRACKS)

We have seen that the probability for thermal neutron capture is very high in certain isotopes and that the resulting nuclear reactions lead to the emission of radiations to which photographic emulsions are very sensitive. Unfortunately such favorable conditions cannot be attained with fast neutrons, as the absorption cross-sections for fast neutrons in most elements are fairly low. Thus we are left with the possibility of utilizing elastic collisions in which the kinetic energy of fast neutrons is imparted to light nuclei, such as those of hydrogen, present in the gelatin, in the base and in the paper wrapping of photographic films. The proton tracks thus produced can be observed by microscopic inspection.

5.8a Neutron-Proton Collisions

From the collision laws of classical mechanics we conclude that the transfer of energy in a head-on collision from one colliding body to another is greater the closer their respective mass values. If we denote the mass of the nucleus by M, that of the neutron by m_N, the energy imparted to the nucleus by E and the kinetic energy of the neutron by E_n, we can write

$$E = \frac{4 m_N M}{(m_N + M)^2} E_n \tag{5.4}$$

Using this equation, in the following Table 5.4 calculated values are given which show the percentage of neutron energies imparted to

TABLE 5.4. Percentage Energy Transferred from Neutron to Particles in Head-on Collision

	Electron	Proton	Carbon nucleus
Rel. mass	1/1850	1	12
Max. energy transfer	0.2%	100%	28.4%

nuclei, if a neutron has a head on collision with an electron, a proton or a carbon nucleus.

It is seen that the head-on collision between particles such as the neutron and the proton, having practically equal masses, is 100 per cent, but even in the case of the carbon nucleus the energy transfer is appreciable. It should be noted however that this calculation is only strictly valid for free and not for bound atoms, in which case a corresponding reduction would have to be made. The chance of a head-on collision is, however, small and we have to consider glancing collisions in which the part of the energy transferred to hydrogen nuclei depends on the angle formed between the direction of the neutron and the hydrogen nucleus. This effect is governed by the relation

$$E_H = E_n \cos^2 \theta \qquad (5.5)$$

where E_H is the energy of the hydrogen nucleus, θ is the angle formed between the direction of the knock-on proton and that of the motion of the incident neutron, the energy of which is given by E_n. A proton projected at a small angle, for instance of 5°, receives almost the full energy of the neutron, as $\cos^2 5° = 0.992$. In the majority of cases, a neutron will not lose all its energy at once but will undergo several collisions until its energy is reduced to that of thermal neutrons. (see p. 35).

5.8b The Proton Track Method

Only a brief review of the proton particle method will be given here, as the reader will find a more detailed discussion on practical measurements for dosimetric purposes in Chapter 6 pages 236-247 and 250.

The number of proton tracks produced in a nuclear particle emulsion is proportional (a) to the number of hydrogen atoms per unit volume of emulsion, (b) to the number of neutrons passing through unit volume of emulsion, and (c) to the scatter cross-section (σ) for a given energy of fast neutrons. Because of item (a), the photographic efficiency of fast neutrons can be increased by surrounding a photographic layer by hydrogenous material, such as polythene or other plastic material, from which protons can be ejected. (J. E. Cook, 1958; R. L. Lehman, 1960). The number of protons per neutron incident on the emulsion decreases with increasing energy of neutrons from 0.25–10 MeV to about 1/12th (see Table 6.3 p. 240). Proton tracks equivalent to 0.25 MeV consist of about 4 grains only, hence below this energy the counting of the number of tracks present in a microscopic field becomes uncertain. The range of protons in nuclear particle emulsions is shown in Figure 14.4 and it is seen that a 10 MeV proton has a range far beyond the normal thickness of a nuclear particle emulsion, unless this is manufactured for specialized cosmic ray work, where emulsions of up to 600–1000 microns thick and also stacked emulsions are used. The average thickness of nuclear particle emulsions used in photographic dosimetry is of the order of 30–50 microns.

5.8c Proton Track Counting and Optical Density Reading

It is of particular interest to realize that the counting of the number of proton tracks per unit area in a nuclear particle emulsion, although much less convenient than the reading of optical densities for instance in a no-screen X-ray film is a much more sensitive method. In order to appreciate this statement the following comparison will be useful. It is found that using 1 MeV neutrons, a dose of 0.1 rem (equivalent to 2.6×10^6 neutrons per cm²) produces about 10^3 proton tracks per cm² in a 50 μ thick nuclear emulsion. This can easily be counted under the microscope by scanning a fraction of this area. In terms of optical density in a fast no-screen X-ray emulsion this would correspond to no more than a density of 0.001 which can hardly be measured with an average densitometer.

5.9 THE RESPONSE TO HIGH ENERGY PROTONS

The photographic response to exposure of protons has been studied recently by K. E. Huff and H. M. Cleare of the Kodak Research Laboratories in cooperation with R. R. Adams of the National Aeronautics and Space Administration, Langley Research Center U.S.A. Although this work will be published in detail in the near future, the writer is privileged to have the permission of the authors to describe some of their main results obtained primarily during the Summer of 1967.

These investigations which are of particular interest in studies of space flight refer to exposures to protons having energies of 10, 17.6, 50, 90, and 130 MeV. The exposures have been carried out on the 160 MeV synchrocyclotron of Harvard University and the 22 MeV cyclotron at Oak Ridge National Laboratory. This study has been undertaken partly with the aim of comparing the response of several different types of film, which are quoted in the Table 5.5, together with the developer, development time and temperature used in the processing of the various types of emulsion.

TABLE 5.5. Photographic Emulsions Used in Proton Exposures

Type of photographic material	Developer	Development time	Temperature degrees F
Kodak Special Solar Recording Film, Type SO-375	D19	8 min.	68
Kodak Panatomic-X Aerial Film (Estar Thin Base) Type 3400	D19	8 min.	68
Kodak Spectroscopic Film Type 103-0	D19	4 min.	68
Kodak Spectroscopic Film Type 103-0, U.V.	D19	4 min.	68
Kodak Plus-X Aerial Film (Estar Thin Base) Type 3401	D19	8 min.	68
Kodak No-Screen Medical X-ray Film (one emulsion coating only)	Rapid X-ray	5 min.	68
Kodak High Speed Pan-chromatic Film (Estar Gray Base) Type SO-166	D19	12 min.	75
Kodak Special High Definition Aerial Film (Gray Base) Type SO-243	D19	8 min.	68
Kodak SWR Film	D19 (1:1)	2 min.	68

In Figures 5.12 and 5.13 the spectral sensitivity is shown in two ways

(a) by plotting the dose sensitivity, i.e. the reciprocal of the number of $rads_{air}$ necessary to produce a net density of 0.5 and

(b) by plotting the fluence sensitivity, i.e. the reciprocal of the number of protons per cm^2 necessary to produce a net density of 0.5.

The net densities refer to gross density minus base plus fog density. In order to compare the proton sensitivity with that obtained with ^{60}Co γ-rays, the corresponding values for the latter are given at the right of Figure 5.12.

It should be noted that the data shown in these graphs should be considered only as representative of the film types and processing conditions used in the experiments. Further, it is emphasized that SWR Film is extremely sensitive to pressure and abrasion. Hence the accuracy of the data for this film is not as good as that for the other films investigated.

Apart from the relative sensitivities of the various types of film investigated, which can be read off from Figures 5.12–13, the following conclusions can be drawn.

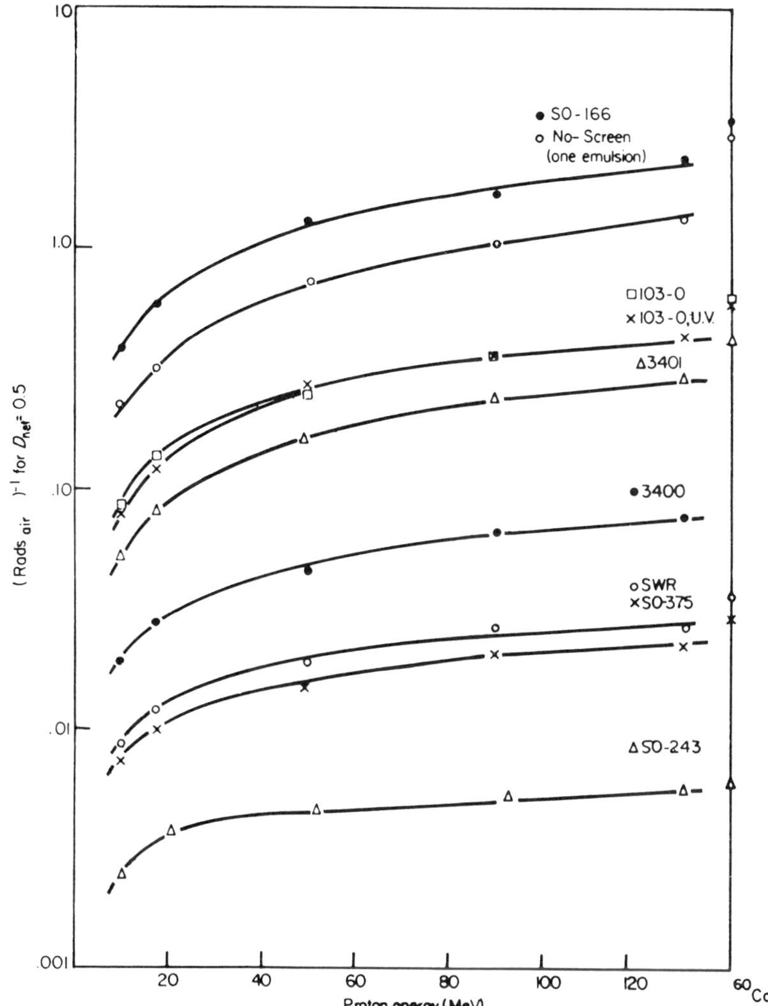

Figure 5.12. Dose sensitivity $(\text{rads}_{\text{air}})^{-1}$ for Density $= 0.5$ as a function of proton energy in MeV. (*After* K. E. Huff and H. M. Cleare.)

1. The dose sensitivity increases for each film with increasing energy, i.e. a smaller number of rads_{air} is required with increasing energy to produce the same density.

2. The relative sensitivity in terms of rads_{air} of the various films to ^{60}Co γ-rays is only slightly but consistently higher than that to protons of 130 MeV.

3. In the energy range from 10 to 130 MeV, the number of protons

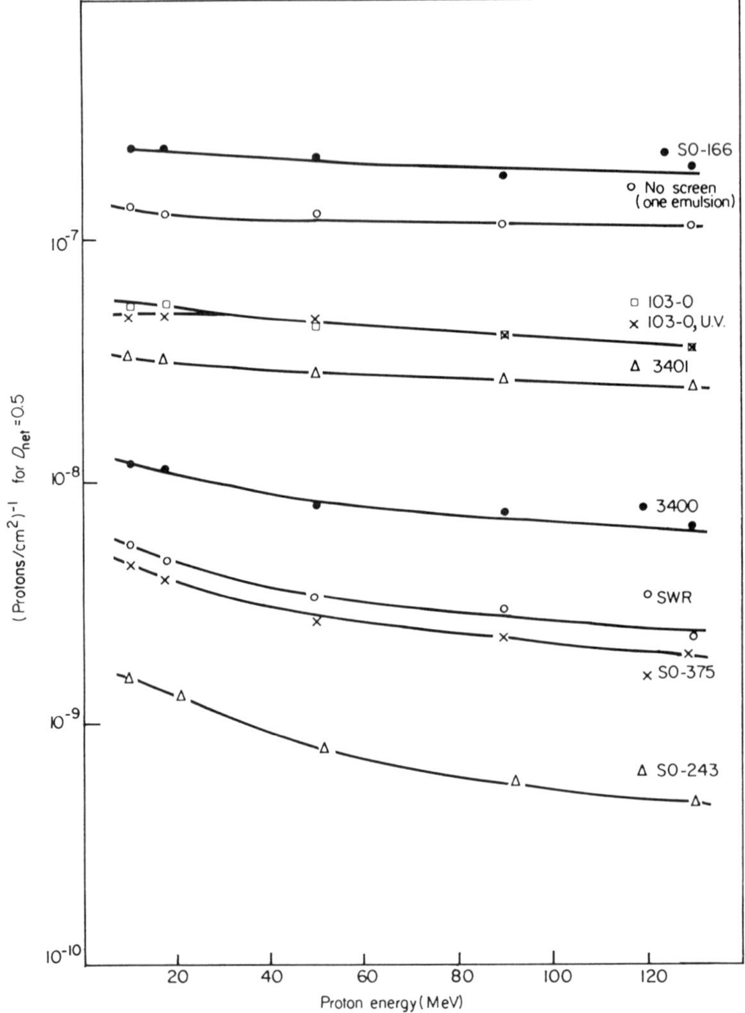

Figure 5.13. Fluence sensitivity (protons per cm²)⁻¹ for Density$_{net}$ = 0.5 as a function of proton energy in MeV. (*After* K. E. Huff and H. M. Cleare.)

per cm² required to produce a net density of 0.5 increases with increasing proton energy.

4. Films of highest fluence sensitivity to protons are somewhat less dependent on proton energy.

It was found that the effect of a uniform 'fogging' exposure to 50 and 130 MeV protons and ^{60}Co γ-rays on the response of these films to light is dependent only on the density produced by the 'fogging' exposure regardless of which of the three radiations was used. The

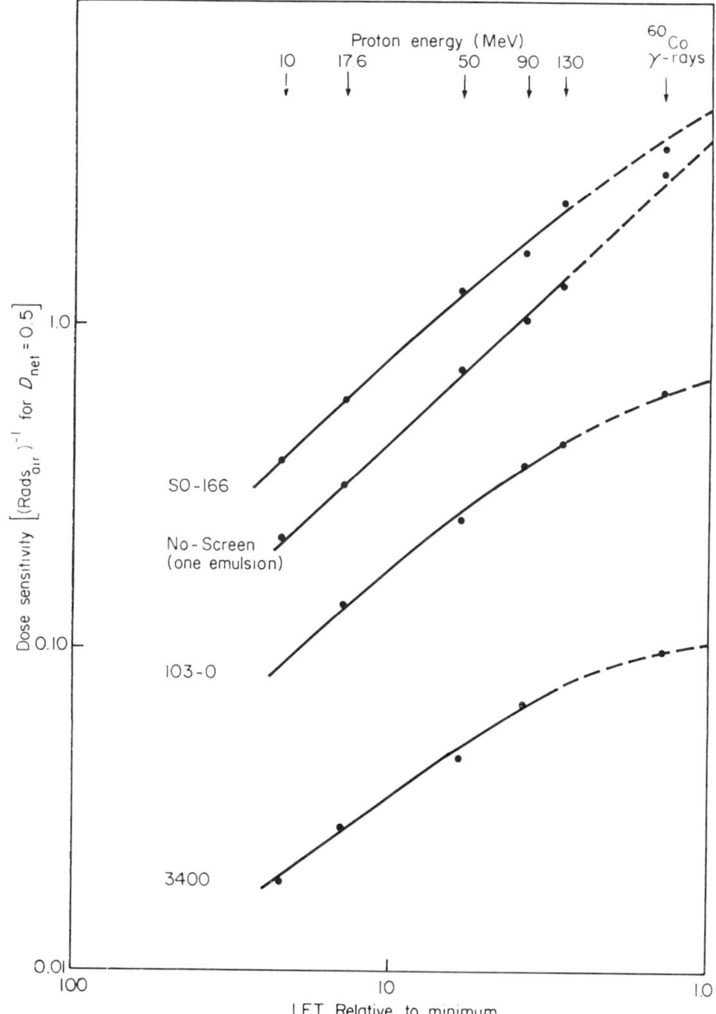

Figure 5.14. (a) Dose sensitivity $(\mathrm{rads_{air}})^{-1}$ for $\mathrm{Density_{net}} = 0.5$ as a function of LET relative to minimum. (*After* K. E. Huff and H. M. Cleare.)

response curve for a given uniform 'fogging' density produced by the ionizing radiations is not the same in every case as that obtained for a uniform light exposure.

In order to facilitate the understanding of the mechanism involved in proton exposures, dose sensitivities are plotted against proton LET (realtive to a minimum) in Figures 5.14a and b. (See the definition of LET on p. 31.)

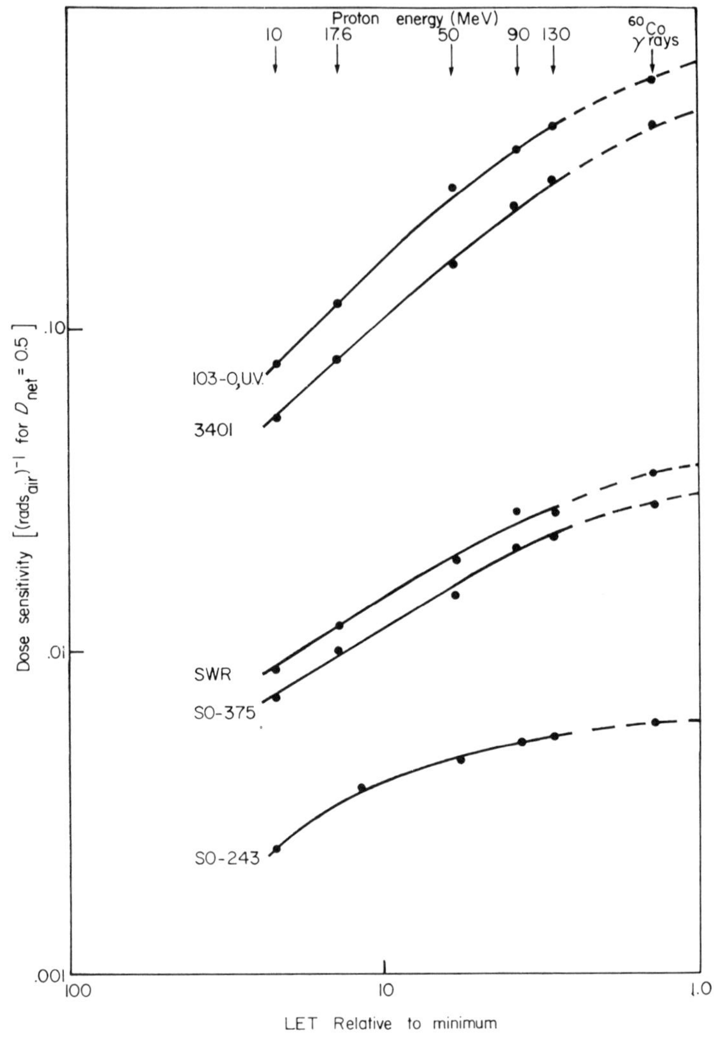

Figure 5.14 (b)

The dose sensitivity values for ^{60}Co are plotted at a relative LET (relative to minimum for a singly charged particle) value of 1.4. This is actually an effective value in water for the spectrum of Compton electrons under electron equilibrium conditions. (D. V. Cormack and H. E. Johns, 1952.) It is assumed that a photographic grain would have the same sensitivity to protons and electrons of equal LET. Because of this assumption and because there actually is a spectrum of Compton

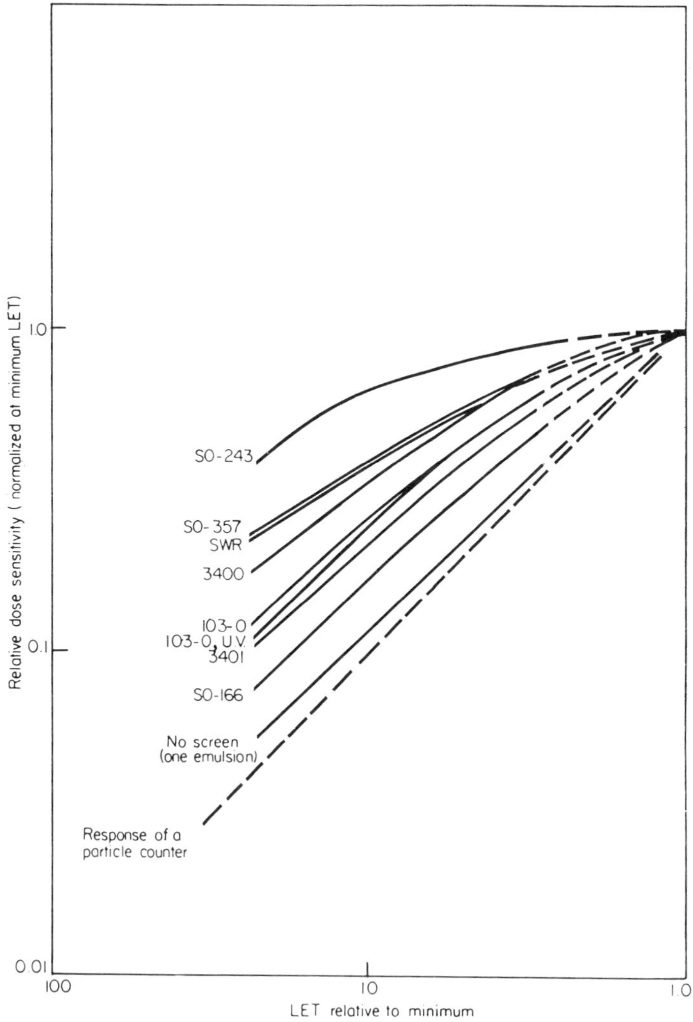

Figure 5.15. Relative dose sensitivity normalised at minimum LET relative to minimum. (*After* K. E. Huff and H. M. Cleare.)

electrons the sensitivity curves are shown as dashed lines in this LET region and must be considered approximate.

Since the mean LET of Compton electrons, resulting from ^{60}Co γ-rays, is equal to the LET of 450 MeV protons and since they are both singly charged particles, one may expect the same degree of film fogging for equal doses.

In Figure 5.15 the relative dose sensitivity is normalized at minimum

LET for the various films and the response of a particle counter is shown as well.

As mentioned earlier, the extrapolations of the sensitivity curves have shown that even for minimum ionization protons, film sensitivity is only slightly higher than for ^{60}Co γ-rays. Hence in applications where the expected proton dose cannot be predicted with great accuracy, film sensitivity to ^{60}Co γ-rays can be used to make conservative estimates of the degree of film fogging that will result from exposure to high energy protons.

REFERENCES

Barkas, W. H., (1963). *Nuclear Research Emulsions.* Academic Press, New York.

Becker, K., (1962). *Film Dosimetrie.* Springer-Verlag, Berlin-Göttingen-Heidelberg.

Brownell, G. L., (1952). Interaction of P-32 beta-rays with matter. *Nucleonics* **10**, 6, 30.

Coates, A. C., and Herz, R. H., (1949). Experiments with nuclear track emulsions sensitive at minimum ionizing power. *Phil. Mag.*, **XL7**, 1088.

Colombie, N., (1959). Étude de l'action des Électrons sur Differentes Émulsions Photographiques. *Sci. Ind. Phot.*, **2**, 30, 165.

Cook, J. E., (1958). Fast neutron dosimetry using nuclear emulsions A.E.R.E. HP/R2744.

Cormack, D. V., and Johns, H. E., (1952). Electron energies and ion densities in water irradiated with 200 keV, 1 MeV and 25 MeV radiations. *Brit. J. Radiol.* **25**, 369.

Demers, P., (1958). Ionography. *Les Presses Universitée Montreal*, Montreal.

Dudley, R. A., (1951a). The measurement of beta radiation dosage with photographic emulsions. Ph.D. Thesis, M.I.T.

Dudley, R. A., (1951b). Photographic detection and dosimetry of beta rays. *Nucleonics*, **9**, 3, 5.

Dudley, R. A., (1954). Photographic detection and dosimetry of beta-rays. *Nucleonics*, **12**, 5, 24.

Dudley, R. A., (1956). Photographic film dosimetry in G. J. Hine and G. L. Brownell, (Ed.), *Radiation Dosimetry*, Academic Press, New York.

Ehrlich, M., (1960). The sensitivity of photographic film to 3 MeV neutrons and thermal neutrons. *Health Physics* **4**, 113.

Ehrlich, M., (1963). Use of photographic film for personnel dosimetry. Basic physical considerations. *In personnel dosimetry techniques for external radiation. Symposium Madrid. Europ. Nucl. Energy Agency.* O.E.C.D.

Fomina, I. A., (1959). Zh. Nauchn. i. Prikl. *Fotogr. i. Kinematogr.* 494.

Frieser, H., and Ranz, E., (1962). Die Wirkung von Tritium auf Photographische Schichten. *Inst. Wiss. Phot.* T. H. München. Forschungsbericht 5.

Frieser, H., and Klein, E., (1958). Die Eigenschaften Photographischer Schichten bei Elektronen Bestrahlung. *Mitteilungen Forschungslaboratorien der Agfa. Leverkusen-München*, Vol. **II**, p. 121. Springer-Verlag. Berlin-Göttingen-Heidelberg.

Frieser, H., Klein, E., and Zeitler, E., (1961). Das Verhalten Photographischer Schichten bei Elektronen-Bestrahlung (II) *Mitteilungen Forschungslaborat. Agfa. Leverkusen München.* Vol. **III** p. 182, Springer-Verlag Berlin-Göttingen-Heidelberg.

Glocker, R., (1960). Das Photographische Schwärzungsgesetz für Elektronenstrahlen verschiedener Energie. *Z. Phys.*, **160**, 568.

Kryachko, V. V., (1960). Zh. Nauchn. i. Prikl. *Fotogr. i. Kinematogr.*, **5**, 34.

Lehman, R. L., (1960). Energy response and physical properties of NTA personnel neutron dosimeter nuclear track film. *5th Ann. Meetg. Health Phys. Soc. Boston.*

Powell, C. F., Fowler, P. H., and Perkins, D. H., (1959). *The Study of Elementary Particles by the Photographic Method.* Pergamon Press, New York.

Ray, R. C., and Stevens, G. W. W., (1953). Reciprocity failure and latent image fading in autoradiography. *Brit. J. Radiol.*, **26**, 307, 361.

Shushkin, N. G., and Kovner, I. A., (1948). Dokl. Akad. Nauk. *SSSR* **62**, 633.

Titterton, E. W., (1949a). Slow neutron health monitoring with nuclear emulsions. *A.E.R.E.*-G/R **36**, 2.

Titterton, E. W., (1949b). Slow neutron monitoring with boron and lithium loaded nuclear emulsions. *Nature* **163**, 990.

Titterton, E. W., and Hall, M. E., (1950). Neutron dose determination by the photographic plate method. *Brit. J. Radiol.* **23**, 465.

Tochilin, E., Shumway B. W., and Kohler G. D., (1956). Response of photographic emulsions to charged particles and neutrons. *Radiation-Research* **4**, 467.

Yamakawa, K. A., (1951) Silver-bromide crystal counters. *Phys. Rev.*, **82**, 522

Part Two

6

Photographic Dosimetry

6.1 INTRODUCTION

This chapter deals primarily with the technological aspects of film badge dosimetry and reference will be made frequently to the chapters on the direct response of photographic material to ionizing radiations, i.e. to photons and particles. This means that an understanding of some of the photographic principles involved in dosimetry will be assumed.

The rapidly growing use of ionizing radiation in medicine, industry and research demands a continuous watch on radiation hazards. Dosimetry in general deals with the quantitative and to some extent with the qualitative measurements of the radiation received by personnel engaged in radiation work.

Although there are many methods available which permit a survey to be made of the radiation dose to which persons have been subjected, the photographic method is probably the most widely used one.

The photographic dosimeter consists of a small X-ray film, usually of dental film size, which is housed in a container (the film holder or, colloquially, the film badge). This container is fitted with one or more absorbing filters. The film badge is worn on the clothing (e.g. breast pocket of personnel) throughout working hours. The evaluation of the dose received by the film is ultimately based on a calibration of the response of the film by means of an ionization dosimeter or other reliable instrument which has been calibrated against an ionization chamber. The main advantages and disadvantages of the photographic dosimeter if compared, for instance, with the pocket ionization dosimeter are:

Advantages

1. It provides a permanent record.
2. It is practically independent of dose-rate.
3. It may provide information on the quality of radiation.
4. It covers a much wider range of exposures than pocket dosimeters.
5. It is small and cheap to use.
6. It is particularly suitable for a large survey of persons engaged in radiation work.

Disadvantages

1. The method does not provide a direct reading.
2. It requires processing and interpretation.
3. It is dependent on the energy of radiation.
4. It is dependent on the angle of incidence of radiation.

The advantages have been regarded as being so great that the photographic method is now very widely applied throughout the world in hospitals, universities, atomic energy research establishments, nuclear power plants, biological research organizations and industry. From this it may be gathered that the film badge system is particularly useful in establishments where a great number of people are subjected to radiation hazards and where a central organization (film monitoring unit) can be set up to deal with the delicate processing procedures required for a quantitative evaluation of monitor films.

In recent years a considerable amount of research has been carried out on dosimeters based on thermoluminescent phenomena using, for instance, lithium fluoride or calcium fluoride as detectors. These dosimeters show high sensitivity to X- and γ-rays, electrons and can also be made sensitive to neutrons. They show an almost linear response with radiation dose over a wide range, but their response is partly wavelength dependent. Although thermoluminescent dosimeters are also suitable for large scale use, they do not supply the same kind of permanent documentary record as photographic materials. In spite of this such instruments may replace photographic dosimeters in some fields of work in the future.

6.2 DOSIMETRY OF PHOTONS

6.2a. General Techniques

In the following section some of the developments towards achieving a satisfactory film badge system will be reviewed. Since the photographic effect is not only dependent on the exposure but also on the energy of the radiation (see p. 121), attempts have been made both to produce energy independent and alternatively energy dependent film badge systems. With the former the aim is to determine directly the correct dose whatever the quality of radiation applied. In the latter case the film badge provides information about the energy of the radiation and this is used in assessing the correct dose. Experience has shown that both attempts have certain limitations and it is difficult to separate one method completely from the other. Most modern film badge systems make use of both the energy independent and dependent methods simultaneously, with the emphasis on one or the other.

6.2b. Energy Independent Methods

(1) *Compensation by Filters*

In this section the energy dependence will be briefly expressed by the 'hardness factor' (German: Härte-Faktor) as first suggested by M. Dorneich and H. Schäfer (1942). This expression is derived from the fact that the sensitivity of an emulsion depends on the hardness of the radiation applied. The hardness factor (HF) is defined by the ratio of the sensitivities of a given film at about 40 keV to that at between 600 and 1000 keV. For most X-ray films the HF is of the order of 20–50, as discussed in Chapter 4, p. 123.

Probably the most obvious method of smoothing out the HF effect is to use suitable compensating metal filters. If the radiation penetrates first a filter and then the film, the softer or low energy radiations will be more readily absorbed by the filter than the hard or high energy radiations. On the other hand, the film is more sensitive to the soft radiation, so that one can expect that a suitable filter or combination of filters will be capable of compensating to some extent for the variation of response with energy.

Indeed, many workers E. Haschè (1939), L. A. Pardue, N. Goldstein and E. O. Wollan (1944), R. B. Wilsey (1951), M. J. Heard, J. E. Cook and P. D. Holt (1960) and M. J. Heard (1962) have shown that by using compensating filters, a nearly constant photographic density can be obtained from the same dose over a wide range of energies. The actual range of energies over which this is possible, and the limits of accuracy obtainable, depend not only on the choice of filter material and its thickness, but also on the type of film in use.

In Figure 6.1 the effect of a typical combination of compensating filters is illustrated. This graph shows the relative sensitivity of a radiation monitoring film against the radiation energy in keV. Curve

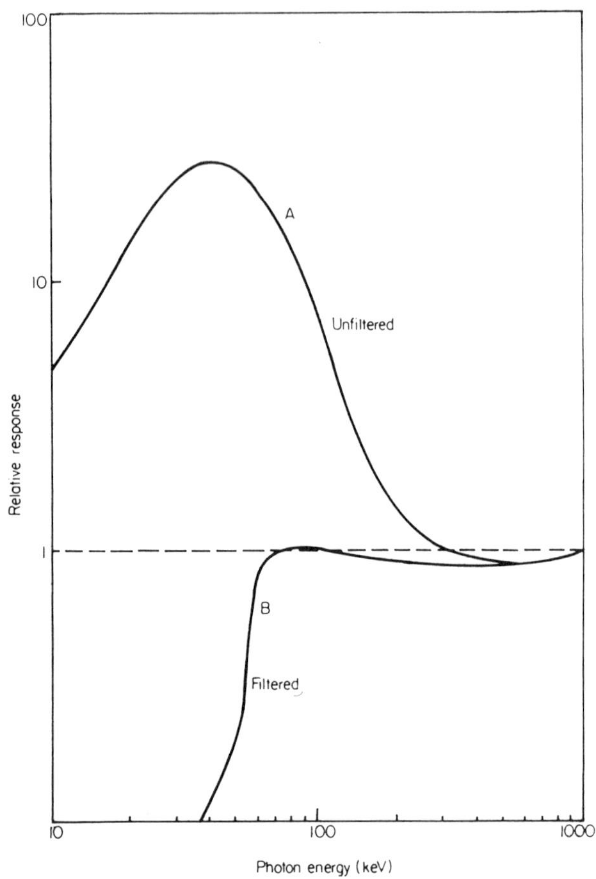

Figure 6.1. Reponse curves as a function of photon energy (keV). A, unfiltered; B, using a typical compensation filter. (*After* M. J. Heard. *J. Phot. Sci.* **13**.32 (1965).)

A shows the strong dependence on energy of the unfiltered film. Curve B is the result achieved by a combination of filters of tin and lead. From the graph it becomes evident that the response of the film below the filters is almost independent of energy over the range from 75 keV to 1 MeV. It should be remembered that the effective thickness of these filters increases with the angle of incident radiation. This effect is smaller for thinner filters and it is therefore an advantage to have as low a value as possible for the HF.

Another method of overcoming the effect presented by the HF is to use two separate films, one of high and the other of low sensitivity in conjunction with a combination of metal filters in the following order. Radiation—low sensitive film—filters—high sensitive film—lead filter. Here the radiation acts first on the slow film and is then more heavily filtered before it reaches the fast film. Thus by using a suitable ratio of the sensitivities of the two films and a suitable filter combination, the total response of the two films can be made to be almost independent of energy, as the unfiltered slow film will decrease in response with increasing energy, whereas the filtered fast film will increase in response, due to the decreasing absorption of the filter.

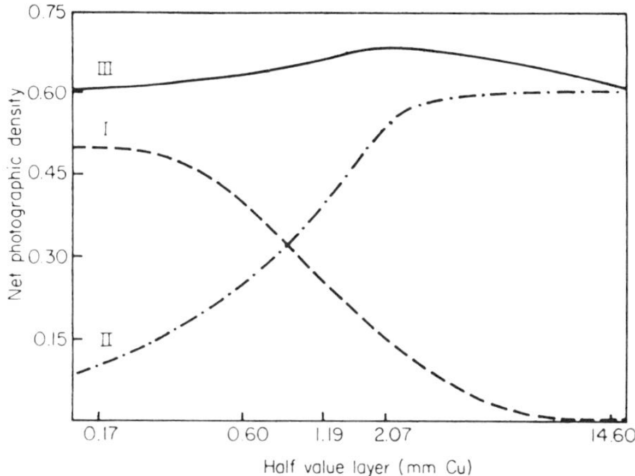

Figure 6.2. Compensation for energy dependence of film response. I. Slow X-ray film; II. Fast X-ray film; III. Sum of the densities of both films. (*After* M. Ehrlich, *The use of Film Badges for Personnel Monitoring*. Safety Series No. 8. International Atomic Energy Agency, Vienna (1962).)

The two effects thus give rise to an almost constant response with energy. The result of such a combination of films and filters, is illustrated in graph Figure 6.2.

Curve I reveals the change of density with half value layer thickness (mm Cu) for the slow film, curve II presents the result of the fast filtered film and curve III shows the summation of the densities of films I and II. It is seen that a fairly constant density over a wide range of radiation energies is achieved. This system was first suggested by A. Allisy (1955) and in a modified form by M. Ehrlich (1957). Several other systems of a similar kind have been proposed by L. H. M. van Stekelenburg (1958), H. Stadelmann (1960) and F. Wachsmann and H. Stadelmann (1961).

(2) *Compensation by Fluorescence*

Another method of reducing the effects of the HF is based on the use of the light emitted by a suitable fluorescent material when it is exposed to ionizing radiation. The response of a photographic material to light depends however on the rate of the light exposure and thus this method is influenced by reciprocity failure effects (see p. 128). This behavior of the photographic emulsion sets strict limitations on the use of light in dosimetry. An interesting example of the use of fluorescent substances was made by H. Hoerlin and his collaborators (1953 and 1956). These authors applied an organic scintillator of p-therphenyl (p,p'-diphenylbenzole), the fluorescence of which varies with the energy of radiation in the opposite direction to that of the photographic emulsion. Hence this scintillator is able to counteract the energy dependence of the film. In the Hoerlin system a special film sandwiched between two terphenyl intensifying screens is used in a holder. The selected film emulsion has a relatively low HF and suffers only slightly from reciprocity failure. It is claimed that this system is energy independent from 40–1000 keV within ± 15 per cent and reveals a failure of reciprocity of ± 33 per cent within an exposure time range from 1 second to 55 hours. The range of sensitivity of this system is however limited from about 20 mR to 20 R. (K. Becker, 1960a, 1960b, 1961a, 1961b, 1962).

(3) *Compensation by Emulsion Properties*

Considerable efforts have been devoted to reduce the HF in the photographic emulsion. Some authors have proposed and successfully experimented on the incorporation of terphenyl into the photographic emulsion. However, it appears that this type of film has not found application in routine use. (K. Becker, E. Klein and E. Zeitler, 1960; K. Becker, 1960a, 1960b; K. Becker, 1961.)

According to K. Becker (1962) who used a series of Agfa emulsions of different mean grain size, but comparable grain sensitivity, it could be shown that the HF decreases with increasing mean grain diameter as illustrated in Figure 6.3.

Any effect which has a tendency to increase the sensitivity to high energy radiation, and keep the sensitivity to low energy radiation constant, will reduce the HF. Thus H. Hoerlin and F. W. H. Mueller (1950) have shown that gold sensitization tends to reduce the HF by increasing the sensitivity to hard radiation. It is also of interest that D. Klein (1961) using dye desensitization on emulsion grains was able to increase the HF considerably.

According to a suggestion by J. R. Greening (1951) it should be possible to reduce the HF by diluting an emulsion, i.e. by increasing the ratio of gelatin to silver bromide content. Hoerlin and also the author have shown that although it is possible to do this, the dilution leads to a considerable reduction of sensitivity.

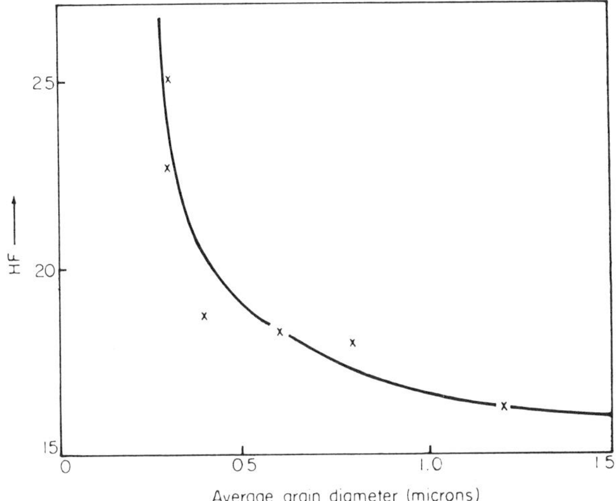

Figure 6.3. Hardness Factor as a function of average grain diameter (microns). (*After* K. Becker, Filmdosimetrie, Springer-Verlag, Berlin-Göttingen-Heidelberg 1962).

A further proposal by H. Hoerlin and his collaborators (H. Hoerlin, R. H. Clark, D. P. Jones, F. J. Kaszuba and E. T. Larson, 1953) is to reduce the HF by means of suitable developers. The authors suggested a developer which does not only act on the surface latent images but also fully develops the internal latent images (see p. 133). In this way they achieved a 30 per cent reduction of the HF.

Summarizing all the results of influencing the emulsion itself in order to reduce the HF, it must be said that none of the various attempts mentioned above is capable of producing even a nearly wavelength-independent emulsion.

6.2c   Energy Dependent Methods

It has been shown in the previous sections that all the methods of achieving energy independence suffer from certain limitations. Hence many film badge systems rely entirely on the energy dependent method, the basic principle of which will be described in this section.

(1) *Filter Analysis*

In this method several filters, either of the same material and different thicknesses, or of different materials and thicknesses, are used adjacently as absorbers in front of the photographic emulsion. The differential absorption by the filters and the ratios of the resulting optical densities below these can be utilized for the evaluation of the dose received, assuming suitable calibration has been carried out. As the relationship between the densities below two or more of the filters is a function of the quality of radiation, the method of filter analysis can take into account the energy of radiation applied. The difference between the densities (i.e. the objective contrast, see p. 112) below two filtered areas depends on the relative thicknesses of the type of filter material used, on the quality and degree of homogeneity of the radiation, and on the inherent contrast which is given by the slope of the characteristic curve of the emulsion in the region concerned. As the degree of homogeneity of radiation (see p. 28) has a very strong influence on the accuracy of the result, the basic response data for the dosimeter is usually obtained with heavily filtered radiation. To increase the accuracy of the determination of dose when using less homogeneous radiation, as is generally the case in practical monitoring it is advantageous to use as many different filter thicknesses as possible. On the other hand, this leads to greater complexity in the evaluation and obviously demands more room in a filter badge, which should necessarily be of small dimensions.

No general recommendations can be given for the choice of filter materials, but it is advisable to choose the atomic number and the thicknesses of the materials to suit the range of radiation quality and type of film for which the filters are to be used. This will be more easily appreciated in a practical case such as in the British film badge (see p. 201) where tin, lead, cadmium, aluminum and plastic are used as filters, whereas in the German film badge (see p. 215) only copper and cadmium are used. The former film badge is based on both energy independent and dependent methods, whereas the latter is entirely based on an energy dependent system.

A material of low atomic number (Z), e.g. aluminum is obviously unsuitable for the differentiation of high energies of radiation, whereas the absorption in lead ($Z = 82$) will increase too rapidly at low energies. Studies on the optimum choice of filters have been made from a theoretical and experimental point of view by many authors, such as E. Tochilin, R. Davis, and V. Clifford (1950), M. Ehrlich and S. Fitch (1951), H. Langendorff, G. Spiegler and F. Wachsmann (1952), E. D. Gupton (1956), M. J. Heard, J. E. Cook and P. D. Holt (1960).

(2) *Secondary Electron Emission*

A further attempt to develop an energy dependent method is based on the use of secondary electron emission from metal foils. (W. V.

Mayneord, 1930; C. W. Wilson, 1941; G. J. Hine, 1952, and G. J. Hine, 1954.) These authors have shown that the number and direction of secondary electrons emitted from metal foils depend on the atomic number of the foil material and on the quantum energy of photons used. This is illustrated in Figure 6.4 according to G. J. Hine (1954) in which the ratio of the number of forward directed electrons

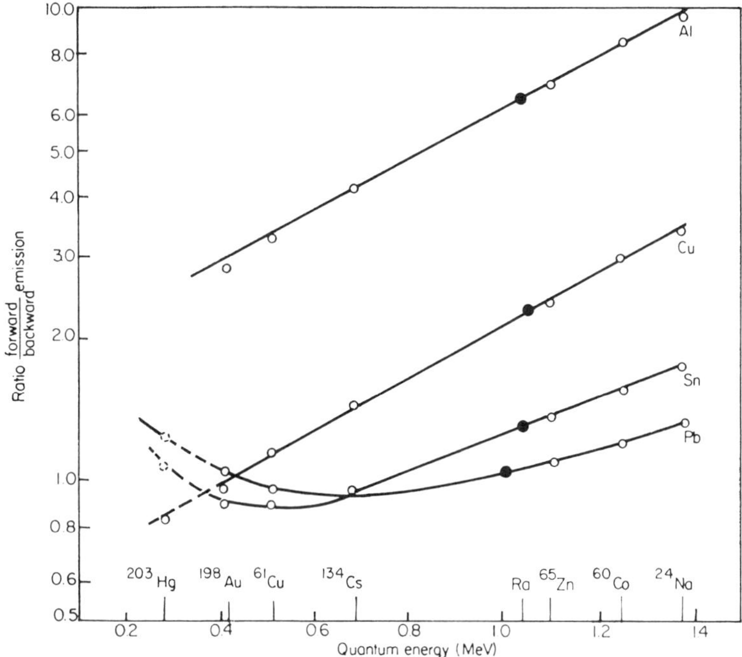

Figure 6.4. Ratio of forward to backward emission of secondary electrons for various absorbers as a function of quantum energy (MeV). The radiations have been generated by the radioisotopes mentioned above the abscissa axis. (*After* G. Hine, *Am. J. Roentgenol.* **72.**293 (1954).)

to the backward directed electrons is plotted against the quantum energy for various types of absorber. The thicknesses of the absorbers have been adjusted to the respective ranges of electrons in the absorbers. Using aluminum at 0.4 MeV about three times as many secondary electrons are emitted in the forward direction (i.e. in the direction of the perpendicularly incident photons) than in the backward direction, whereas this factor reaches a value of 10 at 1.4 MeV according to Figure 6.4. When using material of higher atomic number the direction of electron emission becomes less dependent on the various quantum energies involved. If a film is sandwiched between lead foils it is found that these emit about 50 per cent of electrons in the forward and 50 per cent in the backward direction over the range from 0.4 to 1 MeV.

This principle has been utilized by G. Spiegler (1950, 1951) and also later by B. W. Soole (1956) for the purpose of an energy dependent type of film badge. A simplified form of such a badge was also described by G. Spiegler and R. Davis (1959). Although this method appeared to be promising it has not found wide application in routine use, partly because it is difficult to apply to radiations of mixed type and energy.

6.3   FILM BADGES, MONITORING FILMS AND THE ASSESSMENT OF DOSE

The methods adopted for film dosimetry vary not only from one country to another but even within the same country, where different types of film badges may be used. In some cases the application of very simple badges may be justified, as when the hazard of radiation is limited to one quality of radiation only. As this is not always the case, more sophisticated designs of badge are used in order to cover a wide range of qualities and doses of photons, electrons, neutrons and mixed radiations. At present there is a noticeable tendency towards a standardization of badge design. One example of such a tendency is found in Great Britain, where until quite recently many different types of film badge were used. However, this situation has now been changed by the introduction of the RPS/AERE film badge, which will be called the 'British Film Badge' in the following sections of this chapter. This forms the basis of a British Standard (1963). The voluntary acceptance of one badge throughout practically the whole of Great Britain is mainly due to the combined efforts of two Government organizations (a) the Radiological Protection Service (RPS) and (b) the Atomic Energy Research Establishment (AERE) of the U.K. Atomic Energy Authority. The former covers a country-wide surveillance of film badges; this includes distribution, processing, dose estimation and communication of results to the wearers of film badges in hospitals, research institutions and industry. For a better appreciation of the various functions of a photographic film dosimeter a rather detailed description of one given type is preferable to a general treatment. For this reason most of the operations that can be carried out with the British Film Badge will be discussed in some detail. A considerable amount of research has gone into the development of the British Film Holder and as it is a versatile device and as it has proved its usefulness in practice, it is chosen for this more elaborate treatment in the following sections. The selection of the British Film Holder does not imply any special preference to other well established photographic dosimeters and short descriptions of film badge systems used in other countries will follow in further sections.

6.3a. The British Film Badge

The badge has the advantage that it allows a simple derivation of dose when applied to photon doses of one or more mixed energies.

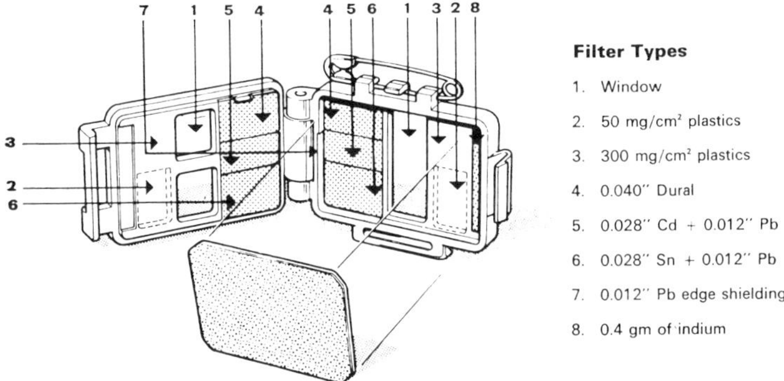

Figure 6.5. General design of the AERE/RPS film holder. (*After* M. J. Heard, A.E.R.E. Report-M 1370 and M. J. Heard and B. E. Jones, *Personnel dosimetry techniques for external radiation. Symposium Madrid* (1963). ENEA-OECD.)

Its versatility also permits the evaluation of mixed doses of photons and electrons of various energies and of doses arising from electrons and thermal neutrons. In the case of the simultaneous incidence of mixed energies of photons and electrons the evaluation becomes more complex. The description of the British Film Badge follows closely the report given by M. J. Heard (AERE) and B. E. Jones (RPS) at the Symposium on 'Personnel Dosimetry Techniques for External Radiation' held in Madrid (1963).

General Design

The badge consists of a holder* moulded in polypropylene and is in the form of a flat hinged box with a strong hinge on one short side and a fastening catch on the opposite side (see Figure 6.5).

---

* Supplied by Loxford Equipment Company Limited, Barking, Essex, England.

The holder is designed for use with a particular film (Kodak Radiation Monitoring Film; manufactured in England); the holder plus film making up the film dosimeter. The holder can, however, be used with other types of film provided that any necessary changes in filter thickness are made to suit the characteristics of the new film. The film holder has a standard blue color which indicates that the filters incorporated match the properties of Kodak R.M. film. Slight asymmetry in the position of the window and a bar across the rear window helps to ensure that the film is inserted with the correct orientation, because there is only one position in which the identification number is clearly visible. The word 'Back' is marked on the rear surface of the holder to indicate the correct position when in use. These steps are necessary because the film pack is not completely symmetrical. The film consists of a base coated on one side with a fast X-ray emulsion and on the other side with a slow emulsion. The fast emulsion faces the white i.e. the front side of the film packet which is of dental film pack dimensions. The badge incorporates six filters and an open window, as seen from Figure 6.5. The filters have the following purposes.

1. The *open window* together with the two plastic filters (see 2 and 3) permit the separation of the effects due to photons and β-radiation within certain energy limits. Part of the area of the film in the open window is utilized for the identification number of the person who wears the badge.

2. The *thin plastic* filter (0.020″ thick) permits practically the same photon response as the open window but attenuates β-radiation to an extent which depends on its energy.

3. The *thick plastic* filter (0.110″ thick) passes all but the lowest energy photons; on the other hand, it absorbs all but the highest energy β's.

4. The *Dural filter* (aluminum alloy spec. British Standard L72) is 0.040″ thick. For the higher photon energies (70–2000 keV) its absorption is low and comparable with that of the thick plastic filter. However, the response under the Dural filter begins to fall below that under the thick plastic filter at energies below 60–70 keV. This is approximately the energy at which response under the tin-lead filter (see next item) falls below 80 per cent of its 1 MeV response. The response under Dural remains measurable down to an energy of about 15 keV, below which this filter is opaque. Thus the Dural filter provides a basis for measurements at energies between 15 keV and the lower threshold of the tin-lead filter. The ratio of the response under the thick plastic filter to that under the Dural filter changes continuously

with decreasing photon energy within the range between 65–15 keV.

5. The *tin-lead* filter (0.028″ tin plus 0.012″ lead) permits a response which is almost independent of energy over a range from 75 keV– 2 MeV.

6. The *cadmium-lead* filter (0.028″ cadmium plus 0.012″ lead) provides the same photon absorption as the tin-lead filter. Its purpose is to detect thermal neutrons. The photographic response under the cadmium-lead filter is indirectly (see p. 214) due to the high cross-section of cadmium for capturing thermal neutrons. The nuclear reaction occurring leads to the emission of very hard γ-rays (approx. 9 MeV max. energy) which contribute to the density of the film. If the response below the cadmium-lead filter exceeds that below the tin-lead filter, the excess is due to thermal neutrons.

The narrow strip of indium 0.4 g (see also Figure 6.5) which may be inserted in the recess in the front half of the holder, does not form part of the film dosimeter proper. It is often omitted and is normally only used when there is a possible criticality hazard. In an accident of this type the indium is activated by thermal neutrons and helps in the identification of exposed persons. The indium may also serve another purpose. If photon radiation below 2 MeV is incident on the front of the badge a negative image along the top edge of the film will appear below the indium, hence a differentiation between front and rear exposures is possible.

As indicated in Figure 6.5 the badge is further provided with a 0.012″ lead edge which shields the tin-cadmium—tin-lead joint and is bent up around the two outer edges of these filters. This prevents low energy X-rays from leaking beneath the γ-ray filters which could cause serious error because the film sensitivity is considerably greater for low than for high energies.

Angular Incidence

The response of a film dosimeter is always dependent upon the angle of incidence. An important point is that the optimum thickness for the filters was calculated for an average angle of the incident radiation of 35°. Investigations suggest that the radiation is received predominantly from the front within an angle of $\pm 60°$ to the normal. A film oscillating through this angle was found to yield the same response as that of a film at a static angle of 35°. Thus the design of the film badge is based on an angle of incident radiation of 35° to the normal, and this angle must be used for any calibrations employing energies which are significantly attenuated by the filters (see also p. 141).

Monitoring Films

(i.) *General Characteristics.* Amongst the most desirable photographic properties of monitoring films are:

1. Adequate speed to X- and γ-rays, to enable small doses to be detected.
2. Uniformity of speed within the area of a film.
3. Minimal variations of speed between various batches.
4. Constancy of hardness factor.
5. Constant and reproducible low fog level.

Further, the film should be capable of covering a fairly wide range of exposures and should possess a reasonably high inherent contrast, so that small differences of exposure can be determined. The two latter requirements are in fact contradictory and a compromise is advisable between not too low a contrast and not too small a coverage of range of exposure. Further, the latent image fading of the films should be as little as possible. Some monitor films are supplied with a coating of the same emulsion on both sides of the base, others have a fast emulsion on one side and a slow emulsion on the other side. Yet another variant is for the holder to contain two separate double coated films of different speeds. In general it can be said that the choice of the type of film depends on the type of film badge for which it is intended. It is important not to put two films in a holder intended for only one (and vice versa), because an important contribution to film density comes from the back filters and this would be completely altered by the second film.

In order to satisfy most of the requirements of a satisfactory monitoring film, the packing of the film is of vital importance and must be considered as well. The dimensions of the packet should be chosen so that it fits easily into the holder without being pressed, since pressure may cause undesirable marks on the film. An emulsion with the least possible pressure sensitivity is most desirable. The packet must be light tight and easy to open, further it must be well sealed to minimize the effect of humidity on the film and its internal wrapping. The total wrapping should not be too thick or heavy, so that electron absorption is reduced to the minimum attainable with a light-tight packet. As an example it may be mentioned that 200 keV electrons are entirely stopped by a paper thickness equivalent to 45 mg/cm^2.

Other properties of monitoring films might be required for special purposes, such as for megaroentgen dosimetry, for fast neutrons and other applications, which will be discussed in further sections (see p. 221).

(ii) *The Kodak Radiation Monitoring Film.* In general it is not intended in this book to describe individual brands of film. An exception is, however, made by the description of the response characteristics and properties of the film recommended in conjunction with the British Film Badge. This is because the design of the badge is matched with a particular type of film. Other types of film could be used in the British Film Badge, but either the filter thicknesses or the dose assessment system or both would have to be altered to match the properties of the other film. The brand of film chosen for the British Film Badge is the Kodak Radiation Monitoring Film. This consists, as already indicated, of a safety base, coated with a fast X-ray emulsion on one side and a slow emulsion on the other side of the base. The fast emulsion should face the front of the badge. The dimensions of the film are $1\frac{1}{4}'' \times 1\frac{5}{8}''$. It is the combination of the two different emulsions which permits a very wide range of exposures to be given. If, for example, an X-ray exposure at $80\,\mathrm{kV_p}$ and 0.4 mm Cu filtration should exceed 1 roentgen, the processed fast emulsion would reach an optical density of about 4, which is rather high for most densitometers to read with accuracy. In this case the slow emulsion is used. The fast emulsion can easily be removed after processing so that the density on the slow one can be read. The emulsion can be removed by immersing the film in water of approximately $120°\mathrm{F}$ or approximately $50°\mathrm{C}$ for about 30 secs and then gently wiping it with a wet swab of cotton wool. Alternatively a small area of fast emulsion can be removed using a small swab dipped in hot water. This leaves the remainder of the film area in its original state. No damage will be done in this way to the slow emulsion which is capable of recording fairly high doses. In fact the Kodak R.M. film is capable of covering a range of exposures of about $10^5:1$. In the daily routine of dosimetry the slow emulsion is hardly used, as the fast emulsion covers a sufficiently wide range for most purposes. However, if a person should accidentally receive a dose considerably in excess of the maximum permissible dose the slow film can be utilized and can be evaluated with the same accuracy as the fast emulsion at lower dose levels. The slow emulsion will also be useful if the dosimeter is placed at special sites where higher doses are to be expected.

Separate characteristic curves for the response of both emulsions together and for the slow emulsion only have to be plotted for a proper calibration of the film if the complete range of exposures is to be utilized. Such characteristic curves of the Kodak R.M. film are illustrated in Figure 6.6. These curves show the net density as a

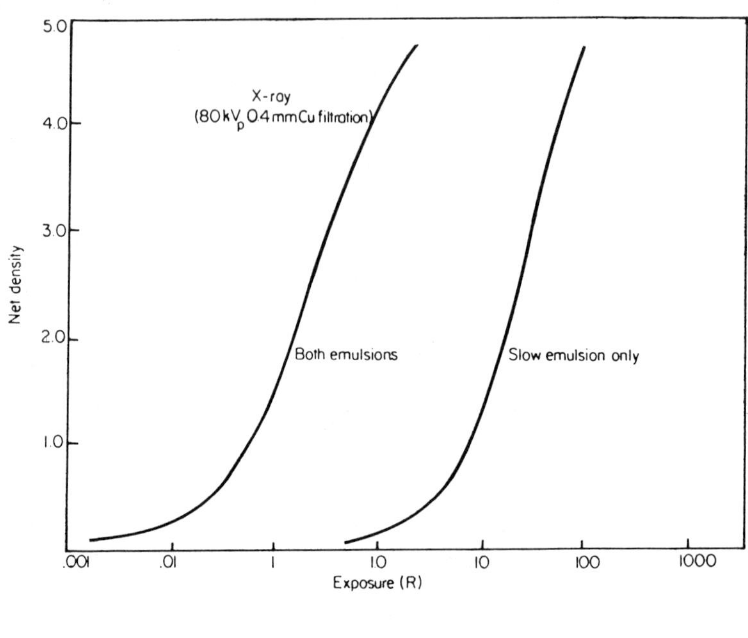

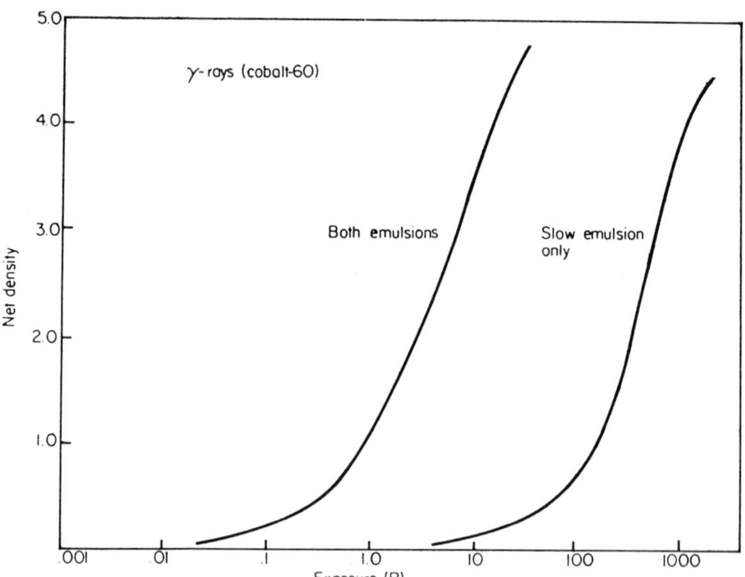

Figure 6.6. Characteristic curves of Kodak Radiation Monitoring film for X-rays (80 kV$_p$, 0.4 mm Cu filtration) and for γ-Rays of cobalt-60. (*By courtesy of* Kodak Ltd., England.)

function of the exposure in roentgens for X-rays generated at 80 kV$_{\rm p}$ 0.4 mm Cu filtration and for γ-rays of cobalt-60. The radiation of 80 kV$_{\rm p}$, filtered by 0.4 mm Cu is almost equivalent to a radiation of 40 keV. The relative sensitivities for X- and γ-rays in terms of the net density obtained for a given exposure to X- or γ-rays can be read from the graphs. Figure 6.6 should only be used for general guidance, since calibration using individual processing conditions for each batch is essential. Details of recommended processing conditions and storage of films are given on pages 234-236.

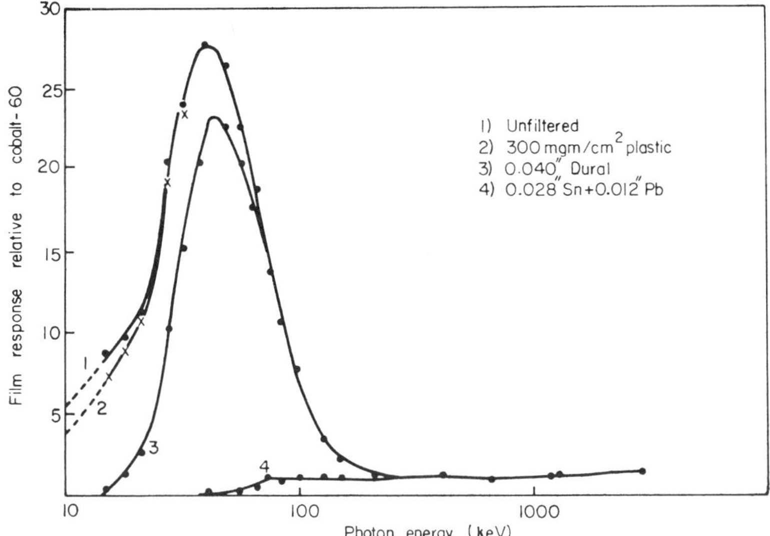

Figure 6.7. Response of the British Film Holder plus Kodak R.M. film versus energy of radition at 35° incidence. (*After* M. J. Heard and B. E. Jones, *Personnel dosimetry techniques for external radiation. Symposium Madrid* (1963). ENEA-OECD.)

(iii) *Response of Kodak R.M. Film in the British Film Badge.* The assessment of dose from a film exposed in a film holder requires a detailed knowledge of its response to radiation of different energies under the various filters.

Response to Photons. The response of the film shielded by the filters incorporated in the British Film Badge is shown in Figure 6.7, according to M. J. Heard (1964). The film response given on the ordinate of this figure is defined as the reciprocal of the dose measured in rads in air to give the same density relative to that obtained using

^{60}Co γ-rays. The curves refer to incident radiation at 35° (see p. 203) to the normal. The radiations used were strongly filtered before reaching the badge, so that energies correspond approximately to homogeneous beams of photons. In the Appendix (p. 609) Table 14.5 shows the filtration for various radiations to approach homogeneity. It is seen from Figure 6.7 that the response of the film below the tin-lead filter is practically constant from 2000 keV down to 75 keV.

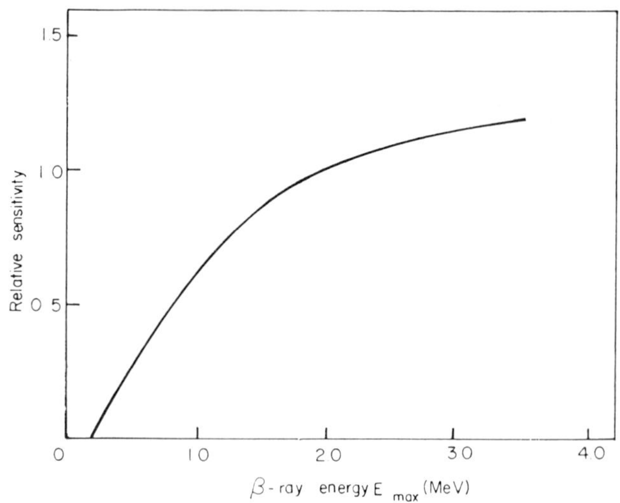

Figure 6.8. Variation with β-ray energy of the relative sensitivity of unfiltered Kodak R.M. Films exposed at normal incidence. (*After* B. E. Jones and T. O. Marshall, *J. Phot. Sci.* **13**.1.12 (1965).)

The minimum radiation dose which can be detected by the Kodak R.M. film is about 0.3 mR at a density of 0.01 above fog for X-rays of 40 keV and about 7 mR for ^{60}Co γ-rays.

The effect of small manufacturing variations in the thickness of the filters on the percentage change of the response is described in the paper by M. J. Heard and B. E. Jones (1963). The manufacturing tolerances are $\pm 0.001''$ for the thicknesses of the metal filters of the British Film Badge.

(iv) *Response to β-Rays.* Details of the measurement of β-rays using the Kodak R.M. film in the British Film Badge were recently published by B. E. Jones and T. O. Marshall (1965). Only the plastic filters and the open window are of practical use when measuring β-rays. The relative sensitivity of the R.M. film in its packet using no filtration as a function of energy in MeV is shown in Figure 6.8.

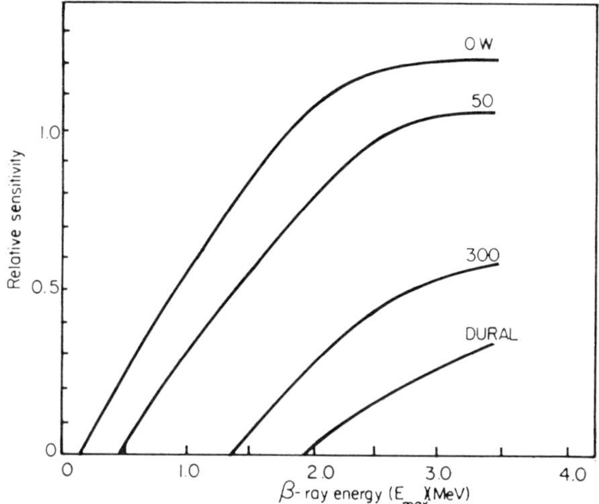

Figure 6.9. Variation of the relative sensitivity of Kodak R.M. Films (unfiltered) with β-energy. (open window) and beneath the 50 and 300 mg/cm² plastic and Dural filters for 35° incidence. (*After* B. E. Jones and T. O. Marshall, *J. Phot. Sci.* **13**.1.12 (1965).)

The relative sensitivity refers to the standard film sensitivity, i.e. to the response to γ-rays of ^{60}Co using the tin-lead filter of the badge. It is seen that the relative sensitivity approaches saturation above 2 MeV, when the apparent sensitivity is about 1.2. This means that the sensitivity of the film for β-energies above 2 MeV is very nearly that to γ-rays of ^{60}Co.

The effects of the filters on the response of the R.M. film to β-rays as a function of β-ray energy is presented in Figure 6.9. The curves shown in Figures 6.8 and 6.9 are the result of exposures by Jones and Marshall using radioisotopes emitting β-rays within the range from 0.17–3.6 MeV max. The isotopes used by these authors are tabulated below:

TABLE 6.1. Radio-isotopes used for β-exposures and maximum energies of β's emitted

Isotope	Max. energy (MeV)
Sulfur-35	0.17
Tungsten-185	0.43
Thallium-204	0.78
Yttrium-91	1.55
Rhodium-106	2.4 (12 %), 3.1 (12 %), 3.6 (70 %)
Strontium-90 + Yttrium-90	0.54 + 2.2

The β-dose was measured by means of a Loevinger type extrapolation chamber (R. Loevinger, 1953).

6.3b.   The Assessment of Dose

(1) *Photon Dose*

It has been mentioned before that the evaluation of dose by means of the British Film Badge is based on the conversion of the measured film densities into 'apparent' doses of the dose-density curve for ^{60}Co or ^{226}Ra γ-rays, the latter being filtered by 0.5 mm Pt. *The apparent dose is thus the equivalent γ-ray dose beneath the various filters* and is symbolized by $D_{tin\text{-}lead}$, D_{Dural}, $D_{thick\ p}$ and $D_{thin\ p}$, the last two standing for doses beneath thick and thin plastic filters respectively.

As the tin-lead filtered film area is approximately energy independent from 75 keV–2000 keV, the dose for radiations in this range of energies may be evaluated directly from this area with an accuracy of about ± 20 per cent. If radiation between 40–75 keV was incident on the badge the densities beneath the dural and the tin-lead filters would be used. Heard, Jones and Marshall have shown that the apparent energy of the radiation as obtained from the ratio of the apparent Dural to tin-lead filter dose is a mean of the individual photon energies weighted according to the magnitude of the dose at each energy. After conversion into apparent γ-ray doses, the ratio of these apparent doses, called the quality ratio, indicates the effective photon energy received by the badge using the associated correction factor (see Figure 6.10), which must be applied to the apparent dose under the tin-lead filter to obtain the true dose. This factor is the reciprocal of the relative sensitivity of the tin-lead covered area for the energy involved.

The relationship between the tin-lead correction factor and the quality ratio (i.e. the ratio of the apparent doses beneath the Duralumin and tin-lead filters) is illustrated in Figure 6.10. Knowing the quality ratio the correction factor can be derived and the true photon dose is then obtained by multiplying the apparent tin-lead dose by this correction factor.

For energies from 20 keV–40 keV the ratio of the film sensitivity below the thick plastic filter to that below the Duralumin filter is used for evaluation of the dose. Similarly for energies below 20 keV the film areas below the open window and the thick plastic area may be utilized. It is important to remember that any measurement using the unfiltered area (open window) may however be in error if electron radiation was also present.

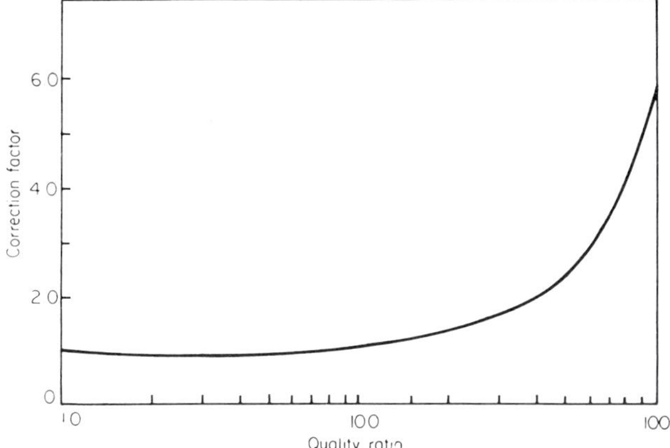

Figure 6.10. Tin/lead apparent dose correction factor against $\dfrac{\text{Dural}}{\text{tin/lead}}$ quality ratio for use with combined emulsions. (*After* B. E. Jones and T. O. Marshall, *J. Phot. Sci.* **12**.6.319 (1964).)

The effective quality of the radiation to which a badge was exposed can be recognized not only in this way, but also, in many cases, from a visual inspection of the film. Examples of films exposed in the badge at various radiation qualities are illustrated in Figure 6.11. It is then only necessary to measure the densities and to evaluate the apparent

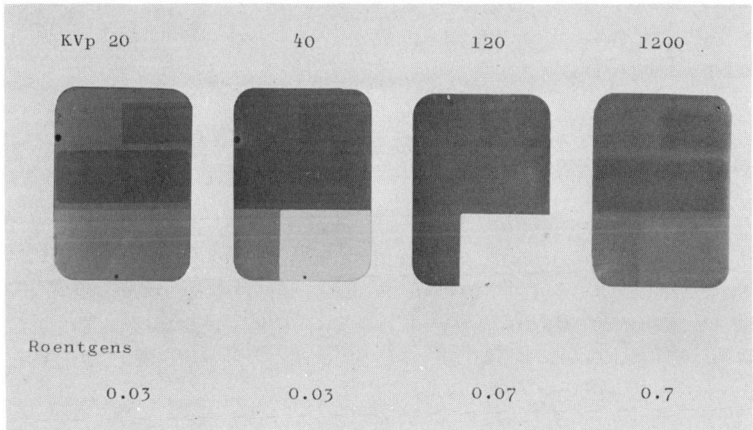

Figure 6.11. Appearance of processed Kodak Radiation Monitoring film after exposure in the AERE/RPS film holder using radiations generated at 20, 40, 120 kV_p and γ-rays from a cobalt-60 source. The number of roentgens measured at the position of the holder is stated below.

doses for the relevant areas. Hence by a subjective selection large numbers of films can be assessed very rapidly.

(2) *β-Dose*

It has been shown in Figure 6.8 that the response of the R.M. film to 2 MeV β-radiation is very close to that of ^{60}Co γ-rays, but the response drops with decreasing β-energy so that it is necessary to derive and to apply β-correction factors. If pure β-sources are used,

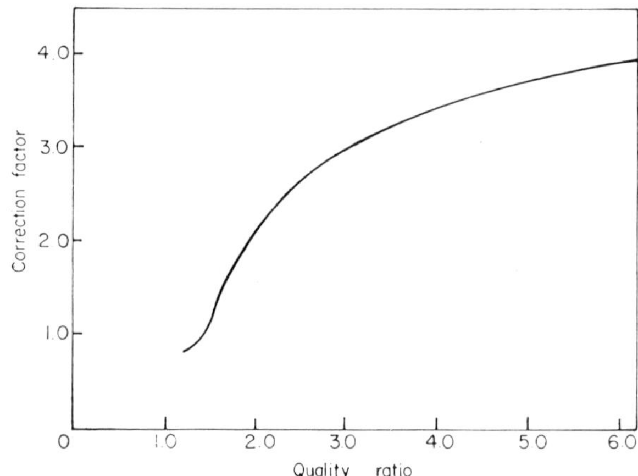

Figure 6.12. Open window (OW) apparent dose correction factor against $\dfrac{OW}{50}$ quality ratio for use with combined emulsions where only β-radiation has been recorded. (*After* B. E. Jones and T. O. Marshall, *J. Phot. Sci.* **13**.1.12 (1965).)

i.e. without any contribution from photons, the ratio of the apparent doses at the open window and the thin plastic area, i.e. $D_{open\ window}/D_{thin\ p}$ is first evaluated. This ratio is called the quality ratio as mentioned previously and the relationship between the quality ratio and correction factor is given in Figure 6.12. The β-dose is then obtained by multiplying the apparent open window dose by the appropriate correction factor. If photon radiation is also present, more elaborate procedures become necessary.

(3) *Mixed Photon and Electron Dose*

In practice a person may receive not only radiation of different energies, but also simultaneous doses from different types of radiation, for example from photons and electrons of various energies. Hence a film badge should be capable of correctly recording the integrated dose of mixed radiation. The evaluation of these mixed doses can be carried out by means of the British Film Badge and the interested reader may find a detailed account of one such procedure of the assessment in the paper by B. E. Jones and T. O. Marshall (1964).

(4) *Mixed Photon Dose*

The assessment of doses produced by a mixture of different energies is fairly straight-forward, if one deals with photons only. It is however, not possible to derive the individual doses and energies composing a

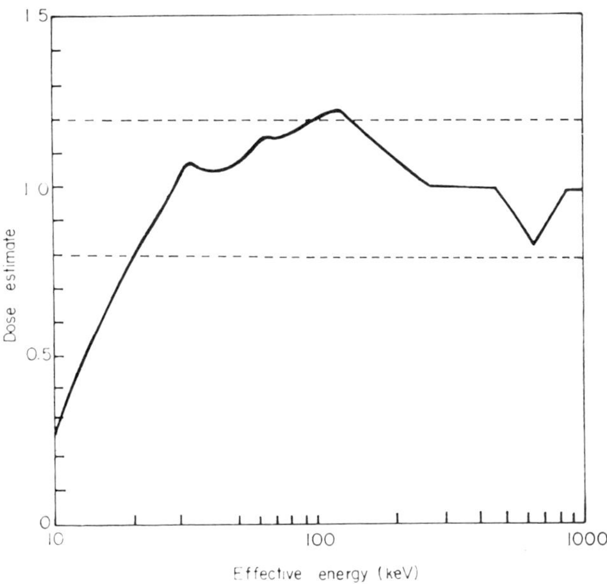

Figure 6.13. Dose estimate as a function of effective photon energy obtained from

$$\text{Dose} = D_{tin/lead} + \frac{D_{Dural}}{50} + \frac{D_{thick\ p} - D_{Dural}}{10}$$

(*After* B. E. Jones and T. O. Marshall, *J. Phot. Sci.* **12**.6.319 (1964).)

mixture. Heard, Jones and Marshall have derived the following simplified empirical expression for mixed photon energies over a range from 20 keV to 2 MeV;

$$Photon\ dose = D_{tin\text{-}lead} + \frac{D_{Dural}}{50} + \frac{D_{thick\ p} - D_{Dural}}{10} \qquad (6.1)$$

where D refers in each case to the apparent ^{60}Co dose for the filter indicated. This expression is a measure of the true photon dose to within ± 20 per cent and is shown in graphical form in Figure 6.13. Other expressions and systems of evaluation which are more suitable for machine computation have also been derived and the reader is referred for more details to the papers by M. J. Heard and B. E. Jones (1963), M. J. Heard (1964a, 1965), B. E. Jones and T. O. Marshall (1964, 1965).

(5) *Thermal Neutron Dose*

Using the British Film Badge the dose due to thermal neutrons can be assessed from the relative densities below the cadmium-lead and the tin-lead filters. Both filters have the same absorption for γ-rays, but any difference between the apparent doses evaluated from the densities below these filters is due to exposure by slow neutrons. The effect of the excess density below the cadmium-lead filter is due to the extra γ-rays produced in the neutron capture reaction in the cadmium as discussed in more detail below.

Certain correction factors have to be introduced in the evaluation of the thermal neutron dose. These factors are caused by special effects, such as neutron activation of the atoms composing the photographic emulsion, to the influence from γ-rays from cadmium on other areas of the film and to the sensitivity of the film to the γ-ray emission from cadmium, in comparison with the ^{60}Co γ-rays below the tin-lead filter. Considering these effects the thermal neutron doses and the γ-ray doses (in presence of neutrons) can be evaluated according to Heard and Jones from:

$$Thermal\ neutron\ dose = \frac{D_{cadmium\text{-}lead} - D_{tin\text{-}lead}}{F}\ rems \qquad (6.2)$$

and

$$\gamma\text{-}ray\ dose = D_{tin\text{-}lead} - \frac{thermal\ neutron\ dose}{3}\ rads\ in\ air \qquad (6.3)$$

where F, the correction factor, is between 2–2.5 and D again refers to the apparent ^{60}Co γ-ray dose below the respective filters. The γ-ray dose is of importance in neutron measurements as neutron exposures are usually accompanied by γ-rays.

Cadmium is often used as a thermal neutron absorber because of its very high absorption cross section $\sigma_a = 20{,}000$ barns for ^{113}Cd, (see p. 38) which is an isotope having an abundance of 12.26 per cent in natural cadmium. In the reaction ^{113}Cd$(n, \gamma)^{114}$Cd γ-rays having a wide spectrum of energies are emitted.

As a general guide in thermal neutron measurements it is useful to know that 9.6×10^8 thermal neutrons per cm^2 is taken as being equivalent to a dose of 1 rem. Further information on the number of neutrons per cm^2 per rem as a function of energy is given on page 239.

(6) *High Energy Photon Dose*

Dosimetry of photons above 2 MeV becomes of practical importance when using high energy linear accelerators or betatrons in medical, industrial and research applications. Apart from this, dosimetry in the vicinity of (gas-cooled) power reactors sometimes has to deal with γ-rays up to about 7 MeV due to the radiation emitted from the excited states of ^{16}O for instance in heat exchangers.

A study of the response of photographic material to doses of photons with energies above 2 MeV was carried out recently by P. D. Holt (1964) and by P. D. Holt and J. A. B. Gibson (1965). These authors found that the dose for photons up to 6 MeV may be overestimated by about 50 per cent when using the British Film Badge and Kodak R.M. Film under electronic equilibrium conditions. The overestimate is about 10 per cent less if electronic equilibrium is not established. Holt and Gibson suggested that the ratio of the readings below the Dural and tin-lead filters could be used for the assessment of the average energy of the radiation applied, if the latter is above 2 MeV. The use of this ratio to derive a correction factor fails, however, if there are also photons of energy below 0.3 MeV present.

Holt has also studied the response of Kodak R.M. Film towards higher energy photons up to 27 MeV (peak energy of spectrum). The average energy for this spectrum is about 12–13 MeV. He found that the sensitivity of the film below the tin-lead filter relative to that for γ-rays of radium increases further with energy and reaches a maximum at about 22 MeV when the curve begins to flatten. This latter result refers both to electron equilibrium and non-equilibrium conditions and changes with the type of film in use.

6.3c. Other Types of Film Badges

Amongst the many different designs of photographic film badges, only a few other typical ones will be described in this section.

(1) *A German Film Badge*

A German film badge due to H. Dresel which is illustrated in Figure 6.14 is designed as an energy dependent system for the measurement of photons, electrons, thermal and fast neutrons. For a detailed description the reader is referred to the publications by H. Dresel (1956, 1960, 1963). The badge consists of two separate units containing two different types of film. The fast neutron compartment

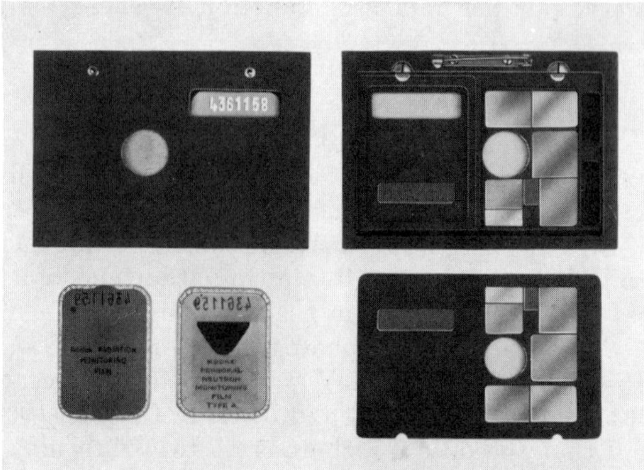

Figure 6.14. German Film Badge with two monitoring films for X- and γ-rays and for neutron dose measurements. (*After* H. Dresel, Freiburgi. Br.)

is provided with a fast neutron nuclear track monitor film, (Kodak Personal Neutron Monitoring Film Type A), and contains an open window for the identification number and a 0.5 mm Pb filter. The latter facilitates the counting of the number of proton tracks per unit area, (see p. 245) if X-rays below 300 keV are present. The other compartment of the badge is provided with seven filters and an open window, as seen in Figure 6.14. The Cu filters of 1.5, 0.5 and 0.05 mm and the open window permit the measurement of photon doses. In brief, the doses are evaluated from the ratios of the apparent doses $D_{F0.5}/D_{F1.5}$, $D_{F0.05}/D_{F0.5}$ etc. where D and F denote dose and filter respectively. A plot of the logarithm of the apparent doses against the filtration in mm Cu is extrapolated to open window doses and these are converted into true doses after multiplying by specially evaluated quality factors. The true dose is obtained by adding the various component doses. The badge also allows the evaluation of β-doses between energies from 0.4 to 2 MeV. The filters of 0.05 mm Cu and 0.18 mm Al are used for the discrimination between high energy β's and soft X-rays (below 30 keV). Both the filters have the same mass per unit area and therefore they stop β-rays to the same extent. If soft X-rays are incident, different photographic densities below the two filters will be recorded. This makes it possible to distinguish β- and X-rays.

The determination of the thermal neutron dose is based on the comparison of the photographic densities obtained below a cadmium and a tin filter, similar to the methods used in the British Film Badge.

(2) *A French System*

Various types of photographic dosimeters are used in France. The following description is of a dosimeter which was designed in collab-

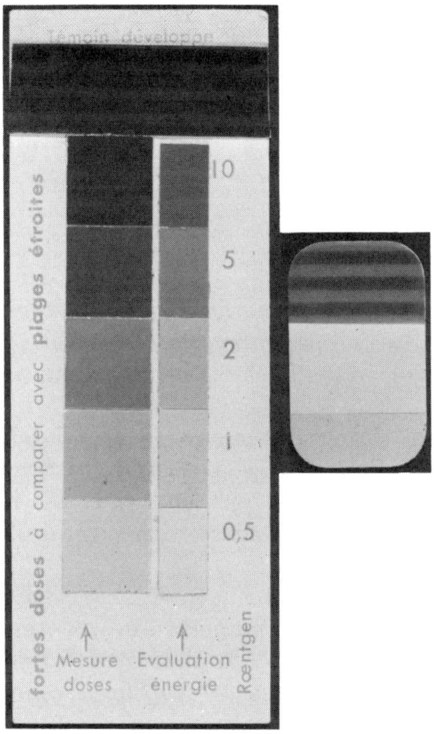

Figure 6.15. Kodak Pathé Paper Dosimeter (on the right), unwrapped; top shows development test strip, the other two strips are for the recording of the radiation dose. Left: A comparison step wedge for evaluating doses. On top is the processed development test strip. (*By courtesy of* Kodak Pathé S.A. France.)

oration of the Commissariat à l'Energie Atomique with the Pharmacist General Chassende-Baroz and Kodak Pathé S.A. The outstanding feature of this dosimeter is that it makes use of photographic paper instead of film. There are several types of such paper dosimeters on the French market which differ in sensitivity.

Design. Each dosimeter consists of a heat sealed and water proof paper packet (of similar, but not the same dimensions as a dental film packet, (see Figure 6.15). Each packet contains three different emulsion strips in adjacent positions and stuck on a common paper base. Only two of the strips are used for the recording of radiation doses, whereas the third one serves as a development test strip. There are three variations of dosimeters available which cover the following ranges (a) 0.01–10 R for X-rays, (b) 0.02–20 R and (c) 0.02–800 R, (b) and (c) to be used in the γ-ray region. The different ranges of sensitivity can easily be recognized by different colored printing on the packets. The radiation recording emulsion strips in the packets are of different widths, the one of 9 mm width covers the larger, the 14 mm wide strip the smaller dose range. Hence in type (a) the wider strip records doses from 10–500 mR, and the narrower strip records doses from 0.5–10 R.

Identification

For identification purposes the name of the wearer can be written on the outside of the sealed paper packet, the writing being transferred through the packet on to the back of the recording paper base.

Plastic container. Each dosimeter packet should be worn in a plastic container which is partly covered by a filter of 0.2 mm Cu. This filter gives an indication of the energy of the incident photons. There are also more complex holders with several filters available.

Processing. The photographic paper may either be developed under standard conditions of temperature and time or, if for some reason this is not possible, the completion of development might be controlled by a development test strip. The development control or test strip is pre-exposed by the manufacturer of the dosimeter in such a way that it will reveal during development four dark and three lighter line shaped bands of density. These appear uniformly blackened to the dark adapted eye when development is complete. The test strip may also be compared with a processed strip supplied with the dosimeter.

Evaluation of Doses. The doses can be assessed by measuring the reflection density or by comparing it with a density step wedge supplied with the dosimeter (see Figure 6.15). Separate step wedges are supplied for each exposure range. Each step has the corresponding exposure in terms of roentgens printed alongside it. The step wedges thus serve as a calibration of the dosimeter for standard processing conditions.

This type of dosimeter is characterized by its simplicity in use and it has the further merit that the results can be evaluated fairly quickly. According to F. Duhamel and G. Soudain (1963) the assessment is so rapid that if amongst 150 dosimeters 2 reveal a dose greater than 20 mR this can be assessed within 2 minutes.

(3) *Two American Film Badges*

The rather elaborate system to be described in the following is restricted mainly to the dosimetry in nuclear energy plants in the United States. Badges are successfully used by persons employed in chemical processing of fissionable material, in nuclear reactor practice and by those engaged in the operation of high energy accelerators and other fields. One example of such a system is presented by the Oak Ridge National Laboratory (ORNL) badge dosimeter, as described by D. M. Ross (1963) and also by F. V. Cipperley and W. P. Gammell (1961). An exploded view of the badge is illustrated in Figure 6.16. In this badge combined use is made of photographic, chemical, phosphate glass and nuclear activation dosimeters. It provides in most cases a double check on the doses received by different methods. Furthermore it is not only suitable for the measurement of small and large doses from photons, β's, thermal and fast neutrons, but it is also capable of giving additional information in

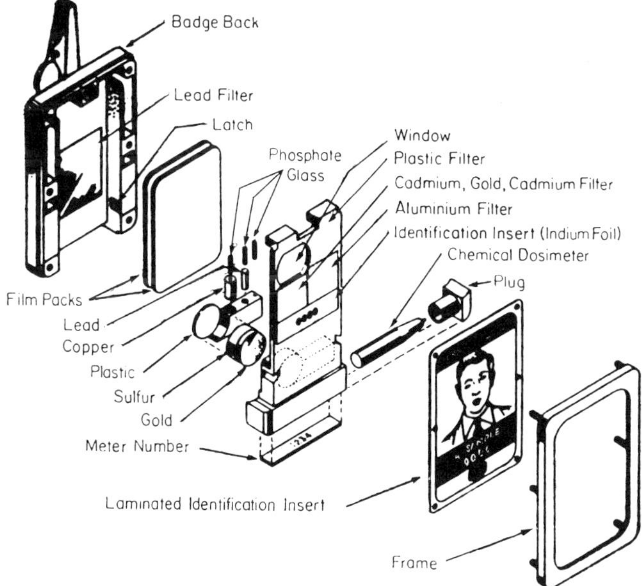

Figure 6.16. ORNL Badge-Meter Model II. (*After* D. M. Ross, *Personnel dosimetry techniques for external radiation. Symposium Madrid* (1963). ENEA-OECD.)

criticality accidents, where operators may have received very high doses.

The ORNL badge incorporates the following features:

1. A DuPont 544 film packet containing a sensitive film (No. 555) for low range exposures and a less sensitive film (No. 834) for exposures in excess of about 5 rads.

2. A Kodak Personal Neutron Monitoring Film, Type A.

3. Four filtered areas including (a) a gold foil sandwiched between two cadmium foils, (b) an aluminum filter 0.04″ thick, (c) a plastic area which is an absorber of about 215 mg/cm², (d) an open window.

4. An indium foil for screening exposed staff in the case of a nuclear incident (by measurement of the radiation from the neutron activated ^{116}In, using a portable contamination meter).

5. A sulfur pellet for the determination of high energy (2.5 MeV) neutron flux.

6. A bare gold foil for estimation of thermal neutrons.

7. Three meta-phosphate glass rods to measure γ-ray doses above 100 rads. (These rods are shielded by lead, copper, and plastic respectively to determine the approximate effective energy.)

8. A chemical dosimeter to permit quick visual identification of the exposed persons after an incident involving γ-radiation and/or thermal neutrons. The visual appearance is based on light fluorescence.

The function of the sulfur pellet, the gold and indium foils is to act as nuclear activators in severe accidents, where the following reactions are utilized

$$^{32}S(n,\, p)\,^{32}P, \ ^{197}Au(n,\, \gamma)\,^{198}Au, \text{ and } ^{115}In(n,\, \gamma)\,^{116}In.$$

The sulfur reaction is used for the detection of fast, the gold for thermal and the indium for thermal and intermediate neutrons.

The β- and γ-radiations emitted by the reaction products can be measured by means of Geiger or scintillation counters.

A similar system is used for the new Hanford Film Badge dosimeter which is issued by the U.S.A. Hanford Laboratories (L. F. Kocher, P. E. Bramson and C. M. Unruh, 1963). In this badge an open window and plastic, iron, and tantalum filters are used for the evaluation of photon doses. The tantalum filter of 0.508 mm thickness and covered by 0.51 mm of plastic, provides a linear density response within ± 10 per cent for photons at any energy between 50 keV and 2 MeV, using a DuPont No. 508 film. The badge also makes use of two separate films, one of low and the other of high sensitivity. These cover doses from 0.015–1500 R for an energy range from 50 keV upwards, and from 0.001–15 R for the energy range from 10–50 keV. β-ray doses can be assessed for energies above 0.5 MeV.

For the evaluation of dose from the film dosimeter the density behind each filter and the open window is measured by a mechanized densitometer which punches an electronic data processing machine card. For a radiation field consisting of β-particles of a single energy, photons of energies exceeding 50 keV and photons of a single energy less than 50 keV, a set of equations is used which define the film density for each filter region. Although these equations contain many unknowns, the latter reduce to a few figures only due to the ingenious experimental design of the dosimeter. This is based on the choice of thicknesses of the iron and other filters which provide equal film densities for uranium β's and also equal densities behind the iron, plastic filters, and open window for photons with energies greater than 50 keV.

The Hanford Film Badge dosimeter also incorporates activation foils for neutron dosimetry in the event of a criticality; furthermore two small silver phosphate glass rod dosimeters (fluorods) are included to provide measurements of high γ doses and an indium foil for prompt sorting of directly involved personnel following a criticality event. The film identification and film exchange are performed mechanically. The film slide is held in the dosimeter by a 'T' lock which is released magnetically for film exchange.

This brief description of the Hanford dosimeter should only provide an idea of the methods involved and for further details the reader is referred to the Hanford Report HW-76944 by L. F. Kocher, P. E. Bramson and C. M. Unruh (1963) and also to L. F. Kocher (1957 and 1962).

6.4 NON CONVENTIONAL DOSIMETRY TECHNIQUES

In the previous sections we have discussed the techniques and problems involved in the more conventional applications of photographic dosimetry. In the following section we want to discuss the less frequently applied methods, such as megaroentgen dosimetry, low dose, accidental radiation and finger-film dosimetry.

6.4a. Megaroentgen Dosimetry

The range of sensitivity using ^{60}Co γ-radiation, that can be covered by various types of monitoring films lies between exposures of about 2 milliroentgens–1000 roentgens. Although there are many slow fine grained films and plates commercially available which permit the range to be extended to about 10^5 roentgens the use of slow emulsions may produce certain difficulties caused by latent image instability and reciprocity failure, as pointed out on pages 130 and 144. Other methods for the photographic recording of extremely high doses have been tested, but most of them suffer from certain disadvantages, particularly if they are required for a great range of radiation energies. These methods can be divided into those in which (i) special development procedures are used, (ii) print out methods are used and (iii) silver analysis is employed.

(i) *Special Development Procedures*

W. L. McLaughlin (1962) has made a special study of the possibilities of evaluating films which have received unexpectedly large radiation exposures. He investigated the following methods: 1. The use of developers which chiefly act on the internal latent image (see p. 133) known as internal developers. 2. The use of developer components which retard the action of conventional developers. 3. The use of weak developers.

1. *Internal Developers:* In examining the performance of internal developers, i.e. solutions which act on the internal latent image by bleaching the surface image, McLaughlin found that there is generally a reduction of the energy dependence, but also an undesirable solarization in the region of about 10^4 R units depending on the type of film applied.

2. *Retarding Developers:* A practical method which is easily applied, but suffers from similar disadvantages as those described in 1. is the use of conventional developers to which 200 grams of sugar per liter is added, (H. M. Cleare and N. H. King, 1961). The application of this developer revealed a rather flat region in the characteristic curve in the exposure region between 10^4–10^5 roentgens and solarization for certain emulsions.

3. *Weak Developer Solutions:* When using slow acting developers for short times, or conventional developers at low concentration, or processing at low temperatures, McLaughlin observed not only an uneven density distribution, but also a tendency for the characteristic curve shapes to vary widely with the rate of exposure in the higher density region. Summarizing it may be said that most of the developing procedures mentioned tend to increase the exposure latitude of the photographic emulsions combined with a simultaneous reduction of contrast or even solarization in the region of 10^4–10^5 roentgens.

(ii) *Print-out Method*

The silver bromide system owes its sensitivity chiefly to the amplification caused by the development process. Although the latent image formed on exposure consists of metallic silver, the minute amount produced per unit area cannot be detected without development. If, however, exposures are used which exceed those required for latent image formation by factors of 10^6 and above, visible blackening appears. Practical use was made of this effect in the early decades of this century for printing positive images from negatives on photographic paper (so called print-out papers) and hence the process is known as 'print-out exposure' (see also p. 63). Because the method appeared to be promising for recording very large radiation doses, print-out papers were experimentally exposed to γ-rays using doses within the range from 10^4–10^6 R (H. F. Nitka and D. P. Jones, 1957 and H. F. Nitka, 1959). After exposure the reflection densities of these papers were measured or were assessed by visual inspection. W. L. McLaughlin (1960) successfully used the same method on commercially available X-ray films.

For a precise quantitative evaluation of the print-out image a special densitometric procedure is required. When measurements are made, diffusion of the light by the aggregates of silver produced by heavy exposures, adds to the inherent turbidity of the emulsion layer. Hence a high degree of scatter, diffraction, refraction and reflection occurs when light is passed through the emulsion. This makes conventional densitometry difficult and even impossible. In McLaughlin's investigations optical densitometry was carried out using light with wavelengths limited to those for which the emulsion was practically insensitive. He used an Ansco-Sweet color densitometer with filters placed between light source and film which transmit either in the range 665–800 mμ (peak transmission 720 mμ) or 550–610 mμ (peak transmission 580 mμ). The choice of filter depended entirely on the type of X-ray film used as it was shown that various emulsions differed in their relative spectral absorption. When various types of industrial X-ray films and monitoring films were exposed to doses varying from 10^2 to about 10^8 roentgens using various exposure rates, it was found that the results were free from reciprocity failure. Furthermore the densitometric experiments showed that the superimposition of three exposed pieces of film instead of the use of one film has great advantages. It also improved the accuracy of measurement due to the greater slope of the print-out characteristic curves of superimposed films especially at low doses. A typical example of McLaughlin's results from print-out exposures is illustrated in the

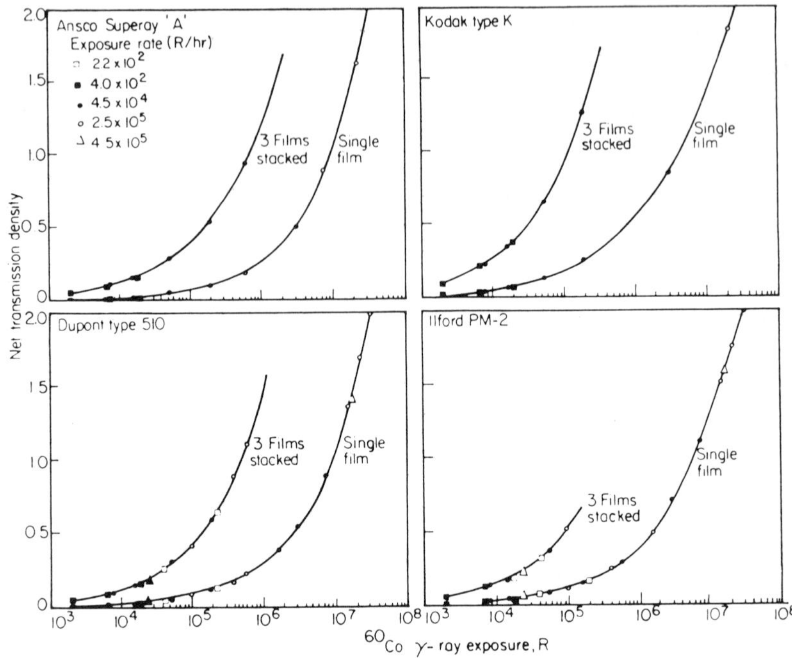

Figure 6.17. Characteristic curves of print-out exposures with ^{60}Co γ-rays, both for single films and for three superimposed films. (*After* W. L. McLaughlin, *Radiation Research*, **13**.594 (1960).)

characteristic curves in Figure 6.17. This shows results for both a single film and three superimposed films; the different rates of exposure applied are indicated on the top left of the diagram. In addition the dependence on energy (Hardness Factor) was found to be considerably greater for print-out, than for conventionally developed images. It is also interesting to note that the print-out effect depends on the environmental conditions of the emulsion. In contradiction to the behavior of most emulsions when conventionally processed, the sensitivity of print-out exposures increased by 50 per cent at 100 per cent relative humidity compared with the sensitivity at room humidity. McLaughlin was able to measure γ-ray exposures from 2×10^4–10^8 R with an accuracy of ± 5 per cent under the appropriate conditions of densitometry, as described.

(iii) *Use of Silver Analysis*

A very different approach to the problem of how to determine high doses of radiation photographically employs silver analysis, i.e. by measuring the amount of silver per unit area of the processed emulsion,

instead of measuring the optical density. The silver in an exposed and
processed area of film can be measured by means of an X-ray fluorescence spectrometer, i.e. by exciting the silver K-radiation for instance,
while the film is exposed to X-rays. The excited characteristic silver
K-radiation which is reflected by a crystal of, lithium fluoride, for

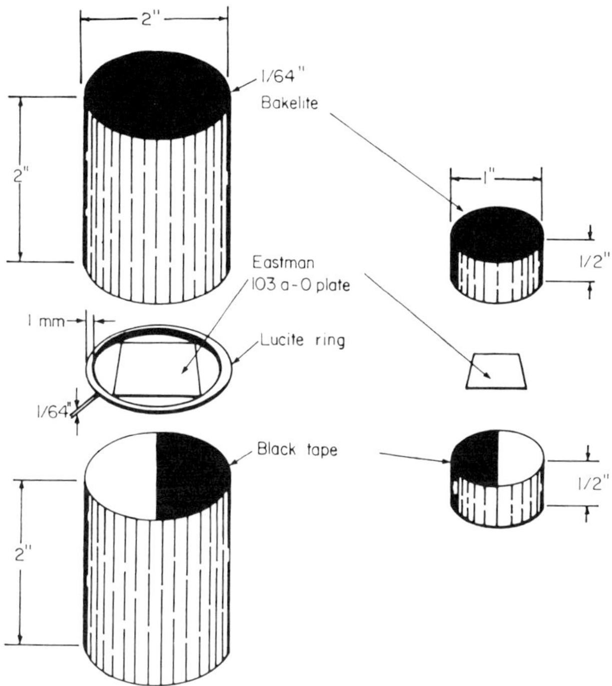

Figure 6.18. Assemblies of photographic plate and Pilot B plastic scintillators.
(*After* M. Ehrlich and W. L. McLaughlin, Nat. Bureau of Standards, Techn.
Note 29 (1959).)

example, is then measured with a Geiger or scintillation counter. This
method has been applied to both high and low doses (I. B. Berlman,
1953; L. S. Birks, 1959; W. V. Baumgartner, 1960).

6.4b. Low Dose Dosimetry

Although some films permit an accurate measure of X-ray doses
down to about 1 mrad, and of ^{60}Co γ-rays doses to about 20 mrads,
attempts have been made to assess doses below 1 mrad (in the
γ-ray region) by sandwiching a light sensitive photographic emulsion
between plastic scintillator cylinders, as shown in Figure 6.18.

M. Ehrlich and W. L. McLaughlin (1959) used a scintillator of diphenyl-stilbene in plastic material. The sandwiched photographic plate responds to the light flashes produced by ionizing radiation. Half of the area of the plate is shielded from the light flashes by black tape in order to provide a distinction between the density obtained by the effect of direct ionizing radiation and that produced by the light flashes.

This system is capable of measuring ^{60}Co γ-ray doses which are accumulated at rates as low as about 0.1 mrad per hour, but as would be expected, the effect is dose rate dependent. In spite of the use of a special photographic plate which is intended for astronomical purposes and therefore is relatively free from reciprocity failure within a limited range, the response was still found to be rate dependent. Furthermore, the assembly is energy dependent owing to the considerable thickness of the plastic cylinders used. This is caused by the change in spatial distribution of the light flashes in the plastic material at various energies.

Similar scintillator dosimeters have been used by other workers, notably by P. W. Henson (1963 and 1964). He employed the dosimeter in pocket form in investigations on the contribution of rock and soil radioactivity to the natural background radiation and their contribution to the gonad and bone marrow doses received by certain sections of the population. Although background dose-rates could be measured, which were 25–30 times lower than those measurable by film alone, errors due to reciprocity failure and exposure temperature dependence were observed. It was also noted that the temperature coefficient varied with the dose-rate. A series of seven two-week exposures was, however, made with a person having no occupational connection with radiation, from which the resultant dose-rate of 108 mrads per year was established with a standard deviation of ± 10 per cent.

For the measurement of low doses an increase of sensitivity can also be achieved by 'latensification' methods, for instance, by post exposure to visible light and special development techniques. W. L. McLaughlin (1965) was able to increase the ^{60}Co γ-ray sensitivity of commercial X-ray emulsions by a factor of 6 in the very low density region of the density-exposure curve.

No specific recommendations for latensification can be given as the result depends on many factors, such as (1) the type of emulsion used, (2) the ratio of the energies absorbed by the emulsion from the exposure to ionizing photons and from the light post-exposure, (3) the relative rates of energy absorption, and (4) the wavelength of the light used in the post-exposure and the type of development. The process might even lead to a desensitization effect under certain circumstances. (Villard effect, see p. 137.)

The best combination obtained by McLaughlin consisted of (a) a post-exposure to light with a wavelength of 580 or 620 mμ and (b) doubling the normal 5 minutes developing time using an addition of 0.5 g/liter of sodium thiosulfate to the developer solution. The result

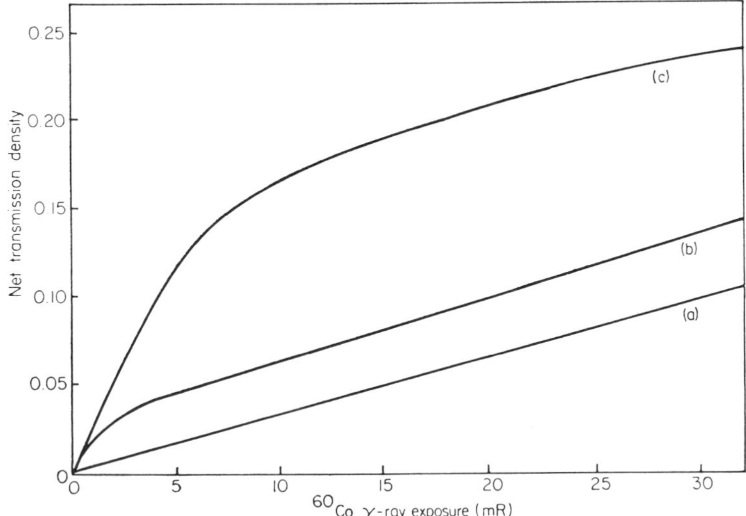

Figure 6.19. Latensification experiments. Density-exposure curves. (a) No post-exposure. (b) 580 mμ post-exposure. (1 m$\mu = 10^{-6}$ mm $= 10$ Å). (c) 580 mμ post-exposure with 10 min development and 0.5 gm/l $Na_2S_2O_3$. (*After* W. L. McLaughlin, *J. Phot. Sci.* **13**.1 (1965).)

of the latensification is shown in Figure 6.19 where curve (a) is due to the conventional procedure, curve (b) due to the post-exposure and curve (c) due to the post-exposure combined with prolonged development and the addition of 0.5 g/liter of sodium thiosulfate to the developer solution. It is seen that the major increase in the sensitivity of the emulsion is within the region of useful densities. No dependence on the exposure rate over the range of 0.4–6000 mR/hour was noted using this latensification method.

6.4c. Dosimetry of Accidental Radiation

'Dosimetry of accidental radiation' is the assessment of the distribution of radiation doses in nuclear accidents. It may happen, for instance, that various types of photographic material are normally stored in commercial stocks in the vicinity of a region where a nuclear accident has occurred. It may then be possible to assess photographically the radiation dose in this particular locality, after the accident. For the assessment one is assumed to have a knowledge of the relative sensitivity of various commercial films to X- and γ-rays of about 1000 keV. It may also happen that pre-exposed or unexposed films have accidentally received a radiation dose, e.g. in a nuclear research laboratory or while being transported. In such cases one wants to know whether the dose received by the films makes these useless for further exposures. Cases have been reported of cine films which have been accidentally exposed to γ-rays during aeroplane flights whilst the plane carried a radioisotope in one of the wings. In such a case expensive studio exposures or outside scene exposures may have been entirely ruined by γ-rays and the damage may involve the loss of enormous sums of money. Although it is true to say that a knowledge of the sensitivity of the film to γ-rays does not prevent the fogging, it is important to be able to assess whether the potential risk of the presence of the isotope may have actually caused the fogging or whether the fogging may have been caused by other circumstances. A knowledge of the relative sensitivities to light and ionizing radiations is obviously of particular importance in considering film fogging during space flights.

G. M. Corney (1960), G. M. Corney and H. M. Cleare (1955), J. S. Moser (1953) and B. K. Boller (1964) have published accounts of the sensitivity of photographic material to million volt X-rays. The effect of this radiation can be taken as being practically identical to that of γ-rays from ^{60}Co.

The following graphs in Figure 6.20 according to B. K. Boller (1964) are the result of exposures of various film samples to known doses of one million volt X-rays followed by a sensitometric exposure to simulated daylight through a stepwedge. The effect on various black and white films is illustrated by characteristic curves (density plotted against log exposure of simulated daylight in terms of meter candle seconds) for no added X-ray exposure and for various X-ray flash exposures measured in roentgens. From these curves the effect of a 1000 kV X-ray exposure on the relative fog obtained at various light exposures can be read. Graph Figure 6.20 shows, for instance, that an exposure of 1 R on a Kodak Panatomic X Aerial film has hardly any influence on the film when exposed to light, whereas a 30 R exposure leads to an additional density of unity within the area of low light exposure.

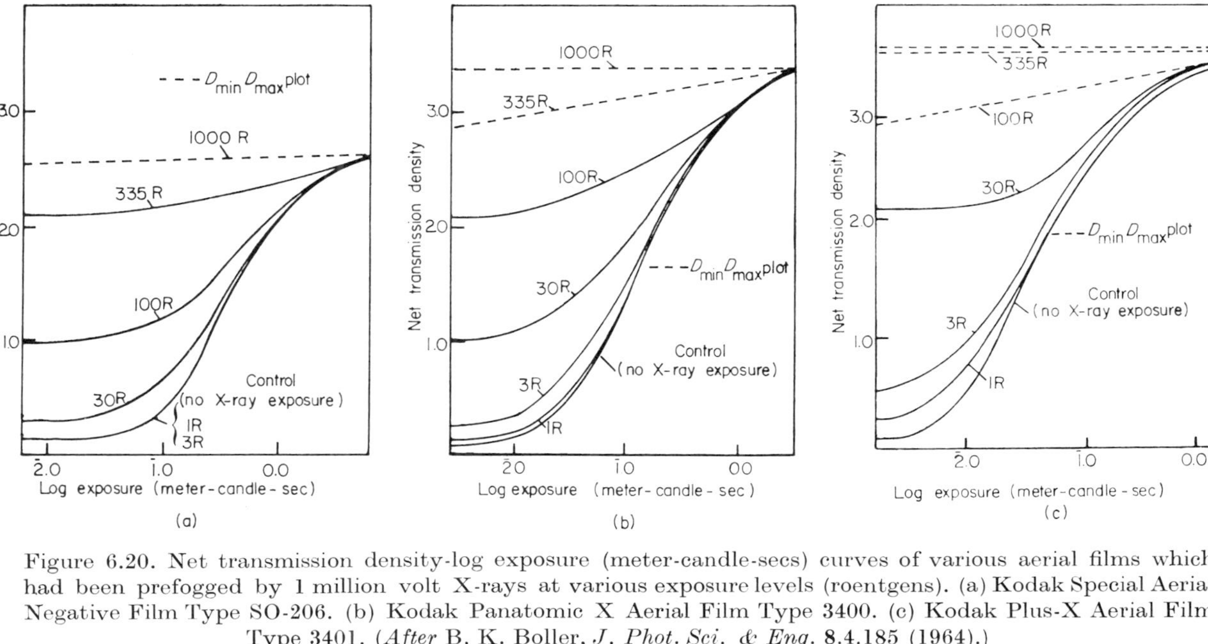

Figure 6.20. Net transmission density-log exposure (meter-candle-secs) curves of various aerial films which had been prefogged by 1 million volt X-rays at various exposure levels (roentgens). (a) Kodak Special Aerial Negative Film Type SO-206. (b) Kodak Panatomic X Aerial Film Type 3400. (c) Kodak Plus-X Aerial Film Type 3401. (*After* B. K. Boller, *J. Phot. Sci. & Eng.* 8.4.185 (1964).)

229

A similar Figure 6.21 due to B. K. Boller shows the effect on a typical color emulsion. Apart from the fog level produced by the million volt X-ray exposures, the image quality is also impaired by the severe graininess which is due to the high energy X-rays.

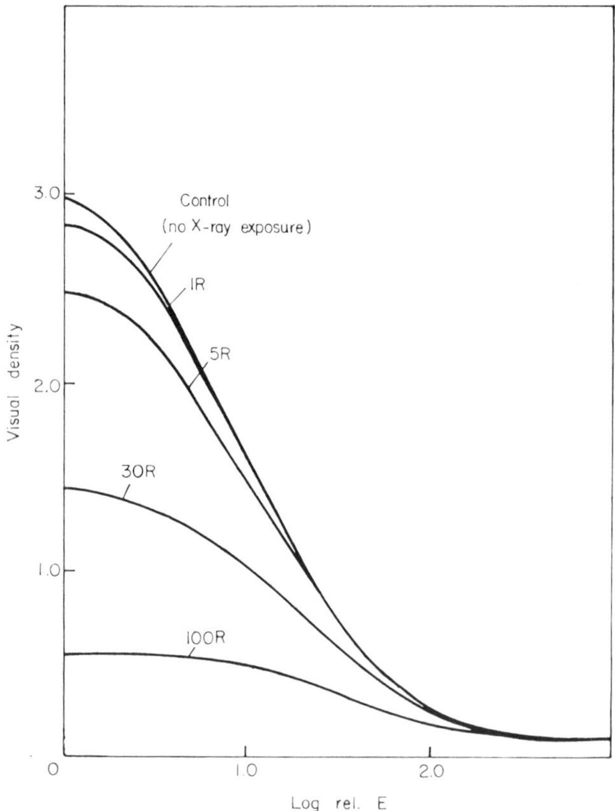

Figure 6.21. Visual density-log exposure curves of Kodachrome II film (for daylight) which has been prefogged by exposure to one million volt X-rays at various exposure levels (roentgens). (*After* B. K. Boller, *J. Phot. Sci. & Eng.* 8.4.185 (1964).)

6.4d.   Finger-Film Dosimetry

In some types of work people may receive considerably higher doses to their fingers, hands and wrists than to their bodies. Such exposures may be received by operators manipulating radioisotopes with their hands, to doctors inserting radium needles into patients (T. M. Robb and R. E. Ellis, 1952), to X-ray crystallographers adjusting their specimens close to the exit of an X-ray tube without proper precautions, and others. The exposure level on the wrist can usually be assessed by conventional film badges worn on a wrist band. However, the dose received by the fingers is preferably measured on the finger tips by small film packets of an approximate area of 1 cm². These packets can be fixed on the finger nail or on the inner side of the finger tip. Plastic finger rings have been constructed, into which a small film may be inserted. S. K. Stephenson (1964) gives details of how such packets can best be prepared. In many applications of radioisotopes it is necessary to handle β-emitters. In these cases the thickness of the packet, which determines the response of the film to β-rays becomes of critical importance. Finger film dosimetry in the region of low β-ray energy and of soft X-rays is not an easy problem. The use of a plastic filter (of about 50 mg/cm²) covering part of the area of the film may assist in the evaluation of β-doses in a similar way as in the British Film Badge. For the precise assessment of the dose received by a finger film an accurate calibration under simulated conditions is an absolute necessity. It may be added that the photographic method is probably not the ideal way of achieving the best results as it is unlikely to compete successfully with lithium fluoride detectors in finger dosimetry.

6.5 CALIBRATION, PROCESSING AND STORAGE OF FILMS

The most essential feature in the treatment of photographic material for dosimetry is the standardization of calibration and processing procedures. Unless strict reproducibility of both conditions is ensured, the percentage error in the assessment of dose might become so great that the wearing of film badges would serve no good purpose. In the following sections an attempt is made to indicate some of the essential and basic recommendations which should be followed in most circumstances.

6.5a. Storage and Calibration

As slight batch to batch variations in sensitivity of the supply of films are unavoidable, it is advantageous to order an amount of fresh unexposed films, of the same batch, and sufficient to last for a period of, say, three months. The films should preferably be stored at a temperature between 6–10°C and at a relative humidity between 40–60 per cent. The films should be protected from undesirable fumes and must be kept free from chemical contamination and from the influence of ionizing radiation. The upper limit of the latter is about

5–10 μR per hour of 1 MeV γ-radiation depending on the type of film used and considerably smaller amounts for low energy radiation.

In general it can be assumed that films supplied in one box are of the same sensitivity but slight variations in speed may occur between films of different boxes, even if they have the same batch number. Hence it is advisable to select films at random from each box for calibration tests.

When a consignment of film is received which has a different batch number than that of the film in current use, it is desirable to make some comparisons between the response of the two batches. Exposures to 40 keV X-rays and ^{60}Co γ-rays are required in order to establish any changes in the minimum and maximum sensitivity, fog level and shape of the density-exposure curve. The sensitivity may be expressed as the reciprocal of the number of roentgens required to obtain a net density of unity. Comparison exposures should preferably be carried out by processing the films from both batches together. The 40 keV radiation, mentioned above is almost equivalent to a heterogeneous X-ray beam which is generated at 80 kV_p and filtered by 0.4 mm Cu. The American Standards Association (1965) recommends X-rays generated at 50 kV_p and a filtration of 0.5 mm Cu + 1.0 mm aluminum.

Comparison of the β-responses should also be made because this is affected by the packet (wrapping) thickness in addition to the emulsion sensitivity. Thallium-204 is a suitable β-emitter for this purpose.

Sensitometric exposures for calibration purposes can be carried out either by a series of exposures of increasing times at a fixed distance or by one single exposure at a range of distances. The latter procedure is usually more convenient, especially for γ-radiation. The films must be exposed in their packets through the filters actually used in the film badges and preferably in their respective badges. In order to avoid scatter from surrounding material, films should be exposed about one meter away from any scattering material. Furthermore the support for the films and the source should be made of low atomic number material. The shortest source to film distance using ^{60}Co γ-rays should not be less than 25 cm, whereas that for X-ray exposures should not be less than 50 cm. The 40 keV X-ray exposures should preferably be carried out at an angle of incidence of 35° with respect to the normal, in the case of the British film dosimeter or otherwise at an angle specified by the design data for other holders. In γ-ray exposures the directional variations were found to be of less importance and films can be most conveniently exposed at normal incidence. Figure 6.22 illustrates the arrangement for the calibration of

Figure 6.22. Arrangement for the calibration of monitoring films in the British Film Badge, using γ-rays from a 100 millicurie radium source, filtered by 0.5 mm platinum. (*By courtesy of* the Radiological Protection Service, Belmont, Sutton, Surrey, England.)

monitoring films using a source of 100 mg radium γ-rays (which are filtered by 0.5 mm Pt) as it is used at the Radiological Protection Service, England. If the 100 mg radium source is not in use, it is stored in a lead container below the exposure table. The source is moved into the exposure position by a pneumatic system which is controlled from outside the room. Thus the films can be positioned in their respective support stands prior to exposure without risk of radiation exposure to the operator.

In the case of radium or of a ^{60}Co γ-ray source, an accurate knowledge of the activity of the source is sufficient for the calculation of the doses received by the films at various distances. In the case of X-ray calibration exposures the dose should be measured for instance by means of a thimble ionization chamber.

6.5b.   Processing

Many of the manual processing units which are used for medical radiography in hospital darkrooms are also suitable as processing units for radiation monitoring films. Accurate temperature control of the solutions is an absolute necessity.

Agitation.   In order to achieve uniformity of development, agitation of the developer solution is a further necessity. This can be achieved either by manually moving the hanger carrying the films for 5 seconds at 1 minute intervals or preferably by nitrogen burst agitation. The latter is carried out by injecting short bursts of nitrogen bubbles at the bottom of the tank so that they rise through the solution. Automatically timed movement of the film hanger from one bath into the next is advantageous. A very convenient hanger for one hundred and fifty monitoring films in which the films are arranged crosswise is supplied by Ilford Limited, England.

Developer and Temperature.   The composition of the developer should be that recommended by the manufacturer supplying the film. The temperature of the developer solution should be kept constant at $20°C$ within $\pm 0.5°C$. The various solutions used in processing, such as the stop bath, fixing bath and washing water should have a temperature not differing by more than $\pm 3°C$ from that of the developer solution. Exact temperature control can be facilitated (particularly in tropical countries) by using an air conditioned darkroom.

Development Time.   Although the development times recommended by the film manufacturers should be adhered to in general, some experts in this field have found that an extension of the recommended time by 50 per cent is advantageous for the following reasons: (1) it leads to a fuller utilization of film speed, (2) a greater spatial uniformity of development within the volume of the tank is achieved assuming constant agitation (3) very slight variations between the shapes of the density-exposure curves for radiation at low and at high energy are reduced. The 50 per cent increase in developing time will on the other hand contribute to a slight increase in fog level.

Topping Up.   Although most manufacturers recommend the use of a replenisher solution for keeping the developer at a constant level of efficiency, in many monitor film darkrooms the developer solutions are topped up (to maintain their level in the tank) by standard developing solutions. The reason for this deviation from usual darkroom practice is that the developer solution does not become sufficiently exhausted during use. This is evident because the majority of monitor films do not reveal a measurable net density or if they do, it is so small that the action of the developer is hardly utilized. Nevertheless, the solutions should be renewed at intervals not exceeding one month depending on the volume of the liquid in the tank and also on the number of films processed. It is recommended that the tank be kept closed by a lid, when not in use.

Stop Bath. As it is essential that the action of the developer is promptly stopped before films are transferred to the fixing solution, the use of an acid stop bath is important. Films should only be immersed for a time of about 10 seconds, using continuous agitation, before transfer to the fixing bath.

Fixing Bath. The fixing time should be not less than twice that for clearing the films from their milkiness and fixing should, in general, take not more than ten minutes. Since the majority of films reveal only a very low net optical density, the action of the fixing bath is utilized at a fairly high rate, as all the unexposed silver bromide has to be washed out from the emulsion. It is therefore important to watch the strength of the fixing solution and to replace it at given intervals, otherwise the storage life of the processed films may suffer.

An electrolytic silver recovery unit might be used with advantage (see p. 394) to maintain the concentration of silver in the fixing bath at a minimum, i.e. below 2 grams per liter. The fixer should then be regenerated in the manner recommended by the film manufacturers.

Alternatively the use of two successive fixing baths is suggested. The films should be cleared from milkiness in the first bath and should then be transferred to the second fixing bath for the same time, as was required in the first. The first fixing bath should be discarded when it fails to clear the films in twice the time needed in a fresh bath. Then the first bath should be replaced by the second bath and a fresh second bath made up.

Washing. The films should be washed in running water for at least 30 minutes at between 18–24°C. At lower temperatures or with inefficient washing the times have to be increased considerably to guarantee the permanence of the processed film material. A final rinse in water containing a wetting agent is useful to ensure uniform drying.

Drying.   The drying should be carried out in a dustfree atmosphere at a temperature not exceeding about 30°C. This can be most easily achieved in one of the commercially available drying cabinets. Spin-drying has been suggested as a much more rapid method.

6.5c.   Storage of Processed Monitoring Films

Processed monitoring films have usually to be kept for several years for future reference. Whatever the storage time for which films have to be kept, it is essential that films are kept free from fumes of either hydrogen sulfide or sulfur dioxide, as these may cause slow deterioration of records. The relative humidity should be kept between 25–60 per cent and high temperatures should be avoided in processed film stores. If films become too dry, they become brittle.

If films are to be kept only for a few years, say five, no particular precautions are necessary apart from those mentioned above. They should be well fixed and washed as described in a previous section. (A. Biermann and Y. Feige, 1965.)

For the storage of films for periods up to about twenty-five years more stringent precautions are required. The films must be so thoroughly washed, that they should not contain more than about 5 μg of residual hypo per square inch. Further spot tests using sulfide for residual silver compounds should prove negative. (J. I. Crabtree and others, 1943.) Air purification and air conditioning will certainly help in the long preservation of processed films. If films are kept in envelopes, these should be free from harmful chemicals and should have a low degree of hygroscopicity. The cabinets in which the films are stored should permit free circulation of air around the films.

6.6   FAST NEUTRON DOSIMETRY

It has been shown on p. 214 that the recording of thermal neutrons by photographic means within the maximum permissible dose rate presents no particular difficulties, as the sensitivity of fast X-ray films is sufficiently high to respond to the β- and γ-rays emitted by certain foils having high cross sections for thermal neutrons. Proceeding further towards the spectrum of higher energy neutrons from about 1 eV to 2.5×10^5 eV, we find that the photographic emulsion is unable to respond to the maximum permissible dose rate. There are two main reasons for this no-man's land as far as the photographic emulsion is concerned. These are (1) that the neutron flux per rem decreases towards higher energies and (2) that no suitable elements having high enough cross-sections have been found within this range of neutron energy. Attempts are being made to overcome this deficiency in the photographic emulsion and they may perhaps be successful in the future. One of the possibilities is to moderate the higher energy neutrons by low atomic material, such as graphite, plastic, wax etc. The thermal neutrons thus formed may then be captured by foils of high cross section or produce nuclear reactions in boron or lithium loaded emulsions or foils containing these elements (see pp. 176 and 492).

For neutron energies exceeding 2.5×10^5 eV, the use of another method of recording the dose in rems becomes possible. This section deals with this method, i.e. with the recording of fast neutrons with energies from about 2.5×10^5 eV up to approximately 14 MeV.

The basis of the method is that when fast neutrons strike hydrogenous material their kinetic energy is partly transferred to hydrogen nuclei (protons). The impact of fast neutrons on nuclear particle emulsions thus causes protons from the gelatin and film base to move through the emulsion, leaving tracks of grains which after processing can be observed by means of microscopic inspection, (see pp. 177 and 243).

The more practical aspects of fast neutron dosimetry will now be reviewed. If the number of proton tracks counted per unit volume of an emulsion can be related to neutron dose, a dose measurement by photographic means becomes possible. The recoil proton method is very tedious and certainly not ideal for routine measurements and it would be advantageous if it could be replaced by a technique in which the dose can be read off from optical densities. Nevertheless no superior photographic recording method has been introduced so far.

The main disadvantages of the proton track method are: (1) that to obtain the number of proton tracks, the individual tracks per unit area have to be recognized and counted under the microscope, (2) that the counting becomes more difficult when the tracks consist only of a few grains, which occurs with neutron energies below 0.5 MeV, (3) that the recognition of tracks becomes more difficult, if there is a background in the emulsion caused by γ-rays, (4) that the number of tracks produced per unit area per rem decreases with increasing neutron energy and finally (5) that photographic proton recording may be accompanied by significant latent image fading. The steps which have been taken to counteract these various drawbacks of the method will be discussed in the following sections.

6.6a. Photographic Material

The photographic material most suitable for the recording of proton tracks is the nuclear track type emulsion. Several varieties of these are available coated both on glass plates or on film base. The

Eastman Kodak Company supplies a special fast neutron recording nuclear track emulsion coated on film base and packed in a similar way to standard Radiation Monitoring films. The emulsion thickness used for dosimetric proton track recording is usually of the order of 30–50 microns when unprocessed, but the processed emulsion shrinks to about $\frac{1}{3}$–$\frac{1}{4}$ of its original thickness.* This is because the unexposed

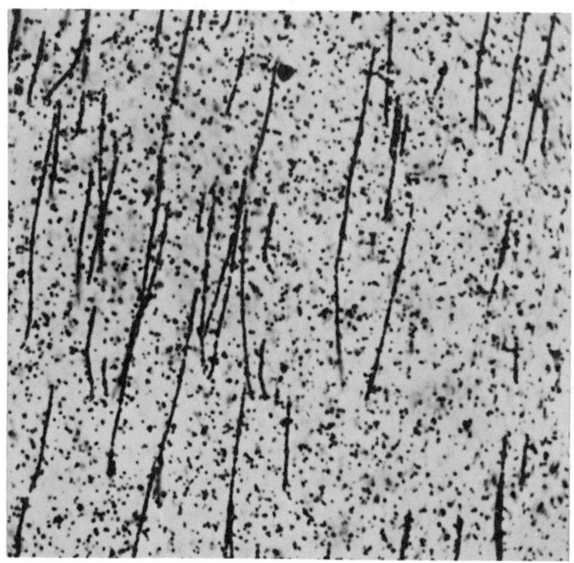

Figure 6.23. Photomicrograph (X 950) of proton tracks with slight γ-ray background. (*By courtesy of* Research Laboratories, Kodak Ltd. England.)

silver bromide will be washed out by the fixing bath. As mentioned on page 60, nuclear particle emulsions are characterized by their high ratio of silver bromide to gelatin, being of the order of 5:1 by weight, hence the grains are packed very closely. The undeveloped grains may be of the order of 0.1–0.3 microns in diameter, and are usually distributed over a very narrow range of sizes. This contributes to an easy recognition of tracks. Proton tracks as seen under the microscope with slight γ-ray background are illustrated in Figure 6.23.

6.6b.   Fast Neutron Dosimeter

The number of proton tracks recorded per unit volume of an emulsion depends on the neutron flux, on the number of hydrogen atoms present per unit volume and on the probability of neutron-proton collisions

* Corrections for shrinkage are important when determining the energy of particles. (See section 6.6d p. 244.)

at various neutron energies. Thus in order to assess the dose in rems for a given neutron energy one has to know (a) the number of neutrons per cm^2 required per rem for various neutron energies and (b) the probability of neutron-proton collisions for various neutron energies.

The number of neutrons per cm^2 per rem as a function of neutron energy is shown in Table 6.2. It is seen that the neutron flux for a

TABLE 6.2. Number of Neutrons per cm^2 per rem as a Function of Neutron Energy

Neutron energy	Integrated flux equivalent to 1 rem (n/cm^2)
Thermal	9.6×10^8
5 keV	8.2×10^8
20	4.0×10^8
100	1.2×10^8
500	4.3×10^7
1 MeV	2.6×10^7
5	2.6×10^7
10	2.4×10^7
14	1.4×10^7
20	1.4×10^7
40	1.0×10^7
50	1.0×10^7
70	9.6×10^6
110	8.8×10^6
190	6.8×10^6
300	4.9×10^6
475	3.3×10^6
650	2.4×10^6
1000	1.6×10^6

The data are taken from the National Bureau of Standards Handbook No. 63. *Protection against Neutron Radiation up to 30 Million Electron Volts.*

constant dose decreases with higher neutron energies. Concerning the second requirement (b) it is found that for high energy neutrons (about 1 MeV) an average of 10^4 protons per cm^2 per rem are produced in a nuclear emulsion of 50 microns thickness. The probability of neutron-proton collisions in the emulsion decreases, however, with increasing energy, as seen in Table 6.3 p. 240, which refers to the relative probability of collision for constant dose.

The effect illustrated by this table makes it necessary to calibrate emulsions for the spectral range of neutrons applied. Generally a wide energy spectrum is used and several attempts have been made to

TABLE 6.3. Probability of Neutron-Proton Collisions

Neutron Energy (MeV)	Relative probability of neutron-proton collision for (fixed) dose in rems
0.25	12
1.0	4
10.0	1

overcome the energy dependency by producing the majority of protons outside the emulsion in layers of materials known as radiators, which are rich in hydrogen, such as polythene. This appears to be promising as the higher the neutron energy, the greater will be the range of protons, and thus a greater number of high energy than of low energy protons will reach the emulsion from the radiator. One would thus expect that a certain compensation may be achieved by using this method. Several attempts have been made in this direction; for instance J. E. Cook (1958) uses a radiator of polythene of 250 microns thickness which is sandwiched between two 50 micron thick Ilford Nuclear Research Plates. The emulsions face the radiator. A 0.04 inch shield of cadmium covers half of the package, front and back, to suppress thermal neutrons which would otherwise contribute to the number of proton tracks on account of the $^{14}N(n, p)^{14}C$ reaction (see p. 173). In this manner the energy dependency has partly been compensated for, as seen from Figure 6.24 in which the number of protons per cm^2 per rem is plotted as a function of neutron energy in MeV for 50 micron thick emulsions using a 250 microns thick polythene radiator. The full curves refer to recoil proton energies above 0.25 MeV whereas the dashed curves present the result for recoil proton energies above 0.5 MeV. The latter curves show a considerable reduction of the energy dependence, as one would expect from Table 6.2, since the very short proton tracks are excluded. As noted on the graph Figure 6.24 one pair of curves refers to a neutron incidence of $0°$, i.e. normal incidence, the other pair to a $90°$ incidence. The probability of energy transfer in neutron-proton collisions is discussed on p. 177.

Another attempt to make fast neutron dosimetry practically independent of energy within the range from 1–14 MeV originates from J. S. Cheka (1953, 1954). This author proposed a film dosimeter in which a nuclear track film is packed in the following manner (see Table 6.4).

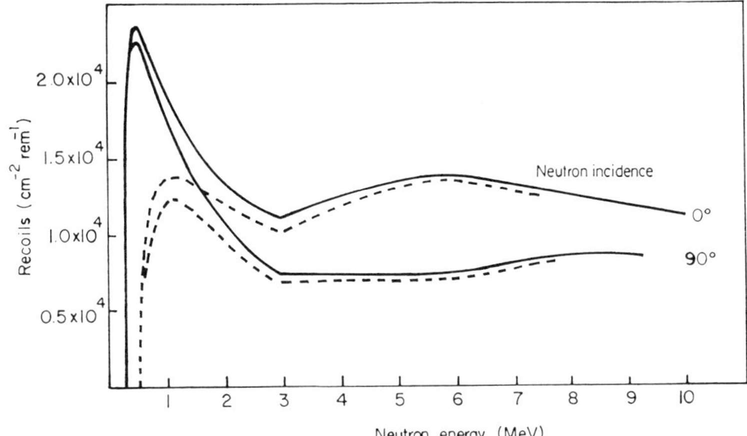

Figure 6.24. Proton recoil yield as function of dose and neutron energy. Proton recoils produced per cm² of emulsion per Rem for a 50 μ thick emulsion. 0.01″ polythene radiator, 50 μ emulsion pack. ————— Recoils of energy > 0.25 MeV – – – – – – – Recoils of energy > 0.5 MeV. The upper two curves refer to 0° neutron incidence, the two lower curves to 90° neutron incidence. (*After* J. E. Cook, *AERE* HP/R 2744 (1958).)

Such a neutron film dosimeter has been commercially available from the Eastman Kodak Company U.S.A. ready packed as a radiation monitoring film, but only the packing components indicated by + were built into the Kodak Personal Neutron Monitoring Film Type B. The original packing reduces the energy dependence, as

TABLE 6.4. Composition of Packing as suggested by J. S. Cheka (1954)

Material	Density in mg/cm²
Cellulose	76
Aluminum	85
Cellulose	24.2
Cellulose (Film wrapper)	+10.3
Aluminum	+27
Base of emulsion	+28.5
Emulsion	
Film Base	+28.5
Aluminum	+27
Cellulose (Film wrapper)	+10.3
Cellulose	24.2
Aluminum	85
Cellulose	76

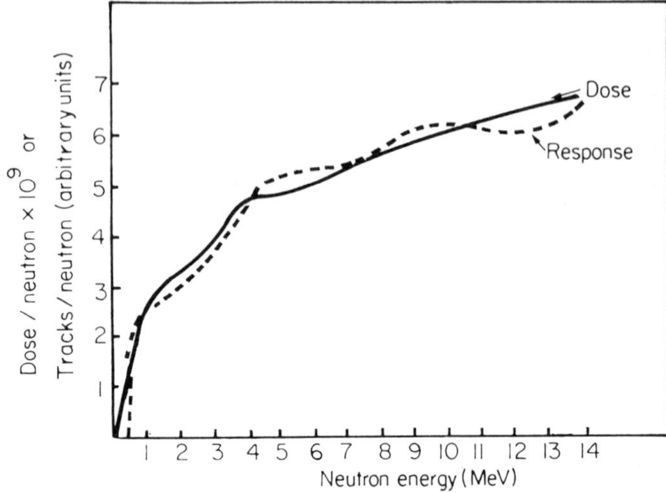

Figure 6.25. Response of Eastman Kodak Fast Neutron Monitoring Film (in the original packing as suggested by J. S. Cheka (1954)) as a function of neutron energy. ——————— Neutron Dose – – – – – – Photographic response. (*After* J. S. Cheka, *Nucleonics*, **12**.640. (1954).)

shown in the graph Figure 6.25, which illustrates the relative number of proton tracks and the relative dose as a function of neutron energy. It is seen that both curves practically coincide. Since some of the components suggested by Cheka are not included in the film packet, the type B emulsion packet was found to be energy independent up to 10 MeV, instead of up to 14 MeV (G. Portal, 1963).

In order to extend the independence to about 14 MeV, Portal uses an ordinary monitoring film in front of the type B film, which acts as

TABLE 6.5. Experiments on Neutron Energy Independence (due to G. Portal, 1963)

Neutron Spectrum	Type B alone	Type B with full package according to Cheka	Type B behind monitor film pack
	Number of proton tracks (10^5/cm² per rad)		
2.5 MeV	1.22	1.20	1.28
Polonium-Beryllium			
source*	1.26	1.24	1.27
14.0 MeV	0.79	1.15	1.62

* See Figure 11.3.

a further radiator. The Table 6.5 on page 242 illustrates his results, using type B film alone, type B film fully packed and type B film with monitoring film packet in front.

With reference to this table one notices the deficiency at 14 MeV for type B film alone (column II), the energy independence (column III) and the over-compensation at 14 MeV (column IV). The latter leads to an over-estimation of the dose of about 25 per cent.

6.6c. Microscopic Examination

Proton tracks in nuclear emulsions are usually examined with an oil immersion ($\times$ 100) objective and a ($\times$ 10) eyepiece preferably with a binocular microscope. (Total magnification $\times$ 1000). The tracks as seen in Figure 6.23 appear as straight rows of developed grains, their range relationship with energy is shown on page 604. For the counting of tracks by the scanning method it is advantageous to use an inclined scanning stage system, for which the inclination used may be about 10°. This facilitates the recognition of tracks which form an angle of dip with respect to the plane of the emulsion and also enables the full depth of the emulsion to be examined without much need for adjusting the focusing level. For dosimetric purposes the counting of tracks per unit volume is usually sufficient. In order to achieve a significant statistical value for the number of tracks per cm^2 a considerable area of the emulsion should be scanned. In a plate of 50 μ thickness exposed to a neutron dose of 1 rem, about 10^4 protons per cm^2 are expected. At least one per cent of these tracks should be counted e.g. those that occur within an area of one square millimeter. Using, for instance, a magnification of 1000 and a microscopic field of 150 microns diameter, 57 individual fields have to be scanned, which takes about 5 minutes for a skilled person. This rather tedious procedure can be improved by semi-automatic devices which move the microscope stage with the emulsion layer progressively in fixed steps over the whole field in one direction. In such a manner overlapping of fields can be avoided. Automatic proton track counting, machines have been developed, which permit rapid assessment of doses (see p. 250).

The number of developed silver grains produced by a proton of given energy depends of course on the size of the grains and on the ratio of silver to gelatin in the emulsion. The latter does not differ very much for different makes of nuclear emulsion, but the grain size depends on the particular type of emulsion chosen. As an average value one might expect that a 0.25 MeV proton track contains about four silver grains. This number of grains per track can be regarded as about the limit of fairly certain recognition, since a row of three grains may be due to a chance accumulation of developed grains for instance by the formation of background fog grains. Artifacts and scratches on the surface of the emulsion can usually be distinguished from tracks quite easily, as the majority of tracks are inclined with respect to the plane of the emulsion. In order to be sure that one deals with a track, its length should be followed by means of the fine focusing knob of the microscope. Occasionally one may find α-tracks in nuclear emulsions either singly or several originating from a point forming stars. These may be due to contamination of the emulsion (or the base in the case of emulsions coated on glass) by natural

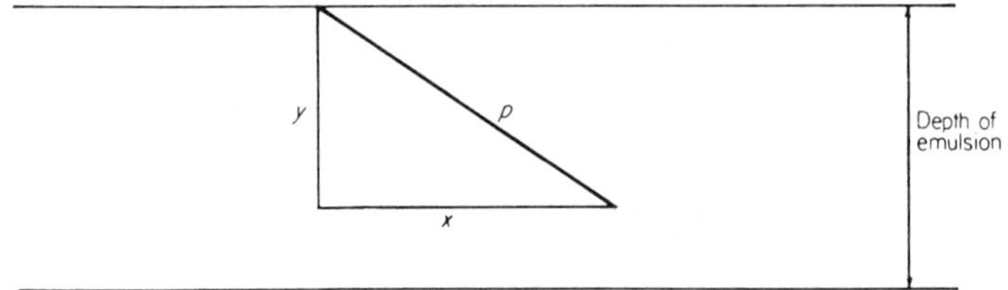

Figure 6.26. Diagrammatic section through nuclear emulsion showing a proton track of length p at an angle with respect to the emulsion plane.

radioactive material, such as uranium, radiothorium and others. These tracks should not be counted, but α-tracks can usually be distinguished from proton tracks by their higher grain density throughout their lengths.

6.6d.   Energy Distribution of Protons in Emulsion

In assessing the energy distribution of protons the following facts have to be considered. (1) The emulsion shrinks after processing, hence the lengths of the tracks appear foreshortened, (2) the actual track length can be evaluated from the projected length of the track. This latter point is explained by reference to Figure 6.26 which shows a section through an emulsion with a track (p) at an inclined angle with respect to the plane of the emulsion. The actual length (p) is found from $p = \sqrt{x^2 + y^2}$, where x is the projected length and y is the depth. Both quantities x and y can be measured with the microscope. The length x can be measured by focusing the microscope on the first and last grain of the track, y is measured by means of the depth screw. The observed value of y has to be corrected by the shrinkage factor (s) i.e. the ratio of the thickness of the unprocessed to that of the processed emulsion. The range of the particle in the emulsion is then found from

$$R = \sqrt{x^2 + s^2 y^2}$$

The respective energies are found from the range–energy curve of protons (see p. 604). It should be noted however that this distribution of proton energies, which can be established in the described manner, is not identical with the distribution of neutron energies and appropriate corrections have to be made based on the probability of energy transfer in neutron-proton collisions (see p. 177).

6.6e.   Latent Image Fading

Many of the results of fast neutron dosimetry can be spoilt by latent image fading (LIF) in the exposed but undeveloped grains forming proton tracks. LIF occurs during the interval between exposure and processing and has been discussed on p. 144. It is shown there that it occurs chiefly as a result of high humidity and temperatures in the environment of the photographic material. LIF effects and their avoidance in proton tracks have been described in detail by J. E. Cook (1958), H. Dresel (1963), K. Becker (1963) and G. Portal (1963).

LIF may occur in all nuclear emulsions, although to different degrees and several methods have been suggested to solve the problem. In Figure 6.27 due to G. Portal (1963) the effect of 50 per cent and 85 per cent relative humidity on the percentage latent image fading of non packed nuclear emulsion film is illustrated as a function of the time interval in weeks between exposure and processing. The graph reveals that after one week at 85 per cent relative humidity only 25 per cent of the number of proton tracks could be observed in the emulsion. The same author however shows how he was able to solve the problem by dessicating the film packs for a week (using silica gel as dessicator) and packing the films into sealed aluminum-polythene bags, without subjecting them to room atmosphere between dessication and sealed packing. It should be stressed that the sealing of the bags fulfils its purpose only if the interior atmosphere of the bag is at low humidity. H. Dresel (1963) and R. H. Herz (1959) suggest an additional precaution by displacing the oxygen within the bag by carbon dioxide (see Figure 12.26).

6.6f. γ-Ray Background

The exposure to fast neutrons is usually accompanied by exposure to γ-rays. Hence the latter contribute to a general background fog which is superimposed on the proton track images. This may lead to difficulties in recognizing proton tracks. Most nuclear emulsions chosen for the recording of proton tracks have a relatively low sensitivity to γ-rays, usually of the order of $\frac{1}{10}$ or $\frac{1}{100}$ of that of fast X-ray

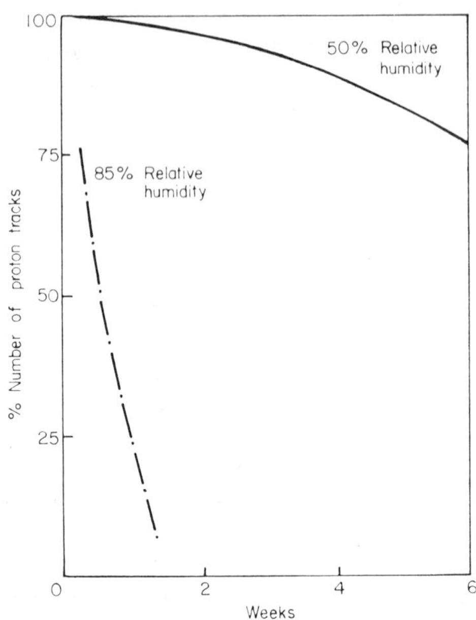

Figure 6.27. Latent image fading of proton tracks at 50 per cent and 85 per cent relative humidity in non-wrapped dosimeter films, as a function of time (weeks) elapsed between exposure and processing. (*After* G. Portal, Personnel Dosimetry Techniques for External Radiation Symposium Madrid (1963) ENEA-OECD.)

emulsions. Nevertheless, the background may be so troublesome that the accuracy in the assessment of dose can be very much impaired. K. Becker (1963 (b)) reports that the number of recognizable proton tracks was reduced to two-thirds of its actual value, when a background due to a dose of 1 rad of ^{60}Co γ-rays was present. The lower the energy of the proton recoils the less accurate will be the count, as the short tracks are more easily overlooked in a strong background fog.

There are basically three different methods which have been proposed to reduce the effects of γ-rays, two of them are based on discrimination by development techniques.

1. Special processing techniques using developers with low activity have been suggested which should help in the discrimination between high and low ionizing radiation, such as between protons and γ-rays respectively. K. Becker (1963b) however, did not find significant differences between conventional processing methods and those recommended by H. Farragi, A. Bonnet and M. J. Cohen (1952). H. G. de Carvalho, M. Muchnik and others (1962) have proposed other processing methods which have apparently not been tested for dosimetric purposes.

2. G. Portal (1963) makes use of underdevelopment. He measures first the degree of γ-ray fog on an ordinary monitoring film which was exposed together with a neutron film. He adjusts his developing time according to the value of the γ-ray dose he obtained, as seen from the table below. Successful methods for the discrimination of different

TABLE 6.6. Reduction of γ-ray fog by under-development (due to G. Portal 1963)

γ-ray dose in rems	Developing time in minutes	Percentage loss of tracks
0–3	15	$\leqslant 10$
3–6	8	$\leqslant 25$
6–10	5	$\leqslant 40$

particle tracks by means of pH buffers have been discussed in detail by G. W. W. Stevens (1951).

3. Another means of selective reduction of γ-ray fog (i.e. induced γ-fading) is to obtain a greater latent image fading of γ-ray fog, than of proton tracks. This was tested by treating the exposed emulsion with oxidizing agents or with acids prior to development. The best results were obtained with HCl which gave a 50 per cent reduction in density due to γ-rays without an appreciable loss of proton tracks. (I. V. Mejdhal, C. R. McCarthy, 1963).

6.6g. Processing of Neutron Monitoring Emulsions

The processing of nuclear emulsions for recording proton tracks is fairly straight forward.

Developer. X-ray developers, i.e. those suitable for the development of X-ray films are applicable, although some workers prefer weaker developing solutions in order to reduce the background due to γ-rays, as mentioned above. The development times for nuclear emulsions are generally recommended to be somewhat longer than those applied for X-ray films, because the emulsion coatings usually have at least double the coating thickness of X-ray films. Hence, sufficient time must be given to the developer solution to diffuse through the emulsion.

A weak developer solution proposed by J. E. Cook (1958) has the following composition.

Hydroquinone	0.5 grams
Sodium sulfite (anhydrous)	10.0 grams
Potassium carbonate (anhydrous)	50.0 grams
Potassium bromide	0.5 grams
Water to make up	1 liter

A stock solution is kept containing all the chemicals except hydroquinone which is added before use. Developing time is 15 minutes at 20°C.

Stop Bath.   After development, the films or plates should be rinsed and then transferred to a stop bath of 4 per cent sodium metabisulfite for 5–10 minutes. It is advisable to remove any surface deposits from the plates or films after the use of this solution. This can be done by gently wiping with the finger or wetted cotton wool.

Fixing Bath.   A non hardening 40 per cent hypo fixing bath is recommended in which the emulsions are left for twice the time to clear. The time required to clear the emulsion in this fixing bath is considerably longer than for conventional films. The reason for this is that the nuclear emulsion contains a much greater mass of silver bromide per unit volume than, for example, an X-ray emulsion and relatively few of the silver bromide grains are converted into silver, which means that most of the original silver bromide has to be removed by the fixer. Depending on the thickness of the emulsion, the fixing time may take up to 1–2 hours for emulsions from 50–100 microns in thickness. The Eastman Kodak Company recommends a fixing time of only 20–30 minutes for their Neutron Monitoring films in a sodium thiosulfate fixer (not a rapid fixer).

Washing and Drying.   Thorough washing is recommended for at least 2 hours in running water. Drying in air at room temperature and room humidity is acceptable for emulsions up to 50 microns thick. Using thicker emulsions, greater care must be taken to avoid distortion and it is recommended that the water is removed by immersion in successively more concentrated solutions of either alcohol or polyethylene glycol containing 5 per cent glycerol. After removal from the last solution the final traces of liquid are evaporated in a strong current of air at room temperature and humidity.

6.7 AUTOMATION

If personnel dosimetry is carried out on a large scale, with say thousands of monitoring films per week, the amount of routine work involved in the standardized processing and evaluation of films becomes a formidable task. It is, therefore, feasible to consider semi- or full automation of the main procedures involved, such as in the coding and processing of films, obtaining densitometer readings and in the assessment of doses including automatic recording of the necessary data. The advantages of semi- or full automation increase with the sophistication of the film badges. For example, it may be necessary to take 6–8 density readings on each film used with certain types of film badges.

Also the measurement of fast neutron doses which is such a tedious operation could be reduced to a minimum of routine work, if automatic proton track counting apparatus were used. In several countries a considerable amount of thought has been devoted to the application of semi- and full automation in both fields, although it is true to say that the use of automation in photographic dosimetry is at present still more an exception than the rule. In the following section only a brief account is given of the kind of apparatus which have been developed, but for a full appreciation of the subject the reader is referred to the literature (J. W. Adams and C. N. Wright, 1960; P. G. Lale, 1960; M. Wilhelmsen, D. Parker, R. W. Coulter and P. R. Boren, 1960; D. H. Peirson, 1963; M. J. Heard, 1964; and on automatic proton counters, S. Becker, 1961; B. J. Wilson, 1961; A. Narath and P. Koeppe, 1962).

6.7a. Automatic Coding

One of the first operations in the handling of monitoring films is to print identification numbers on the packing and on the films. An automatic film coding machine has been developed by the Atomic Energy Research Establishment, Harwell, England, which numbers monitoring film packets and films sequentially. In order to eliminate visual reading of the number on the film itself a coded number is printed on the film through the packing by means of soft X-rays (20 kV_p). The number consists of a six digit identification number which is interpreted and recorded by an automatic recording densitometer. The coding machine together with its X-ray tube operates automatically at the rate of 60 films per minute.

6.7b.   Automatic Processing

An automatic processing unit for monitoring films has been designed and constructed by the United Kingdom Atomic Energy Authority Establishment, Windscale, in conjunction with industry. In this unit racks containing 50 films are taken into the machine at intervals of 2 minutes and pass successively through the various processing baths and eventually into a dryer. The handling capacity of the unit is 1500 films per hour. The baths are thermostatically controlled and nitrogen burst agitation is used in the developing tank together with recirculation of the solution.

6.7c.   Automatic Recording Densitometers

Such units have been described by D. H. Peirson (1963) and by M. J. Heard (1964b) and one type has been developed by the Atomic Energy Research Establishment, Harwell, England in conjunction with an industrial organization. The apparatus exists also in a semi-automatic version. A picture of the fully automatic unit is shown in Figure 6.28a. A schematic diagram of the various operations of the unit is shown in Figure 6.28(b). The unit automatically reads and records optical density measurements of each of the separate areas of the films. It also automatically reads, decodes and records a six digit identification number. Furthermore the density measurements can be converted automatically to equivalent γ-doses, on the basis of a calibration curve. This conversion is adjustable to suit different calibrations. The data, i.e. the identification number and density measurements are presented on punched cards, whilst the equivalent γ-doses and the identification number are recorded on an output writer. This information is for immediate use, but the punch card data can, if required, be used for dose calculation on a computer. The machine also draws attention to radiation doses in excess of a preset level by red print. If abnormalities occur, such as inadequate identification code, incorrect orientation of film, excess density levels etc., a reject gate opens and the film is diverted into a reject magazine for individual examination and the particular fault is indicated by an output code. Films are handled at the rate of one every 14 seconds which means that about 240 films are automatically read and evaluated per hour.

6.7d.   Automatic Proton Track Counter

The basic principle of particle track evaluation by semiautomatic units has been applied by cosmic-ray physicists in the past, e.g. by S. J. Goldsack and H. B. van der Raay (1956), P. V. C. Hough, W.

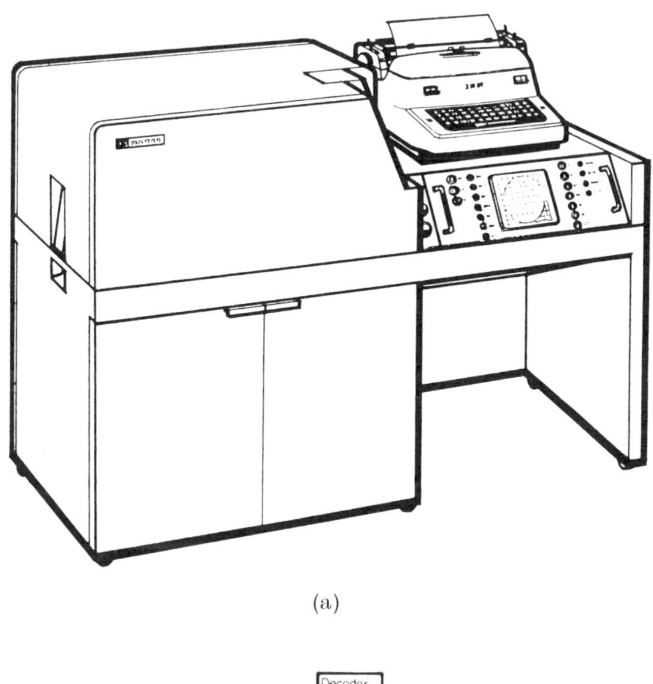

(a)

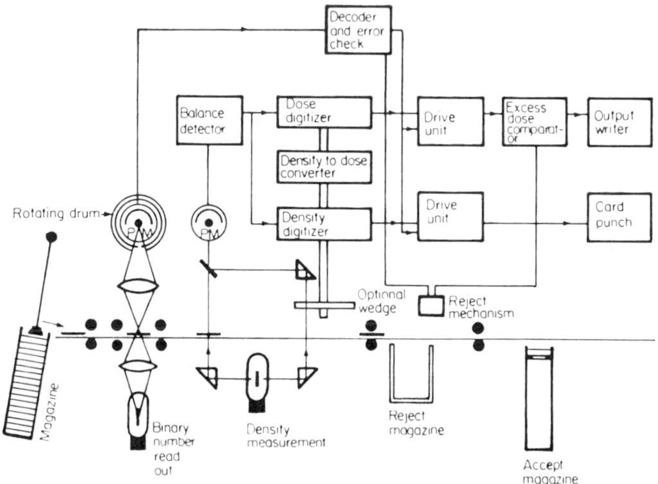

(b)

Figure 6.28. (a) External Appearance of fully automatic recording densitometer.
(b) Diagram (Functional) of the above densitometer. (*After* M. J. Heard, AERE-
M1371 and M. J. Heard, *J. Phot.Sci.* **12**.312. (1964).)

Williams, and R. C. Winder (1958), W. H. Barkas (1960) A. E. Vornonkov and A. I. Glaktinov (1961) and others. Cosmic ray physicists are generally interested in counting the number of grains per unit length of track and in the angular scattering of particles along their paths. In such cases the track or event is presented to the instrument and does not have to be specifically identified. The problem in dosimetry is different, as it is chiefly concerned with the automatic counting of the number of tracks per unit area of an emulsion. In this case it is essential that the instrument should be capable of differentiating between tracks and blobs of grains or small artifacts.

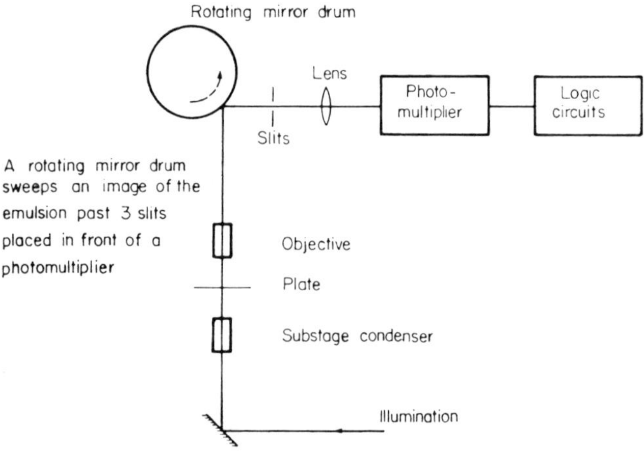

Figure 6.29 (a). Diagram of the basic system used in the Automatic Proton Track Counter. (*After* M. J. Heard, AERE-M1371 and M. J. Heard, *J. Phot. Sci.* **12**.312 (1964).)

Furthermore, all tracks should be recognized and counted whatever their orientation in the volume of the emulsion.

Automatic proton track counters suitable for dosimetry have been developed for instance by F. P. Cowan (1959), S. Becker (1961), W. H. Barkas (1961), P. Koeppe (1961), A. Narath and P. Koeppe (1962), D. H. Peirson (1963), M. J. Heard (1964) and others.

A description follows of an automatic proton track counter (M. J. Heard, 1964) which has recently been designed by the Health Physics and Medical Division of the Atomic Energy Research Establishment, Harwell in collaboration with Southern Instruments Contracts Limited, England. The basic function of the apparatus is as follows. A magnified image of the microscope field in which proton tracks may be seen is projected on to a diaphragm provided with a system of slits. The image of the field is swept across this system of slits by means of a rotating mirror drum (see Figure 6.29(a)). A photomultiplier behind

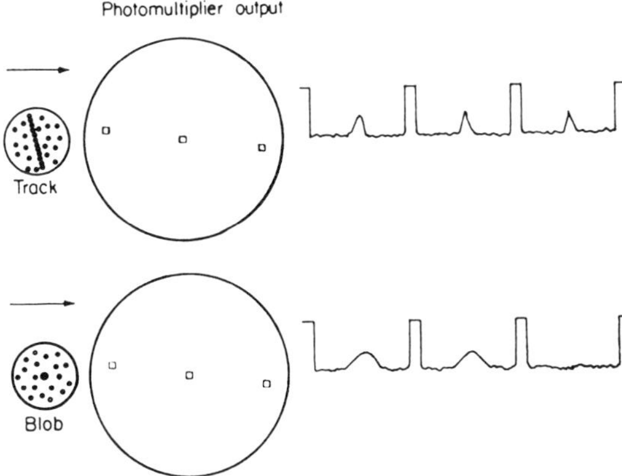

Figure 6.29 (b) Diagram illustrating the projected image, slit system and photomultiplier output of the Automatic Proton Track Counter. The type of signals obtained from a track and a small 'blob' are shown. A track is recognized by (1) The shape of the output pulse of the photomultiplier as an object passes the slit, and (2) the characteristic shape appearing from each slit after a given time interval. (*After* M. J. Heard, AERE-M1371 and M. J. Heard, *J. Phot. Sci.* **12**.312 (1964).)

the screen converts the light intensity passing through the slits into electrical signals. The shape of these signals is analyzed electronically so that tracks can be differentiated from scratches, individual blobs of grains etc. In the diagram (b) of Figure 6.29 the track recognition and slit system is illustrated. The type of signals obtained from a track and also a small blob are shown on the left. A track is recognized by the shape of the output pulse of the photomultiplier (as shown on the right) as an object passes a slit. The slit system consists of three slits, each equivalent to 2 microns in length and 1 micron in width in the plane of the emulsion; they are displaced relative to one another by 2 microns along their length and are spaced across the diameter of the projected field of view (this is illustrated in Figure 6.29(b)). This arrangement corresponds to a segmented slit of dimensions 6 microns by 1 micron. Hence the slit system permits a discrimination between short tracks and small groups of silver grains and it further permits a large acceptance angle. The mirror drum, rotating at 3000 r.p.m. has 45 faces. The emulsion on the stage of the microscope is moved with a steady speed of 7.2 mm per second, so that the image moves 3.5 microns for each pass of a mirror. The total distance scanned in the y direction is 15 mm, at the end of which the stage is stepped 100 microns in the x direction.

Complete scanning of a single plate takes 12.5 minutes. This includes the time required to automatically replace the plate on the microscope by the next plate. One cassette containing 60 plates is scanned in 12.5 hours. Thus the apparatus will run unattended for this period of time. The data are recorded by printing the number of scans and the number of tracks counted in that area.

REFERENCES

Adams, J. W., and Wright C. N. (1960). A modernized film badge system *5th Ann. Meetg. Health Phys. Soc.*, (Abstract) Boston.

Allisy, A. (1955). La mesure des doses de rayons x ou γ à l'aide d'émulsions photographiques. *J. Rad. and d'Electrol*, **31**, 249.

American Standards Association Inc. (1965). Method for Evaluating Films for Monitoring X-rays and γ-Rays having energies up to 3 million electron volts. PH2, 10-1965.

Barkas, W. H. (1958). Equipment and methods for automatic track scanning. *U.C.R.L.*, 8482.

Barkas, W. H. (1960). New facilities and automatic track analysis equipment at the Lawrence Radiation Laboratories. *U.C.R.L.*, 9180.

Barkas, W. H. (1961). Nuclear Research Emulsions, Academic Press, New York.

Baumgartner, W. V. (1960). X-ray spectrometry extends film badge dosimetry, *Nucleonics*, **18**, 8, 76.

Becker, K. (1960a). Probleme und Ergebnisse der Filmdosimetrie ionisierender Strahlen. *Phot. Korr.*, **96**, 83, 115.

Becker, K. (1960b). Ein wellanlängenunabhängig registrierender Dosismessfilm. *Z. Elektrochem.*, **64**, 1102.

Becker, K. (1961a). Beitrag zur Filmdosimetrie energiereicher Quantenstrahlung. Dissertation. Tech. Hochsch. München.

Becker, K. (1961b) Zum Stand der praktischen Filmdosimetrie, *Kerntechnik*, **3**, 120.

Becker, K. (1962, 1966). *Filmdosimetrie*, Springer-Verlag, and *Phot. Film Dosimetry*, Focal Press, London.

Becker, K. (1963a). Capabilities and limitations of the different methods applied in personnel dosimetry. *Personnel Dosimetry Techniques, Symposium Madrid*, 393. E.N.E.A. (O.E.C.D)

Becker, K. (1963b) Fehlerquellen bei der Neutronen-Personendosismessung mittels Kernspuremulsionen. *Atomkernenergie* **8**, 2, 74.

Becker, K., Klein E., and Zeitler E. (1960). Ein wellenlängen-unabhängig registrierender Dosismessfilm. *Naturw.* **47**, 199. A wavelength independent recording dosimeter film. *A.E.C.* Tr. 4408.

Becker, S. (1961). Automatic nuclear emulsion scanner. *Health Phys.*, **4**, 164.

Berlman, I. B. (1953). Determination of photographic film exposure by activating ^{109}Ag. *Nucleonics*, **11**, 2, 70.

Biermann, A., and Feige Y. (1965). Density permanence of processed monitoring films. *Health Phys.*, **11**, 320.

Birks, L. S. (1959). *X-ray Spectrochemical Analysis*. Interscience, New York.

Boller, B. K. (1964). Sensitometric Effects of Million-Volt X-rays on Selected Kodak Films. *Phot. Sci. and Eng.*, **8,** 4, 185.

British Standards Institution (1963). Specification for film badges for personnel radiation monitoring. BS. 3664

British Standards Institution (1965). General recommendations for the testing, calibration and processing of radiation monitoring films. BS.3890.

Carvalho, H. G. de., Muchnik, M., Potenza G., and Rinzvillo R. (1962). Discrimination of proton tracks on heavy gamma-ray background. *Korpuskular-Photographie Inst.Wiss. Phot., Techn. Hoch-Schule,* München.

Cheka, J. S. (1953). Fast neutron film dosimeter. *Phys. Rev.*, **90, 353.**

Cheka, J. S. (1954). Recent developments in film monitoring of fast neutrons. *Nucleonics*, **1, 26,** 40.

Cipperley, F. V., and Gammel, W. P. (1961). Improvements in personnel metering procedures at the national reactor testing station. *Health Physics*, 4, 173.

Cleare, H. M., and King, N. H. (1961). 6th Annual Meeting of Health Physics Society, Las Vegas, Nevada.

Cook, J. E. (1958). Fast neutron dosimetry using nuclear emulsion. *A.E.R.E.* HP/R 2744.

Corney, G. M. (1960). The effect of gamma-ray exposure on camera films. *Phot. Sci. Eng.*, **4,** 291.

Corney, G. M., and Cleare, H. M. (1955) Disaster monitoring with amateur photographic films. *Nucleonics* **13,** 8, 40.

Cowan, F. P. (1959). Techniques of Radiation Protection, in H. Blatz (Ed.), *Radiation Hygiene Handbook*, McGraw-Hill, New York.

Crabtree, J. I., Eaton, G. T., and Muehler, L. E. (1943). The removal of hypo and silver salts from photographic materials as affected by the composition of the processing solution. *J. Soc. Mot. Pict. Eng.* **41.**9.

Dennis, J. A. (1963). Neutron dosimetry using activation techniques. *Personnel Dosimetry Techniques*, Symposium Madrid, 189. E.N.E.A. (O.E.C.D.)

Dorneich, M., and Schäfer, H. (1942) Über die Strahlenschutz-Messung in r nach der photographischen Methode, *Phys. Zt.*, **43,** 390.

Dresel, H. (1956). Filmdosimetrie bei Strahlenschutz-Messungen, *Fortschr. Geb., Röntgenstr.*, **84,** 2, 214.

Dresel, H. (1960). Praktische Bestimmung der Personendosis mit photographischen Emulsionen in Röntgen-, Isotopen- und Reaktor-betrieben, *Kerntechnik*, **2,** 7/8, 239.

Dresel, H. (1963). Eine neue Filmplakette für Strahlengemische verschiedenster Art, *Kerntechnik Isotopentechnik und Chemie*, **5,** 11, 446.

Duhamel, F., and Soudain, G. (1963). Expérience passée et practique actuelle de la dosimetrie par films. *Personnel Dosimetry Techniques. Symposium Madrid*, 77. E.N.E.A. (O.E.C.D.)

Ehrlich, M. (1957). A photographic personnel dosimeter for X-radiation in the range from 30 KeV to beyond 1 MeV. *Radiol.*, **68, 549.**

Ehrlich, M., and Fitch, S. H. (1951). Photographic X- and gamma ray dosimetry. *Nucleonics*, **9**, 3, 5.

Ehrlich, M., and McLaughlin, W. L. (1959). Photographic Dosimetry at Total Exposure Levels Below 20 mr. *Nat. Bureau of Standards, Technical Note* 29, October 1959.

Faraggi, H., Bonnet, A., and Cohen, M. J. (1952). Irradiation et développement des émulsions nucléaires exposées a des flux intenses de neutrons thermiques accompagnés de rayons gamma. *J. Phys. Radium.*, **13**, 105.

Goldsack, S. J., and Van der Raay, H. B. (1956). An automatic scanner for nuclear emulsions. *J. Sci. Instr.*, **33**, 135.

Greening, J. R. (1951). The photographic action of X-rays, *Proc. Phys. Soc.*, (*London*), **B64**, 977.

Gupton, E. D. (1956). A revised technique for film dosimetry at Oak Ridge National Laboratory, *Radiol.*, **66**, 253.

Haschè, E. (1939). Uber die Messung des Strahlenschutzes auf photographischem Wege in Röntgeneinheiten. *Fort. Geb. Röntg., Strahlen*, **60**, 74.

Heard, M. J. (1962) *Review of Present Practice and Past Experience in Film Badge Monitoring for X- gamma- and beta-Radiation and Thermal Neutrons in the Atomic Energy Authority.*, A.H.S.B. (RP) R-19.

Heard, M. J. (1964a). Photographic radiation dosimetry and the development of the A.E.R.E./R.P.S. Film dosimeter. *A.E.R.E.*-M 1370.

Heard, M. J. (1964b). Report on the progress of automation of photographic dosimetry, April 1964. A.E.R.E. M-1371, 1964 and *J. Phot. Sci.*, **12**, 6, 312.

Heard, M. J. (1965). Photographic radiation dosimetry and the development of the A.E.R.E./R.P.S. Film dosimeter. *J. Phot. Sci.*, **13**, 32.

Heard, M. J., Cook, J. E., and Holt P. D. (1960). Photographic emulsion dosimetry and the A.E.R.E. film dosimeter. *A.E.R.E. Report* 3300.

Heard, M. J., and Jones, B. E. (1963). A new film holder for personnel dosimetry. In '*Personnel Dosimetry Techniques for External Radiation*', p. 89. *Symposium Madrid.* E.N.E.A. (O.E.C.D.)

Henson, P. W. (1963). A photographic dosimeter for the measurement of personal doses of environmental radiation. *Phys. Med. and Biol.*, **8**, 4, 423.

Henson, P. W. (1964). The influence of heat on the sensitivity of a photographic film to low intensity light. *J. Phot. Sci.*, **12**, 102.

Herz, R. H. (1959). Methods to improve the performance of stripping emulsions. *Lab. Invest.* **3**, 1, 71.

Hine, G. J. (1952). Secondary electron emission and effective atomic numbers, *Nucleonics*, **10**, 1, 9.

Hine, G. J. (1954). The range of usefulness of photographic film in roentgen dosimetry. *Am. J. Roentg.*, **72**, 293.

Hoerlin, H. (1956). Photographic materials for nuclear laboratory and test work. *Wiss. Phot. Internat. Konf. Köln.*

Hoerlin, H., and Mueller, F. W. H. (1950). Gold sensitization of X-ray films. *J. Opt. Soc. Am.*, **40**, 246.

Hoerlin, H., Clark, R. H., Jones, D. P., Kaszuba, F. J., and Larson, E. T. (1953). Development of a wavelength-independent radiation monitoring film. *Final Report ANL*-5168.

Holt, P. D., The response of photographic film dosimeters to high energy gamma radiation. *A.E.R.E. Report* 4168, 1964.

Holt, P. D., and Gibson, J. A. B. (1965). High-energy gamma-ray dosimetry: Experimental measurements at a nuclear power station. *J. Nucl. Energ. Parts A/B* **19**, 17.

Hough, P. V. C., Williams, W., and Winder, R. C. (1958). Nuclear emulsion scanner for nuclear spectroscopy. *Bull. Am. Phys. Soc.* **2**, 3, 181.

Jones, B. E., and Marshall, T. O. (1964). The dosimetry of mixed radiations, involving more than one energy or type of radiation, with the R.P.S./A.E.R.E. Film dosimeter. *J. Phot. Sci.*, **12**, 6, 319.

Jones, B. E., and Marshall, T. O., (1965). A practical method of personnel monitoring for beta-radiations using photographic emulsions. *J. Phot. Sci.*, **13**, 1, 12.

Klein, D. (1961). Einfluss von Hypersensibilierung und Farbstoff-Desensibilisierung auf die Wellenlängenabhängigkeit der Empfindlichkeit für die energiereiche Strahlung. *Intern. Koll. Wiss. Phot.*, Zürich.

Kocher, L. F. (1957). HAPO adopts a new personnel dosimeter. *Nucleonics*, **15**, 12, 70.

Kocher, L. F. (1962). A personnel dosimeter filter system for measuring beta and gamma doses in mixed radiation fields. HW-71764 March 1962, Unclassified.

Kocher, L. F., Bramson, P. E., and Unruh, C. M. (1963). The New Hanford Film Badge Dosimeter. Unclassified Report. HW-76944 From *Radiation Protection Operation*. Hanford Laboratories, G.E.C., Richland, Washington.

Koeppe, P. (1961) Automatisches Auswertungs-Gerät fur Kernspuremulsionen. *Dissertation D83*. Techn. Univ. Berlin.

Lale, P. G. (1960). A modified method for personnel monitoring. *Brit. J. Radiol.*, **33**, 748.

Langendorff, H., Spiegler, G., and Wachsmann, F. (1952). Strahlenschutz-Überwachung mit Filmen. *Fortschr. Geb. Röntgenstrahl.* **77**, 143.

Langmead, W. A. (1963) External radiation dosimetry practice in the United Kingdom Atomic Energy Authority. *Personnel Dosimetry Techniques. Symposium, Madrid*, 455. E.N.E.A. (O.E.C.D.)

Loevinger, R. (1953). Extrapolation chamber for the measurement of beta sources. *Rev. Sci. Instr.*, **24**, 10, 907.

McLaughlin, W. L. (1960). Megaroentgen dosimetry employing photographic film without processing, *Radiation Research* **13**, 594.

McLaughlin, W. L. (1962). Evaluation of Unexpectedly Large Radiation Exposures by Means of Photographic Film. Nat. Bureau of Standards Technical Note 161, August 1962.

McLaughlin, W. L. (1965). Latensification in Radiographic Exposures. *J. Phot. Sci.*, **13**, 1.1.

Mayneord, W. V. (1930). Secondary electronic emission from metal foils and animal tissues. *Proc. Roy. Soc.*, A. **130**, 64.

Mejdahl, V., and McCarthy, C. R. (1963). Induced gamma fading in neutron films. *Personnel Dosimetry Techniques for External Radiation. Symposium Madrid* E.N.E.A. Organization for Economic Co-operation and Development p. 253.

Moser, J. S. (1953). *The Effects of Nuclear Radiation Upon Photographic Film.* AD-144586, Armed Services Technical Information Agency, Arlington 12, Va.

Narath, A., and Koeppe, P., (1962). An automatic scanner for nuclear emulsions. *Symposium on Neutron Detection, Dosimetry and Standardisation.* I.A.E.A. Harwell, December 1962.

National Bureau of Standards, *Protection Against Neutron Radiation up to* 30 *Million Volts.* Handbook No. 63.

Nitka, H. F. (1959). Photographic methods (megaroentgen dosimetry). *Nucleonics* **17**, 10, 58.

Nitka, H. F., and Jones, D. P. (1957). Photographic method for megaroentgen dosimetry. *Nucleonics,* **15**, 10, 128.

Pardue, L. A., Goldstein, N., and Wollan, E. O. (1944). Technical Information Service Extension AEC, MDDC-1065. 8th April 1944.

Peirson, D. H. (1963). Progress in the automation of photographic dosimetry. In 'Personnel Dosimetry Techniques.' *Symposium Madrid*, 305, *Personnel Dosimetry Techniques for External Radiation.* E.N.E.A. (O.E.C.D.)

Portal, G. (1963). Dosimetrie des Neutrons Rapides Au Moyen D'Emulsions A Protons De Recul in 'Personnel Dosimetry Techniques for External Radiation' Symposium Madrid p. 219. E.N.E.A. (O.E.C.D.)

Robb, T. M. and Ellis, R. E. (1952). The use of small protection films for the estimation of the doses received on the fingers and hands during radium manipulations. *Brit. J. Radiol.* **25**, 100.

Ross, D. M. (1963). External radiation dosimetry practices in the United States in 'Personnel Dosimetry Techniques for External Radiation.' Symposium Madrid p. 443 E.N.E.A. (O.E.C.D.)

Soole, B. W. (1956) Photographic badges for the estimation of the quality of X- and gamma-radiation. *Brit. J. Radiol.*, **29**, 450.

Spiegler, G., 1950. A new method of use of the photographic film as quality indicatior and dose-meter for X- and gamma-rays. *Phot. J.*, **90B.**, 166.

Spiegler, G. (1951) Photographic protection measurements with X-rays gamma- and beta-rays. *Phot. J.* **91B**, 128.

Spiegler, G., and Davis, R., (1959). An improved method and film-holder for personnel monitoring. *Brit. J. Radiol* **132**, 464.

Stadelmann, H. (1960). Kombinations-Filter zur Unterdrückung der Energieabhängigkeit von Dosis Filmen. Diss. Univ. Erlangen.

van Stekelenburg, L. H. M. (1958). New film badge enables cheaper X-ray monitoring. *Nucleonics* **16**, 6, 83.

Stephenson, S. K. (1964). Experience with finger-film dosimetry in work with ionizing radiations. *J. Phot. Sci.*, **12**, 1, 91.

Stevens, G. W. W. (1951). Some Applications of pH buffers in the development of nuclear particle plates. *Photographic Sensitivity Symposium Bristol*, 1950, Butterworth, London.

Tochilin, E., Davis, R., and Clifford, V. (1950). A calibrated roentgen-ray film badge dosimeter. *Am. J. Roentgenol.* **64**, 475.

Voronkov, A. E., Glaktinov, A. I., Murin, I. D., Sukhov, L. V. and Shtranikh, I. V. (1961) Apparatus for the automatic scanning of nuclear emulsions by the television method. *Prib. i. Tekhn. Eksp.* **2**, 68.

Wachsmann, F., and Stadelmann H. (1961). Unterdrückung der Energie-abhängigkeit von Dosisfilmen durch Kombinationsfilter. *Phot. Korr.*, **97**, 83.

Wilhelmsen, M., Parker, D., Coulter, R. W., and Boren, P. R. (1960). Automatic film-badge reader. *Nucleonics* **18**, 4, 84.

Wilsey, R. B. (1951). The use of photographic films for monitoring stray X-rays and gamma-rays. *Radiol.*, **56**, 229.

Wilson, B. J. (1961). Automatic scanning of nuclear emulsions. *An Annotated Bibliography*. A.E.R.E.-Bib., 137.

Wilson, C. W. (1941). The dependence of the secondary electronic emission produced by gamma radiation upon the direction of the radiation. *Proc. Phys. Soc.*, London, **53**, 613.

7
Fundamentals of Radiography

7.1 INTRODUCTION

The term 'radiology' covers the whole field of the use of ionizing radiation in medicine, biology, and industry. 'Radiography' is concerned with the physical and technical principles of recording images from three-dimensional objects by means of X- and γ-rays and other ionizing radiations.

In this and the following chapters the photographically important aspects of 'radiography' in its various applications will be discussed. The most widely applied field of radiography is undoubtedly that of medical diagnosis. As distinct from the medical use, industrial radiography covers another important variety of applications, particularly in engineering investigations of light and heavy metals, plastics, rubber, and other fabricated goods. Apart from medical and industrial radiography other less conventional applications have been developed, such as the examination of paintings, of forgeries and of microscopically small objects in biology and metallurgy (microradiography) or the use of electrons and neutrons as radiation sources in radiography and others (L. Mullins, 1949).

This chapter presents a brief introduction into the elementary principles of radiography which are common to both medical and industrial applications. We shall deal with the geometry of image formation, with the evolution of a radiograph and the basic terms of contrast and definition. In the subsequent chapters medical and industrial radiography will be treated separately. As the processing of radiographs is basically the same in the medical and industrial field, both aspects will be covered together in Chapter 9.

7.2   THE GEOMETRY OF IMAGE FORMATION

7.2a   Central Projection

The radiographic image is caused by the projection from a relatively small point of the origin of radiation (focal spot of an X-ray tube or a radioactive source) through the object to be radiographed on to a two-dimensional plane, the film, see Figure 7.1. Hence the formation of a radiograph has essentially the character of a central or point-projection. As the radiation penetrates the three-dimensional

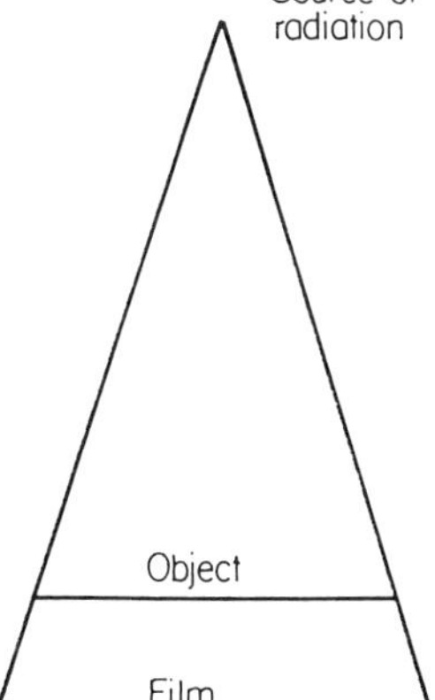

Figure 7.1. Central or point-projection.

object, the radiograph necessarily presents a superimposition of images of an infinite number of planes of the object. The final image is sometimes, although not quite correctly, called a shadow image of the object, as it presents differentiated shadows caused by the differentially penetrated regions of the object.

7.2b   Formation of Penumbra

If ideal projection i.e. from a point source, could be obtained in practice, we would be able to obtain a very high degree of definition or image sharpness. This is because each point in the object would be rendered as a point in the image. Unfortunately this is not so in practice because of the finite area of the so called focal spot of the X-ray tube. A lower limit to the area of the target of the X- ray tube is required to prevent damage by the heat generated under electron bombardment.

The influence of the dimensions of the radiation source on the formation of the radiographic image will be evident by reference to

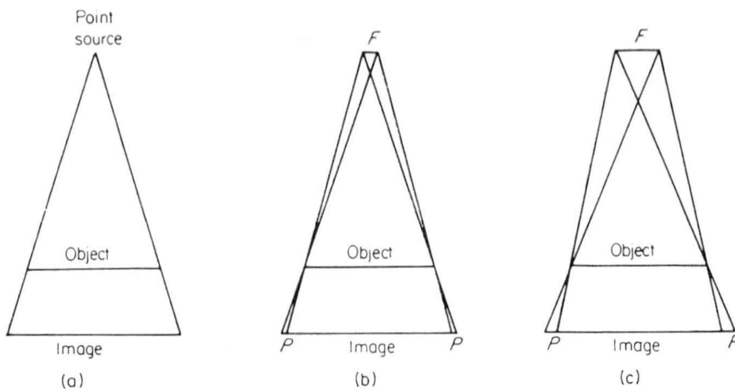

Figure 7.2. The effect of focal spot size on the width of the penumbra.

Figure 7.2 which shows diagrammatically the effect of focal spot dimensions (F) on the outline of the shadow image of the object. The larger the focal spot (F) for the same focus-object-film distances the greater will be the blur around the image, as seen in Figure 7.2(b) and (c). Figure 7.2(a) shows the result attainable if the focal spot were an ideal point source. The blur around the image in (b) and (c) is caused by the so called half shadow or penumbra (p) as indicated in the diagram and this blur impairs the rendering of details in the image. The width of the penumbra depends not only on the dimensions of the focal spot, but also on the focus to object to film distances as seen in Figure 7.3. It is concluded from Figure 7.3 that the width of

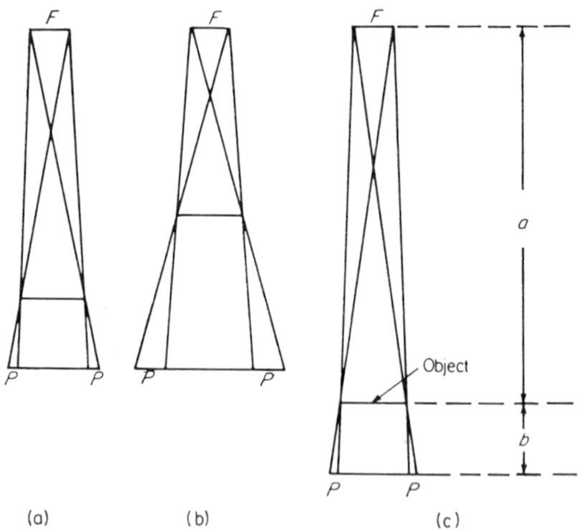

Figure 7.3. The effect of focus–object–film distances on the width of the penumbra
for a given focal spot diameter (F).

the penumbra is the smaller, the greater the focus to object and the
smaller the object to film distances. The change of the width
of the penumbra (p) with focal spot size (F), object-film (b) and focus-
object (a) distances can easily be derived from simple geometry
with reference to Figure 7.3c.

$$\frac{p}{b} = \frac{F}{a}$$

or

$$p = \frac{bF}{a} \qquad\qquad (7.1)$$

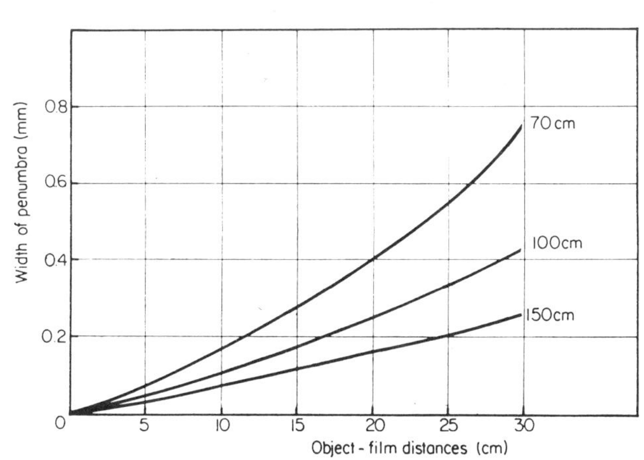

Figure 7.4. Width of penumbra at various focus to film and object to film
distances. The curves are based on a focal spot diameter of 1 mm.

This equation reveals that the width of the penumbra is proportional
to the maximum dimension of the focal area and to the object-film
distance and inversely proportional to the focus-object distance.
Hence in order to keep the width of the penumbra as small as possible,
it is important that the object is placed as close to the film as is practically possible. In an object of great depth those parts of it which are
close to the film will be less blurred than those which are remote from
the film. This is evident for instance in chest radiographs in a P.A.
projection, where the front regions of ribs are better defined than the
regions more distant from the film. The width of the penumbra (p)
which is usually called the geometrical unsharpness of a radiographic
image has been calculated from equation (7.1) for various focus-film
and object-film distances and this relationship is plotted in Figure 7.4
for a constant focal spot size of 1 mm diameter. The dimensions
chosen in this graph cover the most frequently used conditions in

normal practice in medical and industrial radiography. For the use of larger focal spots (as for instance applied in industrial radiography) the penumbra widths (p) have to be multiplied by the width of the actual focal spot in mm to be employed.

7.2c   Objects Smaller than the Focal Spot

In the previous discussion on the formation of the penumbra, the size of the object was not considered as we were concerned essentially with the penumbra as an edge effect. In medical radiography the radiologist is obviously interested in small details of an object, such as the trabecular structure of bones, capillaries of small diameter, roots of a tooth etc. In industrial radiography it may be a hair line crack in a casting, a small diameter void, or a fine wire, which is to be radiographed. All these objects mentioned above may be smaller in width than the diameter of the focal spot of the X-ray tube used in the investigation. The diameter of focal spots in medical radiography ranges from about 0.3–2 mm, whereas those used in industrial radiography, may be as large as 7–10 mm but they can be as small as 1–5 mm (or smaller with some megavoltage equipment).

We shall now consider the geometry of the formation of the penumbra for objects smaller than the focal spot, which is illustrated in the diagram (see Figure 7.5 p. 265) where F is a diameter of the focal spot, Q is the width of the object, d is the focus-object distance and a is the depth of the so called umbra, which is the blackened area in this diagram. Depending on the position of the image plane AA or BB, the image consists either of the width of the umbra and two penumbra shadows H_A' and H_A'' or the width of the pseudo-umbra (dashed area) and two penumbra shadows H_B' and H_B''. The depth of the umbra (a) can easily be derived from the geometry of Figure 7.5 and is found to be

$$a = \frac{Qd}{F - Q} \qquad (7.2)$$

Calculations of the depth (a) are of interest in considerations of image sharpness of small objects.

It is however of greater practical importance to calculate in a given case the actual total width of the image of an object smaller than the focal spot, if the diameter of the latter, the object width, and the focus-object and object-film distances are given. In circumstances where these are very unfavorable it can be useful to predict the extent of the penumbras by calculation in order to check to what extent the image may resemble the object, assuming that the resolving power

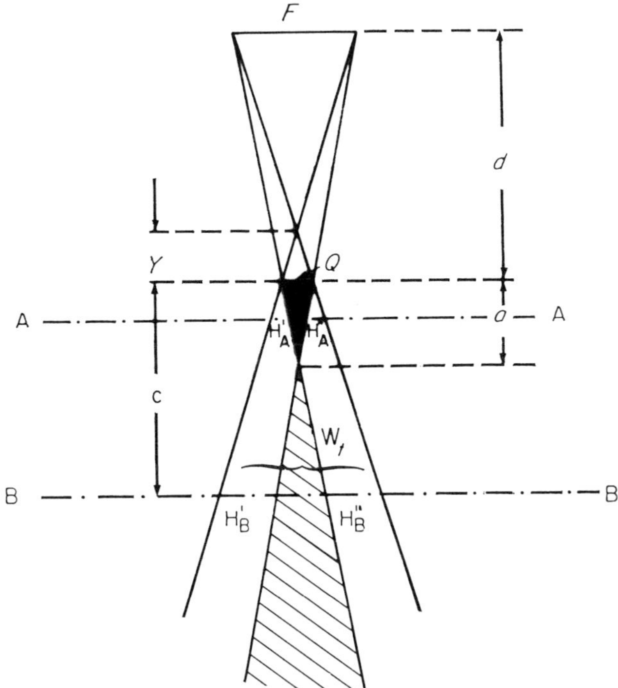

Figure 7.5. Object (Q) smaller than focal spot (F). A–A is an image plane through the umbra (dark area) B–B is an image plane through the pseudo-umbra (shaded)

of the film or screen-film system is adequate (see p. 345). The calculation can be carried out using the following elementary geometrical consideration based on Figure 7.5.

$$\frac{F}{d - Y} = \frac{Q}{Y}$$

$$Y = \frac{dQ}{F + Q}$$
(7.3)

and the total width (W_t) of the image of the object at any given image plane, such as AA, BB or others, follows from

$$\frac{W_t}{Q} = \frac{c + Y}{Y}$$

$$W_t = \frac{Q(c + Y)}{Y}$$
(7.4)

By substituting the value for Y, we have

$$W_t = \frac{Q\left(c + \dfrac{dQ}{F + Q}\right)}{\dfrac{dQ}{F + Q}} = \frac{Q(c + d) + cF}{d} \qquad (7.5)$$

The actual image width in a film is also influenced by purely photographic effects and it usually appears to be slightly smaller than the calculated width for the reasons given on page 278.

7.2d Geometric Enlargement and Distortion

Image Enlargement. As a result of the central projection in radiography only those regions of an object close to the plane of the film will be projected almost in their original size. With increasing distance from the film however, the object will appear enlarged in the radiograph. This enlargement of the object is merely dependent on the ratio of the focus–film to the focus–object distance and the percentage magnification ($M_\%$) can be expressed by

$$M_\% = \frac{\text{Focus–film distance}}{\text{Focus–object distance}} \times 100$$

Hence the greater the focus–object distance for a constant focus–film distance the smaller will be the magnification. The following Table 7.1 will be useful for an appreciation of this effect, as the percentage

TABLE 7.1. Percentage Increase in Image Size for Various Object–Film and Focus–Film Distances in cm

Object–film distance (cm)	Focus–film distances (cm)				
	50	75	100	150	200
		Percentage magnification			
0	0	0	0	0	0
2	4.2	2.7	2.1	1.4	1.0
4	8.7	5.6	4.2	2.7	2.1
8	19.0	12.1	8.7	5.6	4.1
12	32.0	19.0	13.6	8.7	6.4
20	66.7	36.4	25.0	15.4	11.1

linear increase in the image is important for instance for absolute measurements of the dimensions of the human heart in medical diagnosis and in the determination of the true size of a defect in industrial applications.

Image Distortion. The projection of an object on the film plane may be subject to distortions in the rendering of the image with

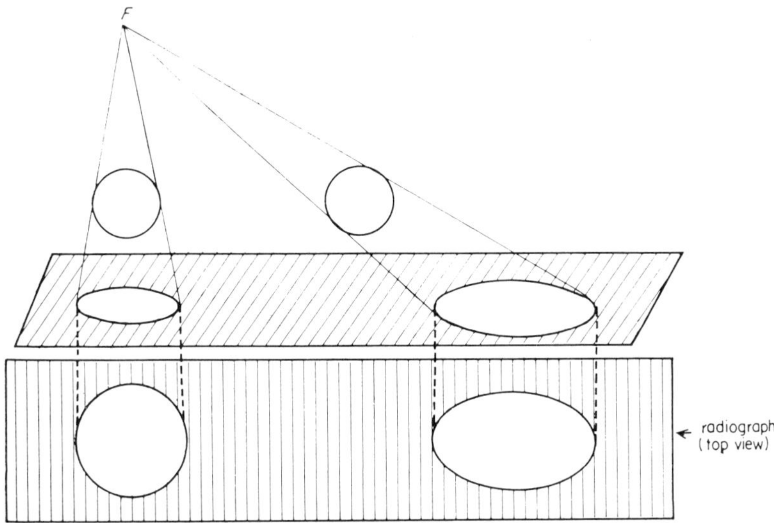

Figure 7.6. Normal and oblique projection of a sphere.

respect to the actual shape of the object. One of the most obvious examples is the oblique projection of a sphere, which will appear as an ellipse, as illustrated in Figure 7.6. Such a distortion may occur if a small cavity is present in a larger object, so that the X-ray beam strikes the small cavity obliquely. Distortion may also occur in the case of the radiographic projection of a rod-shaped body being placed obliquely or perpendicularly with respect to the film plane, as seen in Figure 7.7. The radiologist must be aware that if such a rod-shaped

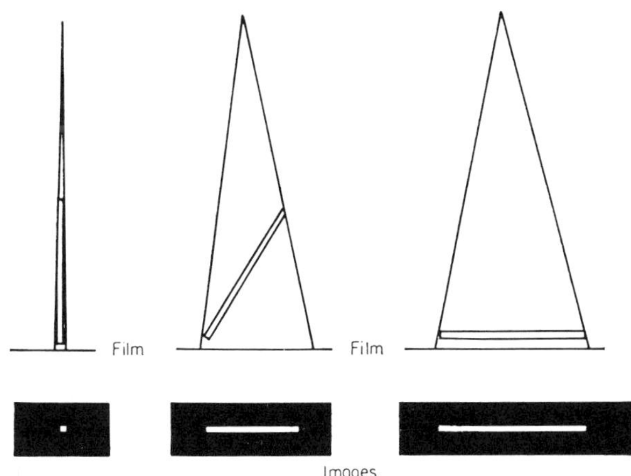

Figure 7.7. Projection of rod-shaped object in normal, oblique and horizontal position.

object forms part of a larger object structure, it is only by taking radiographs of different projections that the true dimensions and shape of an unknown object can be determined.

7.3   THE EVOLUTION OF A RADIOGRAPH

7.3a   The Ray Image

The evolution of a radiograph can be visualized in four stages. (see Figure 7.8.)

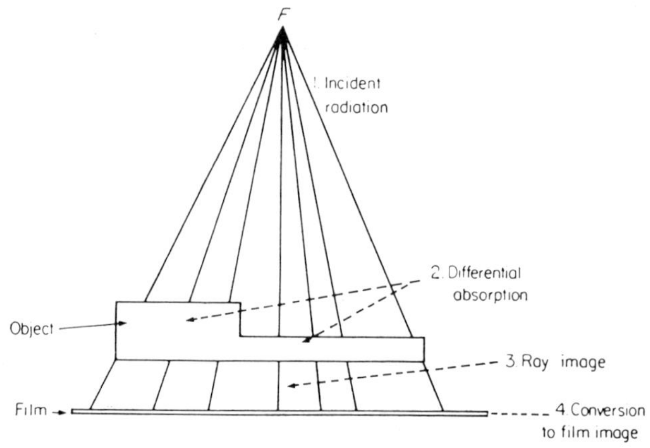

Figure 7.8. The evolution of a radiograph in four stages.

1. The X-ray or γ-radiation arising from a small focal spot or source is incident on the object to be radiographed.

2. The incident radiation is differentially absorbed by the varying composition, structure and thickness of the object.

3. Radiation of an inhomogeneous intensity and wavelength composition, sometimes called a 'ray-image' is transmitted by the object.

4. A small fraction of this radiation forming the ray image is absorbed by the photographic material and is converted into latent images, thus producing by subsequent processing a radiographic image of the object which can be inspected by the eye.

The art of radiography can therefore be specified mainly by two aims: (1) to produce the most suitable ray image of the particular object concerned and (2) to convert this ray image into a radiographic image of optimal quality. What is meant by optimal radiographic image quality is to some extent subject to individual opinion. More recently however, certain methods have been developed (see p. 342) which allow a more objective approach to be made to the informational character of the radiographic image.

7.3b Radiation Quality and the Ray Image

The differentiation between two adjacent regions of an object (see Figure 7.9) to be radiographed is based on the relative absorption of X- or γ-rays by these regions. As shown on page 18 the intensity of the radiation (I_1) transmitted by an object, can be expressed by

$$I_1 = I_0 e^{-\mu t} \tag{7.6}$$

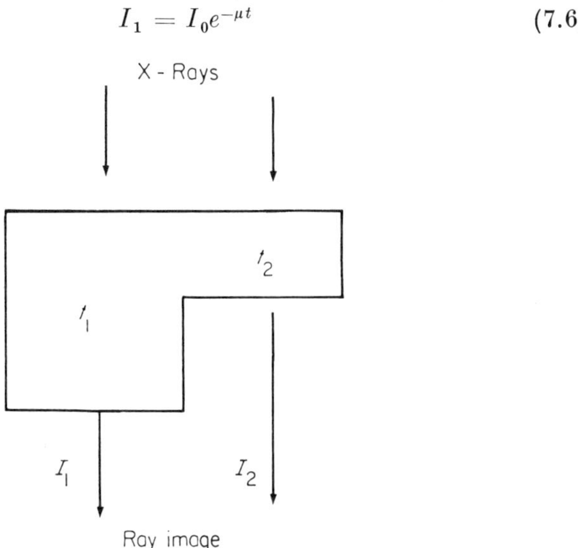

Figure 7.9. Two different thicknesses of a given material are penetrated by X-rays and form a ray image.

where I_0 is the intensity of the incident radiation, μ the linear absorption coefficient of the material concerned (see p. 18) and t its thickness. This equation is strictly valid for monochromatic radiation only, whereas the radiation emitted for instance by an X-ray tube covers a wide range of wavelengths, as shown in Figure 1.1. In the following considerations a close approximation to the use of monochromatic radiation is made, by using a radiation having an equivalent wavelength to that of a given monochromatic radiation. This can be done by using a strong filtration of the X-ray beam and by determining the equivalent wavelength from half value layer measurements (see p. 27). If our object to be radiographed consists of a given material, having two different thicknesses t_1 and

t_2 (see Figure 7.9), and if μ is the linear absorption coefficient for the material and wavelength chosen, then

$$I_1 = I_0 e^{-\mu t_1} \quad \text{and} \quad I_2 = I_0 e^{-\mu t_2}$$

thus

$$I_2/I_1 = e^{\mu(t_1 - t_2)} \tag{7.7}$$

where I_1 and I_2 are the intensities of the radiations after having passed through the object thicknesses t_1 and t_2. By changing the wavelength of the radiation, μ changes as well and hence the ratio of the intensities transmitted. In fact the absorption coefficient increases with the wavelength for the same material (neglecting selective absorption phenomena). Hence by taking radiations of

TABLE 7.2. Change of Radiation Contrast using Two Different X-ray Wavelengths

Wavelength in Å	keV	μ	$I_2/I_1 = e^{\mu(t_1-t_2)}$
0.175	70	6.29	$e^{6.29 \times 0.5} = 23.2$
0.142	87	4.04	$e^{4.04 \times 0.5} = 7.54$

increasing hardness (corresponding to decreasing absorption coefficients) for the two given thicknesses of the same material the exponent $\mu(t_1 - t_2)$ in equation (7.7) will be smaller and so also will the ratio I_2/I_1. In other words *the ray image contrast (often called 'subject contrast') decreases with harder radiation, which is one of the most fundamental aspects of radiography.* A more detailed aspect of this is shown on page 574.

The following example will illustrate this more clearly. Consider (by reference to Figure 7.9) two adjacent sheets of iron of 1.0 and 0.5 cm thickness and the use of two radiations of equivalent wavelengths of 0.175 Å (70 keV) and 0.142 Å (87 keV). The corresponding absorption coefficients (μ) which are taken from published tables are quoted in the Table 7.2. We can then calculate the ratio of intensities transmitted by 1.0 and 0.5 cm iron for the two wavelengths as shown in the Table 7.2.

Hence the ratio I_2/I_1 or the contrast of the ray image has been reduced by a factor of more than three by changing the quality of the radiation from 70 to 87 keV. The specification keV (instead of kV_p) means that the radiation corresponds to an equivalent wavelength to that of monochromatic radiations generated by 70 and 87 keV respectively. Similar calculations have been carried out for two further wavelengths and the result is shown graphically in Figure 7.10.

It should be emphasized that the relative change of I_2/I_1 with kilovoltage will be considerably smaller than indicated in Figure 7.10 if the change is measured with X-rays of increasing peak kilovoltage instead of with keV. This is because of the fact that the radiation emitted by an X-ray tube contains a strong component of

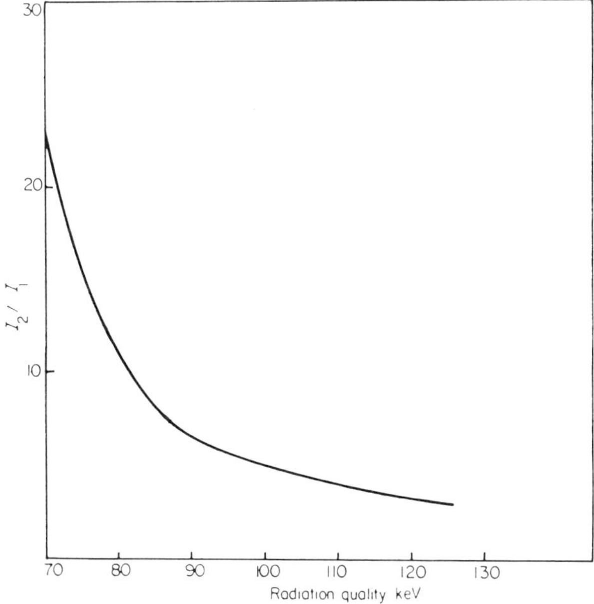

Figure 7.10. The ratio I_2/I_1, i.e. the ray image contrast (or subject contrast) as a function of radiation quality in keV.

soft X-rays leading to a more gradual increase of average penetrating power with kV_p than with keV (see p. 8).

Although in general it can be said that the contrast of the ray image or the subject contrast increases with softer radiation, there are of course limitations to such a statement, which are due to the fact that by further increasing the softness of the radiation, the object might not be penetrated at all.

7.3c   Ray Image and Film Image

So far we have discussed the influence of the quality of the radiation on the ray image and now we have to proceed one step further in relating the ray image to the film image. This can be done for instance by reference to Figure 7.11 in which the characteristic curve of an X-ray film is shown and the object, consisting of two thicknesses of the same material t_1 and t_2 is drawn above the curve. The ray image is indicated by I_1 and I_2.

The densities on the film obtained by the two intensities of the transmitted radiations I_1 and I_2 depend on the slope (see p. 111) of the characteristic curve for given development conditions. If both densities D_1 and D_2 lie within the straight line region of this curve, as indicated in Figure 7.11, then we obtain, using equation (4.8),

$$D_2 - D_1 = \gamma(\log I_2 - \log I_1)$$
$$= \gamma\mu(t_1 - t_2)\log_{10} e \tag{7.8}$$

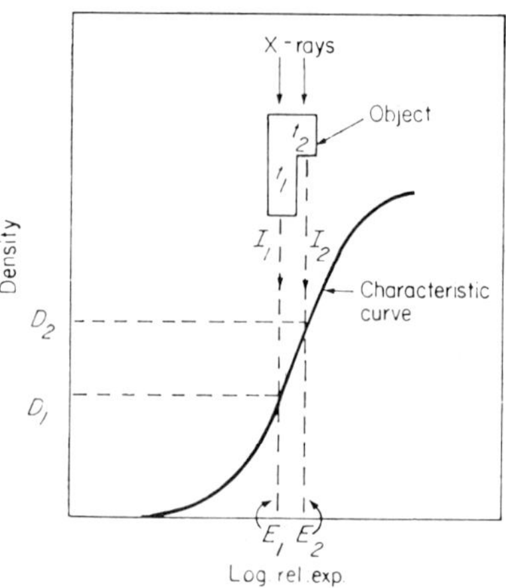

Figure 7.11. The conversion of the X-ray image into a film image.

where γ is the slope of the straight line region of the curve. Hence the objective contrast (see p. 274) $D_2 - D_1$ of the radiograph produced by the radiations transmitted by the object is proportional to the slope of the characteristic curve, the difference between the thicknesses and the absorption coefficient μ. It should be mentioned that the straight line region of the characteristic curve referred to above is an idealized situation and may not necessarily occur in all types of X-ray film, as discussed on page 112. If the slope is greater than unity, the photographic material acts as an 'amplifier' with respect to contrast, i.e. it converts the ratio of X-ray intensities into increasing differences in densities, or, in other words, into higher object or film contrast.

Similarly it can be shown that the contrast for two objects, having different thicknesses and made of different materials, is given by

$$D_2 - D_1 = \gamma(t_1\mu_1 - t_2\mu_2)\log_{10} e$$

where μ_1 and μ_2 are the respective absorption coefficients of the two materials (see p. 574).

7.3d Physical Density and Atomic Number

Physical density. In the previous sections we have dealt with the differential absorption taking place in an object. In this we have made use of the 'linear' absorption coefficient μ which does not explicitly show the effect of the physical density of the object concerned, as does the mass absorption coefficient μ/ρ (see p. 19). It is important however at this stage to draw attention to the significance of the physical density in X-ray absorption processes and a practical example of it will therefore be given.

If we consider, for instance, a radiograph of a chest, we see a fairly high contrast within the region of the lung's structure and this is chiefly caused by the difference in physical density between the tissues and the air in the cells of the lung (alveoli). Although the average atomic numbers of tissues and air are fairly similar, the ratio of physical density of air to that of tissue, which causes the striking contrast, is of the order of 1000:1.

Similarly in industrial radiography the visibility of very small gas cavities can be detected chiefly because of the difference in physical density between the metal and the gas.

Contrast media. In those cases where the ray contrast is too small to be converted into visible density differences in a radiograph, so called contrast media may be introduced in order to render certain regions of the human body opaque or semi-opaque to X-rays. Such contrast media consist of substances containing heavy elements which by their presence in organs or cavities cause high X-ray absorption, thus allowing a contrast to be formed between these regions and their surroundings. This is a well known method in medical radiography where barium meals are swallowed by the patient in order to render visible the contours of the stomach and intestines. Other contrast media, containing for instance iodine, are being injected in order to show radiographically kidneys, bladder, gall-bladder, arteries of the brain etc.

7.4 CONTRAST AND DEFINITION

Definition is an all embracing term referring to the quality of reproduction of detail in an image. It includes specific estimates of detail reproducing ability, such as sharpness, resolving power, modulation transfer and quantum mottle. (The latter terms are considered in some detail on pp. 335–350).

The technical quality of a radiograph is generally judged by two criteria, i.e. by the contrast and by the definition it reveals to the viewer. In recent years another aspect has been added to these criteria, which is associated with the spatial fluctuations of X-ray photons absorbed in the photographic system. This aspect will be presented at a later stage on pages 335–342.

Technical quality refers to the optimum quality of the record of the structural details of the object which the radiologist wishes to study. Any influence due to blemishes in the radiograph, such as developer splashes or fixer spots, fog due to unwanted radiation, scratches etc. are excluded from our following considerations on technical quality. (See also section on image quality on p. 342).

In the preceding sections we have made some reference to the terms contrast and definition. We have seen how the ray contrast is controlled and we have shown how it is converted into the film contrast. In discussing the geometry of image formation we have seen how the penumbra influences the sharpness of a radiograph.

7.4a Objective and Subjective Contrast

'Objective contrast' which can be determined by means of density measurements is defined on page 112 as the difference between two densities of adjacent areas. In radiography contrast measurements are rarely made in practice, (except for certain scientific purposes) and the radiograph is generally viewed or inspected for the necessary interpretation. Hence the 'subjective contrast', i.e. the perception of contrast by the human eye is the important factor. This 'subjective contrast' depends on the ratio of the light intensities transmitted by the corresponding density areas when these are inspected at a viewing box or illuminator. As the objective contrast changes the subjective contrast does not necessarily change in proportion, as the latter depends on many physiological and psychological factors associated with the observer; thus it cannot be expressed by a simple formula. Some of the most significant factors influencing 'subjective contrast' are the brightness of illumination, the size and shape of the image, the rate of change of density from one area to an adjacent one and the influence of these factors may be different for various observers. The 'subjective contrast' is also influenced by the experience of an observer in seeing radiographs, by the degree of fatigue of his vision, by dazzling from the light transmitted by neighboring fields and many other factors. The dazzling can be minimized by suitable masking of the radiograph. The visual sensitivity of the eye to small differences of density as a function of the brightness level was originally studied by A. Koenig and E. Brodhun (1888) and is shown in Figure 7.12.

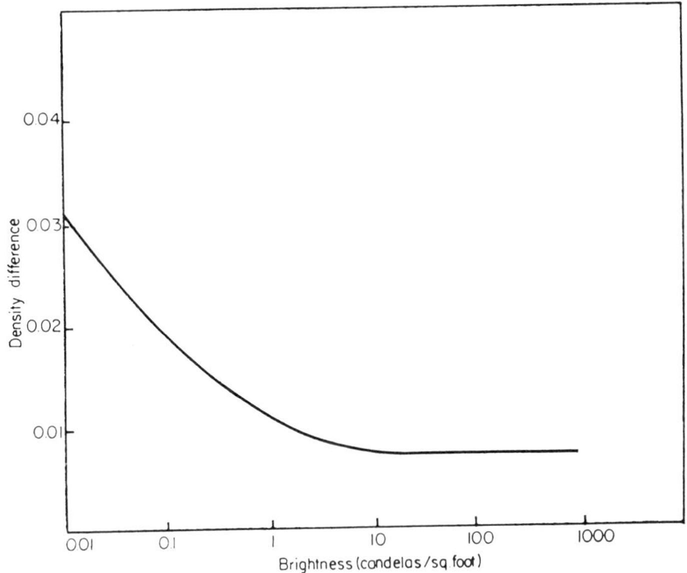

Figure 7.12. The sensitivity of the eye in detecting small differences of density as a function of the brightness level of an illuminator, assuming dark surrounding of the examined field.

On the ordinate of this figure the density differences are plotted and on the abscissa the brightness level of an illuminator in terms of candelas per square foot (see p. 429). The graph indicates that under suitable conditions the eye can detect density differences of the order of 0.01 within a range of brightnesses between about 1 to 1000 candelas per square foot. The limits are set by the loss in sensitivity of the eye at low levels of illumination and by the glare causing loss of contrast at the upper levels. (P. Moon, 1961).

This diagram presents however, only a part of the facts, as the visibility of small differences in densities also depends on the rate of change of density from one area to an adjacent one.

If there is a gradual change, the density difference might not be detected at all by the eye.

It is particularly important to realize that the visibility of small differences of density becomes very low indeed at high density levels, such as in the region of three and above, if viewed at too low levels of brightness. On the other hand, by increasing the brightness the visibility can be very much improved.

As the 'subjective contrast' is influenced by so many factors it is advantageous to make use of illuminators permitting variations of brightness and also of 'masking', so that the area illuminated can be obscured in those regions where the dazzling effect of the surrounding light interferes with the viewing conditions. It is generally found in practice that too little use is made of these facilities.

7.4b   Objective and Subjective Definition

In analogy to our concept of the ray image contrast (p. 270) we can also visualize a ray image sharpness. In Figure 7.11 we saw two adjacent steps of material drawn above the characteristic curve and we can imagine that the width of the boundary between the X-ray intensities transmitted by the two steps will depend on the dimensions of the penumbra. The smaller this boundary the better the sharpness of the ray image.

This ray image will be converted into the film image and the accuracy of its transfer to the film depends on the inherent contrast of the film and on its ability to differentiate between adjacently positioned detail, called the resolving power, (usually expressed as the number of lines per millimeter) which the film or screen-film system is capable of resolving (see p. 349). Where salt fluorescent intensifying screens are used, as is practiced in the majority of medical radiographs the conversion from the ray image to the film image will be strongly influenced by the screen characteristics (see p. 284) and by other effects (see p. 313).

We have thus seen that the rendering of detail in a radiographic image is a complex result of geometrical and photographic properties of image formation. The estimates of such detail reproducing abilities are often summarized as mentioned before by the general term 'definition' which includes considerations on sharpness, resolving power (see p. 349), modulation transfer (see p. 343) and quantum mottle (see p. 335).

In the following figures it is assumed that two adjacent areas of photographic density are separated by a gradual decrease from one density to the neighboring one, see Figure 7.13(b) or are separated

by a very sudden change of density as in Figure 7.13(a). In the
former case we speak of poor definition and in the latter case of good
definition. The visual appearance of this effect is called 'subjective
definition' and the physical assessment of the effect is referred to
as 'objective definition'.

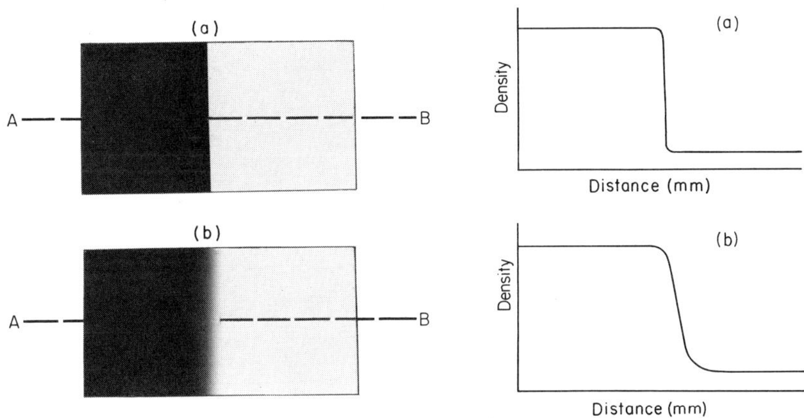

Figure 7.13. (a) Good definition (upper image). (b) Bad definition (lower image).

Objective Definition.   This can be assessed by the measurement of
the breadth of the boundary between two adjacent image elements
of different but uniform density (see Figure 7.13). The breadth of
the boundary may be recorded by making a microdensitometer scan
along the line AB. The smoothed out microdensitometer patterns are
shown on the right of Figures (a) and (b). These patterns reveal an
almost vertical drop of density from the higher to the lower density
level, whereas a gradual drop occurs, if the width of the boundary
between two adjacent density levels is great. In order to give a
quantitative assessment of definition we can measure the breadth
of the boundary on the microdensitometer record.

On page 266 it was mentioned that the total width of the penumbra
of an image of an object smaller than the focal spot is smaller in the
radiograph than derived by calculation. This will be understood by
reference to Figure 7.14 in which the distribution of X-ray intensities
is illustrated by the dashed curve and the corresponding density
distribution in the radiographic film image by the full curve. The
steeper full curve is due to the ability of the film to amplify the
radiation contrast. (W. Bronkhorst, 1927).

The 'subjective definition' may be defined as the more or less sudden change from one light stimulus to an adjacent one, as perceived by the eye. It cannot be approached by a simple quantitative measurement, as the effect varies with the observer and is influenced by similar factors as the 'subjective contrast'. As the subjective definition depends on the rate of change from one light stimulus to an adjacent one, the visual appearance of contrast and definition influence each other and it may sometimes be difficult even for an

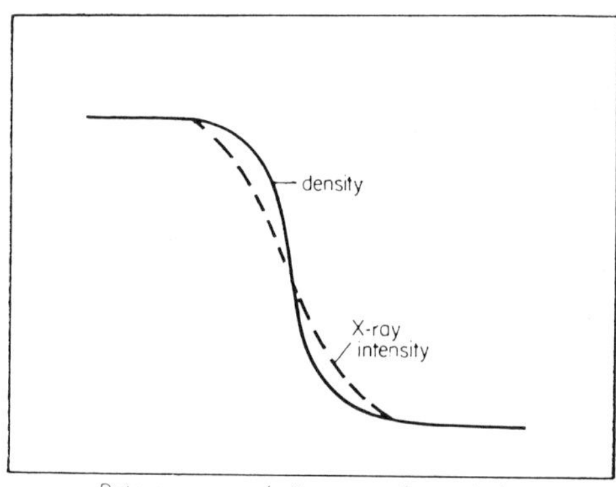

Figure 7.14. Relative rate of change of X-ray intensity and resulting rate of change of density within region of penumbra.

experienced radiologist to distinguish these two image forming factors.

The control of contrast and definition and how these are influenced by the use of intensifying screens will be discussed separately in the chapters on medical and industrial radiography.

In recent years it has become customary to consider the visibility of detail from a statistical point of view and this arises from the statistical fluctuations of the X-ray quanta absorbed in film or in screens. This aspect will be discussed in several sections of this book, notably on pages 296, 335 and 420.

REFERENCES

Koenig, A. and Brodhun, E. (1888, 1889). *Berliner Berichte*, 917, 641.

Koenig, A. and Brodhun, E. (1932). In A. C. Hardy and F. H. Perrin (Eds.), *The Principles of Optics*. McGraw-Hill, London. p. 192.

Moon, P. (1961). *The Scientific Basis of Illuminating Engineering*. Dover Publications, New York. p. 418.

Mullins, L. (1949). Some uncommon applications of industrial radiography. In J. A. Crowther (Ed.), *Handbook of Industrial Radiology*, 2nd Ed. Edward Arnold, London, p. 162.

8

Medical Radiography

8.1 INTRODUCTION

In the preceding chapter an attempt was made to present the geometrical and some of the physical factors of the formation of the radiographic image. A knowledge of these factors is essential for the understanding of the production of medical radiographs.

In the following sections we shall be dealing with the radiographic material, its sensitometric characteristics, the control and assessment of image quality and some of the auxilliary methods of medical radiography. As we are going to discuss chiefly the photographic aspects, it is assumed that the reader is familiar with the operational functions of the X-ray apparatus and X-ray tube and also with the control of kilovoltage (kV), milliamperage (mA) and the product of mA and time, i.e. milliampereseconds (mAs).

8.2 RADIOGRAPHIC MATERIAL

8.2a The Double Coated X-Ray Film

As mentioned briefly in Chapter 3, X-ray films are generally coated on both sides of the base with emulsion of equal thicknesses. In the past the base of most X-ray films consisted of cellulose acetate, but more recently some of the large film manufacturers have introduced film base made of polyesters, which is essentially inflammable, shows less tendency to curl and also has better dimensional stability than cellulose acetate films. On the average, the base of X-ray films is about 150–200 microns thick and the coating thickness on each side varies from 10–20 microns. The emulsions are provided with a protective supercoat of about $\frac{1}{2}$ to 2 microns thickness, the purpose of which is to prevent abrasions of the film surfaces. The majority of commercially available X-ray films have a bluish tinted base. This color gives a more pleasing impression to the eye, when viewing radiographs, than that of clear based films and apparently emphasizes the subjective contrast. It has sometimes been argued that the slightly more restricted wavelength band caused by the bluish tint is important with respect to detail visibility of the eye, but this theory appears to have little scientific basis.

Significance of Double Coating

Since X-rays penetrate both emulsion layers the densities of the images formed in both coatings are added together. Hence twice the respective densities that would be obtained in a single layer will be recorded by the double layer with the result that the contrast is also doubled. This can be seen from the following argument. Suppose two adjacent densities in a single layer are $D_1 = 1$ and $D_2 = 0.5$; thus the objective contrast $D_1 - D_2 = 0.5$. The objective contrast in the double layer is then $D_1 - D_2 = 2 - 1 = 1$. Hence the objective contrast is doubled.

The advantage of the double coated film over a single coated film also becomes evident from the comparison of their respective characteristic curves, as seen in Figure 8.1.

It is seen from Figure 8.1 that the density differences, i.e. the average inherent contrast of the double coated film II is twice that of the single coated film I, assuming that the same range of exposure is considered, as indicated in Figure 8.1.

A further advantage of the double coated film is its increase of speed when referred to film I. For example at a density of 1, film II is 4.5 times and at a density of 2, film II is 10 times faster than film I (see Figure 8.1). This can be verified by taking the antilogarithms of the difference between the respective log exposure values at densities 1 and 2 for both films I and II (see p. 113–116).

We have seen that the double coated film image consists of a superimposition of two equal, but underexposed film images each of low contrast. Let us assume we expose a single coated film so that the ratio of light transmitted by the darkest and the lightest region of the film is 1:10, corresponding to a density difference of 1, for example densities 1.2 and 0.2 respectively. If we then superimpose two single layers (being equivalent to a double coated film), then we obtain a ratio of light transmitted of 1:100 or in the example densities 2.4

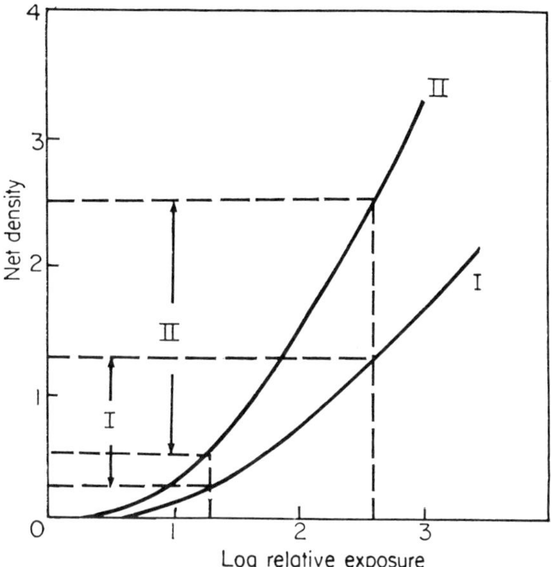

Figure 8.1. Characteristic curves of a single (I) and a double coated (II) X-ray film

and 0.4. Hence the film with the two layers reveals a 10 times increase of the range of brightnesses for the same exposure given to the single layer.

A further advantage of the double as against the single emulsion is that harder radiation can be used with the former, in order to obtain the same image quality, as that with a softer radiation on the single film. This is because of the greater inherent contrast of the double film as against that of the single layer film. Furthermore double films can be used with a 'pair' of intensifying screens (see later sections) and this allows a further reduction of exposure time.

There are generally two basic varieties of double coated X-ray films available, those for direct X-ray exposure (no-screen films) and those for salt-screen exposures (screen films). Reference to their relative properties will be made in section 8.3.

8.2b X-Ray Paper

This consists of a raw paper provided with a layer of baryta, i.e. the trade name for barium sulfate, with added binding material on which an X-ray emulsion is coated. In order to prevent abrasions the emulsion is provided with a thin supercoat. The baryta serves as in any other photographic paper (a) as an efficient reflector and (b) as a chemical separator between the raw paper and the emulsion.

X-Ray paper has only a restricted use, the reasons will be evident from a discussion on its sensitometric properties (see pp. 309–312).

8.2c Salt Intensifying and Fluoroscopic Screens

In medical radiography intensifying screens are indispensable as they permit not only a very considerable reduction of exposure time, when used in conjunction with double coated X-ray films, but also a similar reduction of the radiation dose applied to the patient. Because of the reduction of the exposure time, if compared with film used without screens, image unsharpness due to involuntary movements of the patient and of certain organs, such as the human heart, stomach and intestines can practically be eliminated. The reduction of radiation dose to the patient by the use of intensifying screens is particularly important if several exposures have to be taken of one patient. The majority of the exposures taken in medical radiography are therefore screen exposures. The screens, however, cause a certain degree of blur which impairs the definition in the radiograph (see p. 284) (H. S. Tasker, 1945).

From the point of view of image sharpness, no-screen exposures are preferable, but only as long as the geometric unsharpness of the ray image (see p. 268) permits the full utilization of the improved definition of which the no-screen film is capable. Hence in cases of relatively thin regions of the human body, such as extremities, (legs and arms) X-rays are taken with no-screen films particularly if longer exposure times such as $\frac{1}{2}$–5 seconds do not matter.

Historically the fluorescent screen has a particular significance, as it was the fluorescence of cardboard coated with crystals of barium platino-cyanide which was partly responsible for Röntgen's discovery of X-rays.

Luminescence, Fluorescence and Phosphorescence

The physical properties of such screens underlie the phenomena of *luminescence*. This is a general term denoting the emission of light by the absorption of energy. The excitation of energy may be by electromagnetic radiation or by charged particles, it may also be produced by physical strain, heat, and other effects. In this application it is the absorption of X- or γ-rays which causes luminescence in certain substances, called phosphors. If the light emission lasts only as long as X-rays are absorbed, or within 10^{-8} seconds after the excitation by X-rays, we speak of 'fluorescence'*. The interval of 10^{-8} seconds

---

* This strict definition of the term 'fluorescence' is not always used and one often speaks of fluorescence even if some slight afterglow occurs, but only when this is negligible under practical conditions of exposure.

has been chosen because it is of the order of the life time of an inter-atomic transition on which fluorescence is based. If the light emission still persists after X-ray excitation has ceased then the phenomenon is generally called 'phosphorescence'. The light emission occurring after excitation by X-rays is called 'afterglow'. The distinction between these terms is of some practical importance in the radiographic use of screens, as will be shown later. (G. W. Stanwix and L. Mullins, 1946).

Luminescent phenomena are very complex and only a brief hint on the principles of their mechanism can be given here. During the absorption of radiation by a fluorescent substance an electronic transition occurs and the atoms or molecules are then said to be in an excited state, in which an electron is raised from a lower to a higher energy level. This change constitutes an unstable condition and the electron returns after approximately 10^{-8} seconds to a lower level whereby radiation is emitted of a frequency corresponding to the difference in energy between the initial and final level. It is the radiation emitted in this process which we call fluorescent radiation. Phosphorescent phenomena are explained by the formation of electron traps. In these an electron may be kept for some time before it returns to its ground level by the emission of radiation (afterglow). Phosphorescence, i.e. afterglow, may persist for fractions of a second to hours after excitation and is typical for many inorganic crystalline compounds.

Apart from a limited class of substances pure crystals exhibit luminescent effects only after they have been subjected to *heat treatment*. The degree of luminescence of some substances is dependent on certain impurities (activators) which are incorporated into the material. The activators may be present in an extremely low concentration, e.g. one part in 10,000 or less, and are incorporated during heat treatment of the parent substance. The characteristic color of fluorescence or phosphorescence is related to the properties of the activator.

The fact that many fluorescent materials show a *reduction of emission with increasing temperature* is a subject of considerable theoretical interest and various mechanisms have been proposed to explain this phenomenon. It is believed that the effect of raising the temperature of a phosphor is to render the transfer of the energy of excitation into heat more probable. Thus a drop of the intensity of emission during irradiation is to be expected as the temperature is raised. (L. Levy and D. W. West, 1933, 1934, 1939; E. Hirschlaff, 1938; F. Seitz, 1939; R. W. Gurney, 1940; P. Pringsheim, 1943; G. R. Fonda and F. Seitz, 1948; G. F. J. Garlick, 1949; Luminescence Symposium, 1954.)

Type of Screens

One distinguishes two general types of fluorescent screen in medical radiology: (a) the X-ray intensifying salt screens which, as their name implies, intensify the action of X-rays on the photographic emulsion and (b) the fluoroscopic screens which render the X-ray image visible to the human eye. Furthermore, certain types of fluoroscopic screens are produced for making photographic images; these are called photofluorographic screens and are chiefly used in mass miniature radiography, in cineradiography, and in image amplifier techniques (see pp. 359–366).

Composition of X-Ray Intensifying Salt Screens and the Performance of Screen-Film Combinations

The various types of screen consist of a cardboard or plastic material on which certain inorganic crystalline substances are uniformly coated and held together by transparent binding material. This coating is sometimes separated from the support by a layer of white pigment which reflects the fluorescent light emitted by the crystalline layer when this is excited by X-rays. The fluorescent layer is usually supercoated by a very thin skin of protective layer in order to avoid abrasion marks on screens in handling, which may otherwise show on the radiograph. The thickness of fluorescent layers in modern salt intensifying screens is, generally, between 0.2–0.4 millimeters.

For the production of X-ray intensifying screens there are two main types of substance in use (1) calcium tungstate ($CaWO_4$) and (2) barium lead sulfate ($BaSO_4$:Pb). Calcium tungstate screens are at present by far the most widely used screens. Both substances are practically free from afterglow. Intensifying screens exist as pairs, so that the double coated X-ray film can be sandwiched between them, each emulsion facing one of the screen coatings (see Figure 8.2).

During the exposure X-ray energy is absorbed by the screens and is converted into visible light radiation to which the double coated X-ray film is particularly sensitive. Hence the screens act as intensifiers. Figure 8.2 illustrates a section through a double coated X-ray film sandwiched between a pair of intensifying screens. Each crystal emits light, i.e. it fluoresces by the excitation of X-rays and this light diverges in all directions, thus causing a certain amount of blur on the X-ray film emulsion. This obviously leads to a deterioration of radiographic definition, called screen unsharpness (see p. 332). The diffusion of light from the crystals affects not only the adjacent but even the emulsion remote from the screen (crossover effect), as

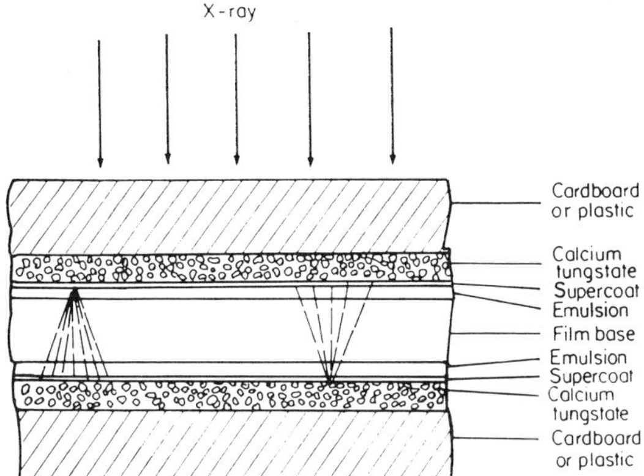

Figure 8.2. Section through double coated X-ray film sandwiched between a pair of calcium tungstate X-ray intensifying screens. (The blue light emitted by the screens is indicated by the divergence of fluorescent light from individual calcium tungstate crystals.)

indicated in Figure 8.2. Furthermore, some screens reveal a certain inherent mottle which influences the image quality. (Structure-mottle). Much thought has been given by manufacturers to counteract the crossover effect by various methods, but each of them suffers from certain disadvantages (E. C. Petri, 1960). The packing of the crystals, the average size of the crystals, the thickness of the screen layer, and the reflecting layer are responsible for the degree of unsharpness obtainable. The smaller the crystal size, the closer the packing, and the thinner the screen layer, the better is the sharpness. The influence of grain size of $CaWO_4$ crystals on the rendering of details has been debated and it was argued that it is not the size but the thickness of screens that is chiefly responsible for the definition attainable.

(H. Schober and C. Klett, 1953). More recently it has been shown by R. Piwonka, G. Voigt, and E. C. Petri (1965), using modulation transfer function techniques (see p. 343), that the grain size has in fact an important influence on the image sharpness. Differences of grain size between 3, 9, and 18 microns were found less striking with respect to changes in definition in thinly coated screens, than in thickly coated ones. The authors found, however, that the effect of screen unsharpness becomes more striking with increase of the thickness of the screen. On the other hand the screen output increases with increasing thickness. Some compromise was found by using very closely packed fine grain crystals in a fluorescent layer of increased thickness. (W. Angerstein and H. Plapperer (1965).)

The thickness of screens is sometimes so adjusted that the front screen, i.e. the screen nearer to the X-ray tube, produces about the same density on the one emulsion as the back screen on the other emulsion layer of the double coated film. Hence the front screen layer is sometimes, but not always, slightly thinner than the back screen layer. The equality of density attainable on both emulsion layers obviously depends on the quality or hardness of the radiation applied. The adjustment of screen thickness by the manufacturer for front and back screen is therefore made at a kilovoltage which is typical for the majority of radiographic exposures.

It may also be mentioned that in general the characteristics of salt intensifying screens are not influenced by prolonged use of the screens, even through several years. It has been observed however by the author that excessive doses imparted to screens, i.e. doses far exceeding those used in normal exposure conditions, may influence the screen characteristics. In practice screens are more likely to become obsolete by surface marks than by the total dose they received during extensive routine use. G. W. Stanwix and L. Mullins (1946) have shown that calcium tungstate screens exhibit afterglow after heavy exposure to X-rays. After a 1 minute exposure to unfiltered X-rays, generated at $140\,kV_p$, $5\,mA$ at $75\,cm$ focus to film distance, the screen was placed in contact with a fast screen film for 15 minutes. This latter exposure revealed a net density of 0.03. A relatively large increase of afterglow was observed by the authors after a screen had been given a series of still heavier X-ray exposures at intervals of 15 minutes.

As the screen unsharpness can be controlled to some extent by the screen manufacturers, three types of calcium tungstate screens for varying degrees of film sharpness are commercially available. These are fine grained, medium and coarse grained screens. Fine grained screens are usually made as thinner layers as well. Unfortunately screens exhibiting improved definition, i.e. fine grained screens, have a smaller light output for a given X-ray dose rate and quality of radiation than coarser grained and thicker screens. The light output for a given X-ray dose-rate and X-ray quality is often referred to as screen speed, which thus increases with crystal size of the fluorescent material and thickness of layer. An improvement of definition has also been achieved by incorporating dyes into the screen layer which reduce the scattered light emission from the remoter regions of the screen facing the emulsion.

Spectral Emission

The peak of the spectral light emission of calcium tungstate screens
lies in the region of about 4300 Å, as seen from the graph Figure 8.3.
It is fortunate that the maximum inherent sensitivity of X-ray screen
films matches the spectral emission of calcium tungstate screens.
The screen film sensitivity to light is shown in this graph as a dashed
line. The peak of the spectral emission of the barium lead sulfate

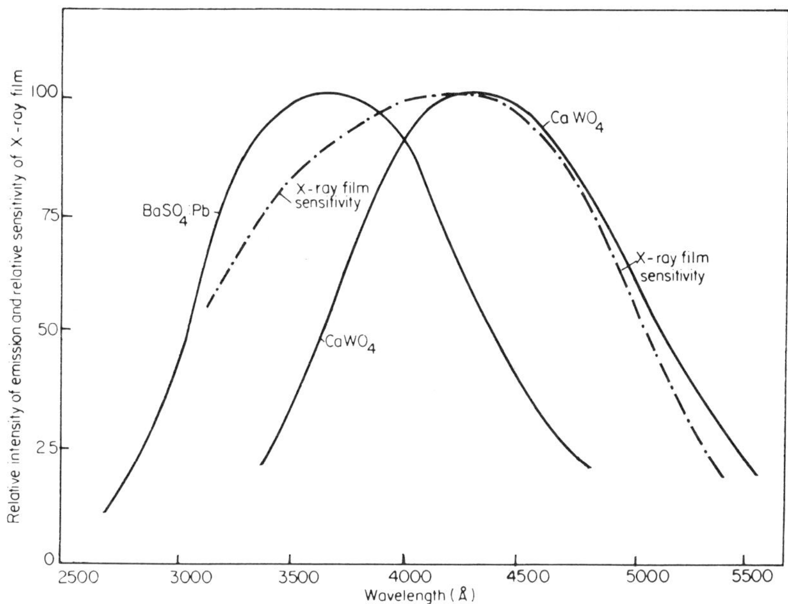

Figure 8.3. Spectral emission of calcium tungstate ($CaWO_4$) and barium lead
sulfate ($BaSO_4$:Pb) screens and spectral sensitivity of X-ray film.

screens lies in the ultraviolet region and is at about 3600 Å (see graph
Figure 8.3). (J. W. Coltman, E. G. Ebbinghausen, and W. Altar,
1947.)

The properties of the two types of screen do not only differ in
spectral emission, but also in their relative speeds. Whereas the
speeds are very close to each other within the range from 50–70 kV_p,
at higher kilovoltages up to about 140 kV_p barium lead sulfate
screens are faster.

It has been established experimentally that the spectral emission
of calcium tungstate screens is not altered by the quality of the
incident X-rays within the limits of qualities tested (M. B. Hodgson,

1918 and G. Knipe, private communication). The author is not aware of any similar experiment carried out on barium lead sulfate screens.

Apart from these two types of screen mentioned above, zinc sulfide (ZnS) has been used as a substance for intensifying screens. A certain type of zinc sulfide screen emits in the blue region and its speed is considerably superior to calcium tungstate in the softer region of the diagnostic range of X-rays. The disadvantage of fast ZnS screens is that they are relatively coarse grained with the effect that they exhibit strong image graininess and furthermore, that they suffer from afterglow. Hence in two consecutive radiographic exposures the afterglow may produce an image of a previous exposure onto the next radiograph. The afterglow can however be quenched between the two subsequent exposures by subjecting the screens to the radiation of an infrared source or a red safelight.

Another substance which appeared to be promising for increasing the speed of screens, namely potassium iodide (KI), containing traces of thallium was suggested by M. M. Ter-Pogossian (1965). KI suffers selective absorption within a range of wavelengths closer to that used in medical radiography than e.g. $CaWO_4$ and ought therefore to exhibit speed in excess of that of the latter. Unfortunately other disadvantages were found which so far made it impracticable to make use of KI screens.

Quantum and Energy Efficiency of Calcium Tungstate Intensifying Screens

In some scientific considerations concerning the efficiency of intensifying screens, it is useful to know the utilization of X-rays with respect to the light output of screens as a function of the incident X-ray quantum energy. Such measurements have been carried out by J. W. Coltman and others (1947), J. Eggert and E. Schopper (1948), and R. H. Herz (1956). The results by Herz are presented in the following, as these refer to a relatively wide range of wavelengths equivalent to X-rays generated between about 40 to 1200 keV. His results were obtained by using a front screen only (typical commercial calcium tungstate screen) and a fast medical X-ray film of the screen type variety (one emulsion coating only).

In any energy considerations concerning the application of screens it is first necessary to know the respective absorption properties of such screens at various wavelengths of X-rays. The percentage absorption of a front screen as a function of X-ray wavelength is illustrated in Figure 8.4.

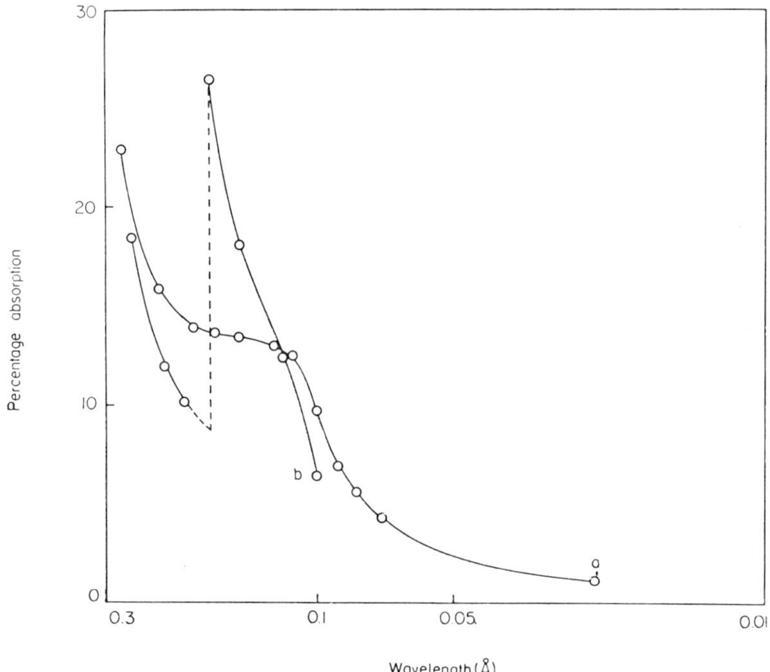

Figure 8.4. The percentage absorption of a front calcium tungstate screen as a function of X-ray wavelength in Å. (a) Experimental and (b) Theoretical. (*After* R. H. Herz, *Brit. J. Appl. Phys.* **7**. 182 (1956).)

Curve (a) shows the results of the experimental data, curve (b) those arrived at by calculation from absorption coefficients. It is seen that the almost flat region of curve (a) is caused by the K-absorption edge of tungsten at a wavelength of 0.178 Å (i.e. about 70 keV). The experimental curve is smoothed out because of the use of heterogeneous, although slightly filtered, radiation, of which the corresponding wavelengths of monochromatic radiation have been established by half value layer measurements and by reference to absorption coefficients. It is further shown that the percentage absorption decreases towards shorter wavelengths, as would be expected.

The quantum efficiency (ϕ), i.e. the number of light quanta emitted by the screen per X-ray quantum absorbed by the screen is shown as a function of X-ray wavelength in Figure 8.5.

From this figure it is evident that at a wavelength of 0.1 Å (124 keV) about 500 light quanta are emitted per X-ray quantum absorbed by the screen, whereas about 8000 light quanta are emitted

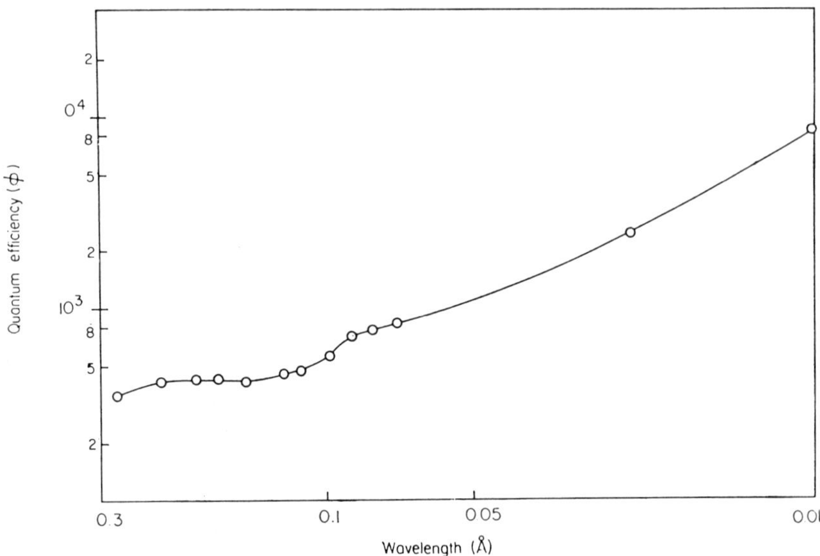

Figure 8.5. Quantum efficiency of calcium tungstate front screen as a function of X-ray wavelength in Å. (*After* R. H. Herz, *Brit. J. Appl. Phys.* **7**. 182 (1956).)

at a wavelength of 0.01 Å (1.24 MeV). The explanation for this increase is thought to be similar to the explanation of the increasing quantum efficiency with increasing quantum energies for no-screen X-ray film (see p. 98), i.e. with the increasing range of electrons released by X-rays in the screen more and more fluorescent centers are excited.

The energy efficiency of a front screen, i.e. the ratio of light energy emitted by the screen to the X-ray energy absorbed by the screen was found to be of the order of 1–2 per cent and to be almost independent of wavelength. This means that the conversion of X-ray energy into that of light energy is practically the same at low and at high X-ray or γ-quantum energies.

The advantage of the screen is explained chiefly by its ability to absorb a greater proportion of the incident X-rays than the X-ray film is capable. Whereas the ratio of percentage absorption of X-rays by the screens over that by the film was found to be 20 at 0.3 Å, it increases to a value of 130 at a wavelength of 0.1 Å.

A further aspect in explaining the merits of the screen is the great photographic efficiency of light exposures if compared with the strikingly low photographic efficiency of direct X-ray exposures, particularly at high quantum energies. This effect is shown in Figure 8.6 which illustrates the ratio of X-ray energy incident on film without screen to the light energy incident on film with screen, to give a standard density of 0.2 above fog as a function of wavelength. It is seen that this ratio increases from 6000 to 400,000 between 0.275 Å and 0.01 Å respectively. These large numerical values are

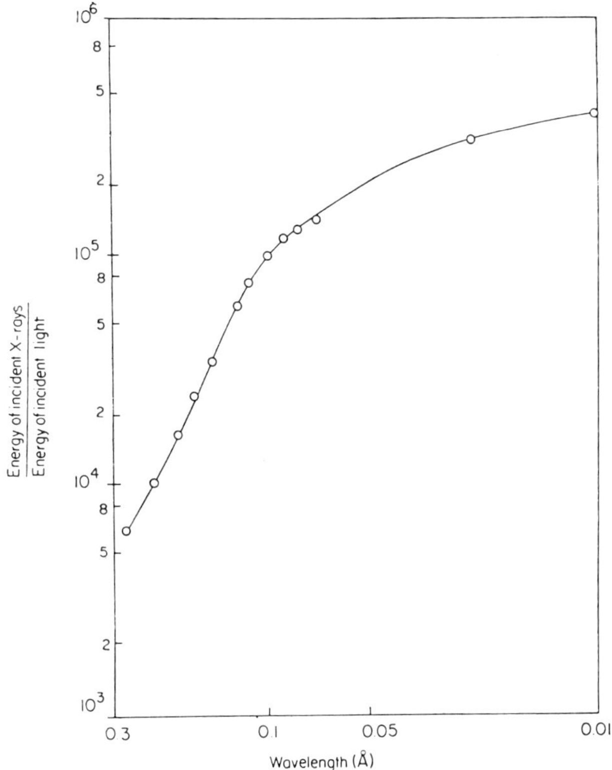

Figure 8.6. Ratio of energy of incident X-rays on film without screen to energy of incident light on film from screen for a density of 0.2 above fog as a function of X-ray wavelength in Å. (*After* R. H. Herz, *Brit. J. Appl. Phys.* **7**. 182 (1956).)

caused by the enormous differences between the respective energies of X-ray and light-quanta, and by the fact that only a small fraction of the X-rays incident is absorbed by the film, whereas a high proportion of the incident light intensity is absorbed by the film. It should also be considered that a very large amount of X-ray energy is wasted, as each X-ray quantum carries much more energy than is necessary to render a silver bromide grain developable, if compared with the sum of the energies of the light quanta required per grain.

Summarizing, the following approximate data have emerged from these investigations when applied to exposures of a screen X-ray film sandwiched between a *pair* of calcium tungstate intensifying screens. The conclusions refer to the range of kilovoltages used in medical radiography, i.e. 60–70 kV$_p$.

1. About 30–40 per cent of the intensity of incident X-rays is absorbed by a pair of screens.

2. About 500 light quanta are emitted per X-ray quantum absorbed in a screen.

3. The conversion of X-ray energy absorbed by a pair of screens into that re-emitted as light energy by the screens is of the order of 1–2 per cent.

4. A relatively high proportion of the light intensity produced by a screen is absorbed by the film.

The intensifying factor of screens is discussed in the section on sensitometry of screen films (see p. 301).

Fluoroscopic and Photofluorographic Screens

Fluoroscopic screens are used by the radiologist for the production of a visible image on a fluorescent screen, i.e. for viewing the fluorescent image with dark adapted eyes during examination of the chest, the intestinal tract etc. Photofluorographic screens on the other hand are available for taking photographs of the screen image by means of a camera as e.g. in mass miniature radiography or in image amplifier techniques. In order to make these screens suitable for the various applications, their color of emission must be adjusted to suit the maximum sensitivity of the human eye, when dark adapted or in photographic applications, to suit the sensitivity of certain types of photographic emulsions.

Both types of screen are generally made of zinc sulfide or cadmium sulfide crystals or mixtures of both held together by a binding medium and coated on a stiff white reflecting cardboard or plastic sheet. Fluoroscopic and photofluorographic screens are in general much coarser grained than intensifying screens and should also be free from afterglow. The natural spectral emission of these screens can be changed by the manufacturer to suit certain applications by adding minute amounts of impurities, so called 'activators' as mentioned earlier. The concentration of these is of the order of 1/10,000 to 1/1,000,000th of the parent crystal and has also a marked influence on the brightness of emission. Thus the concentration is adjusted to the maximum brightness during manufacture. Elements such as Zn, Cu, Ag, and Mn and others are used as activators. The range of wavelengths of emission of phosphors containing different activators, their peak emission, average color appearance and type of application are quoted in the table below.

The disadvantage of afterglow in zinc sulfide types of screen has been reduced to a great extent by the addition of small amounts of nickel (about 1 part in a million) as a so called 'killer' of afterglow (L. Levy and D. W. West, 1939). It has been found that the light emission of these, and also of intensifying screens, is proportional to the tube current of the X-ray tube, assuming that the kilovoltage

TABLE 8.1. Range of Spectral Emission of Certain Phosphors

Luminescent substance	Range of emission in Å	Activator	Peak emission in Å	Colour	Application
Zinc cadmium sulfide	4800–6400	Cu	5300	green	Fluoroscopy & Photofluorography
Zinc sulfide	4000–5400	Ag	4400	blue-violet	Intensification in radiography
Zinc sulfide	5300–6600	Mn	5800	orange-yellow	Fluoroscopy

can be kept constant. This has been verified up to 700 mA tube current using a zinc-sulfide screen, (R. H. Herz, 1936). The brightness of the screens is also increasing with the kilovoltage across the X-ray tube for constant mA within the medical range of kV's. (R. H. Herz, 1936 and J. C. Rockley, 1957.)

Perception of Fluorescent Light

The brightness level of the fluoroscopic screen image is on the average about 10,000 to 1,000,000 times lower than that used for instance in viewing radiographs on an X-ray film illuminator, but how does the eye adapt itself to this low brightness level? The eyes contain two types of light receptors, the rods and the cones. Whereas the perception of the dark adapted eyes at low brightness levels is performed by the use of the rods, the cones are used mostly for normal vision of bright objects. For the selection of screen color for fluoroscopy it is important to realize that the optimum vision for cones and rods changes from about 5700 Å to 5200 Å respectively, i.e. a smaller brightness level is required at 5200 Å than at 5700 Å for rod vision. W. E. Chamberlain (1942)

A great deterioration in detecting details occurs by changing from high brightness levels (cone vision) to those at low levels (rod vision). A quantitative example of this is given by the following comparison. The acuity, i.e. the ability of the eye to distinguish or to separate closely adjacent contours or lines, is considerably worse for rod

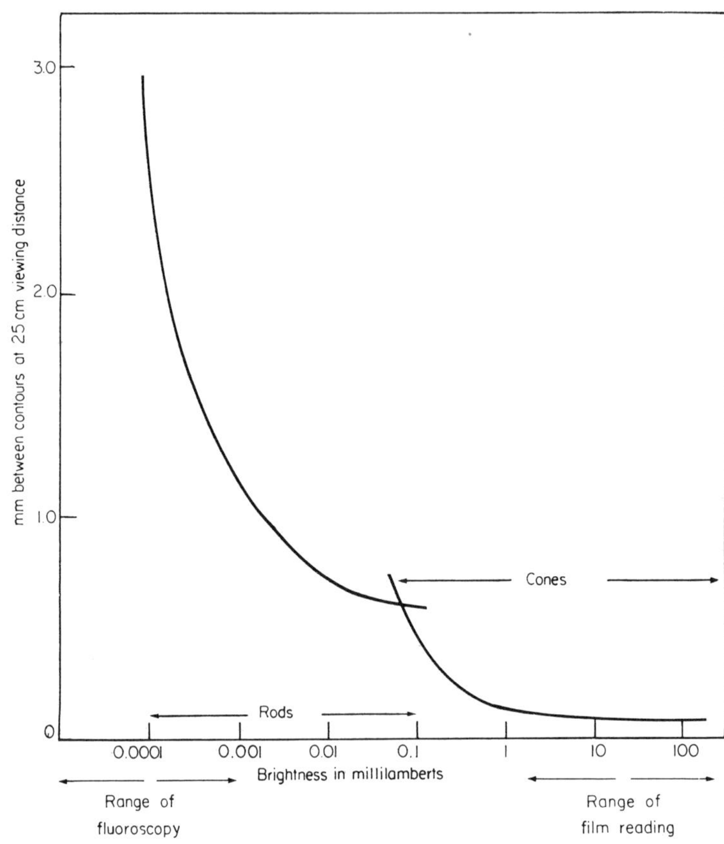

Figure 8.7. Detectable line separation in millimeters as a function of brightness in millilamberts. (See indicated ranges of fluoroscopy and film reading.) *After* G. C. Henny and W. E. Chamberlain in Medical Physics, Vol. 1, edited by Otto Glasser. Year Book Medical Publishers, Inc. *Used by permission of* Year Book U.S.A.

seeing if compared with that of cone vision. Whereas the human eye under favorable brightness conditions is capable of separating lines or contours at a distance of 25 cm, when these are about 0.1 mm apart, only separations of the order of 1–3 mm can be detected at the low brightness level of fluoroscopy (see Figure 8.7).

The quantitative data for rod vision given in the preceding text depend largely on the degree of dark adaptation. A graph showing the relative increase of sensitivity of the eye with the time of dark adaptation is shown in Figure 8.8, from which it is seen that the adaptation time leads to a fairly high sensitivity after about 20 minutes. The adaptation time however depends very much on individual circumstances, and changes from person to person, and with age. Persons who have been in fairly dim light adapt themselves more rapidly

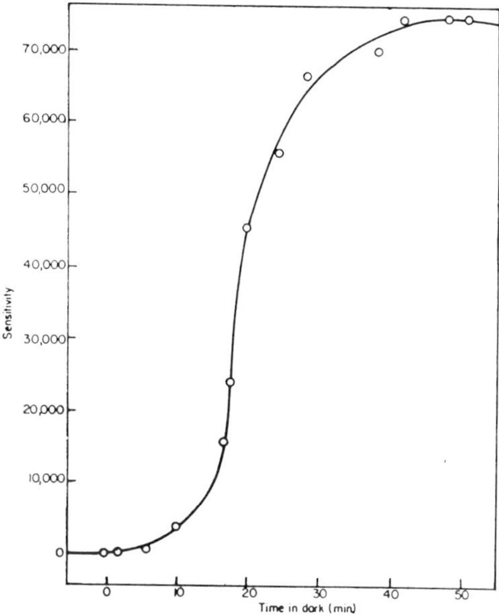

Figure 8.8. Dark adaptation of the human eye. (Sensitivity is defined as the reciprocal of threshold brightness, i.e. the minimum appreciable intensity at each stage.) *After* G. C. Henny and W. E. Chamberlain in Medical Physics, Vol. 1, edited by Otto Glasser. Year Book Medical Publishers, Inc. *Used by permission of* Year Book U.S.A. The original curve is from D. Adams, No. 127 of Special Report Series. Medical Research Council, London. 1927.

than those having come in from bright sunshine (G. C. Henny and W. E. Chamberlain, 1944; R. H. Morgan, 1966). Although one cannot dispense with the adaptation period for good viewing, conditions can be made easier by the provision of red light illumination in the viewing room or by wearing red goggles when white light illumination can be retained.

It is useful at this stage to consider the perceptibility of fluorescent screen images from another point of view. Assuming we have two adjacent image areas on a fluorescent screen differing in brightness, the perception of two distinct areas by the eye does not only depend on the relative brightnesses of these areas, but also on the number of X-ray quanta absorbed by the screen. This can be seen from the following considerations.

The emission of X-ray quanta from the target of an X-ray tube occurs in a random fashion and so does the absorption in the object and in the fluorescent screen. This fluctuation is statistically expressed by the root mean square fluctuation $\sqrt{N}$, where N is the number of quanta or photons and if this is considered as a fraction of the number of photons involved in a given process, it can be formulated by

$$\sqrt{N}/N = 1/\sqrt{N}$$

Hence if $N = 10^6$ photons are absorbed per unit area of the screen, the average fractional fluctuation would be $1/\sqrt{N}$, i.e. $1/10^3$ or 0.1 per cent; if 10^4 photons are absorbed, it would be 1 per cent and if only 10^2 photons are absorbed, it would be 10 per cent.

Thus if the number of X-ray quanta absorbed by the screen decreases the percentage fluctuation of quanta increases. This result is of importance when viewing low contrast images.

If for instance a small area of a screen is only slightly brighter than the background and the brightness is caused by the absorption of only 100 quanta, the fluctuation of 10 per cent may interfere with the visibility of the small area against the background. This is particularly so if the percentage fluctuation of quanta exceeds the percentage brightness difference.

On the other hand if 10^6 quanta are absorbed by the small area, a brightness difference of a few per cent may easily be detected, because the fluctuation is only 0.1 per cent and thus would not interfere with the visibility.

These considerations have shown that statistical fluctuations of photons must be appreciably smaller than the difference between the number of quanta absorbed by two adjacent areas in order to render the contrast visible. This relation between average fluctuations of photons to contrast has been formulated by A. Rose (1948) in another context by the expression

$$C_{\min} = K/\sqrt{N}$$

where $C_{\min}$ is the minimum perceptible contrast and K was found to be of the order of 3–5. Similar considerations apply to the spatial fluctuations of quanta, as observed in quantum mottle (see p. 335). The quantum fluctuations in fluorescent screens are often referred to as 'quantum noise'. It should be noted that the brightness contrast of a screen is influenced by a great number of other factors, such as sharpness, resolving power of the screen, the kilovoltage applied etc.

Further arguments of statistical nature referring to radiographic records are given on page 420.

8.3 THE SENSITOMETRIC CHARACTERISTICS OF RADIOGRAPHIC MATERIAL

The principles of X-ray sensitometry have been reviewed in Chapter 4, page 100 and in the following sections they will be applied to screen and no-screen X-ray films and to X-ray paper. The main purpose of this section is to describe the general characteristics of sensitized X-ray materials and other special characteristics connected with the use of screens, such as the intensifying factor, reciprocity failure, temperature dependence of screen exposures etc. Details about the influence of development conditions on the sensitometry of X-ray material will be discussed in Chapter 9, p. 376.

8.3a Screen-Type X-Ray Films

Screen-Film Emulsions

A screen type X-ray film is usually a double coated film with emulsions which are mainly sensitive to the bluish light emitted by salt intensifying screens. The natural spectral sensitivity of an undyed emulsion is in the blue region of the visible spectrum and the match of wavelength sensitivity of X-ray film with the spectral emission of the screens has been illustrated in Figure 8.3. The emulsion coatings of screen films are sometimes thinner than those of no-screen films and most manufacturers supply screen type films of various speeds. In general those of higher speed have a shorter shelf life, as their fog levels rise faster than those of films of lower sensitivity.

Screen type X-ray emulsions are not only used in conventional radiography, but also in some of the specialized applications of medical radiology, such as tomography, angiocardiography, stereo-radiography etc.

Characteristics of Typical Screen Films

The characteristic curves of two different screen X-ray films (without reference to any specific brand) exposed with salt screens are illustrated in Figure 8.9 and the practical use of such curves will be discussed in this section. The curves shown in Figure 8.9 were

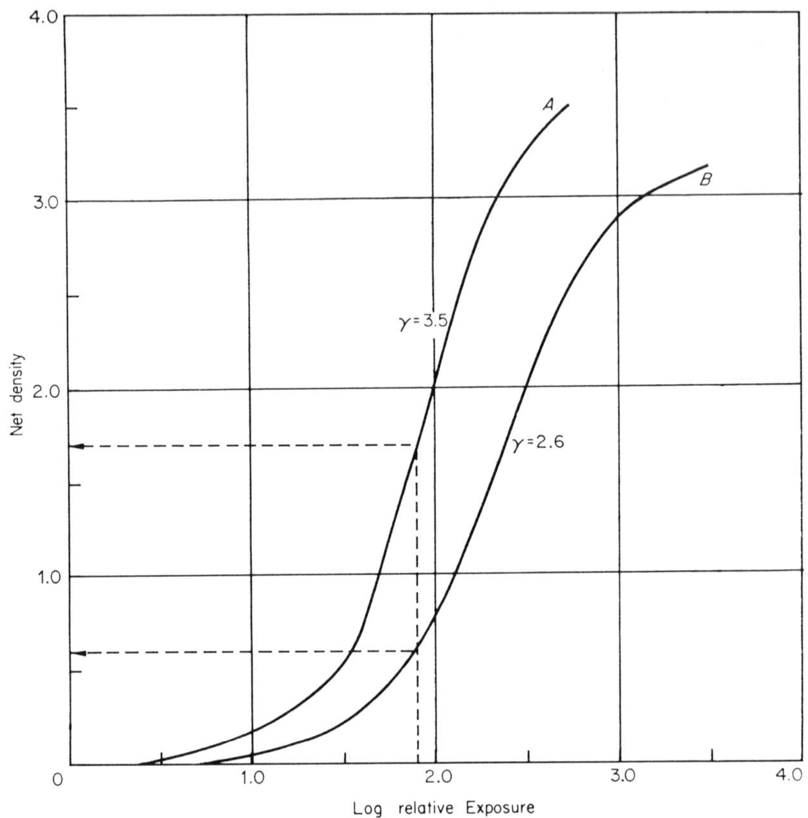

Figure 8.9. Characteristic curves of two different types of X-ray screen film exposed in conjunction with calcium tungstate screens.

obtained from films developed under the conditions as recommended by the manufacturers.

In Figure 8.9 the net density is plotted against the log relative exposure. The net density is that from which the density due to fog has been subtracted. A few of the properties of these films can be seen at once from Figure 8.9 before going into any quantitative details of the curves. These are:

1. That the maximum gradient for both films is obtained at about a density of $D = 1$ above fog.

2. That film A is faster than film B, as for the same log relative exposure, A shows a higher density than B. (See example indicated by dashed vertical line at log relative exposure of 1.9.)

3. The increasing horizontal separation between the two curves towards higher densities indicates that the speed difference between film A and film B becomes greater with increasing exposure.

4. It is seen that the curve of film B is flatter than that of film A; this means that the latter shows more inherent contrast than film B in the region from a density of about $D = 0.6$ onwards.

5. It is further seen that the maximum density of film B tends to be lower than that of film A.

Furthermore we can derive quantitatively how much faster film A is than film B at a given density level. Let us assume we intend to take a radiograph on film A at the same density of 1.2, as that obtained on film B for a given region of the object. Hence we read from Figure 8.9 that $\log E$ is 2.2 at $D = 1.2$ for film B, whereas the corresponding $\log E$ is 1.75 at $D = 1.2$ for film A. The difference in $\log E$ is 0.45 and the antilogarithm of this difference is 2.8. Hence the exposure of film B, which was 20 milliampere-seconds, can be reduced to 7.2 milliampere-seconds in order to obtain a density of 1.2 on film A.

Another quantitative example might be useful. Assuming that a radiograph has been taken on film A and that the important region of the object lies at a density of 0.5 where the contrast (see the respective slope of the curve of film A at $D = 0.5$) is still very low. Hence it is desired to increase the density of the region of importance to a density level of 1.2. The following $\log E$ values for film A can be read from Figure 8.9. At $D = 0.5$, $\log E = 1.45$ and at $D = 1.2$ $\log E = 1.75$. The difference in $\log E$ is 0.3 and the anti-logarithm is 2. Hence the original exposure has to be doubled in order to raise the density in the important region of the object from 0.5 to 1.2 above fog.

Contrast. With regard to the inherent contrast of films A and B, it is seen that the 'gamma' value, i.e. the contrast within the straight line region is 3.5 for film A and 2.6 for film B. The 'gamma' values of films A and B are in general close to those revealed by average modern X-ray screen films. (See definition of 'gamma' value p. 111, Chapter 4, which is here applied to screen X-ray films.) How the gradient $\Delta D/\Delta \log E$ changes with density, for instance for film B, is shown in Figure 8.10. Such a plot may be useful for an easy check on the inherent film contrast at various densities. The flat part of the curve corresponds to the straight line region of the characteristic curve, i.e. to the maximum contrast revealed by film B. It is important to realize that *the shape of the characteristic curve due to screen exposures*

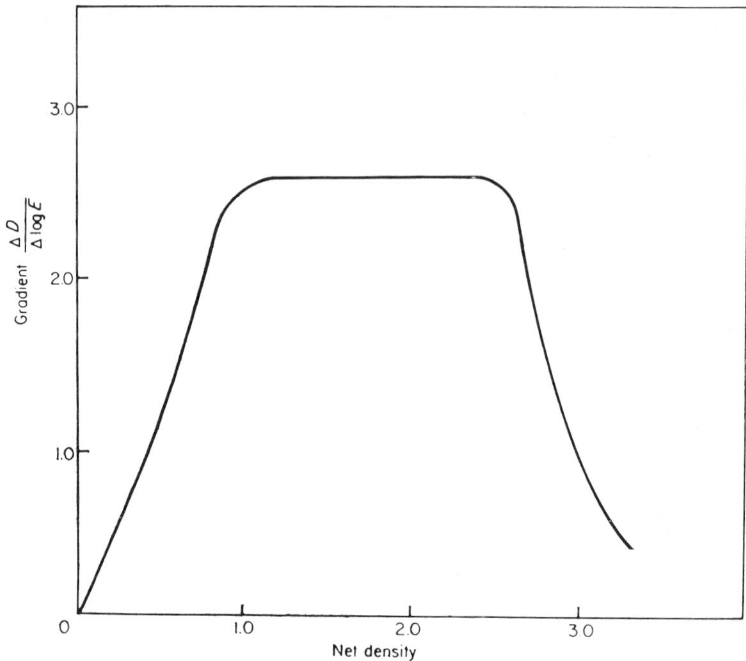

Figure 8.10. Relationship between gradient and net density of film *B* in Figure 8.9.

is not influenced by the kilovoltage at which the incident X-rays are generated, unless reciprocity failure interferes.

Speed.   The speeds of the screen films as derived from Figure 8.9 are relative speeds and no absolute values are quoted by the film manufacturers. One of the main reasons for this is the fact that no practical need has arisen for knowing absolute film speed values. The explanation for this is that the situation in radiography is entirely different from that in light photography where films of such a wide variety of speed exist and the photographer has to adjust the aperture of his camera lens in accordance with the speed of the film applied, and with the light conditions affecting his film. In radiography on the other hand, at least the flux of X-rays can be controlled essentially by means of kV, mA and focus to film distance.

The American Standards Association (1964) has published certain standards for the evaluation of speed of screen and no-screen films in order to give a guide to those who are anxious to carry out their own measurements (see list of references page 369).

The main factors influencing the speed of screen films are the characteristics of the film, the spectral emission and efficiency of the screens applied, the quality and quantity of the radiation incident on the screens, the exposure rate, temperature of screens and film during exposure and the processing conditions (i.e. composition and degree of exhaustion of developer, time and temperature of development, and degree of agitation). It is seen that an accurate reproducibility of all these factors is quite a task.

For a determination of an absolute speed either a standardized screen or a specially defined light source would be required. In addition the contribution of direct X-rays on the film would have to be determined as well. If a standardized screen is used the exposure may be based on roentgen units and on a selected keV. The complexity of all these considerations has not warranted the introduction of absolute speed figures.

On the other hand, it is possible to simulate the light emission of a screen by using an incandescent lamp and a suitable blue filter, which is so chosen that the spectral transmission is practically identical with that of the screen emission. Sensitometric measurements of 'speed' and other characteristics of the screen film can thus be carried out on both sides of the film through neutral step wedges. The very slight influence of direct X-rays on the film which occurs with screen exposures is however neglected, using the conditions described. This influence is negligible in most cases, as only about 2–3 per cent of the density obtained on the film is due to direct X-rays.

Workers in various fields of research may however like to have an approximate guide to an absolute speed, if this is related to an exposure value which is easily reproducible in a laboratory. The absolute speed (S_s) may thus be defined as D_s/R, where D_s is the reference density using a particular screen film combination and R the exposure in roentgens. *For an average modern medium fast screen-film combination S_s is found to be of the order of 1000 for a net density of unity and for an exposure time of 0.1 second*, using average conventional developing conditions. This value refers to a radiation which is generated by a full wave generator at 80 KV_p and filtered by 0.3 mm Cu, thus corresponding to a half value layer of about 0.2 mm Cu. In other words this result means that 1 mR is required to obtain a density of $D = 1$ above fog.

The Intensifying Factor

The intensifying factor (IF) is defined as

$$\text{IF} = \frac{\text{exposure required without screens}}{\text{exposure required with screens}} \tag{8.1}$$

in order to obtain equal density above fog level on the same type of film (with and without screen) and for constant processing conditions. If the exposure without screens is 3 seconds for a density of 1 above fog and that with screens 0.1 second for the same density, using the same quality of radiation, the IF = 30. The intensifying factor may be quoted for an exposure taken without or with absorber (or filter) between X-ray tube and film cassette. In the latter case a different value will be obtained than without absorber because of the hardening effect of the absorber acting as a filter (see p. 28). This

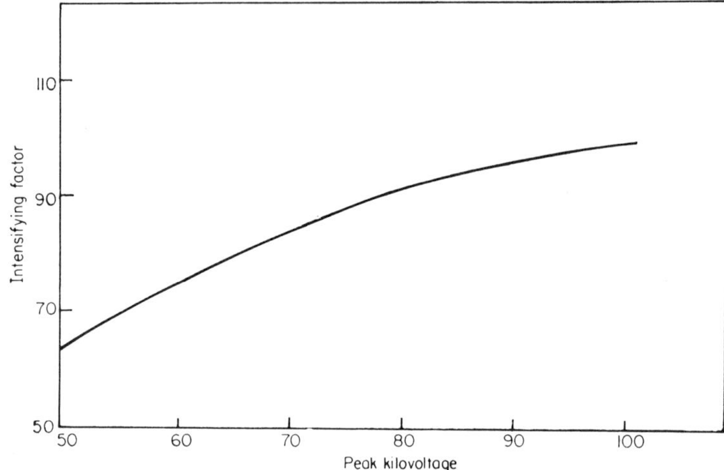

Figure 8.11. Intensifying factor as a function of the quality of X-rays generated at increasing kilovoltage. No filtration and no absorber between X-ray tube and film system. (Medical screen X-ray film with and without calcium tungstate screens.)

difference is due to the fact that the intensifying factor changes with the quality of radiation, as shown in Figure 8.11.

This figure illustrates the increase of the IF with increasing kilovoltage within the medical range of kV's *using no absorber* between X-ray source and screen film combination. It should be noted that the value of an intensifying factor using filtered radiation or an object simulating the human body would be much closer to practical conditions than that derived without absorber.

For the less experienced reader it should be mentioned that the IF leads to considerably higher values if the curve Figure 8.11 were plotted for monochromatic radiation (equivalent to keV) instead of for heterogeneous radiation (as indicated by kV_p as in Figure 8.11).

The reason for this is evident when considering that for instance unfiltered radiation generated at $90\,\mathrm{kV_p}$ corresponds roughly to about $40\,\mathrm{keV}$. The IF at $100\,\mathrm{keV}$ for fast screens was found to be 420 against only 95 for $100\,\mathrm{kV_p}$.

The increase of the IF with kilovoltage is mainly due to the ratio of relative X-ray absorption in screens to that in film, as explained on page 290. In order to derive the IF experimentally, sensitometric exposures, e.g. time scale exposures are carried out simultaneously with and without screens using the same film. Such stepped exposure strips are shown in Figure 8.12(a) below and the characteristic curves derived from the stepped density strips are illustrated in the graph in Figure 8.12(b). The intensifying factor (IF) can be derived by determining the horizontal separation between the two curves at any desired density. As an example, if we draw a parallel line to the abscissa axis at a density of unity, we find the log E values for

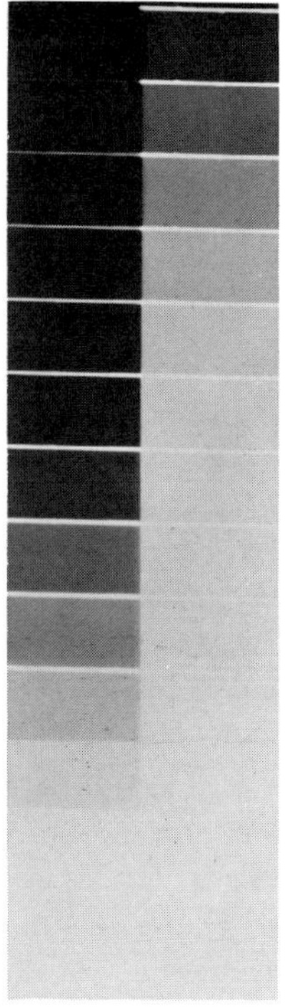

(a)

Figure 8.12.(a) Sensitometric density strip of a screen X-ray film. Left exposed in conjunction with calcium tungstate screens, right without screens.

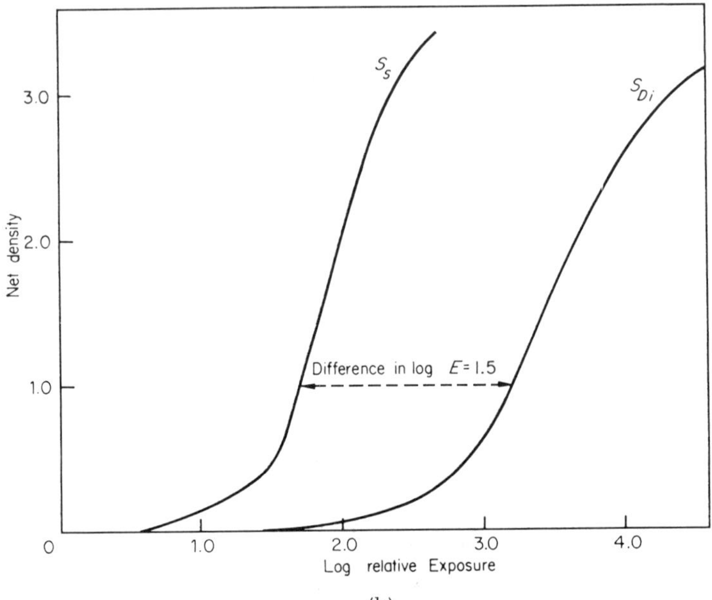

(b)

Figure 8.12(b) Characteristic curves plotted from sensitometric strips (left). Screen exposure (S_s). Direct exposure (S_{Di}).

$S_s = 1.7$ (i.e. for the screen exposed film) and for $S_{Di} = 3.2$ (for the direct exposed film). The difference between the two log E figures is 1.5 and the antilogarithm is 31.6. Hence the IF is 31.6. It is seen from Figure 8.12 that the characteristic curve of the directly exposed film has not only a much longer toe than that of the screen exposed film but is also much flatter. Hence it is to be expected that the IF increases with density. The IF for a density of 2.5 is 63 according to the same graph.

In the opinion of the author the importance of the IF is often exaggerated in discussions on radiographic technology or even in radiographic examinations. In references to the IF it is rarely mentioned whether it applies to exposures with object between tube and cassette or even to what density the IF refers; hence the data are useless. On the other hand it cannot be denied that average figures of intensifying factors, if given with full specification, may be of interest as an approximate guide to the exposures required with screens as against those taken without screens. It should be noted that it is of greater practical interest to determine the intensifying factor for screen film with screens and for a no-screen film without screen, instead for a screen film exposed without screens. The latter is rarely used without screens, because it usually has a lower inherent contrast than typical no-screen films.

Reciprocity Failure with Screen Exposures

In Chapter 4 page 128 it has been shown in some detail that the speed of a photographic emulsion exposed to light depends on the rate of exposure in contradiction to direct exposures with X-rays for which such a dependence practically does not exist. A screen exposure may be carried out at 1 mA for 10 seconds or e.g. at 100 mA for 0.1 second, hence both products of mAsec are the same, but the density obtainable in both cases (assuming the quality of the radiation is kept

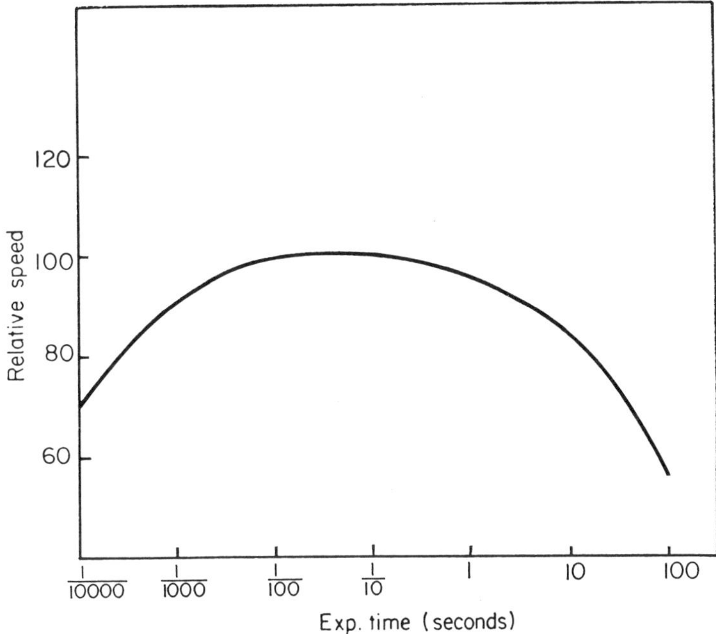

Figure 8.13. Reciprocity failure of an X-ray screen film exposed with calcium tungstate screens. The curve shows the relative speed against exposure time in seconds for a constant product of light intensity and time.

constant) need not be the same. This is because reciprocity failure may have occurred (see p. 129). It is therefore of interest to know to what extent reciprocity failure may influence the speed of screen films in the practice of medical radiography.

Different brands of screen films reveal this effect to varying degrees and therefore a result for a particular film may not necessarily apply to another brand of film. Nevertheless it was thought useful to show the effect of reciprocity failure of a typical commercial screen film, when exposed to screens. Such a result is illustrated in Figure 8.13

which shows the relative speed as a function of exposure time in seconds and it is seen that there exists a maximum of speed in the region between $\frac{1}{100}$–$\frac{1}{10}$ of a second and that the speed drops towards the short and the long exposure times. It can be concluded for the particular brand of film chosen that the effect is negligible for the exposures most likely to occur in medical radiography, as these cover a range from about 1/100th to 5 seconds. It becomes appreciable however for exposures of the order of 1/10000th of a second and for those exceeding 100 seconds and is thus of great significance where such exposure times are used, as for example in industrial radiography (see pp. 451–452, see also R. H. Morgan (1944b)).

The Effect of Temperature upon Screens

Although most screen exposures in medical radiography are carried out at room temperature there may be cases where exposures have to be taken either at lower or at higher temperature, the latter for instance in tropical conditions. It is therefore of interest to realize that the screen efficiency decreases with an increase of the ambient temperature of the screen during exposure. This effect is only slightly compensated for by the opposite trend of the change of film speed with temperature, when exposed to blue light. These effects have been studied by the author for the range of temperatures most likely to occur in practice. The results of this investigation are shown in Figure 8.14 which shows the relative speed of a fast X-ray film exposed in conjunction with calcium tungstate screens for a range of temperatures between 4–44°C. It is concluded from Figure 8.14

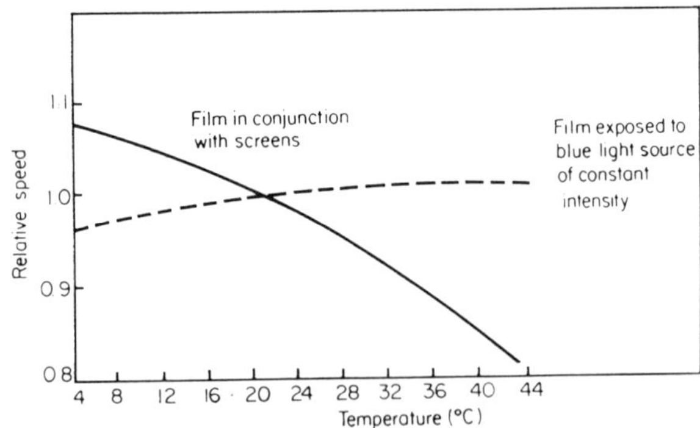

Figure 8.14. Full line curve: Relative speed of X-ray screen film (exposed with calcium tungstate screens) as a function of temperature during exposure. Dashed curve: The same type of X-ray screen film when exposed to a blue light source of constant emission as a function of temperature. (All exposures refer to 0.1 second exposure time.)

that the change of speed over the complete range is about 25 per cent. Similar measurements have been carried out with barium lead sulfate screens, which led to fairly similar results. The variation of speed with temperature for the screen film when exposed to a blue light source of constant intensity is shown in the dashed curve in Figure 8.14. (See also R. H. Morgan 1944a.)

8.3b   No-Screen X-ray Film

No-screen X-ray films are commercially available mostly as double coated films and are used for direct X-ray exposure, i.e. without screens. They are often supplied in individual paper envelopes which facilitates the handling of the film, as it saves the operation of inserting the film into the cassette in the darkroom before exposure. The emulsion coating thickness of the fastest no-screen films is generally greater than that for screen films in order to achieve as much absorption of X-ray energy as possible. The thickness of coatings has its technical limitations caused by the prolonged time of diffusion of the developer solution. The sensitivity to direct X-rays for most no-screen films is higher than that for screen films; on the other hand their sensitivity to blue light is usually considerably smaller, than that for screen films. Most manufacturers supply different varieties of no-screen films, some of these are chiefly used in industrial radiography. The various types differ in speed and in graininess. The fast and coarser grained films in general possess a more restricted shelf life, than the slower varieties. The increase of fog of the fast films with storage time is influenced considerably by the effect of various components of cosmic rays. In fact measurements have shown that films kept below an equivalent thickness of 60 m of water showed only half the fog after one year's storage than those stored under normal atmospheric conditions. (R. H. Herz (1951).)

Characteristics of No-Screen Films

A series of characteristic curves of no-screen films of different speeds is shown in Figure 8.15, usually the fastest no-screen film is used for medical purposes. A curve of a screen film is shown for comparison. One of the significant differences between screen and no-screen films is the long toe exhibited by the characteristic curves of the latter, which indicates less inherent contrast within the low densities than attainable with screen films. A further difference is the fact that the curves of the no-screen films in general reveal no straight line region, as the gradient increases further with density up to very high densities. The maximum density of no-screen films is usually

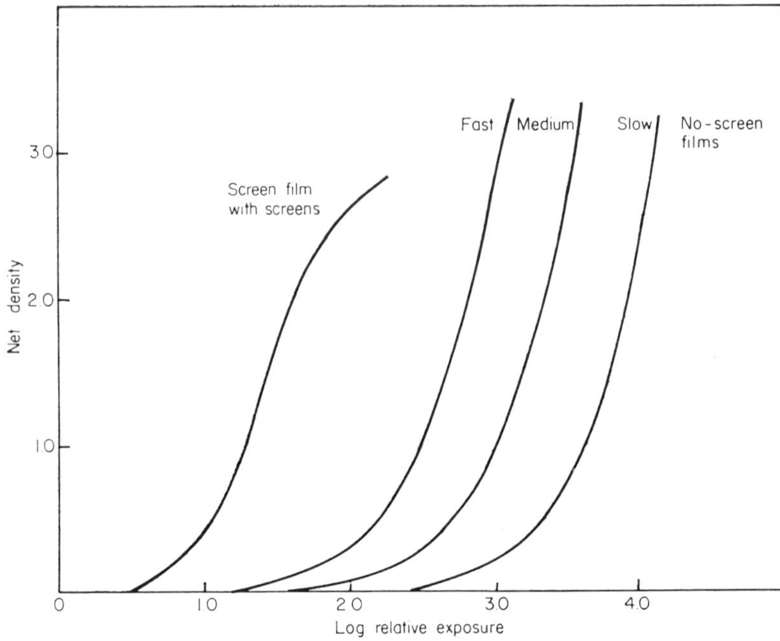

Figure 8.15. Comparison between the characteristic curves of a fast X-ray screen film exposed with calcium tungstate screens and typical no-screen X-ray films of different speeds.

considerably higher than that of screen-films and reaches values of about 12 for the double coating. Hence, if required, the no-screen film may be viewed on a high intensity illuminator, where it may reveal important details in the high densities, thus utilizing the increasing inherent contrast. This aspect is not utilized so much in medical radiography, as the range of thicknesses to be covered is not so great. The shape of the characteristic curve remains the same within the range of kilovoltages applied in medical radiography. In supplementing what has already been said about the inherent contrast revealed by some typical no-screen films, the relation between $\Delta D/\Delta \log E$, i.e. the gradient as a function of the net density is plotted in Figure 8.16. If this curve is compared with the corresponding curve for screen exposures (see Figure 8.10) the difference appears rather striking.

Speed of No-Screen Films

Typical absolute speed figures for fast no-screen films are expressed e.g. by D/R, where D is a density of unity and R the exposure in

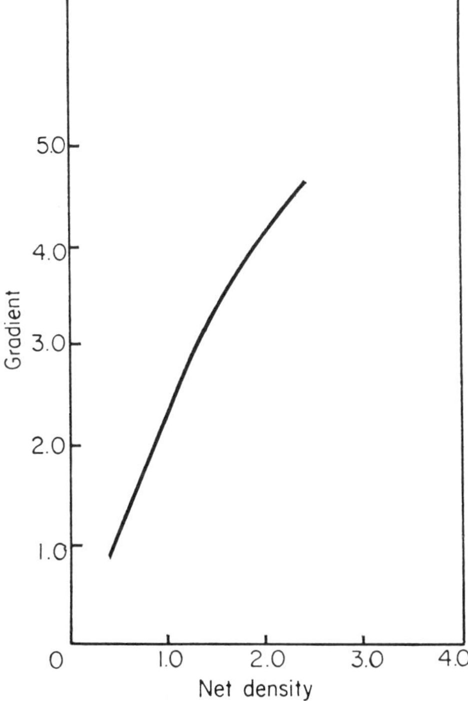

Figure 8.16. The relationship between gradient and net density for a typical no-screen X-ray film. (Note that the gradient at $D = 1$ is 2.3 in accordance with equation 4.13. p. 118)

terms of roentgens. For a fast X-ray film this is of the order of $D/R = 50$. Hence for a net density of unity an exposure of 20 mR is required. This value refers to an exposure to X-rays generated at 80 kV_p and filtered by 0.4 mm Cu. Using other qualities of X-rays the speeds will differ, as has been discussed in detail under the heading of 'Spectral sensitivity' in Chapter 4, page 121. The maximum speed of the no-screen films, exposed to direct X-rays lies at approximately 40 keV and this radiation can be simulated fairly closely by using 80 kV_p, filtered by 0.4 mm Cu. Other more stringent conditions to simulate monochromatic radiations of various keV's are quoted in Table 14.5 in the appendix. It should be mentioned again that direct X-ray exposures obey the reciprocity rule, i.e. no reciprocity failure is involved. (For exceptions see p. 130.)

8.3c X-Ray Paper

The basic composition of X-ray paper has been briefly discussed on p. 281. X-Ray paper is made for use with one intensifying screen, preferably with a back screen and it is advantageous to let X-rays first strike through a back of the paper and then through the screen, depending however on whether A.P. or P.A. projection is required. (A.P. = Antero-Posterior; P.A. = Postero-Anterior).

Viewing Conditions

The basic differences in the viewing conditions between X-ray paper and double coated X-ray film are chiefly responsible for the characteristics of the two types of material. When viewing the double coated film the incident light intensity is attenuated by passing

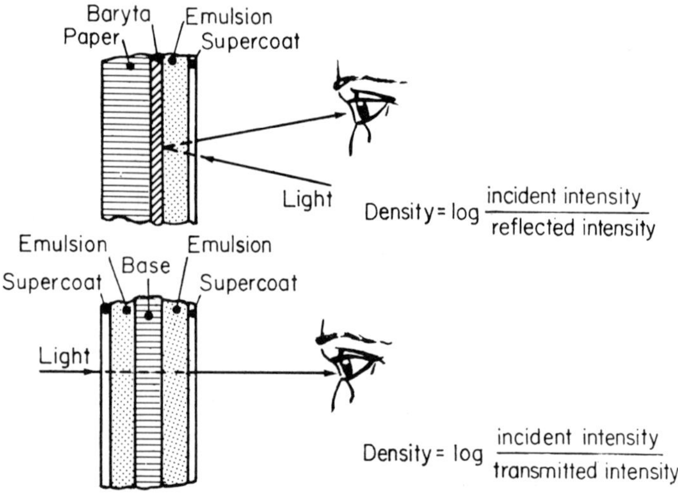

Figure 8.17. Viewing conditions of radiographs taken on X-ray paper (as seen by reflected light) and on double coated X-ray film (as seen by transmitted light).

through two emulsions, supercoats, and base before it reaches the eye. In the paper the light passes through supercoat, emulsion layer and is then reflected back by the baryta (barium sulfate) layer and passes through the emulsion and supercoat a second time, before it reaches the eye. These conditions are illustrated in Figures 8.17 and 8.18.

In Figure 8.17 we see the conditions for viewing double coated film and X-ray paper and in Figure 8.18 the results of the viewing conditions. The latter figure shows on the top left a double density step wedge from 0–3 of a double coated X-ray film viewed against the light of an ordinary viewing box. On the top right of the same figure we see the light sources directed towards the single coated X-ray paper step wedge with its incorporated baryta reflecting layer. On the

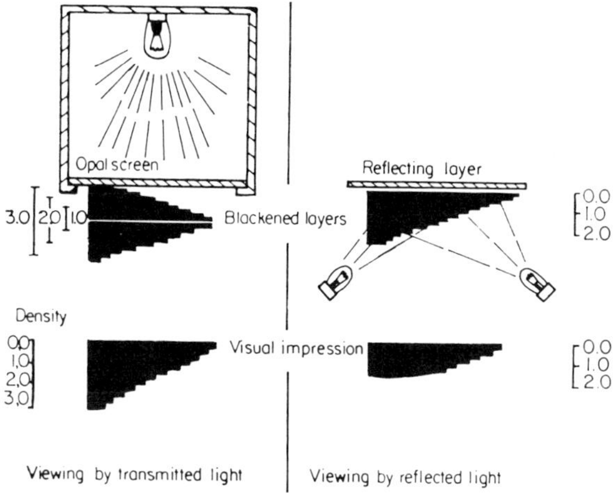

Figure 8.18. Conditions of viewing a density stepwedge of double coated X-ray film (left) and single coated X-ray paper (right).

bottom left the visual impression of the transparent X-ray film is indicated and on the bottom right the visual impression of the paper image. Whereas on the left the eye is capable of distinguishing each step up to a density of 3.0, on the right the paper can only reveal densities up to about 1.6. The reason for this discrepancy is that in the paper image on the right the light reflected by the surface of the paper diminishes the maximum reflection density below that which would be expected from the density of the emulsion layer. In other words the limitation of the paper image is caused by 'surface reflection'. (R. H. Herz, 1933.)

Sensitometric Characteristics

The effect on the relative characteristic curves of X-ray film and X-ray paper is shown in Figure 8.19. These characteristic curves of double coated X-ray film and X-ray paper, are actually not commensurable, because of the different methods of density measurement, transmission and reflection densities respectively. It is seen from the curves that the disadvantage of the paper, if compared with the film, is mainly the restricted range of densities of the former and also the reduction of the inherent contrast. The lower contrast of the paper refers to the whole range of exposures.

The use of X-ray paper without screens is even more inferior to film because of the long toe of the characteristic curve, as exhibited by direct X-ray exposures. The speed of X-ray paper with screen is

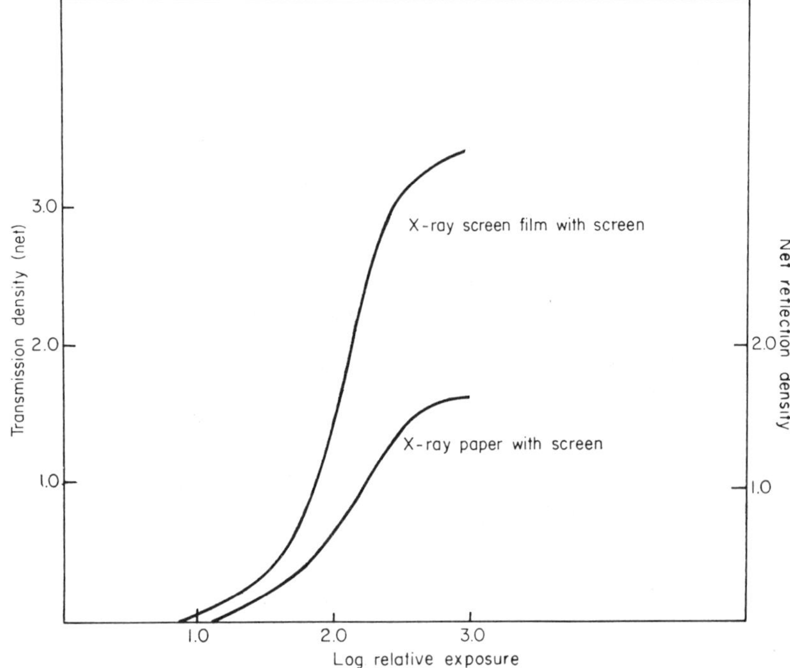

Figure 8.19. Comparison of characteristic curves of an X-ray screen film (transmission density) and an X-ray paper (reflection density).

in general about 50 per cent lower than that of the slowest screen X-ray film combination.

These considerations have shown that X-ray paper is radiographically inferior to the film, although it may be used successfully in certain circumstances, where only a restricted range of densities is required, as for instance in bone alignment, coarse fractures of bone, barium meals etc. For the reasons mentioned relatively little use is made of X-ray paper in medical radiography; its main advantage is that it is cheaper than X-ray film.

8.4   IMAGE QUALITY AND ITS ASSESSMENT

In the preceding sections we have discussed the elementary principles and the photographic tools of radiography, apart from processing. We now intend to survey the many factors which influence the image quality. In the first section the control of contrast and definition will be summarized. A special section will then be devoted to the effects of quantum mottle, a phenomenon which arises only with high speed systems and the importance of which on image quality has only been realized more recently. Having discussed the factors influencing image quality, ways and means of its assessment will be discussed.

8.4a   The Control of Contrast and Definition

In the following diagram a survey is given of the main factors on which contrast and definition in medical radiography depend. As some of these factors have been discussed before, only passing reference will be made to them in order to avoid repetition. Most of the factors mentioned in this diagram are interdependent in their influence on the final image and it would be unwise to give preference to one factor without relating it to the effects on others. The following discussions will essentially be based on the sequence shown in this diagrammatical review.

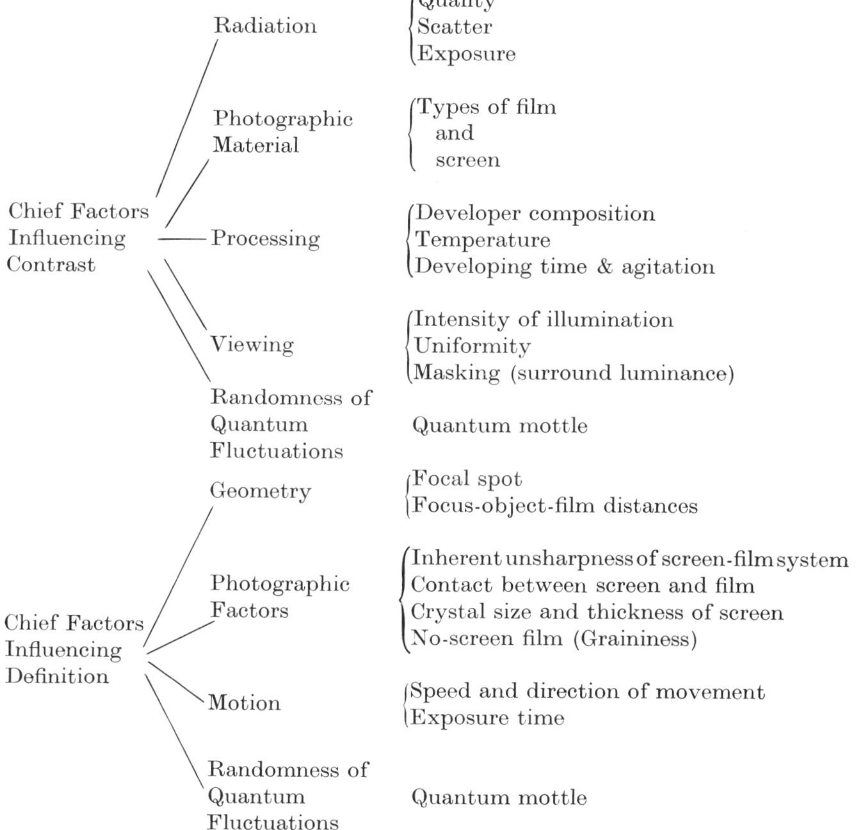

8.4b   Chief Factors Influencing Contrast

(1) *Radiation*

In medical radiography the range of kilovoltage which is at the disposal of the radiographer is generally between 40–120 kV$_\mathrm{p}$. Special high kV techniques ranging up to 150 kV$_\mathrm{p}$ and even higher have been introduced into some hospitals in recent years, partly because of the relatively smaller dose required for exposure.

When taking a radiograph of a patient, the first consideration is generally the choice of a suitable quality of radiation, i.e. the adjustment of the peak kilovoltage. The selection of the kilovoltage is certainly the most effective tool influencing the radiographic image contrast. (This refers to objective and subjective contrast.) The lower the kV$_\mathrm{p}$ the higher the contrast attainable, but this is limited by other considerations (see p. 271). If the object shows a very wide range of thicknesses, such as in an antero-posterior projection of the dorsal spine, low kilovoltage radiation may be completely absorbed by the thicker and bony regions of the patient's body. Unless harder radiation is used in such a case, undesirable and exaggerated contrast will result. Experience must guide the operator in selecting the most suitable kilovoltage in each case, depending not only on the region of the human body, but also on the physique of the patient and on the type of examination. An X-ray exposure table is shown in the appendix to give a guide to the selection of kilovoltage and other exposure conditions for various regions of the human body Other quantitative data on exposures are given in Figures 8.23–8.26. and pages 320–325

Furthermore it is important to realize that the calibration of an X-ray unit in terms of kV$_\mathrm{p}$ is often based on a given tube current in terms of milliamperes. The choice of another higher tube current may cause a drop in the kilovoltage across the X-ray tube due to changes in the distribution of magnetic fields in the secondary coil of the high tension transformer. As long as the operator is sufficiently aware of this possibility, it will not surprise him if he obtains too high a contrast or an underexposed image owing to the drop of kV. A corresponding adjustment can be made by increasing the kV$_\mathrm{p}$. Most modern X-ray units are however provided with automatic corrections for the drop of kilovoltage so that the situation mentioned above is avoided.

The calibration of kilovoltage of X-ray units of different manufacture may differ considerably. Hence the use of a particular kilovoltage control on one X-ray unit may not necessarily lead to the same result on another unit. (R. H. Herz, 1936.) Even if the calibration of kV_p is the same for two units the wavelength distribution may differ (a) because of variations in inherent filtration of the primary beam (see p. 28 and next section), (b) because of differences in the electrical circuit and (c) because of the conditions of the X-ray tube, (tungsten deposit, state of focal spot, etc.).

Filtration. The primary X-ray beam emitted from an X-ray tube must pass through the glass wall of the tube, and sometimes through a layer of oil and the aperture of the tube shield before it emerges from the X-ray tube. The filtration by these materials is called the inherent filtration and this is equivalent to about 0.5–2.0 mm aluminum on the average. An additional filtration of about 1 mm aluminum is usually added in order to eliminate the softest components of the beam which would otherwise be appreciably absorbed by the body of the patient.

Scattered Radiation. One of the most disturbing factors influencing image contrast is that due to scattering of X-rays within the human body. Scattered radiation is very troublesome, because it acts as a diffuse, non-image forming radiation which spreads almost uniformly over the film, and thus deteriorates the quality of the image formed by the primary beam.

The intensity of scattered radiation is largely dependent on the amount of matter traversed by the primary beam and can be weakened significantly by reducing the diameter of the primary beam. In fact, measurements in water phantoms have shown that the intensity of scattered radiation in large volumes of about 10 inches thickness, may be of the order of 5–10 times greater than that of the intensity of the primary beam reaching the film. This occurs with large beam diameters which are generally required in medical radiography.

Scattered radiation is made up of modified and unmodified scatter, but the contribution of the latter is relatively small for low atomic material and within the range of kilovoltages applied in medical radiography. Although the scattering coefficient decreases with harder radiation, experimental findings in practical radiography show that the effect of scattered radiation increases with the kilovoltage applied. The explanation of this apparent discrepancy is the fact that scattering produced by harder radiation will be less easily absorbed by the patient's body and thus reaches the film more readily. There are several means by which the effects of scatter can be reduced. These may be divided into those which minimize the formation of scatter and those which tend to prevent the once formed scattered radiation from reaching the photographic emulsion.

(2) *Reducing the Formation of Scattering*

Cones and Diaphragms. One of the simplest methods of reducing the formation of scatter is to use a cone of lead or lead glass attached to the X-ray tube. The cone restricts the field of X-rays reaching the surface of the object (see Figure 8.20) and thus the effect of scatter from neighbouring tissues. Cones of different diameter may be used

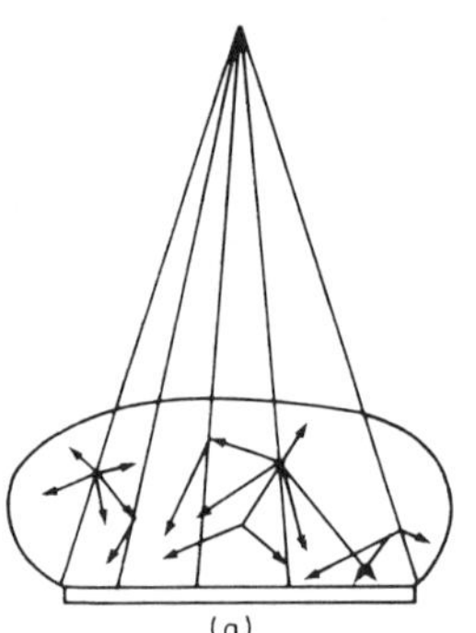
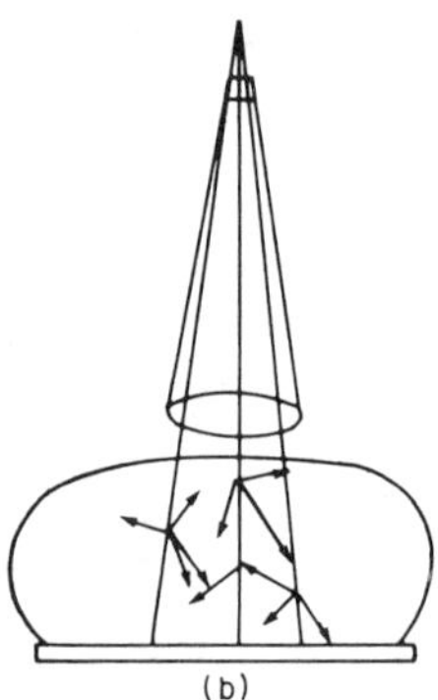

Figure 8.20. (a) Multiscattered radiation in human body. (b) Use of cone to reduce scattered radiation from outside the field of the primary X-ray beam.

to limit the region of the human body to be exposed. The smaller the diameter of the cone, the less the effect of scatter. Modern X-ray tubes are provided with diaphragms, the size of which can be controlled. This device usually consists of a box having a lead diaphragm near to the tube and another one which is more remote from the tube, thus restricting the aperture of the beam according to any cross-section required. Cones may be used in addition to this device to prevent air scatter.

Compression. A further means of reducing the formation of scatter which is chiefly restricted to regions of the abdomen is the compression of soft tissues, thus reducing the volume of the human body to be irradiated. The compression is usually carried out with the aid of a compression band.

(3) *Preventing Scattered Radiation from Reaching the Film*

Potter–Bucky Grid Principle. The most efficient method of preventing scattered radiation from reaching the film is the Potter–Bucky grid principle, so called after the inventors. The grid diaphragm is made up of a series of lead strips interspaced between radiolucent strips. This system is arranged in such a manner that when centered below the X-ray tube between patient and film a portion of the primary radiation can pass between the lead strips, whereas most of the scattered radiation is absorbed by them. Hence a high fraction of the scattered rays does not reach the film (see Figure 8.21).

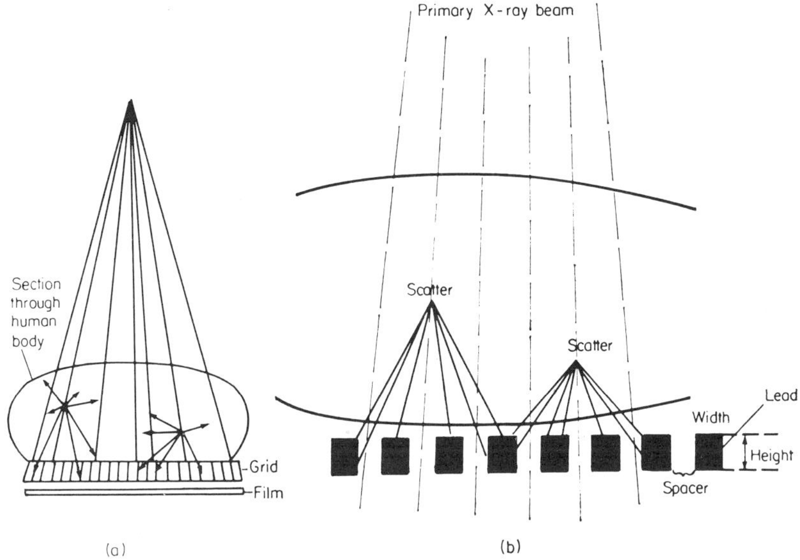

Figure 8.21. (a) Principle of Potter-Bucky grid. (b) Enlarged section of center part of grid.

Stationary Grid System.   That part of the primary beam which is absorbed by the lead strips forms a shadow image on the film, if the grid system is stationary (Lysholm grid system). In a stationary grid the lead and the radiolucent strips are very small in width so that their image is hardly noticeable when viewed at a distance of 1–2 feet. Some types of stationary grids can be placed into the cassette and some even form an integral part of the cassette.

Movable Grid System.   In order to eliminate entirely the formation of shadow images of the lead strips on the film, the grid system is moved during exposure (Potter–Bucky grid). This means that any part of the film will be obscured from radiation only by a fraction of the time of exposure and a uniform image will be obtained. The movement of the grid system may be reciprocal or in one direction only, and it may be carried out by manual or by electrical means from the control board of the X-ray unit. In most modern X-ray apparatus the exposure control is linked with the control of the grid system. The movement of the grid commences before the start of the exposure and ends after the termination of the exposure so that the grid is never stationary whilst the exposure is carried out (H. E. Seemann and H. R. Splettstosser, 1954; J. G. Bonenkamp and W. H. Boldingh, 1959, 1961).

Stroboscobic Effect.   Care has to be taken to avoid stroboscopic effects with moving grids. If these occur, shadows of the lead strips may appear on the film. Stroboscopic phenomena occur if the speed of the motion of the grid is in a certain critical relation to the frequency of the pulses of X-ray emission. (This effect is similar to that observed on a cinema screen when the wheels of a moving cart appear to be stationary.) If this effect happens, the rate of movement of the grid system has to be changed in order to eliminate the effect (O. Mattson, 1955).

Increase of Exposure with Grid.   The movable grid system was introduced shortly after the first world war and is referred to as the Potter–Bucky grid, whereas stationary radiographic grids were first introduced by Lysholm and Schönander in the mid thirties.

Whatever the grid system in use, the exposure time has to be increased in comparison with an exposure without a grid. This is because of the partial absorption of the primary radiation and the elimination of scatter, which would otherwise contribute to the overall density of the radiograph. The increase of exposure with a grid as against no grid may be as high as 4–6 times.

Grid Ratio.   Many types of grid systems are available; some grids consist of parallel lead strips (see Figure 8.22), some have the lead strips placed at an angle (see Figure 8.21(a)). In the former case the primary radiation is cut off at a distance d from the center of the cassette and a greater focus to film distance has to be chosen than that indicated in the diagram, in order to avoid the cut-off of primary radiation. In Figure 8.22 the distance d at which cut-off occurs is found from

$$\frac{a}{d} = \frac{h}{w} \quad \text{or} \quad d = \frac{aw}{h} \tag{8.2}$$

where a is the focus-cassette distance, h the height of the lead strip and w the width of the spacing. The ratio $\dfrac{h}{w}$ is called the grid ratio.

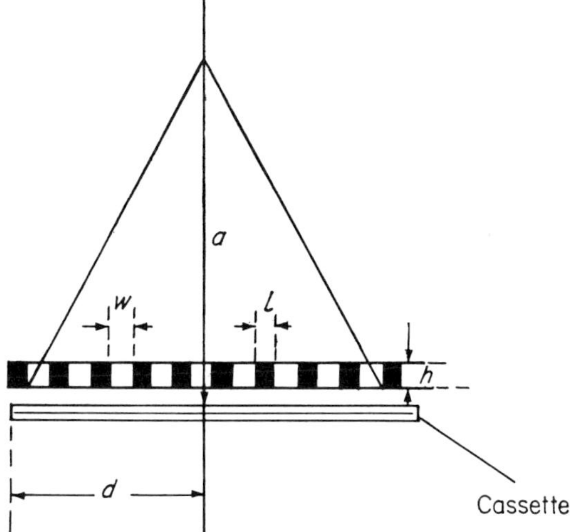

Figure 8.22. Section through grid system with parallel lead
strips. (Grid ratio h/w.)

If $a = 120$ cm and the grid ratio is 8:1, then in the parallel grid,
Figure 8.22, cut-off occurs at 15 cm from the center of the cassette.
The ability of the grid system to eliminate scatter depends on the
grid ratio and on the width of the lead strips (l).

If the lead strips are placed at varying angles with respect to the
plane of the grid as shown in Figure 8.21(a), careful positioning of the
X-ray tube at a given distance is required and then no cut-off
should occur. (Focused grid.)

There are also crossed grids available in which two grids are superimposed at right angles. Their efficiency in eliminating scatter is
high, although relatively long exposures are required. Detailed
information on crossed grids is found in the publication by O. Mattson
(1955).

Movable grid systems of various grid ratios are used within a range
from 4:1 to 20:1. The higher this ratio the greater the elimination of
scatter to be expected for a given kilovoltage and the greater will be
the exposure time. High grid ratios between 10:1 and 16:1 are used
successfully in high kV radiography, where voltages between 100
and 150 kV_p are applied.

The relative efficiency in removing scattered radiation using various
grid ratios is seen from the following example. If an A.P. radiograph of
a pelvis is taken of a patient who necessitates 85 kV_p X-rays for the
exposure, 45 per cent of the density is found to be due to scatter,

using a 4:1 ratio; only 17 per cent is due to scatter using a 10:1 grid, and only 7 per cent using a 20:1 grid ratio. These data have been taken from a graph due to Smit Roentgen Co, Holland, published by H. E. Johns (1964). See also R. B. Wilsey's early work on scatter in radiography (1921a, 1921b, 1922a, 1922b, 1934).

Use of Lead Support.   A further method of preventing scattered radiation reaching the film consists of the insertion of a sheet of lead between table support and film. In this way the primary radiation would be absorbed by the lead sheet and no back scatter would arise from the table, although some characteristic radiation from lead may interfere at higher kilovoltages. This method refers chiefly to the use of no-screen film wrapped in paper envelopes, as most cassettes and exposure holders have lead foils incorporated behind the film.

(4) *Removal of Scatter by Object-Film Distance (Air-Gap Technique)*

Some radiologists have proposed to reduce scatter in chest exposures by having the film at a distance of about 15–30 cm from the patient. Such an exposure would undoubtedly suffer from unsharpness, unless the focus to film distance chosen is considerably greater than usual, and the focal spot small, to compensate for the increased object to film distance (G. M. Ardran, 1964).

(5) *Exposure*

The influence of the amount of exposure on contrast becomes evident by referring to the characteristic curve of a given film. If the exposure is too low, the important region of the object may lie within the toe of the characteristic curve, or if overexposed, it may be within the shoulder of the curve. In both cases the contrast would be reduced, if compared with a correct exposure, where the important region should lie within the more or less straight line region of the curve.

The general relationship between exposure time and kV, mAs and distance can be expressed by

$$\text{Exposure time} \propto \frac{(\text{distance})^2}{\text{mA} \times \text{kV}_{\text{p}}{}^n} \times K \tag{8.3}$$

where K is a factor depending on (a) the object, (b) the photographic conditions (screen, film, processing etc.), (c) whether a grid is used, its grid ratio or cone etc., (d) the filtration used at the X-ray tube and (e) the circuit of the X-ray apparatus.

The power n in equation 8.3 is of the order of 4 for screen exposures in the normal range of peak kilovoltages used in medical radiography, i.e. from 40–100 kV$_p$ and gradually decreases to about 3 in the range of high kV radiography, i.e. from 100–150 kV$_p$.

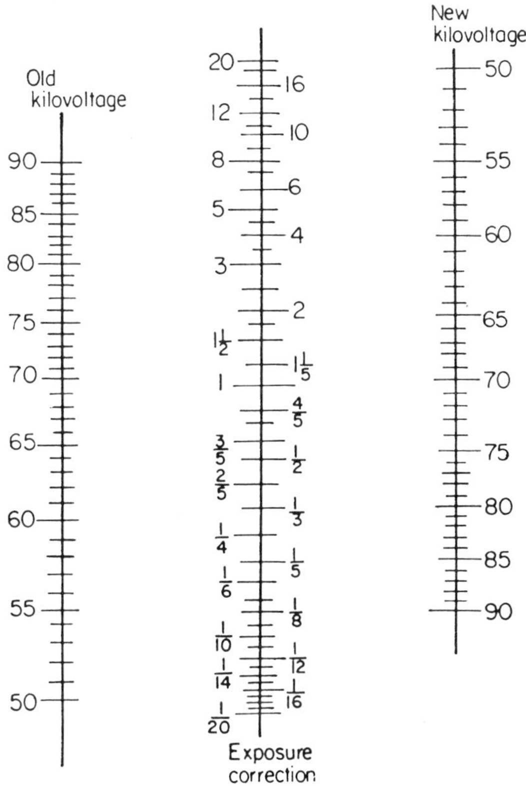

Figure 8.23. Nomogram showing exposure corrections required for changes in kilovoltage. (*By courtesy of* Kodak Ltd., England.)
Example: If the previous exposure was taken at 70 kV$_p$ at an exposure time of 0.1 seconds and the kilovoltage should be reduced to 55 kV$_p$, then the new exposure time is found to be three times greater, i.e. 0.3 seconds.

A nomogram showing the exposure correction to be made, if the previous kilovoltage is to be changed to another kilovoltage, is illustrated in Figure 8.23. This nomogram refers to exposures carried out with intensifying screens and the values derived are subject to some alterations caused by the variable filtration of the subject to be radiographed.

In general the choice of exposures in medical radiography (see appendix) is basically the result of experience in hospital practice combined with simple calculations relating distance, mAsec, film-screen speed, object absorption, processing conditions etc. It would be a fairly complex and probably useless task to derive exposures mathematically from first principles, considering number of quanta incident, absorption coefficients when using heterogeneous radiation and the many other factors involved, such as the region of the human body, physique of the patient etc.

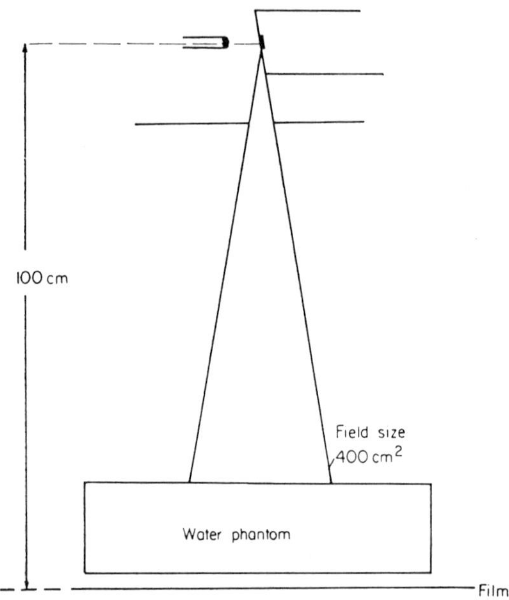

Figure 8.24. Experimental arrangement used in the tests described below.

If we replace the patient by a phantom of water or other material simulating human tissues the relationship indicated in the formula, equation (8.3), can be approached by measurement and can be presented in a fairly precise form for the type of X-ray apparatus and for the particular set up in use. In order to give the physicist and technician, who are less familiar with medical radiographic techniques, a guide for assessing similar relationships of an unknown character the following graphs may be helpful. The experimental arrangement on which the graphs Figure 8.25 and 8.26 are based is shown in Figure 8.24.

Although the calibration of X-ray units and the output of X-rays vary from one set to another and the various types and brands of films and screens differ, the following diagrams approximate average practical conditions.

Figure 8.25 illustrates the number of milliampere-seconds (mAsecs) required with increasing kV_p to give a density of unity above fog with medium fast calcium tungstate intensifying screens and medium fast X-ray screen film, using a water phantom of 20 cm thickness made of a plastic container (see Figure 8.24). The graph refers to a

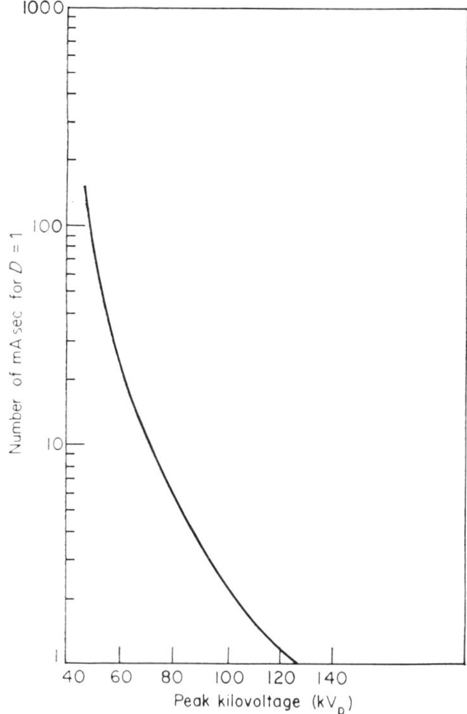

Figure 8.25. Relationship between the number of milliampere-seconds required to obtain a photographic density of 1 above fog and peak kilovoltage, using medium fast calcium tungstate screens and medium fast X-ray screen film. Depth of water = 20 cm; inherent tube filtration = 1 mm aluminum.

depth of water of 20 cm, and to a field size on the water surface of 400 cm^2 (20 $\times$ 20 cm). It should be noted that the values obtained are greatly influenced by the field size due to scattering of X-rays. The X-ray apparatus used was based on a four valve fully rectified circuit. As seen from Figure 8.25 the low exposure (mAsec) values above 100 kV_p indicate the potential advantage of using high kV radiography as far as exposure time is concerned, but this will partly be offset by the necessity of using Potter–Bucky grids of high grid ratios (see p. 319).

A similar diagram is shown in Figure 8.26 for fast no-screen X-ray film, using the same water phantom with a water depth of only 10 cm. This graph shows again the mAsec product required to obtain a density of unity above fog for various peak kilovoltages. No direct comparison between the screen and no-screen exposures is possible

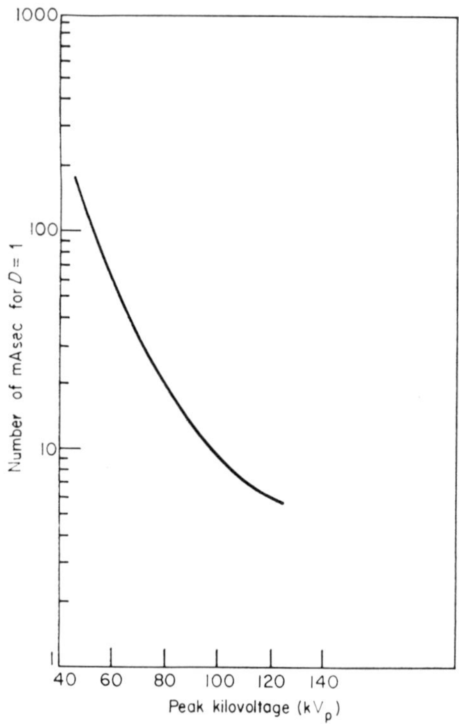

Figure 8.26. Relationship between the number of milliampere-seconds required to obtain a photographic density of 1 above fog and peak kilovoltage, using fast no-screen X-ray film. Depth of water = 10 cm; inherent tube filtration = 1 mm aluminum.

as these refer to different depths of water. A smaller depth of water was used for no-screen film (a) because the necessary mAsec values would tend to be beyond the loading capacity of the X-ray tube for a depth of 20 cm and (b) because the conditions should simulate cases occurring in practice.

Any quantitative considerations in this field will be assisted by a knowledge of the average output of an X-ray tube in terms of

milliroentgens per milliamperesecond with increasing peak kilovoltage. Such a relationship is shown in Figure 8.27 which illustrates
the number of mR obtained per mAsec against kV_p. The milliroentgen
values were measured in air by an ionization chamber at a distance
of 100 cm from the focal spot of a rotating anode tube, using no
absorber apart from the inherent tube filtration of 1 mm Al and an
X-ray unit having a four valve fully rectified circuit. It is seen from
this graph, which is typical for conditions of medical radiographic

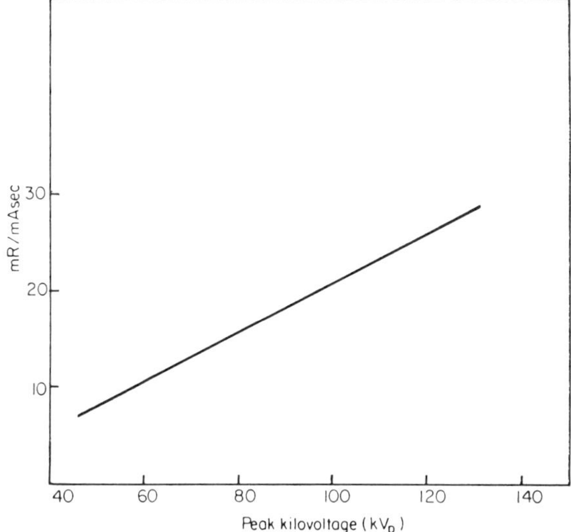

Figure 8.27. X-ray output in milliroentgens per milliamperesecond as a function
of kV_p. (Measured in air at a distance of 100 cm between focal spot and ionization
chamber.)

techniques, that the X-ray output increases steadily with increasing
peak kilovoltage and is about three times as great at 125 kV_p than
at 60 kV_p. It has been shown by O. Mattson (1955) and others
that further proportionality exists between output and kV_p at least
up to 180 kV_p.

(6) *Photographic Material*

Types of Film and Screen. The influence of the inherent contrast
of the film with and without screens as measured by the slope of the
characteristic curve has been discussed in several sections of this book
and no further reference to this subject is required.

(7) *Processing*

The inherent contrast of a photographic emulsion is strongly dependent on the composition of the developer in use, on its temperature and on the development time and agitation of the film in the solution. These items are discussed in the section on the 'control of processing' pages 376–385.

(8) *Viewing*

Much can be done during the inspection of a radiograph in order to utilize fully the objective contrast of the film by means of proper viewing conditions (see p. 275). Large viewing boxes where for instance 50 radiographs are hung up for inspection certainly do not provide optimum viewing conditions, unless these *have individual provision for masking and changing the intensity* of illumination.

8.4c   Chief Factors Influencing Definition

(1) *Geometry*

Focus-Object-Film Distances and Focal Spot. The geometric factors influencing the width of the penumbra, such as the size of the focal spot and the focus-object-film distances have been discussed on pages 260–266.

The *uniformity* and the *shape* of the focal spot are equally important, when rendering fine details in a radiograph. The uniform emission of X-rays from an X-ray tube is often disturbed after excessive use of the tube by roughening and pitting of the target which may result in a split of the focal spot. This may lead to the recording of two adjacent images instead of one image. This can easily be shown in radiographs taken from wires at a distance from the film.

Line-Focus. In order to appreciate the effect of the shape of the focal spot on definition a brief outline is first given of the line-focus principle which is almost universally applied in the construction of modern diagnostic X-ray tubes (see Figure 8.28).

In Figure 8.28 a side view of the anode of an X-ray tube is shown. A stream of electrons from the cathode bombard a circular disc of tungsten which is embedded in the anode. The anode is made of copper and its face is inclined at an angle of about 20° in a direction perpendicular to the axis of the tube as can be seen in Figure 8.28. The area bombarded is usually a rectangle, the length of which is about 3–4 times its width. Due to the inclination of the anode the

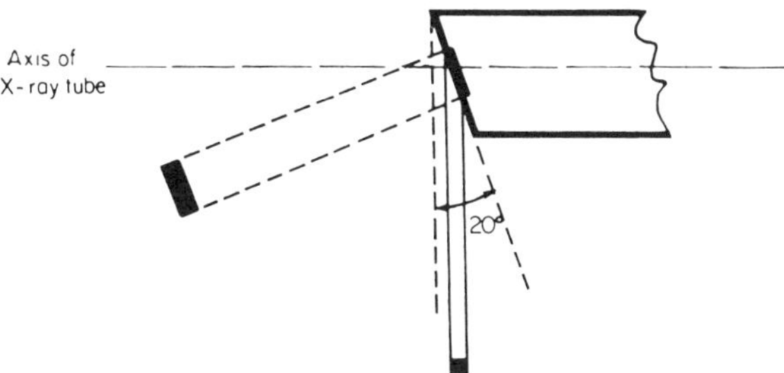

Figure 8.28. Principle of line-focus. (*Due to* Goetze.)

projected area of the focal spot as seen from the direction of the central beam is thus foreshortened to a square. Thus the optically effective focal spot is an area of about $\frac{1}{3}$–$\frac{1}{4}$ of that of the actual rectangular area which is bombarded by electrons. The purpose of the line-focus is thus fulfilled, since the dissipation of heat by the electron bombardment takes place over a larger area than the optically effective focal spot area which should be as small as possible. This principle of the line-focus is also applied to rotating anode tubes.

Oblique Beam.   We can now appreciate that an image taken with an oblique X-ray beam, i.e. one which is not normal to the axis of the X-ray tube, suffers from the disadvantage that it is obtained from a focal spot which differs in two dimensions. Hence the width of the penumbra will be greater in one than in the other direction of the image (see Figure 8.28).

Heel Effect.   An important effect connected with the inclination of the target is the decrease of the intensity of X-rays from the cathode towards the anode, as indicated in Figure 8.29. This shows a section of the cathode–anode assembly and the relative intensity distribution of X-rays is given for the radiation emitted at various angles with respect to the target surface. It is seen that the intensity of X-rays diminishes gradually towards the cut-off of the X-ray beam. The decrease towards the cut off of the anode is partly due to the fact that X-rays produced at a small depth in the target have to pass through a greater thickness of tungsten target (heel effect) and thus are attenuated. Furthermore the intensity of the X-ray beam reaching the film drops away at both sides from the central beam, because the distances between focal spot and film increase from the central beam axis towards both sides (Inverse square law, see p. 21.)

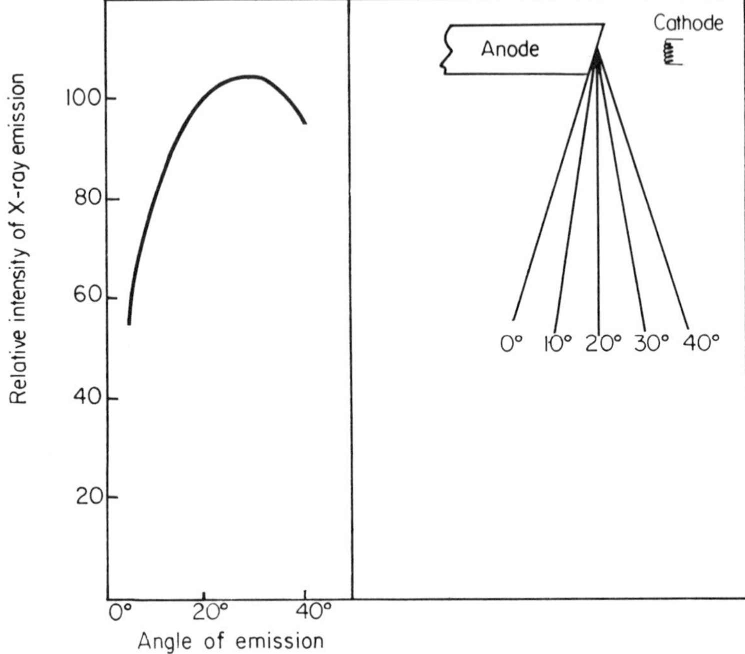

Figure 8.29. X-ray intensity distribution as a function of the angle of emission
from cut off 0° to 40°. (Heel effect.)

Pinhole Radiography.   The shape, size and uniformity of the focal
spot of an X-ray tube can be determined photographically by the
method of pinhole radiography. A thin lead plate containing a small
hole of about 0.5 mm diameter is placed midway between the focal
spot and a film; an anode-film distance of about 1 foot is usually
convenient. The width of the resultant image (see Figure 8.30) is to
be corrected for by twice the diameter of the pinhole according to
$F = i - 2d$, where F = width of the focal spot, i = width of the
pinhole image and d = diameter of pinhole. Hence the width of the
pinhole must be known accurately. The above equation can easily be
derived by reference to equation (7.5), and pages 265 and 266 Figure 7.5
for equal distances between focus to pinhole and pinhole to film,
and if Q in this latter figure is considered to be the pinhole. Typical
focal spot images are shown in Figure 8.31.

If a focal spot image is required which is very nearly of the same size
on the photograph as the focal spot, a pinhole of much smaller dimension, for instance of 0.05 mm diameter, (i.e. 50 microns) should be

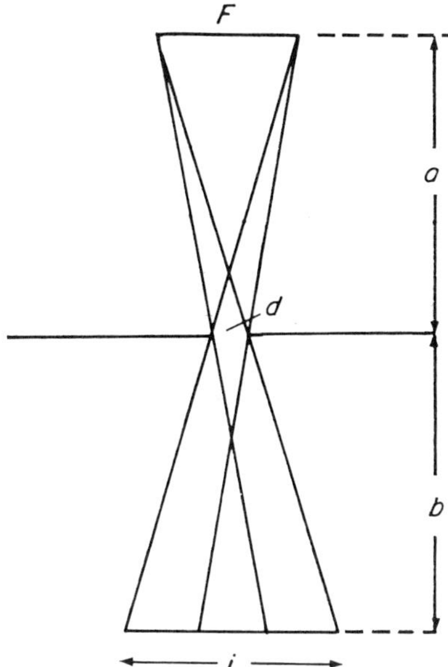

Figure 8.30. Geometry of pinhole image formation ($F = i - 2d$, if $a = b$).

F Focal spot diameter
i Pinhole image
d Diameter of pinhole
a Focus to pinhole distance
b Pinhole to image distance.

taken. Pinholes of 50 and 100 microns diameter in gold–platinum alloy plates of 1 mm thickness are commercially available. The focal spot image should preferably have a density of about 0.6–1.0 above fog. Such an image gives additional information on the structure and homogeneity of the focal spot. (See also I.C.R.U. Report, Brit. J. Rad. 396.784, 1960 on 'Method of focal spot image and measurement for diagnostic tubes up to 150 keV'.)

Definition Test. A further very useful method to obtain information on the image forming properties of a focal spot is to radiograph a metal mesh gauze which should be placed at an angle of about 40° to the film plane. Radiographs taken on no-screen films may then be made at various focus to film distances. These will reveal the definition attainable at various distances of wire mesh to film, if the frame holding the gauze in position is provided with a scale indicating the object to film distance, so that the numbers of the scale, preferably made

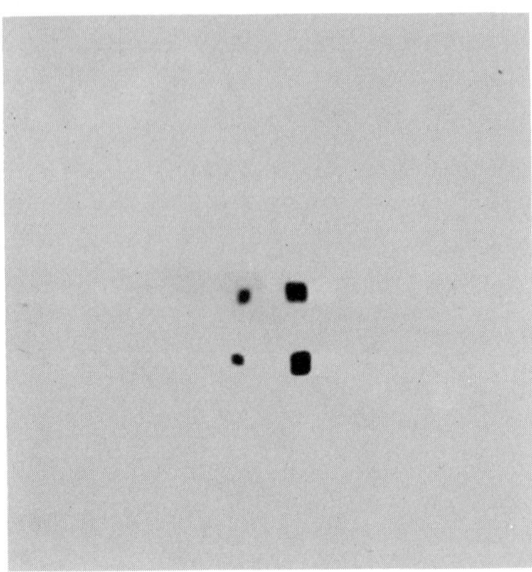

Figure 8.31. Pinhole images of fine and coarse focal spots of a rotating anode
X-ray tube.

of lead, can be recognized on the film. There are many possible
modifications of such a test to suit particular requirements.

(2) *Photographic Factors*

Types of Screen.   The fact that definition is very much dependent
on the type of screen used has been discussed on pages 284–286.
Intimate contact between the whole area of screens and film is of
utmost importance to achieve uniform image quality. The uniformity
can be checked in the following way. A wire mesh gauze of the size
of the cassette to be tested is placed in uniform contact with the
front of the cassette and a radiograph is taken at a distance of about
100 cm between tube and film. The radiograph of the gauze shows,
after processing, whether the contact between screen and film is
uniform over the whole area.

Speed of Screen Film System.   The faster the screen plus film
system, the greater in general is the chance of the formation of
quantum mottle which is detrimental to image quality as explained
in detail on pages 335–342

No-Screen Films.   Definition in no-screen X-ray films may be
influenced by the film itself and it is the graininess of the emulsion

and the parallax effect, which are the possible causes. The parallax effect is caused by two superimposed, but separated coatings. Both effects are in general negligible in most medical radiographs, as these effects are usually smaller than the effects due to other geometrical conditions. The parallax however becomes just noticeable, when viewing wet films, owing to the considerable swelling of the emulsions. This latter effect obviously refers to screen exposed films as well.

(3) *Movement of the Object*

The degree of blur due to movement of the object during exposure depends on the rate of movement and on the exposure time to be used. Blur due to movement is also influenced by the direction of movement with respect to the plane of the film and by the focus-object-film distances. Exposure times should be short enough to prevent blur due to involuntary movements of organs (heart, stomach, intestines). In order to minimize the effects of movement of the patient during exposure a comfortable position of the patient is important.

Check on the Accuracy of Timing Device.   In case of doubt whether the exposure times used are correct, a spinning top can be used to check the exposure time photographically. Depending on whether the frequency of the mains electrical supply is known or unknown, two types of measurement can be recommended. (a) a frequency dependent method and (b) a frequency independent method.

(a) The spinning top consists of a circular disc of lead of about 10 cm diameter and 2 mm thickness having a rod shaped axis leading through the center. The lead disc is provided with a hole near its periphery. For the measurement the disc is centered below the X-ray tube on a cassette loaded with film and screens. By setting the spinning top in motion during exposure, a series of density spots lying on an arc below the area covered by the rotating hole is obtained on the film. These spots are due to the pulsed X-ray emission (see Figure 8.32(a)).

The number of spots obtained on the film is a measure of the exposure time if the mains frequency is known. Example: If a four-valve, i.e. fully rectified X-ray unit is used, and if the frequency of the mains is known to be 50 cycles per second, then each 1/100th of a second corresponds to one half cycle of the X-ray emission.

(b) If the frequency of the mains is not known or subjected to fluctuations, or if the exposure is due to a condenser discharge, the rate of the rotation of the spinning top must be known and its movement should be uniform. From the relationship between the rate of

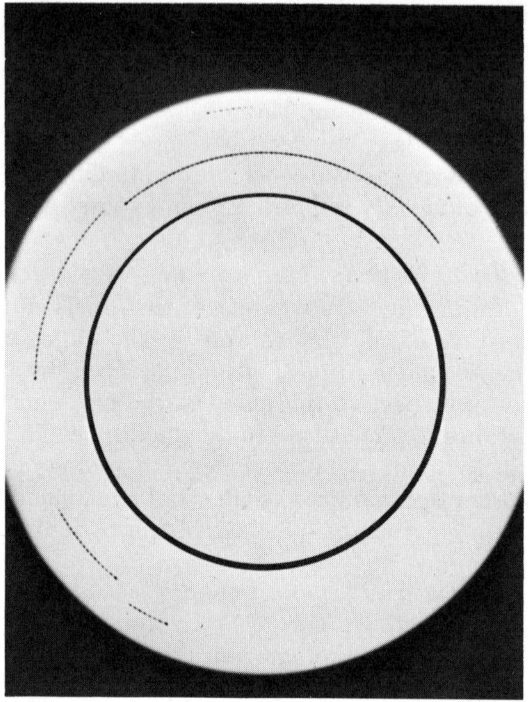

Figure 8.32. (a) Spinning top image taken to check four different exposure times using a fully rectified four-valve X-ray unit, 50 cycles per second.

rotation and the length of the arc of spots obtained, the exposure time can be determined with accuracy (see Figure 8.32(b)). Instead of a hole, a slit may be used and in one version of a spinning top the lead disc is provided with six holes from near the periphery of the disc towards the center, so that six different exposure times can be checked on one film. The holes not to be subjected to exposure must be subsequently covered by lead, an operation which is carried out by a shift of the lead disc in the particular version of spinning top described by R. H. Herz (1931).

(4) *General Considerations on Definition*

The unsharpness of a radiographic image is as we have seen caused by

$$U_g = \text{geometric unsharpness}$$

$$U_s = \text{screen unsharpness}$$

$$U_m = \text{movement unsharpness}$$

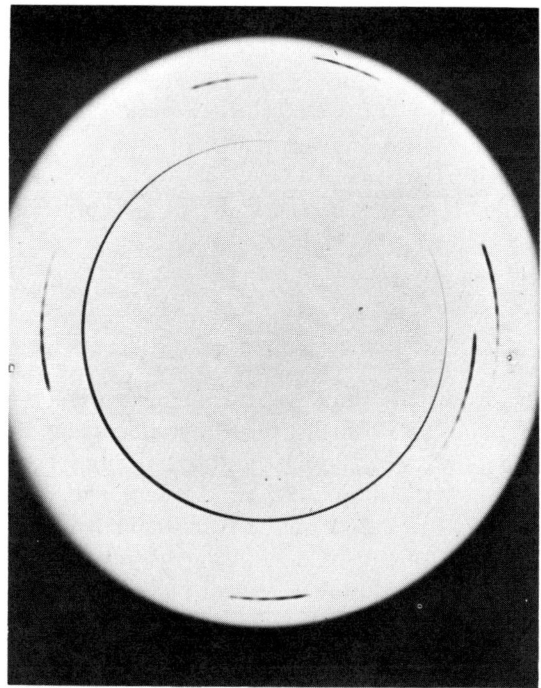

Figure 8.32. (b) Spinning top images of various exposures using a condenser discharge unit. The variations in density along each arc are due to the sinusoidal filament current.

The numerical value of the unsharpness due to geometry is the width of the penumbra, as has been discussed on p. 261. This width is further increased by the unsharpness due to the screens and due to movement. It has been shown by R. R. Newell (1938) that the total unsharpness U_T can be expressed by the square root of the sum of the squares of the individual unsharpnesses, hence

$$U_T = \sqrt{(U_g{}^2 + U_s{}^2 + U_m{}^2)} \qquad (8.4)$$

M. A. Klasens (1947) established that the cube root of the sums of the cubes of the unsharpnesses provides a closer approach to the total blur obtained in a radiograph. In the following, use is made of Klasen's expression

$$U_T = \sqrt[3]{(U_g{}^3 + U_s{}^3 + U_m{}^3)} \qquad (8.5)$$

It is often considered whether an improvement in the rendering of detail is possible by the application of shorter exposure times or by the use of finer grained screens. This can either be checked by experiment or by the following considerations.

Let us assume a chest radiograph is to be taken at 150 cm distance (focus to film) with a focal spot diameter of 2 mm. The tissues to be rendered sharp are found to be 12 cm from the cassette in the A.P. position of the patient. Using the expression (equation 7.1) for the calculation of the penumbra the width of this is found to be $U_g = 0.18$ mm. Let us further assume that the unsharpness due to the pair of medium grained screens in use is $U_s = 0.20$ mm and that due to movement $U_m = 0.25$ mm. Hence

$$U_T = \sqrt[3]{(U_g{}^3 + U_m{}^3 + U_s{}^3)} = 0.309$$

and

$$U_T = \sqrt[3]{(U_g{}^3 + U_m{}^3)} = 0.278$$

It is concluded from this that the unsharpness due to the screen has little effect on the total unsharpness. Hence very little would be gained by using a finer grained pair of intensifying screens for example of $U_s = 0.15$ mm instead of 0.2 mm. In fact the theoretical result turns out to be $U_T = 0.293$ instead of 0.309 when using the finer against the medium screens. Considering however that the exposure time has to be almost doubled (assuming the same milliamperage is used) with a pair of fine grained screens if compared with that of medium grained screens, the final unsharpness value would be even greater because unsharpness due to movement would be greater as well.

Considerations of this kind may be useful when contemplating alterations in technique.

Subject or Absorption Unsharpness.   The reader should also be aware that an apparent unsharpness occurs in most medical radiographs which is due to absorption effects. This is because most anatomical structures are characterized by rounded edges. This means that even under optimum conditions of the three types of unsharpness mentioned above, one would expect unsharp boundaries, which are in fact due to the radiation being attenuated differentially by the varying thickness of the rounded edges through which X-rays have to pass. This is not to be considered as a true unsharpness, but it is sometimes called an absorption or better subject unsharpness.

No-Screen Film Definition.   An important consideration on definition attainable in medical radiography is connected with the use of no-screen X-ray films. In this case one of the three factors of unsharpness, namely that due to the diverging light emission of the screens, is eliminated and furthermore no scattering of light in the emulsion takes place. Scattering due to X-rays within an emulsion is extremely small and hence the definition attainable with no-screen X-ray exposures (mainly of extremities) is usually of a high order, assuming that the geometrical conditions of exposure are favorable. It has been shown experimentally, by letting X-rays pass through fine slits placed on no-screen films, that X-ray scattering within emulsions is almost negligible. (See also MTF measurements p. 345).

8.4d The Effect of Quantum Mottle

In the previous sections it has been shown that exposures with salt intensifying screens in conjunction with screen film permit a considerable increase in speed and contrast compared with no-screen

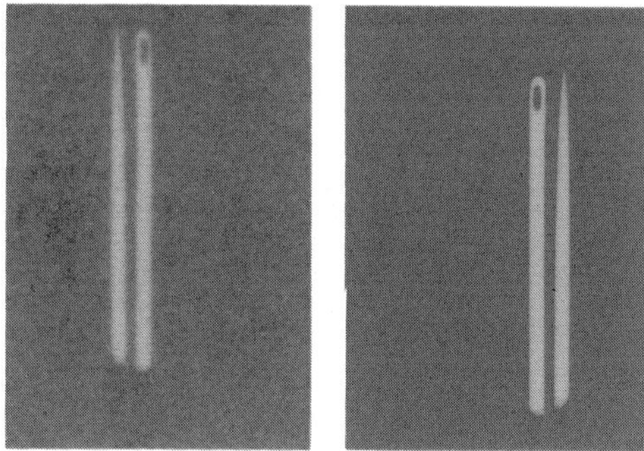

Figure 8.33. Radiographs of a test object consisting of two parts of a needle. *Left:* with X-ray intensifying screens. *Right:* without screens. (*After* K. Rossmann, *J. Phot. Sci.* **12.** 5. 279 (1964).)

film exposures. This advantage is partly offset by impaired definition and in certain cases by a mottled image which is associated with the spatial fluctuations of X-ray quanta absorbed by the screens. The two Figures 8.33 which are due to K. Rossmann (1964) illustrate radiographs of a pair of needles taken with and without intensifying screens. A great difference in the appearance of the background between the two images is noticeable. The screen image appears grainy, mottled and unsharp, whereas the no-screen image has a very uniform background and is sharp. The loss of sharpness of the screen image is caused by the divergence of the screen light from each individual fluorescent crystal in the screen and by the crossover effect (see p. 285). The mottle appearing on the radiograph of the screen image is primarily due to the random fluctuations of X-ray quanta absorbed by the screens and this type of mottle is referred to as 'quantum mottle'. The randomness of the impact of X-ray quanta can be compared with that of raindrops falling on a pavement.

Sometimes an additional mottle has been observed which is caused by a random clumping of the fluorescent crystals of the screens. This effect generally called 'structure mottle' is usually found to be small in modern screens compared with quantum mottle, as will be shown later.

The origin and nature of quantum mottle will be discussed first in an elementary form, then evidence will be given of its occurrence and finally, a theoretical interpretation of this effect follows.

The Origin and Nature of Quantum Mottle

In this section it is attempted to give a pictorial description of the causes and of the nature of quantum mottle i.e. of the spatial distribution of absorbed X-ray quanta in the screens.

Let us assume that we are taking two X-ray exposures on the same type of screen film in conjunction with calcium tungstate screens. One exposure is taken with a pair of fast screens, the other with a pair of slow screens. The fast screens are supposed to be three times faster than the slow ones. It is further assumed that the thickness (coating weight) of both pairs of screens is identical and thus the percentage absorption of X-rays in both screens is about the same. The higher speed of the fast screens is achieved by coarser calcium tungstate crystals, each emitting, on the average, three times as many light quanta per absorbed X-ray quantum than in the slow screens.

Thus the fast screens require only one third of the number of X-ray quanta to be absorbed as do the slow screens, in order to obtain the same number of silver bromide grains to be rendered developable. Although the final number of grains rendered developable in both films is the same, the distribution of the grains differ. This is because the distribution of developed grains in each film is spatially correlated with that of the fluorescent light emission centers in the screens and the latter are the sites of the X-ray quanta absorbed by the screens. As the number of these sites in the fast system is only one third of that in the slow system, the randomness of the distribution of silver grains becomes more striking to the eye, as is shown, somewhat exaggeratedly in the two grain patterns of Figure 8.34.

The figures present two simulated radiographs of a tiny image taken through a triangular aperture of a radio-opaque sheet of material. The left image is taken with the fast system, the right one with the slow system. The total number of grains in both images is the same,

Figure 8.34. Drawings of developed grains to illustrate the effect of quantum mottle. The number of developed grains in (a) and (b) is equal but the distribution of the grains is different. The grain picture in both cases originates from an exposure through a triangular aperture.

(a) Fast system (mottle effect)
(b) Slow system.

but the distribution of grains (left) is clustered and appears more random and mottled than in the image on the right. This is due to the smaller number of X-ray quanta absorbed (left), each producing, however, a greater number of developed grains than in the slow system on the right. It is this appearance of randomness which we call 'quantum mottle'. As is seen in these figures quantum mottle interferes with the formation of images. The triangular shape of the image can only be recognized on the right and not on the left. This effect is even more marked when using a vacuum cassette, as in general this improves detail rendering of the screens. Similar, although not exactly the same effects can be obtained by using faster X-ray films, instead of screens, as the former tend to permit a smaller number of quanta to be absorbed by the screens in order to achieve the same response as would be obtained on a slower film.

We have thus seen that the quantum mottle is caused by the spatial distribution of absorbed X-ray quanta and may severely interfere with image formation. This leads to serious limitations in a further enhancement of film speed in such systems.

The example which we have chosen to explain the causes of quantum mottle can of course be modified. Instead of using screens of the same thickness, we may employ two screens of different speed and different thickness. Such an example is described in Figure 8.35 where we employ on the left a fast and thick front screen, on the right a slow and thin front screen, and both are used in conjunction with the same type of fast screen film. The use of front screens only is shown for simplicity, but does not interfere with the argument.

If for instance 900 X-ray quanta are incident on the fast screen and one third is absorbed, we have 300 X-ray quanta absorbed in

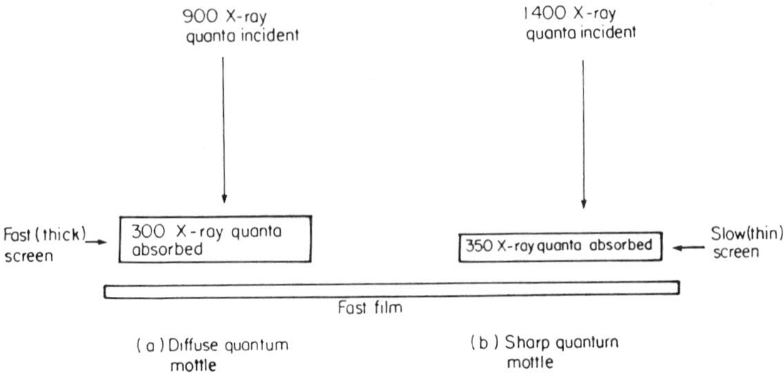

Figure 8.35. Different appearance of quantum mottle.

the screen. In the slower system say 1400 incident quanta are required and as the screen is thinner only one quarter of the X-ray quanta, i.e. 350, is absorbed. Hence almost the same number of quanta is absorbed in both screens, but the appearance of the mottle will differ, as the mottle also depends on the image forming effects, in this case on the light diffusion properties exhibited by the screens. The coarser thicker screen produces a more diffused quantum mottle on the left (see Figure 8.35) and the thinner screen a sharper mottle on the right. The latter mottle is usually considered to be more disturbing in appearance than the former.

Evidence of Quantum Mottle

Although the explanation given for quantum mottle may appear to be plausible to the reader, no evidence has been given that our present fastest screen film systems have in fact reached the speed level required for the phenomenon to be significant in practice. The question for instance arises, how do we know that the effect is not due to a structural mottle or pattern inherent in the screen? Such a structure mottle has indeed been observed in the past on some rather inferior types of screen.

Two different kinds of test have been carried out to show that structure mottle is negligible.

1. If the mottle observed in modern fast screen exposed films were due to a structural pattern of the screen, subsequently flash exposed images of the same area of the screen should reveal the same pattern of image, when these are enlarged and superimposed. H. M. Cleare, H. R. Splettstoesser, and H. E. Seemann (1962) have shown experimentally that the images could not be brought into register, thus

indicating that the patterns were not alike. Hence they concluded that the mottled image must have been caused by the statistical distribution of quanta absorbed by the screen and not by structural properties of the screens.

2. In order to give further evidence that structure mottle is negligible or very small in most modern screens, radiographic images have been taken of a few plastic beads using the same medium speed screens, but in the image on the left of Figure 8.36 with a fast screen film and in the image on the right with a 200 times slower screen

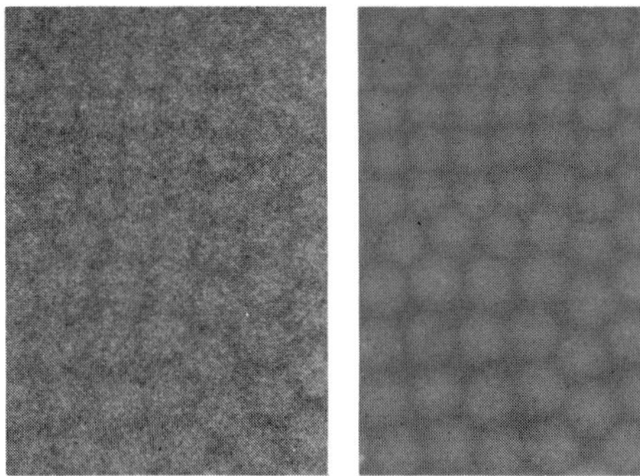

Figure 8.36. The effect of radiographic mottle on the visibility of low contrast detail. Radiographs of plastic beads both taken with the same medium speed X-ray intensifying screens, but in the left image a fast screen film was used; in the right image a screen film 200 times slower was used. (*After* K. Rossmann, *Am. J. Roentgenol.* Radium Therapy and Nuclear Medicine XC.4. (1963).)

film. By comparing the two images it is seen that the left one appears mottled and the right one reveals a uniform background density. If structure mottle were responsible for the mottled appearance of the left radiograph then it would be expected to show as well on the right picture.

The mottle shown in the left picture (Figure 8.36) cannot be due to inherent graininess of the film, as this has been checked separately. Hence there remains no other explanation for the effect of mottle than the one given before. The effect of quantum mottle is to be expected, and it is well known that it occurs in other fast systems, such as for instance in the application of image intensifiers, where it is referred to as 'quantum-noise'.

Factors Affecting the Magnitude of Quantum Mottle

After having discussed the origin, nature, and evidence of quantum mottle, it is useful to consider the main factors affecting the magnitude of this effect.

Quantum mottle becomes more striking if one or more of the following changes occur.

1. Increase of the speed of film.
2. Increase of the gradient of film.
3. Increase of the quantum yield of the screen phosphor, i.e. the number of light quanta emitted per X-ray quantum absorbed.
4. Increase of image sharpness.

As a reduction of these factors will make the disturbing effect of quantum mottle less striking, one has to be aware of their other practical implications. For instance a decrease of film speed and of quantum yield (of the screen) will lead to higher doses for the patient and may interfere with movement of the patient. A lower gradient permitting increased image latitude may be desirable only in special cases. Furthermore a decrease of image sharpness can not usually be tolerated.

It is thus seen that very little can be done to reduce or even eliminate quantum mottle, if the speed of the screen plus film system is not altered. The radiologist who feels troubled by quantum mottle may however balance the various possibilities of reducing the effect. Often a reduction in speed of the screen-film system might be permissible in certain radiographic applications in favor of less mottle and indeed some radiologists and radiographers prefer slower films for this reason.

Theoretical Considerations

In a simple mathematical derivation, K. Rossmann (1963) has shown that the standard deviation of quantum fluctuations increases as the number of X-ray quanta absorbed by the screens decreases. He argues that the mottle apparent in screen radiographs is due to two components, the film graininess and the quantum mottle. This may be described by the standard deviation of the corresponding density fluctuations $\sigma(D)_{grain}$ and $\sigma(D)_{mottle}$. Since both are physically independent phenomena, the total radiographic mottle resulting from their simultaneous appearance can be expressed by

$$\sigma(D)_{total} = \sqrt{\sigma^2(D)_{grain} + \sigma^2(D)_{mottle}} \qquad (8.6)$$

The number of X-ray quanta absorbed per unit area of the screen is denoted by n_x and the number absorbed in area a of the scanning or viewing aperture is $n_x a$. This number fluctuates randomly about an average number $\bar{n}_x a$, according to a Poisson distribution and the resulting standard deviation is given by

$$\sigma_x = \sqrt{\bar{n}_x a} \qquad (8.7)$$

The corresponding fluctuations of the photographic density $\Delta(D)_{mottle}$ are found by combining the last equation with the expression of the characteristic curve; then, for small exposure changes, G (which depends on E), may be regarded as nearly constant.

$$D = G \log E + C \tag{8.8}$$

where G is the gradient, E the exposure, and C is a constant. Then

$$\Delta D = 0.43G \, \frac{\Delta E}{E} = 0.43G \, \frac{\sigma_x}{\bar{n}_x a} \tag{8.9}$$

and by substituting σ_x from equation (8.7) we obtain

$$\sigma(D)_{mottle} = 0.43G \, \frac{1}{\sqrt{\bar{n}_x a}} \tag{8.10}$$

If we insert this expression into that of equation (8.6) we can write

$$\sigma(D)_{total} = \sqrt{\sigma^2(D)_{grain} + (0.43G)^2 \, \frac{1}{\bar{n}_x a}} \tag{8.11}$$

Hence equation (8.11) shows that $\sigma(D)_{total}$ increases as the number of X-ray quanta absorbed by the screen decreases and this is exactly what we experience in practice. Hence it is to be expected that in general the mottle increases with the speed of the screen plus film system.

It was only a few years ago that quantum mottle was noticed in screen radiographs and that we became conscious of the fact that our screen-film combinations had reached a speed level where the disturbing influence of quantum mottle may impair the image quality. No remedy has been found so far to avoid the effect, apart from limiting the speed level (i.e. speed in terms of absorbed quanta) of the screen-film combination. In practice the quantum mottle is noticed by the impression of a kind of image graininess and is the more striking for a given screen-film combination the more intimate the contact between screen and film. Hence it would appear at its maximum in vacuum cassettes. On the other hand the definition would suffer if the contact between screen and film were reduced. Summarizing, this means 'the poorer the image is recorded optically, the less objectionable is the mottle.'

Quantum mottle is (theoretically) not necessarily restricted to screen exposures, but may also occur in no-screen exposures whenever the number of X-ray quanta absorbed per unit area reaches a limiting level. In general the present speed of fast no-screen films does not appear to be great enough for the effect to show, unless exposures are carried out with X-rays generated at higher kilovoltages (see p. 139).

Reference to the quantitative detection of quantum mottle will be given in the next section on 'The assessment of image quality'.

8.4e   Assessment of Image Quality

"Image quality" may be described as a property of a radiograph to reveal as much diagnostic information to the experienced radiologist, as is possible from the inherent structure of the object under examination. This maximum of information is to be achieved by delivering as small a dose as possible to the patient during exposure.

Image quality is influenced by a variety of factors which may be summarized as being due to

(a) the structural properties of the object to be radiographed,

(b) the ability of the radiographic system to record the information impressed upon it,

and

(c) some physiological and psychological properties which are associated with the observer.

In the following we shall be concerned with part (b) of the radiographic system which deals with the recording system. Such a system may consist of a direct X-ray film (i.e. no-screen film) or of a screen film combination. As the latter is most widely used in medical radiography, methods will be described which permit an assessment of image quality of screen-film combinations. Using these methods an attempt is made to establish the ability of screen-film techniques in recording the image impressed upon it.

In the conversion process of the ray image (see p. 268) into the recorded image several factors are involved which lead to an impairment of the image quality. These are mainly the light diffusion from the screens, the crossover, i.e. the diffusion of light from one screen to the non-adjacent emulsion layer of the double coated film, see Figure 8.2, light reflection, screen-structural mottle, poor contact between screen and film, graininess of film and quantum mottle. The parallax, occurring because the base separates the two emulsion layers, is an additional effect influencing the viewing conditions.

Any method of assessing image quality must be capable of covering all the items mentioned above and in addition it must take account of the type of image pattern to be recorded. This may be a patchy image of low density and low contrast or an image showing fine structural detail of higher contrast or other types of pattern. It will be shown that none of the methods to be described in the following has yet given a fully satisfactory answer to the problem and that image quality can probably never be expressed by a single number.

Two methods of approach to the assessment of image quality will be discussed. The aim of one of these is to describe the relation between the contrast in the image of a set of equidistant parallel lines and the separation between the lines. If the luminous distribution follows a sine curve the graph presenting such a relationship is called the 'modulation transfer function'. As the modulation transfer function (abbreviated by MTF) does not take account of quantum mottle, a further method will be briefly described, which is based on the so called 'Wiener Power spectrum'. The other approach is the visual evaluation of a purely experimental method which consists of radiographing suitable test-objects. All these methods have been applied in medical and in industrial radiography.

Elements of the Modulation Transfer Function

In light photography the modulation transfer function (MTF) describes, as already mentioned, the ratio of the contrast between lines and spacings of the recorded image to that of a test chart following a sinusoidal intensity distribution. In radiographic work a line test chart can obviously not be applied because of the penetrating power of X-rays. Wire meshes of varying numbers of wires per inch are also unsuitable because of the different absorption across the width of wires which changes the spectral distribution of X-rays. R. H. Morgan and others (1964) used a sinusoidally shaped metal test object but this also causes changes in the spectral distribution of X-rays. Another approach to the solution of this problem was found in using a slit system through which X-rays pass on to the screen film system. Such a method has been described as well by R. H. Morgan and others (1962 and 1964) and by K. Rossmann (1962a, 1962b, 1962c, 1964a and b). The slit must have a width which is very narrow relative to the width of the image spread, when screens are used; in fact it has to be so fine that by making it still finer, the image spread on the film would not change anymore. Rossmann used a slit of 10 microns width which is made up by two platinum jaws of 2 mm thickness. Platinum is used in order to avoid penetration of X-rays. The slit is in intimate contact with the screen film combination, which itself should be in a vacuum cassette to assure perfect contact. The

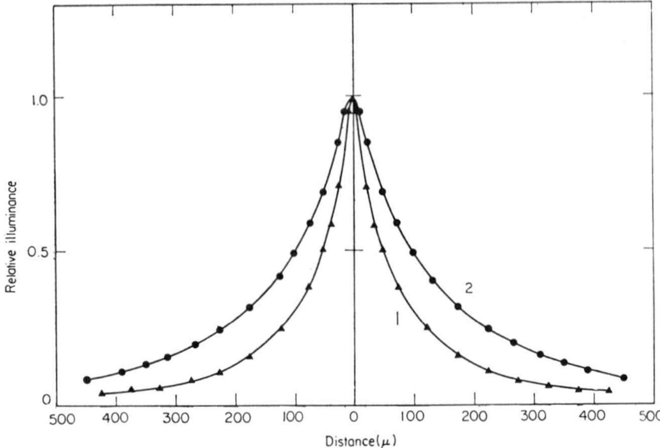

Figure **8.37**. Line spread function of two radiographic systems containing calcium tungstate screens. 1. Two medium speed screens. 2. Two fast screens. (*After* K. Rossmann, *J. Phot. Sci.* **12**.5.279 (1964).)

result of exposures taken with this slit by Rossmann is shown in Figure 8.37 which illustrates smoothed out microdensitometer tracings (made with a long narrow scanning slit) of the two exposures using different screen film combinations. One of these screen-film systems is a high speed, the other a lower speed one. Figure 8.37 shows the relative illuminance against the width of the spread of the image which is called the 'line spread function'. This graph indicates that the image of system 2 is less sharp than that of system 1. It is well known that any periodic function can be analyzed into a series of sinusoidal functions, i.e. sine and cosine functions by means of what is known as a Fourier transformation. If the line spread function is thus analyzed by this method of Fourier, sinusoidal curves of various frequencies will be obtained. By plotting the ratio of contrast between line and spacing in the record over that of the maxima and minima of the intensities of X-rays impressed upon the system against cycles/ mm, a graph of the 'modulation transfer function' is obtained. This is illustrated in Figure 8.38 showing the respective MTF's of the line spread functions of Figure 8.37. The faster screen film system 2 shows a poorer MTF than the slower system 1, as the latter reveals a better resolution. In this way the image degradation in a screen film system can be assessed. The results of MTF measurements on various screen and no-screen films are taken from graphs obtained by

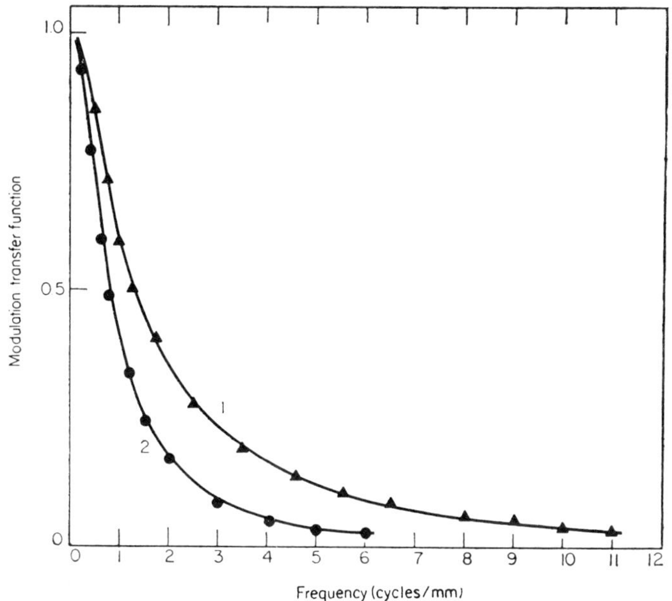

Figure 8.38. Modulation Transfer Functions (MTF's) calculated from the line spread functions of Figure 8.37. (*After* K. Rossmann, *J. Phot. Sci.* **12**.5.279 (1964).)

R. H. Morgan, L. M. Bates, U. V. Gopalarao, and A. Marinaro (1964) and are quoted below.

These data refer to modulation transfer functions of 50 per cent and 10 per cent contrast (see Table 8.2).

TABLE 8.2. Results of MTF Measurements

	Cycles or lines per millimeter	
Types of screen	MTF 50%	MTF 10%
High speed screen	2	5.5
Medium speed screen	2	7.5
High Definition speed screen	3	15.0
No-screen film	20	60.0

Elements of Wiener Power Spectrum

In the previous considerations no account was taken of the quantum mottle, which may be present due to the spatial fluctuations of the X-ray quanta absorbed by the screens. Thus the quantum mottle has to be measured separately, as the slit system used in MTF work has to be narrow (but of large enough area) to suppress the relatively coarse mottle produced by quanta. This leads us to the introduction of the Wiener power spectrum, which permits a measurement of the quantum mottle.

A Wiener spectrum or power spectral density measurement of a screen film combination can be obtained by scanning a previously screen exposed film. The fluctuating light which is transmitted

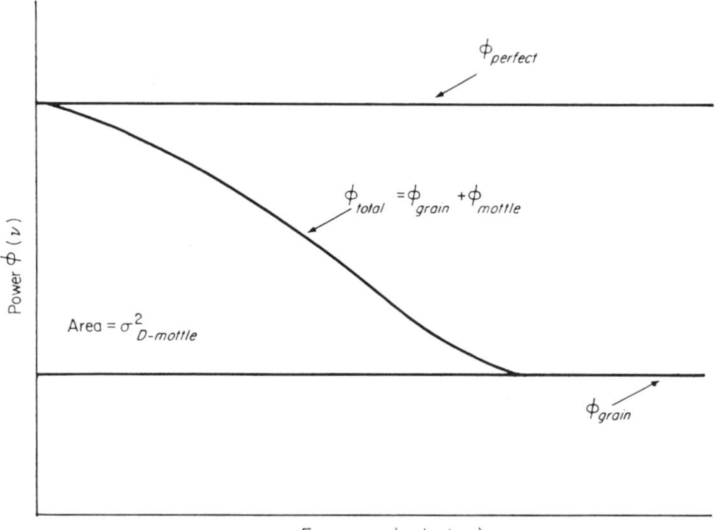

Figure 8.39. Schematic illustration of power spectra of grain patterns.
$\phi_{perfect}$ for a perfect screen film system (theoretical)
ϕ_{total} for a practical screen film system
ϕ_{grain} for a film grain pattern
(Slightly modified illustration.) (After K. Rossmann, *J. Phot. Sci.* **12**.5.279 (1964).)

through a small aperture then falls on a photomultiplier tube. The signals of the latter are recorded by an RMS voltmeter and the content of the fluctuation at different frequencies is examined electrically by means of a variable band-pass electronic filter. The Wiener spectrum is thus given by the squared RMS values of the fluctuations in a narrow frequency band divided by the width of the frequency band and plotted as a function of spatial frequencies in cycles per millimeter.

A schematic presentation of the Wiener spectrum as shown by Rossmann is illustrated in Figure 8.39. The upper horizontal line is the spectrum of mottle as it would be recorded by a perfect imaging system, i.e. all frequencies contained in the so called white noise input of X-ray quanta absorbed by the screen are recorded equally well. The imperfect screen film system, however, which does not transmit all frequencies contained in the input system, is represented by the total density fluctuations $\phi_{total(\nu)}$, which depend not only on the quantum fluctuations, but also on imaging properties of the screen film system as expressed by the MTF and by the graininess of the film. In fact

$$\phi_{total(\nu)} = \phi_{mottle(\nu)} + \phi_{grain(\nu)}$$

The lower horizontal line indicates the spectrum of the inherent graininess of the film. The difference between $\phi_{total(\nu)}$ and $\phi_{grain(\nu)}$ is a measure of the contribution of X-ray quantum fluctuations, i.e. of quantum mottle.

The Use of Test Objects

A less complicated and more direct approach to the problem of assessing image quality may be obtained by the use of test objects, particularly if these simulate the structural pattern of the object to be radiographed.

A test object must permit a quantitative evaluation of the visual assessment of image quality. Wire meshes are unsuitable for the reasons mentioned before. Sheets of lead having holes of diminishing diameters and depths, step wedges made of plastic or metal containing so called penetrameters (see p. 422) or image quality indicators as used in industrial radiography may be useful test objects. Another test object has been described by H. Schober (1957), using Landolt rings with slits, as indicated in Figure 8.40.

In one direction the rings are of the same thickness, but of decreasing diameters, in the other direction the rings are of constant diameter, but decreasing thicknesses. The slit widths are in constant relationship to the diameters of the rings. The rings are mounted on rotatable discs and are placed in random positions. The observer who is trying to evaluate a screen film exposure of such a Landolt ring system, has to indicate the clock position of the slit. If he fails, he has shown that he was unable to see the particular ring properly. Hence the semi-quantitative evaluation of such a system is practically independent of subjective errors.

Test objects of the type described in the preceding section may be suitable for the evaluation of threshold visibility of certain patterns. How careful one has to be with one's judgement of image quality,

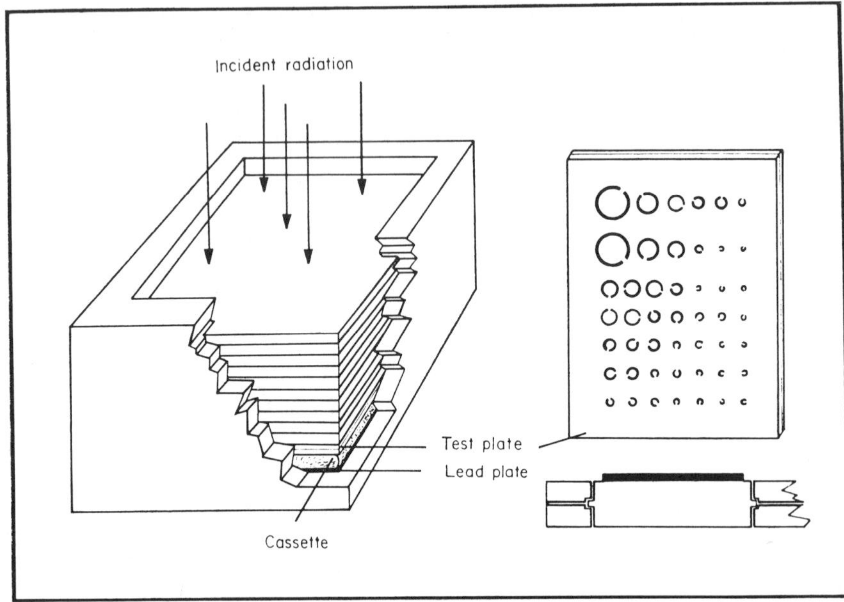

Figure 8.40. Landolt ring phantom test-object. (*After* H. Schober and C. Klett. *Photographie and Kinematographie.* Kurt Wesemeyer, Hamburg (1957).)

however, may be seen from Figure 8.41 which reveals that one type of pattern is seen better in one screen film system, than another one in a different screen film system. The Figure 8.41 which is due to K. Rossmann (1965b) shows that no single figure can be used to indicate image quality.

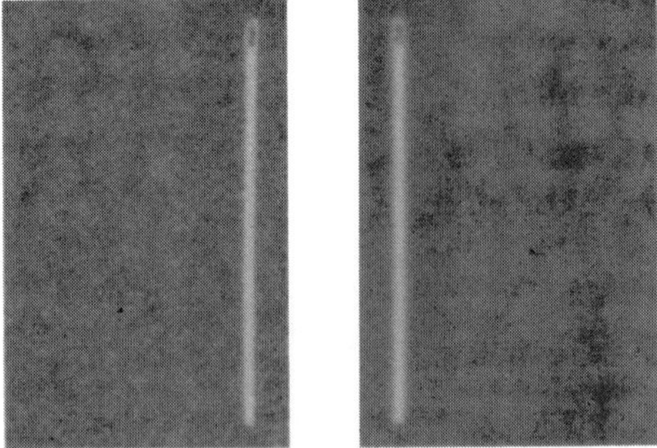

Figure 8.41. Radiograph of a test object consisting of plastic beads and a needle. (*After* K. Rossmann, Diagnostic Radiologic Instrumentation. Edited by R. D. Moseley and J. H. Rust. Charles C. Thomas. (1965).)

In this radiograph (Figure 8.41) use is made of a test object consisting of plastic beads and needles. It is seen that the needle image is superior on the left radiograph if compared with that on the right. On the other hand the reverse is true for the visibility of the beads, which are better imaged on the right radiograph. The test object was radiographed with two screen film systems of equal exposure speed, but the right system absorbed 2.5 times as many X-ray quanta as the left system. Thus the resulting light bursts are less sharply imaged on the right radiograph. The needle is imaged better on the sharp system (left), whereas the beads, which are inherently unsharp and of low contrast are shown better on the image having less quantum fluctuation (right image).

In another investigation K. Rossmann (1965a) has shown that various systems of assessing image quality, such as MTF, wire meshes and hole test objects need not lead to the same judgement of image quality for the same screen film system, because the judgement depends on the type of pattern to be viewed. It is evident from the examples given that there is still more work required in this field before a satisfactory solution of the determination of image quality is found. A discussion on the assessment of contrast will be given in Chapter 10, pages 409 and 416–422

Resolving Power and Unsharpness

In previous sections frequent use has been made of the terms 'resolving power' and 'unsharpness' in radiographic screen and no-screen images. It may be useful at this stage to summarize the various means by which resolving power and unsharpness have been measured and also to indicate the relationship between the two terms.

On page 343 it was shown that a slit image obtained by an X-ray exposure may be recorded by means of a microdensitometer tracing, thus showing the line spread function. A Fourier transform of this line spread function is the modulation transfer function from which the resolving power in terms of cycles or lines per mm was derived. The 'resolving power' is then determined at any desired contrast between line and interval.

Another method of measuring 'resolving power' consists of radiographing a test object which may for instance be made of a thin metal sheet having a number of grooves per millimeter. From the radiograph thus obtained one may also derive the number of lines per mm that can be resolved with a particular film or film-screen combination. This derivation is, however, entirely dependent on the construction of the test object, whether it is built of square or sinusoidally waved or other shaped grooves.

On the other hand 'unsharpness' may be determined by means of a radiograph of a sharp edge of metal or other suitable material. The width of the edge on the radiograph has been taken as a measure of unsharpness due to film, screen but also due to geometrical (penumbra) conditions.

E. W. H. Selwyn (1965) has shown theoretically that the resolving power (R) and 'unsharpness' (U) are related by

$$R \propto \frac{1}{U}$$

and this relationship has been verified by reference to experimental values of R and U.

Image Quality in Chest Exposures

The previous discussions on 'image quality' may have given the impression to the reader that no particular difficulty exists in arriving at an optimum image quality for various regions of the human body. The situation in practice is however very different and the problems involved for instance in establishing a desired image quality in the radiography of the human lung will therefore be indicated as an example in this section.

There is probably no type of medical radiograph which has given more cause for controversial discussions with respect to its image quality than the radiography of the lung. There are several reasons for this. First of all chest exposures form one of the most frequently taken radiographs of the human body; secondly there exists such a wide variety of possible modifications in techniques which may influence the quality of the image, and thirdly personal judgement and previous experience involved in diagnosing lung radiographs are still the factors dominating the controversial discussions on image quality.

Because of this difference in opinion it was thought opportune to discuss some of the many aspects of this problem.

What then are the points of controversy?

Kilovoltage.   Much argument concerns the quality of radiation to be used, whether 50–80 kV$_p$ or perhaps high kV, such as 120–150 kV$_p$ (and even higher) or a radiation lying between the two extremes is preferable from a diagnostic point of view. Some of those favoring the high kV method may do so because of the apparently lower dose given to the patient. Those favoring the low kV argue that initial stages of lung disease can be more easily detected with soft radiation.

The majority of chest exposures are however taken between 60–90 kV$_\mathrm{p}$ depending on the physique of the patient. The question of dose delivered to the patient has certainly aggravated the general problem because of the aim to give the patient the minimum of dose without reducing the information content of the image.

Scattering.   Sometimes it is argued that the sharpness of the image suffers from scattered radiation in the high kV technique in spite of the use of high grid ratios, and furthermore opinions differ as to the grid ratio of Bucky diaphragms to be used in high kV work. The technique of reducing scattered radiation by increasing the chest-film distance is preferred in some quarters. It has been suggested that patients of greater physique should be exposed with softer instead of with harder radiation, in order to reduce the greater tendency of scattering taking place in the heavier patient. Many radiologists insist on the visibility of some structural details to be seen in the heart shadow, otherwise they do not consider the radiograph to be acceptable.

Sharpness.   Several experts take the sharpness of the front rib structure in P.A. radiographs as a criterion for the image quality.

Photographic material.   Some arguments are still expressed with regard to the most suitable shape of the characteristic curve for chest exposures, whether a short or wide toe in the bottom of the curve is to be preferred or where the curve should bend in the shoulder region.

Reproducibility.   Naturally the reproducibility of exposure and processing conditions can play a decisive role in the interpretation if the development of a disease has to be followed up over long periods of time. Phototimers and other devices are used to achieve this aim. An average optical density of $D = 0.8$ within the important region of the lung radiograph is often taken as a standard to be aimed at using a photo-timer. (The photo-timer is a device which works either on an ionometric or photocell principle and stops the exposure when the predetermined dose is obtained.) On the other hand there is some discrepancy as to the most important region in the lung for which an average density is valid.

Fashion.   Perhaps it is not even an exaggeration to speak of certain fashions in lung radiography. From a study of chest exposures in various countries one comes to the conclusion that in certain countries fairly flat radiographs are favored, in others, ones with high contrast.

It should be pointed out that the difference of opinion amongst radiologists with respect to image quality in chest radiographs is by no means a new development, this controversy has already existed for many decades. It may well be that the whole discussion presented in this section should be regarded only as an apparent problem which is artificially created because the term 'image quality' may have in fact no unique meaning. In spite of the many efforts by radiologists and physicists no precise definition of this term has been given yet.

As the detectability of a shadow in an image is, plainly speaking, the result of contrast and sharpness in the record, a very great deal of study has been devoted in the past to the physical, physiological, and psychophysical factors which all contribute to the eye's and brain's perception of the final radiographic information contained in a radiograph. In the author's personal view perhaps too much has already been done on the physical, photographic, and psychophysical aspects of this problem with relatively little reward. It may be more desirable in future investigations if fundamental criteria could be assessed for the medical diagnostic characteristics of radiographs of the lung. This might enable radiologists and physicists to work together on an assessment of a desirable image quality of the radiography of the human lung.

There is an alternative, but not necessarily less optimistic view, which may be expressed here. This is the possibility that the discrepancy of opinions will never be solved, because there exists no unique meaning of image quality to be aimed at and that any attempt of standardizing certain aspects of lung exposure techniques may be a useless undertaking which would deprive radiologists of their individual approach to lung diagnosis based on their personal experience.

8.5 AUXILIARY METHODS OF RADIOGRAPHY

In the course of the evolution of medical radiography, a period of roughly seventy years, many supplementary methods and techniques have been introduced, all contributing either to a greater perfection in diagnosis or to the facilitation of X-ray diagnostic operations.

Although only few of these methods require specialized photographic procedures, their basic principles will be briefly described, as all of them are essentially modified techniques of image recording. Changes in the photographic techniques will be emphasized where this is thought to be necessary, but they will not be discussed in detail. There is no significance to be attached to the order in which these techniques are described.

Stereoradiography

A three dimensional impression of certain regions of the human body can be obtained by means of stereoradiography. Essentially two radiographs of the same object are taken in succession by shifting the X-ray tube after the first exposure into a position for the second exposure. The displacement of the tube should be close to the inter-pupillary distance, i.e. on the average 6.5 cm. This assumes that a focus-to-film and eye-to-film distance (for viewing) of 65 cm is used.

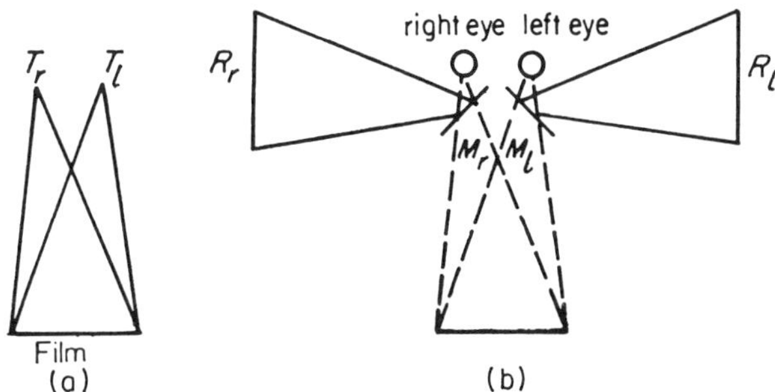

Figure 8.42. Principle of Wheatstone two mirror stereoscope. (a) Conditions of exposure: T_r and T_l indicate X-ray tube in right and left position. (b) R_r and R_l Radiographs taken in right and left tube positions. M_r and M_l right and left mirror.

Under practical conditions, i.e. if no measurements in space are to be taken, the tube shift need not be exactly the average distance between the pupils, but the aim should be a ratio of tube shift to focus-to-film distance of about 1:10. The two radiographs, after processing, can be viewed with a Wheatstone mirror stereoscope or with stereobinoculars. The Wheatstone stereoscope consists of two mirrors at right angles to each other, which are situated midway between two illuminators, as seen in the diagram of Figure 8.42.

The mechanically coupled mirrors are movable so that the viewing conditions can be adjusted. It is essential that the right eye views the image taken with the X-ray tube in the right position and the left eye that taken from the left tube position, as illustrated on the left of Figure 8.42. Using an auxiliary device in combination with the Wheatstone stereoscope, measurements in space can be taken if the viewing conditions are properly related to the exposure conditions.

Prismatic stereobinoculars are used to view stereoradiographs on an ordinary X-ray illuminator when placed side by side. The prisms act as reflectors and also refract the rays of light in assisting convergence. Some people are capable of viewing two radiographs three-dimensionally, i.e. by crossed eyes without the aid of special instruments. There exist several modifications in the methods of viewing stereoradiographs to suit the radiologist's ability of seeing in three dimensions. Stereoradiography is greatly assisted by good radiographic techniques, i.e. by high definition and fairly high contrast radiographs and by exercise in stereoviewing of radiographs taken from simple objects. (M. Cohn and W. Barth, 1931; F. Luft, 1937; A. Hasselwander, 1937; J. F. Brailsford, 1939; D. F. Winnek, 1944; L. E. Etter, 1951.)

Methods for the radiographic localization of foreign bodies are described by K. C. Clark (1939) and in chapter 10 in industrial radiography pages 465–470.

Tomography

If the structural details to be studied in an object are close to the film during exposure usually no difficulties occur in obtaining an image of good definition. If the details of interest are more remote from the film they do not only suffer from lack of definition, but may also be obscured by unwanted shadows, thus adding to the difficulties in the diagnostic evaluation. For instance a cavity or a tumour near the center of the chest may not easily be recognized in normal P.A. exposures, but may be diagnosed with greater certainty in a tomographic exposure, in which only the plane in which the cavity is situated will be recorded sharply and all the other planes will be blurred.

Various techniques have been devised and various names have been given to them, such as 'tomography,' G. Grossmann (1935), planigraphy, J. R. Andrews (1936, 1937), stratigraphy, laminagraphy, J. Kieffer (1938), A. Vallebona (1950, 1952) or layer radiography, but the basic principle is the same for all of them. Tomography and the other techniques do not only permit an image to be recorded of any layer of the human body of interest, but also a quantitative assessment of the distance of the selected layer from the surface of the body. This is achieved by a correlated movement of X-ray tube and film cassette about the stationary object, so that all unwanted object planes appear blurred in the radiograph and only the selected plane is rendered sharp. It should be noted that the formation of penumbra, cannot be eliminated by tomography.

The essential equipment necessary for tomography consists of a stand for a movable X-ray tube which is connected by a rod to the film carriage. The rod is telescopic so that the distance between the focal spot of the tube and film plane can remain constant at all positions of the movement. The rod can be moved about a pivot, so that tube and film move in opposite directions. The position of the pivot can be adjusted so that the radiologist can select the layer of the region of the human body of interest, as required.

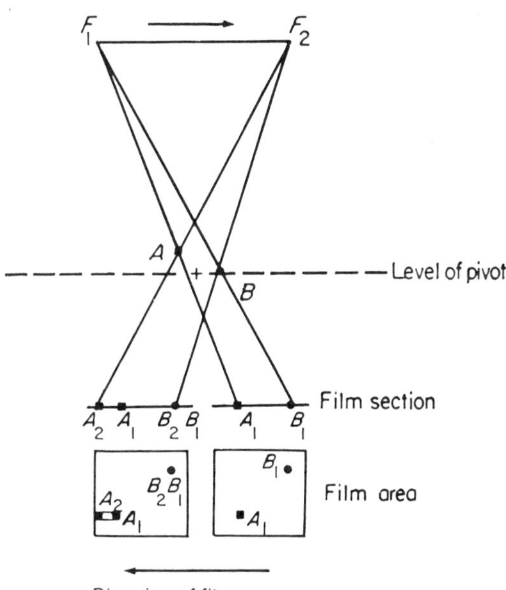

Figure 8.43. Tomography.

The principle of tomography is explained by reference to Figure 8.43. We assume that the pivot is situated at the position of $(+)$ in Figure 8.43 and we follow the projection of point B, which lies in the selected plane of the pivot and of point A which lies outside the selected plane. At the start of the movement of the X-ray tube at F_1, the point B will be projected at B_1 and the point A at A_1, as seen in the section of the film and in the area of the film on the right of the figure.

If the tube has moved during exposure from F_1 to F_2 and the film in opposite direction, we see that point B will be projected on exactly the same position of the moving film throughout the exposure, i.e. B_1 and B_2 coincide. Point A_1 however, being outside the selected plane, has shifted during the exposure towards the edge of the film, as indicated in the section and area of the film (left). Hence the projected image of point A is blurred from A_1 to A_2. This blur refers to all points lying outside the selected plane and the further the selected plane (whether below or above this plane) the greater will be the amount of blur in the tomographic image.

Tomography usually incorporates the use of Potter–Bucky diaphragms, employing, in addition, a cone in order to minimize the amount of scatter produced. Various types of tomographic equipment differ in the way in which the movement is carried out. Some of the apparatus are designed for spiral, others for sinusoidal movement instead of the more conventional rectilinear motion (W. Watson 1962). Differences occur also in the angle of movement and in the speed of the movement of the tube, on which the time of exposure depends. In some cases of horizontal layer tomography the tube is fixed and the X-ray beam is incident through a slit, whilst patient and film rotate synchronously about vertical axes. (A. Weinbren, 1946.)

Simultaneous Multi-Layer Tomography

This technique (M. d'Abreu, 1948; W. Watson, 1951) is often regarded as being more advantageous than conventional tomography, as it allows a series of tomograms of various selected planes to be made in one single exposure, and in the same phase of the involuntary movement of organs. Exposures are carried out simultaneously on several screen film combinations which are superimposed in a pack, by applying a magazine for holding the various screens and films. The magazine might also be used in connection with a Potter–Bucky diaphragm.

The principle of the method is explained by reference to Figure 8.44 in which two films are indicated which are spaced from each other by a distance which is in a certain relation to the separation of the planes in which the points A and B lie, as will be explained later. (See also Figure 8.45 in the next section.) If the pivot is at the position of the $(+)$ then the point A which is lying in the plane of the pivot will be rendered sharp in film section I, as indicated by A_1 and A_2 after the X-ray tube has moved from F_1 to F_2. Point B will however be blurred as shown by B_2B_1 on the left film section I. Since film section II is separated from I by a distance which allows all points in the layer of point B to be recorded sharp, point A will appear blurred in film II, as seen in Figure 8.44. It should be noted that the relative distances of A_1A_2 and B_1B_2 from the edges of the film

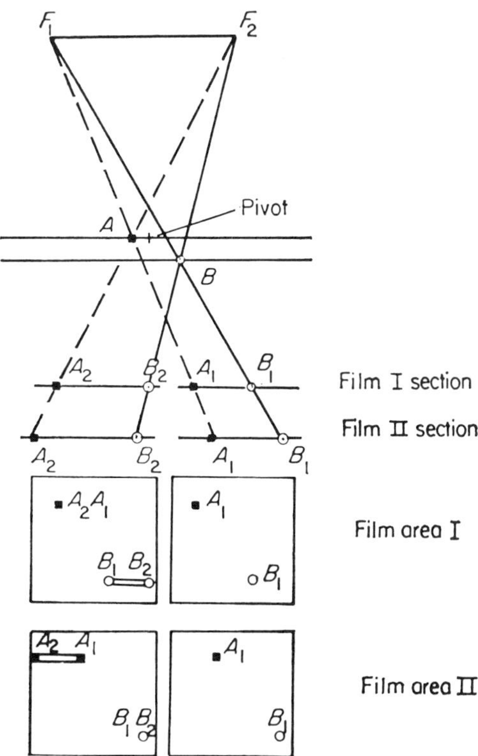

Figure 8.44. Simultaneous multi-layer tomography.
■▬■ indicates blur of point *A* in Film II
○▬○ indicates blur of point *B* in Film I

sections can be seen at once from Figure 8.44. The corresponding positions of the sharp points and the blur of $B_2 B_1$ and $A_1 A_2$ are shown in film areas I and II respectively.

In practice several films, sometimes five, are exposed simultaneously and the screen film combinations are separated from each other by spacers made of a material having a low absorption of X-rays. The aim is that all films should receive about equal exposures. Since the screens absorb X-rays to a considerable extent, the second film will receive less than the first and the third will receive still less than the second etc. In order to compensate for the effects of absorption and distance, screens of increasing speed from top (closer to X-ray tube) to the bottom should be used. The optimum arrangement of the set of screen film combinations depends on the kilovoltage to be used,

TABLE 8.3. Screen-Film Combination for Simultaneous Exposure of Five Layers

Position	Film-screen combination
1st film (nearest x-ray tube)	1 high definition front screen
2nd film	1 medium back screen
3rd film	1 pair of high definition screens
4th film	2 fast front screens
5th film	1 pair of fast screens

but within the range from 70–90 kV$_\mathrm{p}$ the combination shown in Table 8.3 above has been found useful.

Each screen film combination is spaced 1.5 cm from the next one and the whole packet must be under even pressure to ensure intimate contact between screens and films. The same type of film can be used for all the combinations and positions quoted in the table above. The exposure required for this set of film-screen combinations is about 3 times greater than in conventional radiography. If the results should not be satisfactory with regard to the equality of density levels between the various films, this can be improved by a proper adjustment of development times for each film.

The Relation between Separation of Body Layers and Films (*in Multilayer Tomography*). It should be noted that the spacings between the layers of the object to be radiographed are not the same as those between the films, if optimum image sharpness of the respective body layers is to be obtained. This will be seen from Figure 8.45, where p = pivot, a = focus-pivot distance, c = separation between layers, e = separation between films, d = pivot—first film distance, x = distance between X-ray tube positions, y = distance as indicated in the diagram (corresponding to the distance moved by the first film during exposure). From Figure 8.45 it follows, that

$$\frac{a}{x} = \frac{d}{y} \tag{8.12}$$

and

$$\frac{a + c}{x} = \frac{d + e - c}{y} \tag{8.13}$$

By substituting the value of x from the first equation into the second, we find

$$e = \frac{c(d + a)}{a} \tag{8.14}$$

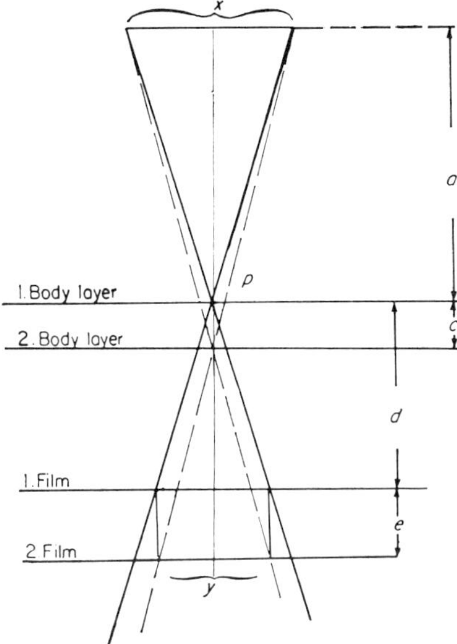

Figure 8.45. Diagram explaining the relation between film spacing and body
layer spacing in simultaneous multi-layer tomography.

Using this equation the film separation needed to give the required
body layer separation can be calculated. It is found that the body
layer separation is slightly smaller than the spacing of the films.

To obtain a body layer separation of 1 cm the film separation should
be 1.08 cm and 1.2 cm for focus–pivot distances of 150 and 60 cm
respectively, using 12 cm as the distance between the first body layer
and the first film.

Mass-Miniature Radiography

This comprises the photography of the fluorescent screen image
which is also called 'fluorography'. The technique was evolved
basically in the early days of the development of radiology, when it
was attempted to carry out cinematography of the screen image of
moving objects. Fluorography has been introduced rapidly in the
mid thirties in mass surveys of the population for the detection of
tuberculosis and other diseases of the chest (H. F. Potter, 1940;
K. C. Clark and others, 1945).

In the fluorographic equipment use is made of a fluorescent screen
(see p. 292), an optical system, and a fine grained photographic
material. In the early days of the development of the technique, the
equipment consisted of a light-tight tunnel fixed to a stand (with

height adjustments) and provided at one end with the fluorescent screen covered by lead glass. At the other end of the tunnel a camera is fitted, e.g. a 35 mm camera having a large aperture of the order of f/1.0. Film lengths of up to 100 feet were accommodated in the reels. Subsequently, larger dimensions of films have been used. A stationary grid (see p. 317) is often used between patient and screen. Due to the necessary enlargement of the images for inspection, the image quality is inferior to normal sized radiographs.

Considerable progress in performance and in resolution was achieved by the introduction of the Bouwers' concentric mirror system (Odelca cameras). It was pointed out by Schmidt, that if a diaphragm is placed at the center of curvature of a concave mirror, the definition is the same all over the image. He corrected the residual spherical aberration by an aspheric correcting plate. A. Bouwers (1951, 1953, 1965) used a correcting device in which the surfaces were spherical. Such systems have been developed by him to yield high definition, high aperture, and a wide field. The Odelca Unit makes use of either 70 mm film in roll form (especially suited for mass miniature radiography) or of 100×100 mm cut sheet film. The cameras have apertures of f/0.63 and f/0.65 respectively. These wide apertures of the mirror system permit a speed increase of approximately 4 times compared with that of the use of refractive lens systems.

In Figure 8.46 the optical arrangement of the 100 Odelca camera is illustrated, showing Bouwers' concentric mirror system. This comprises (1) a spherical concave mirror, (2) the image plane, (3) a cone lens, (4) a meniscus lens, (5) a fluorescent screen, (6) a stationary grid, (7) a cardholder, and (8) a roller separator. The film is held in the curved focal plane (2) and the card holder (7) serves for the provision of identification cards of the patient. The introduction of the cone shaped lens assists in the improvement of the resolving power of the system when designed for extremely large apertures.

The function of the roller separator (8) is as follows.

The unexposed films are in a container of a removable supply magazine, as seen in Figure 8.46. During use the first film in the magazine is pushed forward so that it is separated from the remainder of the batch of films. A hook then engages the curved film and slides it to the set of rollers incorporated in the film guide. The rollers move the film in front of a pressure plate where it is pressed to the required curvature. After the exposure has been taken the pressure plate moves back and the film is gripped by the next set of rollers and is directed into the take-up magazine.

As mentioned before the choice of film for this purpose depends on

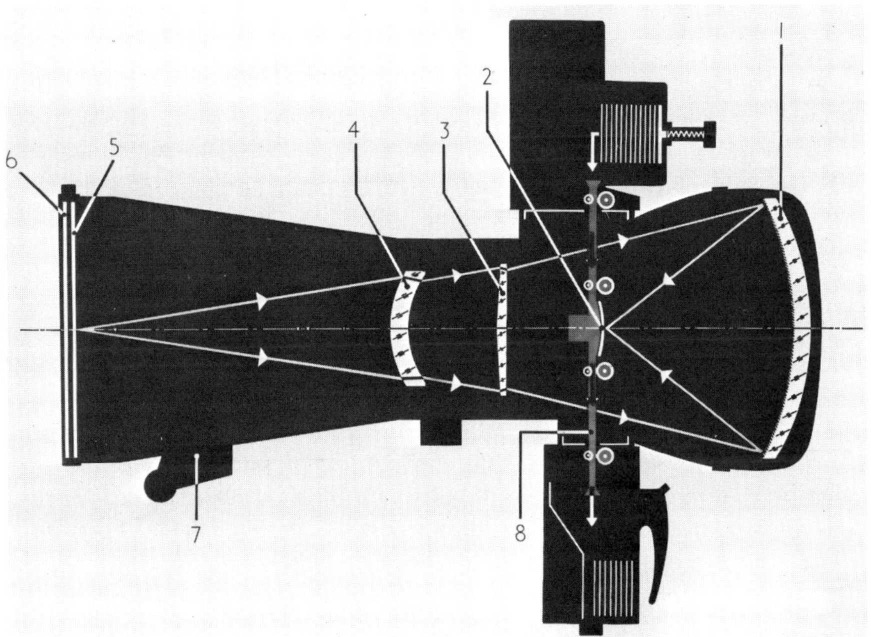

Figure 8.46. Type 100 Odelca mirror system camera. (*By courtesy of* N. V. Optische Industrie, *De Oude Delft*, Delft, Holland.)

the spectral emission of the fluorescent screen and suitable types of film are available either in roll form or in cut sheet. Due to the relatively large film size used with the concentric mirror system of the Odelca cameras, less magnification is required for viewing purposes than with 35 mm film. This and the refinement of the optical system led to the good resolution of this type of radiograph which is also frequently used outside the field of mass-surveys. Exposures with this technique are often controlled by automatic photo-timers which may be incorporated in the equipment. The detector of the timer may be either an ionization chamber or a photoelectric cell. If the former is used the effect of X-rays is detected directly. In the case of a photoelectric cell as a detector, the fluorescent screen light is used. When the necessary exposure has been given a thyratron switches off the X-ray unit automatically. The detector is often used in the apical region of the lungs. The chief advantage of the photo-timer is that in most instances it takes care automatically of the various thicknesses of the patients to be radiographed in mass surveys.

Cineradiography

Attempts to radiograph the movements of the human body (cineradiography) have been made very early in the history of radiography by both direct and indirect methods. In the direct method use was made either of a roll film which is moved between a pair of intensifying screens and is pressed in contact at subsequent intervals for exposure, or use was made of a series of cassettes (loaded with screens and film), which were mechanically moved in rapid succession. As both these techniques require fairly complex mechanisms, the indirect method i.e. the cinephotography of the fluorescent screen image of moving organs, appeared to be more promising. All these methods suffer however from the drawback that the patient receives fairly high doses of X-rays during the exposure, whilst a number of frames per second are taken. With the introduction of image amplifier tubes, cineradiography became a practical possibility without the necessity of giving excessive doses to the patient. A brief discussion of the principle of image amplifiers should precede that of modern cineradiography (R. J. Reynolds, 1934).

Image Amplifier Tubes

The main purpose of the image amplifier is to increase the brightness of the fluorescent image. When the radiologist screens the patient, this is usually carried out with 3–5 mA tube current in a darkened room. Higher tube current is not advisable in the interest of radiation safety and would be against official recommendations. It may be mentioned here that in general a patient receives larger doses in screening (fluoroscopy), than in radiography (G. M. Ardran and others, 1957). This is, in spite of the low milliamperage, because of the relatively long screening times involved in this operation. The dark adapted eyes perceive the screen image under these conditions at such a low brightness level that only 'rod vision', with its reduced acuity, is possible (see p. 293). In order to increase the level of brightness to 'cone vision' an increase of about 1000 is required and this has been achieved by the introduction of image amplifiers, using a tube current considerably below that required for screening. An image amplifier or image intensifier in its simplest form consists of a highly evacuated glass tube, see Figure 8.47 at one end of which is a layer of fluorescent material called the input phosphor, coated at the inside of the tube. A photo-emissive layer consisting of a compound of e.g. cesium, antimony, potassium or sodium is placed adjacent, to the input phosphor and acts as a photocathode. This photo-emissive layer emits low energy electrons when exposed to the light

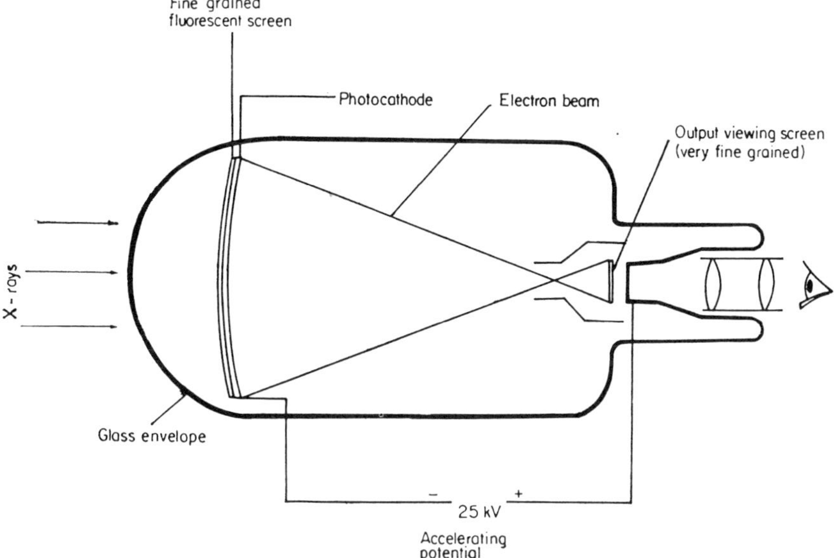

Figure 8.47. Principle of image amplifier.

of the fluorescent screen; hence the fluorescent screen image is converted into an electron image. The number of electrons emitted by the photocathode is directly proportional to the number of light quanta emitted by the fluorescent screen.

The electrons are accelerated by a potential of about 25 kV and brought to focus by an electrostatic lens near the anode on to a small, very fine grained fluorescent output screen, thus forming a fluorescent image. The size of the output screen may be of the order of 1/100th of that of the input screen. The output image is many times brighter than the one formed on the large screen. This is (a) because of the increase of energy acquired by the electrons and (b) because of the reduction of fluorescent image size. A few quantitative data may illustrate how this gain is achieved.

For every 100 light photons emitted by the input screen about 20 electrons are released by the photocathode. The high voltage acceleration of the electrons gives a gain of about 10–15 and the reduction of the size of the output screen image (if 1/100th of the area of the input screen) provides about a 100 times brightness amplification. By these means a gain of about 1000–1500 is achieved. Furthermore the resolving power of the output screen is about of 10 times better than that of the input screen, so that the output screen does not greatly degrade the image.

The viewing screen image which may be of the order of 1″ or ½″ in diameter must now be magnified to a practical size for viewing conditions, by either monocular or binocular microscopes providing a magnification of the order of 10 times whereby again a drop of brightness gain occurs. In another version of image amplifier a system of lenses and mirrors is used to permit simultaneous viewing of the image by two observers, R. E. Sturm and R. H. Morgan (1949), and W. F. Niklas (1961).

The high brightness amplification leads to many advantages, such as

1. A great reduction of X-ray dose to the patient is achieved, if compared with conventional fluoroscopy.

2. The image need not be viewed in the dark because of its high brightness level.

3. The image is perceived by the cone instead of the rod-receptors of the eyes and thus permits an improved visual acuity.

4. It offers the feasibility of cineradiography.

The size of the input screen of image amplifiers originally only 5″ in diameter has been increased in recent years up to 9″ and 12″.

In one type of unit the input phosphor is outside the intensifier tube. The screen image is reflected by a mirror inclined to the screen by 45° and projected onto a photocathode of a small light intensifier tube by means of a Bouwers' concentric mirror system having an extremely high light transmitting capacity, see Figure 8.48.

With the electronic optical system it is possible to attain a two fold magnification during fluoroscopy. The intensified fluorescent image may be viewed by television fluoroscopy and be recorded by a cine-camera. The observation of image amplifier screens on closed television (vidicon and image orthicon systems) introduces the advantage that several persons can view the image simultaneously and remotely, e.g. in another room, if necessary. Furthermore, contrast and brightness level can be changed independently of kilovoltage and current.

There exist many modifications of image amplifiers systems on the American and European markets; in some of them a lens system focuses the X-ray screen image on to the photocathode of a special image orthicon television pick-up tube. This tube converts the optical into an electrically charged image which is scanned off by a scanning electron beam. The tube output signal is then amplified and fed to a number of television monitors for either viewing, film recording or tape recording. (J. Tol, W. J. Oosterkamp and J. Proper, 1955; C. B. Banks, 1958; R. B. Holmes and D. J. Wright, 1962.)

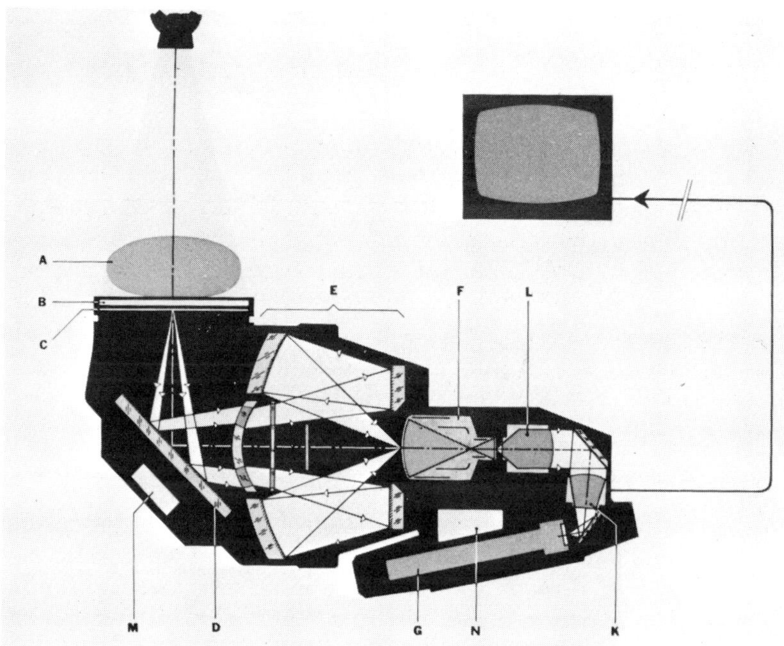

Figure 8.48. Section through Delcalix 12½″ electro-optical image intensifier.
A. Patient; B. Secondary grid; C. Fluorescent screen; D. 45° inclined plane
mirror; E. Bouwers' concentric mirror system, (f/0.68); F. Image intensifying
tube; G. Image orthicon; K. Special "Old Delft" lens f/1.4; L. Optical system for
either 35 mm or 100 mm single shot camera; M. High tension feed unit for image
intensifier tube; N. Amplifier control unit. (*By courtesy of* N. V. Optische Industrie
De Oude Delft, Delft, Holland.)

Cineradiography with Image Amplifiers

Image amplifiers are generally provided with either 35 or 16 mm
cine-cameras. Provision is made so that viewing of the screen image
is possible during cineradiographic exposures, which is an important
requirement. The focusing conditions can be fixed as the distance
between output screen and camera is constant. The number of frames
may be as high as 60 per second and orthochromatic films of medium
and high speed are frequently used for this purpose.

Considering the X-ray dose to the patient and the safe loading of
the X-ray tube it is important to avoid continuous X-ray exposure.
This can be achieved by synchronizing the exposure with the
mechanism of film transport so that each successive film frame receives
a separate exposure. It is essential that the X-ray pulse is timed so
that mainly the peak of the brightness of the screen image coincides

with the opening of the camera shutter. During the period of the decay of the phosphor (afterglow) the shutter should remain closed. This prevents the succeeding frame from being influenced by the afterglow of the previous image.

Cineradiography is mainly used in the study of physiological phenomena, as they occur in gastrointestinal and circulatory processes where the movements are often too rapid to be followed by fluoroscopy. Slow motion projection of the cine film assists in the investigations of these processes in greater detail, which would otherwise escape observation (A. Lignon and P. Bolden, 1958; P. C. Hodges and R. D. Moseley 1959).

In some instances an image orthicon tube may be connected to the image amplifier, when cineradiography is taken from a television monitor.

Kymography

This is a method chiefly for investigations of the heart, diaphragm, oesophagus, and stomach. The essential equipment is a lead sheet larger than the size of the film to be used and provided with a number of parallel slits of equal distances apart. Each slit is about 0.5 mm wide and the spacing between the slits is about 12 mm. During an exposure of the heart, the lead grid is placed between cassette and chest of the patient and is moved in a direction perpendicular to the slits by a distance equal to the spacing between the slits (see Figure 8.49). The resulting kymographs will then show a number of serrations around the heart contour, which may give valuable information on the function of the heart (P. Stumpf, 1937).

The actual motion of the heart can also be simulated by making a reproduction of the lead grid on film and moving this film slowly upwards when superimposed on the kymograph. This rather primitive kind of cineradiography of the motion of the heart or of peristaltic movements as seen on the screen can be carried out by using only two sheets of film, the kymograph and the copy film of the lead grid. Figure 8.49 shows a positive print (drawing) of a kymogram of a human heart (left) (P. Stumpf, 1937), in which the heart and its contours have been blacked out, against a white background. On the right a print of the grid is shown. If the reader wants to simulate the heart movement, he requires only a duplicate print of the grid picture on film (as seen on the right of Figure 8.49). By moving the latter across the kymograph image from bottom to top the motion can be simulated.

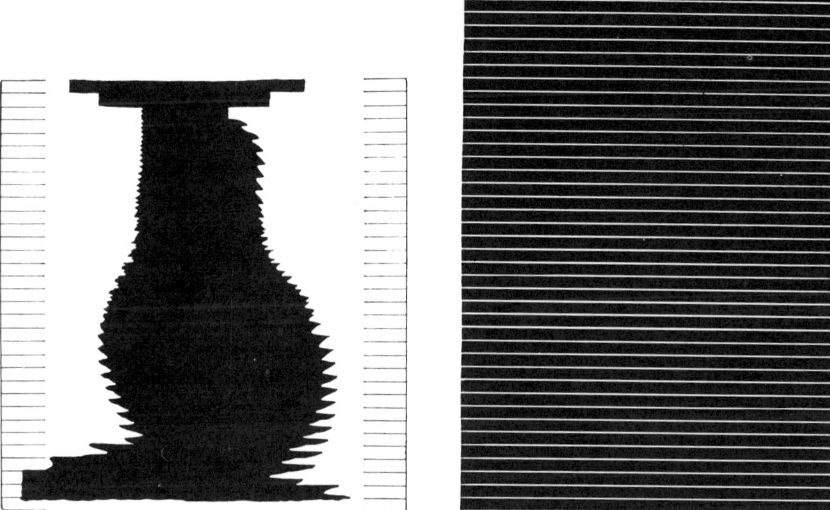

Figure 8.49. *Left:* Positive print (drawing of a kymogram in which the heart and contours are blackened). *Right:* Print of grid. By superimposition and slow upward movement of a transparent duplicate copy of this grid image on the kymogram (left) the motion of the heart can be simulated. (*After* P. Stumpf. Zehn Vorlesungen über Kymographie. Georg Thieme Verlag, Leipzig (1937), now Stuttgart.)

Only a few of the auxiliary methods used in medical radiography have been selected which are of particular interest in recording radiographic images by various photographic and physical techniques. There are many others in current use which have been omitted, such as, for instance, dental radiography, angiography, mammography, contrast media injection and barium meal techniques.

In the following section a method is described which is based on an interesting and different photographic principle.

Polaroid Radiography

In recent years the well known Polaroid-Land system has been introduced into some fields of radiography. This system makes use of the so called 'diffusion transfer process' in which the unexposed region of the negative is utilized to form a positive image on a suitable receiving medium to which it is transferred by diffusion.

The radiographic Polaroid Pack consists of a light tight envelope containing a negative film coated with a silver halide layer, a paper mask, developer pods, and a receiving sheet. The latter is not sensitive to light or X-rays and contains thiosulfate and also silver and silver sulfide particles which serve as nuclei of the positive image. As a preparation for the exposure the pack is inserted into a special cassette by sliding the black paper covered sensitized sheet between a single X-ray intensifying screen and the cassette front. When exposed to X-rays a latent image is formed as in conventional radiography. After exposure the cassette is placed in a special Polaroid processing unit, so that the protruding paper tab of the envelope is fed between rollers which are incorporated in the processor. By a single operation of pushing the starter button of the processor, the rollers pull the packet out of the cassette, the pods are broken and the viscous processing reagents are spread evenly between the negative and the receiving sheet. The highly alkaline developer, containing a silver halide solvent, starts to act immediately producing a negative image. At the same time it converts the unexposed region of the negative to soluble compounds which diffuse to the receiving sheet where they are reduced to form a positive silver image.

After 10–15 seconds of processing the negative film is peeled off and discarded, leaving a positive radiograph. This is treated with a varnish which neutralizes the alkalinity and stabilizes the image. It contains furthermore a drying agent which speeds up the drying process.

Application.   As the cassette can be loaded in daylight and since no darkroom is required for the Polaroid Pack processing, it is a useful equipment in the operating theatre. This is particularly so because of the very short time elapsing between exposure and viewing of the image which may be of the order of only 30–40 seconds. Exposure times are usually shorter than with conventional X-ray film, but the image quality is inferior with respect to graininess. An overexposed image appears too light and an underexposed image appears too dark. The image can be viewed by transmitted, as well as by reflected light, but the radiologist has to become accustomed to a positive, instead of the usual negative image (E. H. Land, 1947; O. W. Kincaid, G. D. Davis, and J. L. Mengis, 1964).

BIBLIOGRAPHY

Bronkhorst, W. (1927). *Kontrast und Schärfe im Röntgenbilde*, Georg Thieme Verlag, Leipzig.
Chesney, D. N. (1965). *Radiographic Photography*. Blackwell, Oxford.

Eggert, J. (1951). *Einführung in die Röntgenphotographie.* S. Hirzel Verlag, Zurich.

Files, G. L. W. *Medical Radiographic Technic.* Charles Thomas, Springfield, Illinois.

Fuchs, A. W. *Principles of Radiographic Exposure and Processing.* Charles Thomas, Springfield, Illinois.

Glasser, O. (Ed.) *Medical Physics.* Year Book, Chicago.

Glasser, O., Quimby, E. H., Taylor, L. S., Morgan, R. H., and Weatherwax, J. L. *Physical Foundations of Radiology.* Pitman, London.

Jaundrell-Thompson, F. and Ashworth, W. J. (1965). *X-ray Physics and Equipment.* Blackwell, Oxford.

Meredith, W. J. and Massey, J. B. (1968). *Fundamental Physics of Radiology.* John Wright, Bristol.

van der Plaats, G. J. (1961). *Medical X-ray Technique, Principles and Applications.* Philips Technical Library.

Schall, W. E. (1961). *X-rays.* 8th Ed. John Wright, Bristol.

Seemann, H. E. (1968) *Physical and Photographic Principles of Medical Radiography.* John Wiley, New York.

Spiegler, G. (1957). *Physikalische Grundlagen der Röntgendiagnostik.* Georg Thieme Verlag, Stuttgart.

Stieve, F. E. (Ed.) (1966). *Bildgüte in der Radiologie. Symposium Herrenchiemsee.* G. Fischer, Stuttgart.

REFERENCES

d'Abreu, M. (1948). Theory and technique of simultaneous tomography. *Am. J. Roentgenol.*, **60**, 668.

van Allen, W. W., and Morgan, R. H. (1946). Measurement of resolving power of intensifying screens. *Radiology*, **47**, 166.

Andrews, J. R. (1936, 1937). Planigraphy, introduction and history. *Am. J. Roentgenol.*, **36**, 575, 1936. **38**, 145, 1937.

Angerstein, W., and Plapperer, H. (1965) X-ray screens: Image quality and dose requirement depending on size of phosphor crystals. *Röntgen-Praxis*, **18** (1), 16–21.

American Standards Association Inc. (1964). *Method for the Sensitometry of Medical X-ray Films*, **PH2**, 9.

Ardran, G. M., Emrys-Roberts, E. and Kemp, F. H. (1957). The duration of fluoroscopic examinations. *Brit. J. Radiol.*, **30**, 381.

Ardran, G. M. (1964). The reduction of scatter fog in chest radiography. *Brit. J. Radiol.*, **37**, 477.

Banks, C. B. (1958). Television pick-up tubes for X-ray screen intensification. *Brit. J. Radiol.*, **31**, 619.

Bonenkamp, J. G., and Boldingh, W. H. (1959, 1961). Quality and choice of Potter-Bucky grids. Parts I, II, III and IV. *Acta Radiol.*, **51**, 479, 1959. **52**, 149, 241, 1959. **55**, 225, 1961.

Bouwers, A. (1951). A new X-ray camera, f. 0.75 for 7 × 7 cm Film. *Brit. J. Radiol.*, **14**, 516.

Bouwers, A. (1953). Improvement of resolving power of optical systems by a new optical element. *Appl. Sc. Res. Sec. B.*, **3**, 147.

Bouwers, A. (1965). Le Role de l'Optique en Radiodiagnostique. *J. Radiol. Electrol. Med. Nucl.*, **46**, 10, 652.

Brailsford, J. F. (1939). Simple radiographic methods for the localization of foreign bodies. *Brit. J. Radiol.*, XII, **134**, 65.

Chamberlain, W. E. (1942) Fluoroscopes and fluoroscopy. *Radiology.* **38**, 383.

Clark, K. C. (1939). Radiographic depth localization of foreign bodies. *Radiography*, **V**, 59.

Clark, K. C., Hart, P. D., Kerley, P., and Thompson, B. C. (1945). Mass miniature radiography of civilians for the detection of pulmonary tuberculosis. *Med. Res. Council, Spec. Rep. Ser.* 251.

Cleare, H. M., Splettstoesser, H. R., and Seemann, H. E. (1962). An experimental study of the mottle produced by X-ray intensifying screens. *Am. J. Roentgenol.*, **88**, 168.

Cohn, M., and Barth, W. (1931). Lehrbuch der Röntgenstereoskopie. G. Thieme Verlag, Leipzig.

Coltman, J. W., Ebbinghausen, E. G., and Altar, W. (1947). Physical properties of calcium tungstate X-ray screens. *J. Appl. Phys.*, **18**, 530.

Eggert, J., and Schopper E. (1948). Zur Energie Ausbeute der Röntgenstrahlung bei der Fluoreszenz von Kalziumwolframat. *Ann. Phys.*, **6**, 270.

Etter, L. E. (1951). Essentials of stereoscopic roentgenography. *Am. J. Roentgenol.*, **66**, 248.

Fonda, G. R., and Seitz, F. (1948). (Eds.) '*Properties and Characteristics of Solid Luminescent Materials.*' John Wiley, New York.

Garlick, G. F. J. (1949). Luminescent Materials. Oxford University Press, London.

Grossmann, G. (1935). Tomographie I and II. *Fortschr. Geb. d. Röntgenstr.*, **51**, 2.

Hasselwander, A. (1937). Das Röntgenraumbild. *Das Raumbild*, **3**, July.

Henny, G. C., and Chamberlain, W. E. (1944). Fluoroscopy. In O. Glasser (Ed.), Medical Physics. 1st Ed. Year Book, Chicago, p. 1292.

Herz, R. H. (1931). Ein handliches Gerät zur exakten Ausmessung von Belichtungszeiten unabhängig von der Periodenzahl. *Röntgen-Praxis*, **3**, 18, 859.

Herz, R. H. (1933). Photographische Bemerkungen zum Vergleich von Röntgenaufuahmen auf Film und Papier. *Röntgen-praxis*, **5**, 4, 260.

Herz, R. H. (1936). Die Anwendung von Photoelementen in der Röntgentechnik. *Fortschr. Geb. d. Röntgenstr.*, **54**, 6, 616.

Herz, R. H. (1936). Ein neuartiges Messgerät zur Bestimmung der Röntgenstrahlen Qualität und-Quantität und der photographisch wirksamen Dosis. *Fortschr. Geb. d. Röntgenstr.*, **57**, 2, 199.

Herz, R. H. (1942). A photoelectric instrument measuring quality and quantity of X-rays of radiographic exposures. *Brit. J. Radiol.*, **XV**, 172, 110.

Herz, R. H. (1951). On the influence of cosmic rays on X-ray films during storage. *Phot. J.*, **91, B.** 134.

Herz, R. H. (1956). The spectral quantum and energy efficiency of calcium tungstate X-ray intensifying screens. *Brit. J. Appl. Phys.*, **7,** 182.

Hirschlaff, E. (1938). *Fluorescence and Phosphorescence*. Methuen, London.

Hodges, P. C., and Moseley, R. D. (1959). Cine fluorography employing split image television type image amplifiers. *Radiology*, **73,** 548.

Hodgson, M. B. (1918). Physical characteristics of X-ray fluorescent intensifying screens. *Phys. Rev.*, **12,** 431.

Holmes, R. B., and Wright, D. J. (1962). Image orthicon fluoroscopy of a twelve inch field and direct recording of the monitor image. *Radiology*, **79,** 740.

Johns, H. E. (1964). *The Physics of Radiology*. (2nd Ed.) Charles C. Thomas, Springfield, Illinois.

Kieffer, J. (1938). The laminagraph and its variations, applications and implications of the planigraphic principles. *Am. J. Roentgenol.*, **39,** 497.

Kincaid, O. W., Davis, G. D., and Mengis, J. L. (1964). Application of the Polaroid-Land method of roentgenography to angiocardiography. *Am. J. Roentgenol.*, **91,** 1364.

Klasens, M. A. (1947). The blurring of X-ray images. *Philips Tech. Rev.*, **9,** 364.

Land, E. H. (1947) A new one-step photographic process, *J. Opt. Soc. Am.*, **37,** 61–77.

Levy, L., and West, D. W. (1933). A new and much more rapid intensifying screen allowing great alterations in radiographic techniques. *Brit. J. Radiol.*, **6,** 85.

Levy, L., and West, D. W. (1934). The elimination of afterglow and latent phosphorescence from fluorazure (ZnS) intensifying screens. *Brit. J. Radiol.*, **7,** 344.

Levy, L., and West, D. W. (1939). The differential action of nickel, iron and cobalt upon the fluorescence and phosphorescence of certain zinc sulphide phosphors when excited by X-rays. *Trans. Faraday Soc.*, **35,** 128.

Lignon, A., and Bolden, P. (1958). Image intensification cinefluorography. *Proceedings of 1st Annual Symposium on Cinefluorography*. Dpt. Radiol., University of Rochester, U.S.A.

Luft, F. (1937). Prüftafeln für Röntgen-Stereoskopie, *Fortschr. a. d. Geb. d. Röntgenstr.*, **56,** 452.

Luminescence Symposium. Electronics Group, Inst., Physics (1954). *Brit. J. Appl. Phys.* Supplement No. 4.

Marshall, F. H. (1947). Microsecond measurement of the phosphorescence of X-ray phosphorescent screens. *J. Appl. Phys.*, **18,** 6, 512.

Mattson, O. (1955). Practical photographic problems in radiography. (With special reference to high voltage technique). *Acta. Radiol.*, Suppl. No. 120.

Morgan, R. H. (1944a). Quantitative study of the effects of temperature on the sensitivities of X-ray screens and films. *Radiol.*, **43**, 256.

Morgan, R. H. (1944b). Reciprocity law failure in X-ray films. *Radiol.*, **42**, 471.

Morgan, R. H. (1949). The sensitometry of roentgenographic films and screens. *Radiol.*, **52**, 832.

Morgan, R. H. (1956). Screen intensification. A review of past and present research with an analysis of future development. *Am. J. Roentgenol.*, **75**, 69.

Morgan, R. H. (1962). The frequency response function. *Am. J. Roentgenol.*, 88(1), 175.

Morgan, R. H. (1966). Visual perception in fluoroscopy and radiography. *Radiology*, **86**, 403.

Morgan, R. H., Bates, L. M., Gopalarao, U. V., and Marinaro, A. (1964). The frequency response characteristic of X-ray films and screens. *Am. J. Roentgenol.*, **92**, 426.

Morrison, A., and Perry, K. J. (1950). Temperature and intensifying screen effects in radiography with cobalt-60. *Nondestructive Testing*, Summer 1950, p. 9.

Mott, N. F., and Gurney, R. W. (1940). Electronic Processes in Ionic Crystals. (1st Ed.) Clarendon Press, Oxford.

Nawyiyn, A. (1963). Modulation transfer and quality of mirror cameras for photofluorography, *Röntgen-Blätter*, **16**, 7, 3.

Newell, R. R. (1938). Sharpness of shadows in radiography of the lungs. *Radiology*, **30**, 493.

Niklas, W. F. (1961). X-ray image intensification with a large diameter image intensifier tube, *Am. J. Roentg. Radium Therapy and Nuc. Med.*, **85**, 2, 323.

Petri, E. C. (1960). Der Röntgenfilm, Veb. Fotokinoverlag Halle.

Piwonka, R., Voigt, G., and Petri, E. C. (1965). Der Einfluss der wirklichen Korngrösse des Calcium-Wolframat-Leuchtstoffes auf die Detail-wiedergabe von Röntgen-Verstärkerfolien. *Röntgen-Blätter*, **18**, 79.

Potter, H. F. (1940). The miniature X-ray. *Radiology*, **34**, 283.

Pringsheim, P. (1943). *Fluorescence and Phosphorescence*. Interscience, London.

Reynolds, R. J. (1934). Cineradiography, *Brit. J. Radiol.*, **7**, 415.

Rockley, J. C. (1957). Fluoroscopy. In W. J. Wiltshire (Ed.) 'A Further Handbook of Industrial Radiology', Ch. 7. Edward Arnold, London.

Rose, A. (1948). Sensitivity performance of the human eye on an absolute scale. *J. Opt. Soc. Am.*, **38**, 196.

Rossmann, K. (1962a). Recording of X-ray quantum fluctuations in radiographs. *J. Opt. Soc. Am.*, **52**, 1162.

Rossmann, K. (1962b). Modulation transfer function of radiographic systems using fluorescent screens. *J. Opt. Soc. Am.*, **52**, 774.

Rossmann, K. (1962c). Detail visibility in radiographs. An experimental and theoretical study of geometric and absorption unsharpness. *Am. J. Roentgenol.*, **87**, 387.

Rossmann, K. (1963). Spatial fluctuations of X-ray quanta and the recording of radiographic mottle. *Am. J. Roentgenol.*, **90**, 863.

Rossmann, K. (1964a). Some physical factors affecting image quality in medical radiography. *J. Phot. Sci.*, **12**, 5, 279.

Rossmann, K. (1964b). Measurement of the modulation transfer function of radiographic systems containing fluorescent screens. *Physics in Medicine and Biology*, **9**, 551.

Rossmann, K. (1965a). Comparison of several methods for evaluating image quality of radiographic screen film systems. *Intern. Radiol. Congr.* Rome.

Rossmann, K. (1965b). Effect of quantum mottle and modulation transfer on the measurement of radiographic image quality. In (R. D. Moseley and J. H. Rust. (Eds.),) *Diagnostic Radiol. Instrumentation.* Ch. 19. Charles C. Thomas, Springfield, Illinois.

Rossmann, K., Lubberts, G., and Cleare, H. M. (1964). Measurement of the line spread function of radiographic systems containing fluorescent screens. *J. Opt. Soc. Am.*, **54**, 187.

Rossmann, K., and Seemann, H. E. (1961). Detail visibility in radiographs. Theoretical study of the effect of X-ray absorption in the object on the edge sharpness of radiographic images. *Am. J. Roentgenol.*, **85**, 366.

Schober, H. (1957). *Photographie und Kinematographie.* Verlag Kurt Wesemeyer, Hamburg.

Schober, H., and Klett, C. (1953). Untersuchungen über die Zeichenschärfe von Verstärkerfolien. *Röntgen-Blätter*, **6**, 214.

Seemann, H. E., and Splettstosser, H. R. (1954). Some physical characteristics of Potter–Bucky diaphragms. *Radiology*, **62**, 575.

Seitz, F. (1939). An interpretation of crystal luminescence. *Trans. Faraday Soc.*, **35**, 74.

Selwyn, E. W. H. (1965). Combination of lens and film. In R. Kingslake (Ed.), Appl. Optics and Optical Engineering, Vol. **II.** Academic Press, New York. p. 165.

Stanwix, G. W., and Mullins, L. (1946). Comparison of the afterglow properties of X-ray intensifying screens. *Phot. J.*, **B86**, 38.

Stumpf, P. (1937). *Zehn Vorlesungen über Kymographie*, Georg Thieme Verlag, Leipzig.

Sturm, R. E., and Morgan, R. H. (1949). Screen intensification systems and their limitations. *Am. J. Roentgenol.*, **62**, 617.

Tasker, H. S. (1945). Some properties and uses of X-ray intensifying screens. *Phot. J.*, **B85**, 75.

Ter Pogossian, M. M. (1965) The reduction of patient dose by diagnostic radiological instrumentation. In (R. D. Moseley and J. H. Rust (Eds.)) *Diagnostic Radiol. Instrumentation.* Charles C. Thomas. Springfield, Illinois.

Tol, J., Oosterkamp, W. J., and Proper, J. (1955). Limits of detail perceptibility in radiology particularly when using image intensifier. *Philips Res. Rep.*, **10,** 141.

Vallebona, A. (1950). Axial transverse lamina. *Radiology,* **55,** 271.

Vallebona, A. (1952). Trattatodi, Stratigraphica, Milan.

Watson, W. (1951). Simultaneous multisection radiography, *Radiography,* **17,** 221.

Watson, W. (1962). Rotascanography, *Brit. J. Radiol.,* **35,** 847.

Weinbren, A. (1946). *Manual of Tomography.* H. K. Lewis, London.

Williams, F. E. (1956). Fluoroscopic image intensification, *Am. J. Roentgenol.,* **75,** 77.

Winnek, D. F. (1944). Three dimensional roentgenography. In O. Glàsser, (Ed.), *Medical Physics,* Year Book, Chicago. p. 1324.

Wilsey, R. B. (1921a). The intensity of scattered X-rays in radiography, *Am. J. Roentgenol.,* **8,** 328.

Wilsey, R. B. (1921b). The effect on scattered radiation in radiography, *Am. J. Roentgenol.,* **8,** 589.

Wilsey, R. B. (1922a). Some practical results with a Potter–Bucky diaphragm, *Am. J. Roentgenol.,* **9,** 441.

Wilsey, R. B. (1922b). The efficiency of the Bucky diaphragm principle, *Am. J. Roentgenol.,* **9,** 58.

Wilsey, R. B. (1934). Scattered radiation in roentgenography of the chest, Preliminary Report, *Radiology,* **23,** 198.

9

The Processing of Radiographs

9.1 INTRODUCTION

The fundamentals of processing have been reviewed in Chapter 3 p. 75. The purpose of this chapter is to discuss the practical aspects of the processing of radiographs. As there is basically no difference between the processing of medical and industrial radiographs this section may also be read by those interested in industrial radiography.

With the general increase of population and the mounting number of radiographs taken per day in hospitals, the manual operations involved in processing require an ever increasing number of staff. Hence the introduction of automatic and time saving processing methods has become an economic proposition, which has been accepted already by a high percentage of modern hospitals and also by industrial departments. The following section will be devoted first to the manual and secondly to the automatic processing methods.

9.2 MANUAL PROCESSING

9.2a. The Preparation of Processing Solutions

The proper function of processing solutions depends on the correct preparation of these solutions. In the following, recommendations are given on the necessary precautions to be taken when preparing solutions from powder chemicals, as they are supplied by photographic manufacturers.

1. Read and follow the instructions attached to the chemicals supplied.

2. Adjust the recommended temperature of the water in which the chemicals are to be mixed.

3. The vessels for mixing and dissolving chemicals should be made of stainless steel, hard glazed earthenware, glass, plastic or other corrosion resistant materials. The use of chipped enamelled containers or those made of aluminum, copper, tin or galvanized iron or soldered tanks should be avoided, as the reaction of the solutions with these materials will lead to a deterioration of the processing solutions and to fogging of the photographic material.

4. Measure the actual volume of the tank to be used, as the solution should have the accurate concentration as recommended by the manufacturer.

5. Clean the containers thoroughly before use.

6. It is essential that the constituents of developer and fixer are dissolved in the correct order, otherwise chemicals might not go into solution properly. Manufacturers usually list the contents in the order in which they should be dissolved.

7. Each chemical should be dissolved completely before the next one is added to the solution.

8. Stir the solution thoroughly until the chemicals have been properly dissolved. Use of an efficient stirrer made of stainless steel or plastic is recommended.

Processing Formulae

Those recommended by Kodak Ltd. England are given in the Appendix. These refer to typical X-ray developer, replenisher, stop bath, fixer and other solutions. The formulae given in the Appendix refer to solutions which are to be made up from powdered chemicals. Many manufacturers supply concentrated liquid developers and fixers in order to save the operator the work involved in dissolving the chemicals and only a given quantity of water has to be added to make the solutions ready for use.

9.2b.   The Control of Processing

After the solutions have been prepared, processing can be carried out, assuming that the temperature of the solutions is adjusted as recommended and that solutions are thoroughly agitated before the film is immersed. The latter is necessary in order to avoid local concentrations in the solution. Although it is most important that the temperature of the developer solution corresponds to that recommended, it is advantageous that the temperature of the other processing solutions is at least close to that of the developer. If great differences in temperature occur, then reticulation may appear, which leads to a net-like appearance of the emulsion layer. This is less likely to happen, however, if hardening fixer is used.

In general it is advisable to use developer tanks of large volume (5–10 gallons), but the surface area of the solution should be small in order to minimize aerial oxidation of the developer solution.

Temperature, development time and agitation are the most important controlling factors influencing the final radiograph for a given developer solution. We shall deal with these three factors separately.

Development Time

The effect of the development time on a given X-ray film at constant temperature of the developer solution can be studied from the following experiment.

We start with the development of a series of equally exposed sensitometric strips (see Figure 8.12a) of a double coated X-ray film exposed in conjunction with a pair of intensifying screens. The temperature of the developer is kept constant at 20 °C (68 °F). Then we remove one strip after 1 minute, the next after 2 minutes, the third after 3 minutes, and so on up to 10 minutes developing time. After complete processing and drying of these film strips, the densities of each are measured by means of a densitometer and the characteristic curve of each strip is then plotted, as shown in Figure 9.1.

Figure 9.1 shows that the average slope of the curves and the speed increase with development time up to about 4 minutes. Beyond this time the slope remains practically constant, although the speed still increases but at the expense of further growth of the fog level. Hence a compromise must be made between unduly high fog and the increase of speed. In this case an optimum development time of 4 minutes might be chosen, which gives a reasonably high speed with a fairly low fog. The photographic manufacturers usually base their recommendations on this kind of sensitometric result.

Temperature of the Developer Solutions

If we carry out a similar test as that in the previous section, but keep the development time constant at 4 minutes and change the temperature of the solution for each sensitometric strip from 15–24 °C (i.e. from 60–75 °F), we obtain a family of characteristic curves, in which each curve corresponds to a particular temperature. The results of such an experiment are shown in Figure 9.2(a) and (b). Both figures refer to X-ray screen film exposed in conjunction with screens.

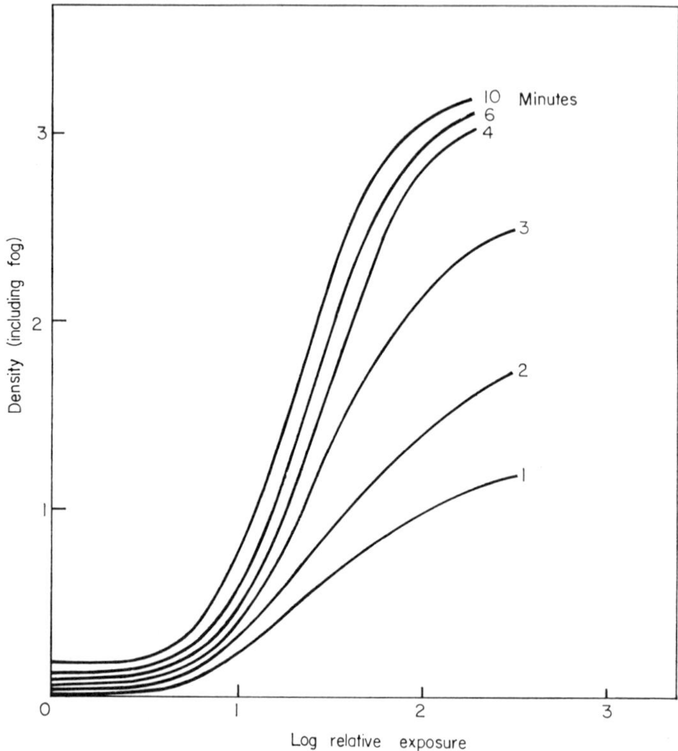

Figure 9.1. Characteristic curves of an X-ray screen film exposed in conjunction with screens, for various development times in minutes at a constant temperature of 68°F.

In Figure 9.2(a) the change of relative speed at $D = 1$ and the fog level are shown as a function of temperature. Similarly we see the change of gradient at $D = 1$ with increasing temperature in Figure 9.2(b), all for a constant development time of 4 minutes. It is found that the relative speed increases with temperature from 8 to 13 for 60° and 75°F respectively, but fog increases as well from 0.07 to 0.25. The inherent contrast at $D = 1$ remains almost unchanged within the same range of temperature. Below a temperature of 60°F the activity of the hydroquinone in the solution is severely affected at pH values below 12, and at a temperature above 75°F softening of the emulsion occurs which makes it difficult to handle the film without damage as the emulsion becomes tacky. It is therefore recommended by most manufacturers to use a temperature of 20°C (68°F), at least

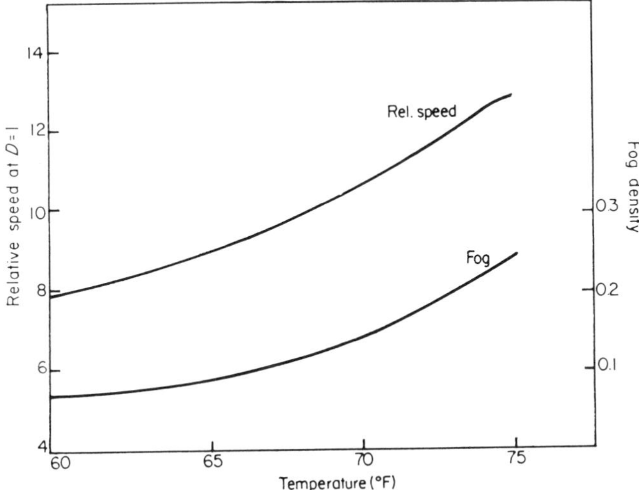

Figure 9.2.(a) Relative speed at $D = 1$ above fog and fog level, both as a function of the temperature of a standard X-ray developer solution. Constant development time of 4 minutes.

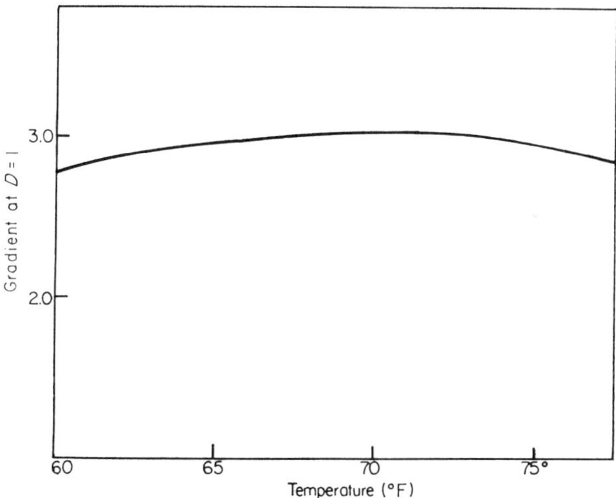

Figure 9.2.(b) Gradient as a function of the temperature of a standard X-ray developer solution. Constant development time of 4 minutes.

for standard manual processing. (See also note on tropical processing p. 385.)

We have seen that the activity of a developer increases with temperature, but development is also time dependent and thus, we can, within limits, compensate for the variations of development rate, by adjusting the time of development accordingly. This can be done by reference to the time-temperature relationship. (See Figure 9.3.) Whatever the temperature the correct adjustment of development time will lead practically to results which would be obtained at the recommended times and temperature.

It should not be forgotten that the shorter the development time, the more difficult it is to obtain even and uniform development. The chart shown in Figure 9.3 is typical for a given film and developer solution, but does not necessarily refer to another type of film and

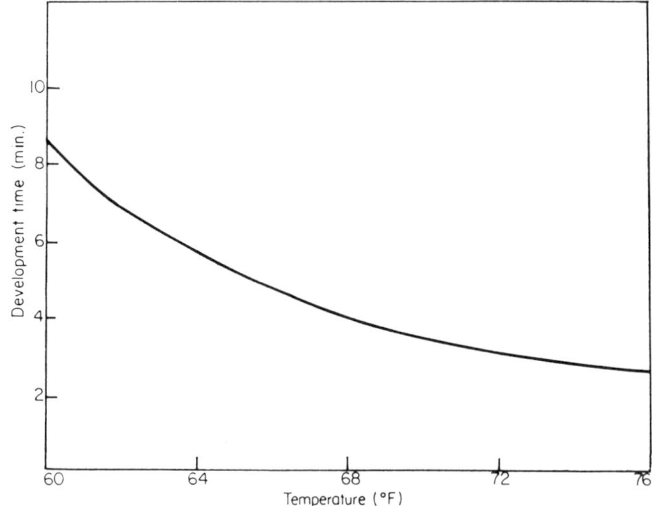

Figure 9.3. Typical time-temperature chart for a standard X-ray developer solution.

developer. In practice it is of less importance whether the recommended temperature is adhered to exactly but it is important in the interest of reproducibility of results from one film to the next, that the temperature chosen should be regulated to within about $\pm 1°C$.

Agitation

It is essential that films are agitated vigorously during the first 10–15 seconds after immersing the film into the developer solution, in order to prevent air bubbles attaching themselves to the film. These may otherwise produce artifacts such as light marks with dark surroundings on the finished film. In order to minimize uneveness of development over the area of a film during development, it is further suggested that agitation of the film is repeated every minute for about 5 seconds.

It is well known that lack of agitation results in film artifacts which are known as bromide streamers. These show as streaks of low density below heavily exposed areas. This effect is caused by the local exhaustion of the developer at the strongly exposed area of the film. Thus the developer has become partly restrained due to the excess bromide formed in this region. On the other hand at a region of a fairly light exposure the local concentration of the developer remains practically unaltered with the result that the region below will appear higher in density.

Defects of this kind are particularly frequent in a film showing wide variations between areas of low and high density. Such artifacts may be very disturbing in the interpretation of radiographs. Thorough agitation ensures that fresh developer is supplied to the entire area of a film at regular intervals in order to prevent any serious loss in speed and contrast.

Replenishment

Continuous replenishment of a developer solution permits the standardization of development to become a practical possibility without changing development times. The quantity of replenisher solution required for topping up a developer depends (a) on the number and size of films to be developed, (b) on their average density and (c) on the time the film is allowed to drain over the developer tank at the end of development. The time for draining should be kept constant, as far as this is practically possible, and should be about 10 seconds. The composition of replenishers of some manufacturers are based on the recommendation of a very short draining time (of only 2–5 seconds). If films are insufficiently drained, too much developer will be carried over into the following stop bath. The result is that too much replenisher has to be used, leading to a too highly concentrated developer solution, which may cause high fog. On the other hand under-replenishment may occur if too much developer is drained back into the tank with the effect of a reduction of contrast and low emulsion speed. A balance between these opposing effects will be found either by using higher or lower concentration of replenisher solution.

In many departments replenisher is added to keep the level of the developer solution constant. This may be carried out several times during the day. Such a technique assumes that the replacement of the volume lost by drainage compensates exactly for the loss of activity of the developer. This may be correct in many cases, but it depends on the method of drainage used, when the film is lifted from the developer tank.

A fairly accurate assessment can be carried out by exposing a series of sensitometric strips. If this should be inconvenient, an X-ray exposure of an aluminium step wedge on a $10 \times 8''$ X-ray film can be made. From this film a series of strips of $\frac{1}{2}''$ width may be cut. These are then processed at regular intervals, e.g. immediately, and after every third day. The densities of the strips need not be measured, but mere comparison will show whether the relative densities have changed. Depending on the results achieved, the replenishment technique can be adjusted accordingly. It should be mentioned that the outcome of this test might be invalidated, if the films used should suffer from strong latent image fading (see p. 144). Some brands of film show this effect stronger than others, some are practically free from fading. The effect can sometimes be minimized by storing exposed films for about two weeks before use after which the fading is not so rapid.

Manufacturers usually suggest that the developer solution should be discarded at regular times, i.e. if replenishment is carried out in the correct manner, a developer may be used to about 100 per cent replenishment. This means that a volume of replenishing solution equal to that of the original developer solution should be used, before the developer is to be discarded.

Acid Stop Bath

After the film has been removed from the developer solution it should be rinsed, preferably in an acid stop bath instead of in water, for about 10 seconds. The acidity of this bath must be checked at regular intervals in order to make sure that it fulfils its function (see p. 78). A formula for an acid stop bath is found in the Appendix, page 612. Ready made liquid stop baths are available which indicate by colour change, when the acidity has been neutralized. However, running water may be used as a less efficient substitute.

Fixing Bath

Following the brief rinse in a stop bath the film is immersed in a fixing solution. This is made up from powders or from liquid fixers, as supplied by the manufacturers. It is advisable to agitate the film strongly when first immersing in the fixing solution, in order to neutralize quickly the alkaline developer still present in the emulsion. This is particularly important if the fixing bath is nearing exhaustion, as otherwise so called 'dichroic fog' may be produced which usually shows as an area insufficiently or not at all fixed and is easily recognizable by a stain appearing greenish-blue by reflected light. It looks pinkish when viewed by transmitted light. In order to minimize the tendency to form dichroic fog and to maintain hardening, boric acid is added to the formula for fixing baths as seen in the Appendix.

Whilst the fixer dissolves the unexposed and undeveloped silver halides, the emulsion becomes clear. The dissolved halides form complex silver compounds and although the emulsion appears clear, some of the compounds may still be insoluble and may not be washed out. It is therefore necessary to extend the fixing time for twice the time taken for the emulsion to clear. Considering this precaution, fixing times should be of the order of 5 minutes in fresh solutions, depending on their composition.

Exhaustion of Fixing Bath

With increasing number of films fixed, the bath becomes gradually exhausted. This is because (a) the fixing bath is used up in forming silver complexes with the unexposed silver halides. If the concentration of these complexes exceeds a certain level, they cannot be completely removed during the following washing procedure and are retained by the emulsion, thus causing stain. (b) The reduction in acidity by the alkaline developer influences the hardening properties of the solution and tends to produce dichroic fog. (c) Soluble halides which are formed as a result of the fixing action delay the rate of fixation and (d) the concentration of the fixer is reduced by the water carried over from the rinse. A fixing bath is regarded to be exhausted, if the time to clear is about twice that when the bath was fresh. The useful life of a fixing solution can be extended by certain silver recovery techniques and fixer regeneration (see pp 394–396).

Washing

After the fixation is complete the emulsion is saturated with chemicals of the fixing bath and silver complexes. These have to be removed by thorough washing of the films.

In the washing process chemicals are diffused out of the emulsion and are removed by circulating the water. By continuous replacement of the water by fresh, contamination of the water is gradually eliminated. In general X-ray films should be washed for 10 minutes, assuming that the water is subject to four complete changes per hour and that its temperature is not less than about 10°C (50°F).

Obviously the necessary washing time depends on the effectiveness of the preceding fixation of the films. The conditions recommended should be adequate for most films which are to be stored for about ten years. If radiographs have to be stored for much longer times, i.e. for archival permanence, washing times have to be prolonged considerably (see p. 236). Inspection of films before washing should not be unduly extended because of the possibility of partial drying, as the hardened gelatin will contract during the drying period and will not swell again; hence the subsequent washing loses in efficiency.

Drying

During the drying process care has to be taken that films are not subjected to mechanical damage to the emulsion and that they are dried in a dust free atmosphere.

The films may be suspended on clips or kept in the hangers in which they were held during processing and they may be dried by natural evaporation. In order to speed up the drying time by about 25 per cent they may be immersed in a final rinse containing a wetting agent. This solution reduces the surface tension of the water and permits a more even drying than is possible without the wetting agent.

In most modern departments drying is carried out in special drying cabinets, in which warm dry air circulates over the films.

Rapid Drying

Continuous driers are available for quick drying. Before introducing films into these, they have to be freed from clips and hangers. They are then fed into the drier through a pair of squeegee rollers. The films are then conveyed between a series of opposing rollers, through a drying chamber and the dry films emerge from the machine and fall into a collecting tray. Drying time in such driers may be about 1–2 minutes.

If such machines are not available, in 'exceptional' cases films can be dried quickly by chemical methods, if they are immersed for 2 minutes in a 70 per cent solution of industrial methylated spirit. After this bath the films can be dried within about 5 minutes by evaporation.

Standardization of Processing

After having discussed the main procedures of processing, it is useful to reemphasize the importance of the standardization of processing in a radiographic department. The main reasons for this are that there are so many technical factors involved which influence the technical quality of a radiograph, that it is advantageous to keep at least one series of factors, namely the processing conditions, constant. It is often quite rightly said that the consistency in processing forms the basis for the standardization of exposure conditions.

Ultra Rapid Processing

During certain surgical operations and certain industrial applications (e.g. pipe line welding) radiographs have to be produced quickly and ultra rapid processing becomes a necessity. Some of the manufacturers recommend their replenishers for ultra rapid development. As these are provided with a fairly high concentration of developing agents and alkali and do not contain a restrainer, development times of 30 seconds to 1 minute can be obtained at a temperature of 20°C (68°F). Continuous agitation is most essential in order to avoid streaks and uneven development.

Fixation may be carried out in ultra rapid fixing baths, such as those recommended by various manufacturers. A fixing bath of this type will allow clearing of an X-ray film in about 30 seconds to a minute after a rapid rinse in an acid stop bath. After fixation a very brief rinse removes the excess fixing solution and the radiograph can then be inspected within just over a minute processing time. After inspection the film should be refixed and thoroughly washed if it is to be kept and stored as other radiographs. Ultra rapid processing may be carried out in dishes instead of in tanks, but special care must be taken not to scratch the films touching the bottom of the dish and special frames might be used which keep the emulsion away from the bottom.

Processing in Tropical Climates

If the processing solutions can be kept at a temperature below 24°C (75°F) generally no special precautions are required, assuming that a hardener fixer is used and that the developing times according to the charts provided by the manufacturer (see Figure 9.3) are adhered to.

For temperatures exceeding this limit the X-ray emulsion becomes increasingly swollen and tacky and may finally melt, unless special precautions are taken. Thus for temperatures between 24°C (75°F) to a maximum of 32°C (90°F) development only becomes possible by adding sodium sulfate (105 grams (cryst.) or 45 grams (anhydrous) per litre of the developer solution) and the modified development times are quoted below, as recommended by Kodak Ltd, England in connection with D19 developer.

Using the short development times as quoted for the higher temperatures vigorous agitation is necessary in order to obtain uniformity of development. It is also important to leave films immersed in the solution during development, since the chances of formation of aerial fog are increased when inspecting films during development. After development, films should be agitated for 30–45

TABLE 9.1. Time-Temperature Relation

Development in D19		Development in D19 with added sodium sulfate	
Temperature	Time	Temperature	Time
55°F	12 min		
60°F	9 min	75°F	6 min
65°F	6 min	80°F	$4\frac{1}{4}$ min
68°F	5 min	85°F	3 min
70°F	$4\frac{1}{2}$ min	90°F	2 min
75°F	3 min		

seconds in a special hardening bath made up to a formula as given in the Appendix, p. 612 and then left for 3 minutes in this bath before immersing into an acid fixing bath with hardener. Renewal of the hardening rinse bath at frequent intervals is important in order to avoid scum marks on the films.

9.2c.   The Darkroom and its Equipment

The planning of a sound and practical layout of a darkroom for processing radiographs is most important for proper processing procedures. Suitable darkroom facilities are often neglected in favor of the expensive X-ray equipment, whereby it is usually overlooked that the quality of the radiographic output of the X-ray equipment is entirely dependent on the darkroom practice. Good planning depends on the amount of work to be handled in a darkroom and is, of course, partly a matter of personal opinion. Nevertheless certain practical recommendations can be given which may be considered when first planning a darkroom or when considering alterations of an older darkroom.

Location

The darkroom should preferably be not too far from X-ray diagnostic rooms and in many cases it is advantageous to place the darkroom adjacent to these. In the latter case provision for a cassette hatch may be made in the wall separating the two rooms, so that cassettes can be passed from the diagnostic room into the darkroom and vice versa, thus saving unnecessary walking between both rooms. For darkrooms situated adjacent to diagnostic rooms, special precautions against X-rays will be necessary to protect the staff working in the darkroom and the sensitized material kept in the latter. The protective measures may consist of lead sheeting or barium cement blocks for the wall of the adjoining X-ray room.

Planning and Basic Requirements

It should be decided before planning whether the darkroom is to be used for any specialized work, such as enlarging, printing etc., apart from processing radiographs and if this is so, adequate provision for other operations should be made. The arrangement of the equipment should be such that it facilitates the logical sequence of operations. Certain basic requirements for the general planning of darkrooms are summarized in the following.

The darkroom has to be entirely light proof and be prcperly screened from external ionizing radiation sources; it should be free from fumes or contaminating gases from neighboring laboratories, it should be provided with a good ventilating system and be heated adequately for healthy and comfortable working conditions. Running cold and hot water should be provided together with proper draining systems. There must be access to the darkroom without interference with the operations of processing, loading and unloading cassettes etc. and without admittance of light from outside. This may be done by a labyrinth entrance, the walls of which should be painted with a matt black paint. A white line of about 2 inches width should be painted along the inside of the light trap at eye level to guide the operators when entering and leaving the darkroom.

Floors

The floors of the darkroom should be resistant to staining by chemicals, they should be durable and should not become slippery when wet; hence highly polished or waxed floors are not recommended. Asphalt, hard rubber sheets and natural clay tiles or plastic sheeting may be used as floor material but wood flooring and linoleum are unsuitable.

Walls and Ceiling.

The walls of a darkroom should be provided with a hard smooth finish; light coloured oil bound paints or enamels are very suitable.

The walls close to the processing units may be covered with ceramic tiles which can easily be washed.

Ventilation and Temperature

The temperature of the darkroom should be maintained at approximately 18–20 °C (65–68 °F). This may be achieved by air conditioning or by means of special ventilation systems. The latter should be arranged so that the air in the darkroom is changed 8–10 times per hour and it should be possible to control the air flow by fans or by means of louvres which must be baffled to prevent entry of light. Whatever ventilation system is used, it is important that fresh and not stale air from outside corridors is taken in.

Safelighting

The darkroom should be provided with as high an intensity of lighting of the correct spectral composition as it is consistent with the safety of the particular sensitized materials to be processed. Photographic manufacturers recommend certain safelight filters for their various types of film; these transmit only light to which the particular film is least sensitive. Care has to be taken that the recommendations are followed for the correct brightness and distance from the safelight permitted, otherwise fogging of films is almost unavoidable.

In addition to the direct safelighting close to the loading bench or processing unit, it is advisable to use indirect safelighting, e.g. at the ceiling, thus providing general reflected safelight spreading over the whole of the darkroom. The switches for indirect safelighting are preferably near the entrance of the darkroom, whereas those for direct safelighting may be placed locally. In order to minimize the accidental switching of white light it is suggested to place the white light switch above the normal level of reach.

Testing Film for Safelight Illumination

The sensitivity of X-ray film exposed to safelight is greater for the exposed, than for the unexposed film and is less for the wet film in the state of processing. It is important to test the safelight sensitivity of X-ray films before the darkroom illumination comes into general use, as safelight fogging is one of the most frequent causes of film fog.

In order to check the safelight sensitivity for both an exposed and an unexposed film the following test is suggested.

A 10″ × 8″ X-ray screen film may be flash exposed in conjunction with a pair of X-ray intensifying screens in such a way that half of the film is covered by a sheet of lead of 2 mm thickness. The exposure may be carried out for instance by using X-rays generated at 60 kV$_p$ 0.3 mm Cu filter, 2 mA secs at a focus–film distance of 100 cm. The loading and unloading of the film should be carried out in complete darkness. After exposure the film should be placed below the safelight lamp at the usual working distance from the lamp. Some cardboard larger than the film should then be placed over the film so as to cover 1″ of both X-ray exposed and unexposed region of the film, see Figure 9.4.

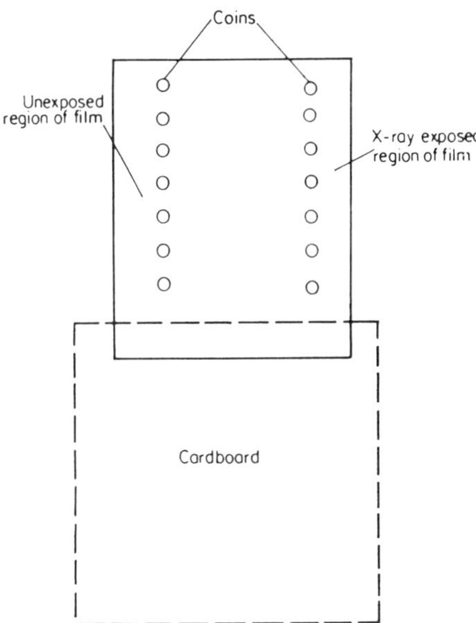

Figure 9.4. Arrangement for safelight testing.

Then the exposed and unexposed regions should be covered at equal distances with coins in the manner indicated in Figure 9.4. After switching on the safelight, the cardboard should be shifted at successive intervals of 1, 2, 3, 4, and 5 minutes to cover another pair of the coins each time. The safelight should then be switched off and the film processed under standard conditions. After processing the permissible safelighting times can be checked from the fog levels obtained for the exposed and unexposed regions of the film.

Layout

The most important division of a darkroom is that the equipment for the dry operations (such as loading and unloading of cassettes, providing films with data of identification etc.) are separated from the equipment for the wet operations, such as processing. Thus it follows that one or two sides of the darkroom as in Figure 9.5 should be provided with a suitable loading bench (or benches), another side with the processing unit and sink, as seen in Figure 9.5.

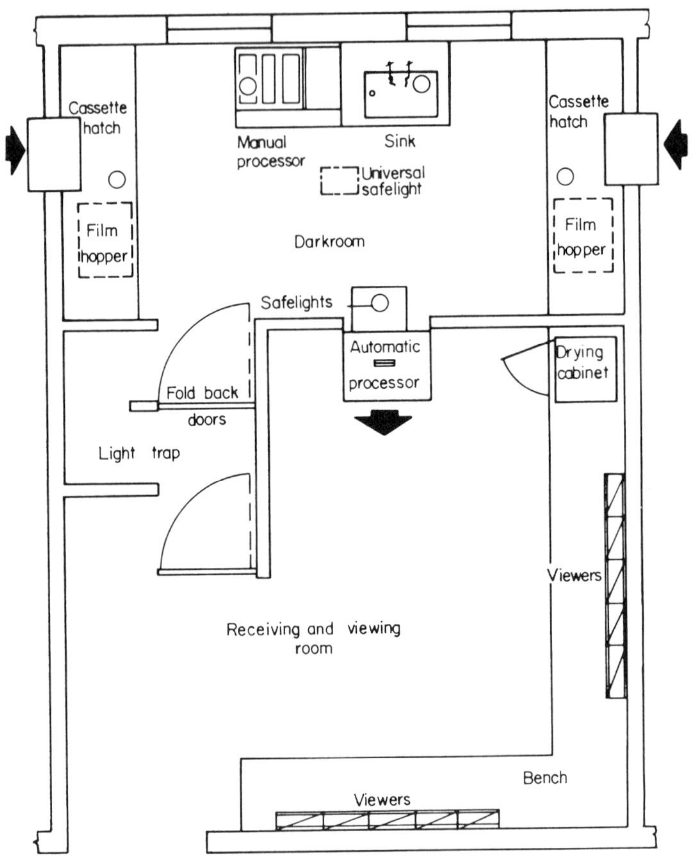

Figure 9.5. General layout of an average medium sized X-ray darkroom designed for manual and automatic processing. (*By courtesy of* Mr. R. Samuels, Kodak Ltd. London.)

This sketch gives an example of the general layout of an average medium sized X-ray darkroom also provided with an automatic processor. The loading bench and the processing unit will be treated separately in the following sections.

Loading Bench

The bench should be 3 feet (90 cm) high, at least 2 feet (60 cm) wide and for each operator on the average about 4 feet (120 cm) of length of bench would be required. The bench should be provided with a suitable dark and washable covering material, such as plastic, formica, cork lino or other fabric. The reason for the choice of a dark material is (a) to perform a contrast between the unprocessed film and the table and (b) to prevent unnecessary reflection of safelighting on to the rear emulsion of the film.

The space below the loading bench should be utilized to provide, for instance, a film hopper, containing X-ray films of various types and sizes arranged in separate compartments. Although the hopper should be clearly marked with a label indicating that it should not be opened in white light, it should be fitted with an interlock switch, which automatically switches white light off when opening the hopper.

The space below the bench may further be used for a container for waste paper, a cupboard for storing film boxes, a shallow light-tight drawer for the temporary storage of exposed films which are awaiting processing. It is useful to have on or near the bench a sectional rack for cassettes and room for an identification printer. Film hangers for processing films should be suspended e.g. at the wall in easy reach from the bench.

Cassettes

The essential requirements of cassettes for radiography with film and screens can be summarized as follows.

1. The pressure exerted by the cassette lid should not only be firm but also uniformly distributed over the total area of the film, thus providing intimate contact between screens and film.

2. It should be completely light-tight (also at edges) and the film-screen combination should be kept perfectly flat, unless a curved cassette is intentionally required.

3. The back, usually lead lined, should prevent scattered radiation reaching the film and the front of the cassette should provide a low degree of absorption of X-rays.

4. The internal dimensions of the cassette for the given standardized sizes of film should have small tolerances, so as to allow a perfect fit for the screen-film combination in the cassette.

5. It should be provided with clips or other means for holding the lid firmly to the cassette front after loading it and these should permit easy handling.

6. It should be made as flat as possible, it should have light weight and the thickness of the cassette should be as small as practically feasible so that the cassette fits easily into the space which is usually provided for it when in use in couches, chest stands etc.

7. Cassettes should be rigidly built, so that they can stand the continuous use in a busy diagnostic department.

Most cassettes are made of an aluminum or magnesium alloy, which ensures a fairly rigid construction. The first mentioned requirement, i.e. the perfect and uniform contact between screens and film is not easy to fulfil and much thought has been given to refined constructions of cassettes in which the air is pressed out whilst the cassette is being closed. Although an intimate contact can be ensured by using a well constructed vacuum cassette, this is often regarded as not being practical, as it adds a further operation, namely that of connecting the cassette to a vacuum pump.

Exposure Holder

For no-screen exposures light-tight exposure holders are available which are either made of cardboard or of thin plastic. These holders are usually provided with a thin lead foil in the back to reduce backscatter from the support. Films may be inserted with or without interleaving paper. The front of these exposure holders usually provides less filtration than the front of a metal cassette.

Cassette Hatch

The installation of a cassette hatch is useful, as mentioned before and it should possess a dividing wall to separate exposed from unexposed films in their respective cassettes. The hatch should preferably be near the loading bench and must be provided with adequate shielding against X-rays. It should have interlocked doors to prevent ingress of light into the darkroom.

Processing Units

The installation of a commercially available unit is probably the best choice for manual equipment. In the following a relatively small processing unit is described. This consists of a 5 gallon (22.5 liters) developer tank usually made of stainless steel and provided with a lid, a rinsing compartment which itself is provided with water sprayers; it further consists of two fixing tanks each of 5 gallons volume. The whole unit, i.e. developer tank, rinsing and two fixer tanks are contained in a larger tank. This outer tank is fitted with an immersion heater and forms a thermostatically controlled water jacket to keep the solutions at constant and preset temperature. The water jacket is provided with inlet and outlet valves for cleaning purposes. A further essential part of the processing unit is the washing compartment. This should hold a great number of film hangers and each is supported and separated from its adjacent hanger by stainless steel slotted channels. The washing section is provided with an overflow pipe and a drain for emptying it.

Obviously a sink must be regarded as an accessory to the X-ray processing unit and this should be close to the latter, but separated by a draining board. The sink may be used for washing film hangers and for making up processing solutions.

The provision of two fixing tanks is recommended so that one of them contains comparatively fresh fixer solution for preliminary fixing, after which films are transferred to the second fixing bath. In the initial stage when both fixing solutions are fresh, films are fixed in the first tank until they have cleared and are then transferred to the second tank for completion of fixing. In this way the second fixing solution remains fairly fresh and only after the solution of the first tank is exhausted and discarded, the second tank is then used for preliminary fixation. The first tank after being provided again with a fresh fixing solution serves then for the final fixing bath. Thus the cycle of fixation is repeated. The discarded fixer which contains valuable amounts of silver may be subjected to a silver recovery process (see p. 394).

Materials Used for Processing Equipment

The essential requirements for the selection of material for the construction of processing equipment are (a) that it is not chemically affected by the processing solutions and (b) that it has no undesirable effects on the photographic solutions or emulsions. Furthermore it must have the necessary mechanical strength to withstand the operations at which it is used and finally its availability and relative cost are of major importance.

Special types of stainless steel are regarded most suitable for use in the construction of processing equipment, such as for instance one which is composed of 18 per cent chromium and 8 per cent nickel, containing also 2–3 per cent molybdenum and a small fraction of about 0.1 per cent carbon. Hard rubber coated metals are useful as long as the rubber is undamaged, otherwise the base metal would be exposed to the solutions. Certain types of polythene and poly-propylene and rigid polyvinyl chlorides have also been found suitable. These latter are, however, less durable and show tendencies to swell and warp.

Film Hangers

Before processing commences films have to be suspended in suitable metal frames, called film hangers. Their purpose is to support the films so that they can be handled without touching the wet emulsion and further to keep the films apart in the processing and fixing tank, if several films are to be processed simultaneously. There are chiefly two types of hanger available (a) the tension clip type hanger and (b) the channel hanger; both should be made of stainless steel. (a) The tension hangers require more time and care in suspending the films, but they have the advantage that films can be dried in these, usually without causing marks on the films. (b) The channel hangers, although they allow greater ease in loading films, are still not suitable for drying, as the wet films stick easily to the channels whilst being dried; furthermore they are less easy to clean. It is necessary to remove films from channel type hangers for drying, they are then usually transferred to metal bars carrying clips for holding the films.

Miscellaneous Equipment

It is not intended to give a detailed description of the many darkroom accessories, as the reader will find a great variety in the catalogues of the manufacturers. Hence only a few of the more important accessories will be briefly mentioned. A more detailed description of them is found in the book by D. N. and M. O. Chesney (1965).

Immersion heaters and thermometers are usually a standard equipment of the processing units. Drying cabinets which provide a rapid flow of clean, warm, dry air circulating over the surfaces of films form a standard equipment of most hospital darkrooms. Continuous driers already mentioned which dry the processed films within about 90 seconds are even superior to the ordinary drying cabinets. Carriers for the transport of wet films are very useful in busy departments. Stirrers of plastic or stainless steel are required when preparing processing solutions or when stirring solutions before processing commences. There are many other devices recommendable to facilitate darkroom procedures, such as hanger racks, identification markers, film corner cutters, fine brushes for dusting screens, darkroom viewing boxes, processing timers, devices for silver recovery etc.

9.2d.   Silver Recovery and Fixer Regeneration

It has been shown on page 393 that a double fixing bath, i.e. the use of two fixing tanks does not only provide a sound method for complete fixation, but it also utilizes the bath fairly economically. The life of a fixing solution may however be extended by the use of silver recovery techniques. The advantage of the latter is (a) to keep the silver content of a fixing solution at a minimum in order to facilitate washing and (b) to recover a high percentage of the silver formed which may subsequently be sold.

An exhausted fixing bath may contain on the average about 8 grams of silver per liter of solution and using an electrolytic silver recovery unit, on the average 30 grams of silver may be recovered from about 60 square feet (or 5.57 square meters) of processed radiographs. These figures obviously depend on the type of radiograph to be fixed. If the films contain an unusually high average density over the total area, most of the silver would be retained by the films. The efficiency of the silver recovery naturally depends also on the amount of silver carried into the wash water.

If an electrolytic silver recovery method is applied whilst the fixing bath is in use, the latter can be regenerated by adding at suitable intervals acid fixing salts, hardener and preservatives either in the form of powder or liquids. The specific gravity of the fixing bath which can be checked by means of a hydrometer, must be kept constant.

A silver recovery unit consists of two electrodes (cathode and anode) which are placed in the fixing tank and are connected to a control box. This provides the necessary potential and current control and is connected to the mains electrical supply. The electrodes may either be placed at opposite sides of the fixer tank or in some units the anode is rod shaped and surrounded by a large area cathode, on which the positive silver ions are deposited. Some recovery units work at low, some others at high currents. Whereas the former are in operation whilst the fixing bath is in use, the latter require strong agitation of the solution and are operated at night when the solution is not in use. The proper function of a unit depends largely on whether the silver is adequately removed at periods of maximum use of the fixing bath.

Certain precautions have to be taken to ensure smooth functioning of the electrolytic process. Whilst the silver recovery unit is in operation the concentration of the preservative (sodium sulfite) is gradually reduced at the anode. This may result in a sulfurization of the fixing agent, furthermore developing agents carried over by the film may decompose at the anode. If a fixing bath contains about 5 grams silver per liter, proper silver deposition (plating) occurs, if there is a minimum concentration of sulfite of about 8 grams per liter. With lower concentration of sulfite, the silver turns into a brownish color (staining the radiographs) and at very low silver concentrations (e.g. 2 grams per liter of fixing bath) the thiosulfate ions may be attacked and sulfiding of the fixing bath occurs. The effects can be counteracted by the addition of sodium or potassium metabisulfite to the fixing solution at each regeneration. Apart from keeping the specific gravity constant, the maintenance of the correct pH value is also necessary.

If no silver recovery unit is used, the silver rich fixing bath may be sold to silver refiners, who collect exhausted fixing baths and extract the silver from them.

9.3 AUTOMATIC PROCESSING

The introduction of automatic processors signified a revolutionary step throughout the organization of radiographic departments. Although automation does not necessarily imply speed of operation, it is one of the additional assets which the most advanced processors provide. Essentially a film is fed into the processor in the darkroom and is taken out at the other end of the unit (usually in a lighted room); this film is developed, fixed, washed, and dried and this whole operation takes only 7–10 minutes. More recently a system has been introduced which permits automatic processing including drying in only $1\frac{1}{2}$ minutes.

The advantages of automatic processing are not only the rapidity, great uniformity and standardization of processing conditions, but also the fact that it permits a faster reporting system resulting in less waiting time for patients and doctors. There are no wet films to be handled and no spilling of solution during processing can occur. Furthermore the average standard and reproducibility of processing quality is far advanced when compared with manual processing. Although the initial cost of automatic processors is high when compared with manual processing units, the saving of personnel, speeding up of intake and release of patients and smoother departmental operation are certainly all beneficial in a long term economy. Hence it is not surprising that due to the advances in automatic processing such units have already displaced manual processing units in practically all larger and even in many smaller hospitals, and also in large industrial radiographic departments.

Many automatic processors of different design are now available and only the general principles of these will be described, as it would be practically impossible to discuss the many constructions in detail without trespassing the purpose of this book.

From a point of view of darkroom planning it is advantageous to install an automatic processor through a wall separating the dark and lighted room, so that films are fed into the processor in the dark, but are taken out dry in the light for inspection, reporting, filing etc.

The more advanced automatic processors may usually be provided with the following sections.

(a) a film feeding system

(b) a film transport mechanism which takes the films automatically through the various stages of processing.

(c) an automatic replenishing system.

(d) a temperature controlling system.

(e) a system which recirculates the processing solutions and provides vigorous agitation.

(f) an air circulating system in the drying cabinet.

The various automatic processors differ not only in basic design, but also in their capacity of film handling per unit time or in the time for a complete processing cycle. The construction of automatic processing units differs chiefly with respect to the type of film transport system adopted. The units are provided either with a roller system (introduced by Eastman Kodak Comp., U.S.A. in 1957) or with a mechanical transport system analogous to manual processing units. The latter method is often referred to as a 'dunking system'.

9.3a. Roller Systems

In this system no film hangers are required as the films, after unloading from the cassette, are fed directly into the processor's feeding compartment. Each film is transported through the solutions and dryer section by a series of small driven rollers, usually arranged in offset positions. The action of the rollers produces a vigorous and uniform surface agitation which combined with an elevated temperature and recirculation of solutions, achieves a considerable reduction in processing time. The solutions, containing special chemicals, operate under rigid time-temperature conditions. As the surface liquids are removed from the film by the squeegee action of the rollers as each film passes from one solution to the next, the solutions are not unduly degraded. Replenishment of the solution is carried out automatically and is controlled by each individual film being processed. In this way consistency of solution activity is maintained for long periods of time. In Figure 9.6 a section through one of the Kodak 'X-Omat processor' models is shown. This picture reveals the film feeding station, the roller transport system, the film processing and dryer section and the receiving bin for the finished radiographs. At the bottom of Figure 9.6, the replenisher tanks are shown. More than 100 sheets of film of various sizes can be processed per hour

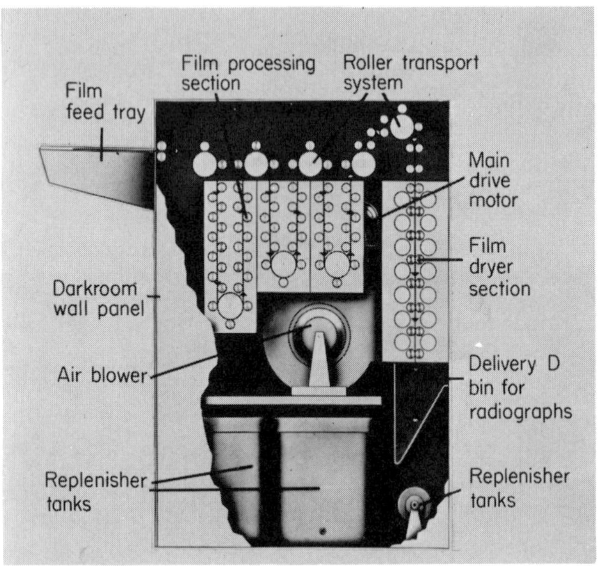

Figure 9.6. Section through a Kodak 'X-Omat' automatic processor model. (*By courtesy of* Kodak Ltd. London.)

with the larger units available, while the smaller more compact processors have a lower capacity. Other processors using the roller system are e.g. Pakoroll XF, Pakoroll XM, Williamson 220, Agfa-Gevaert Compact 100 and the Pixamatic processor.

9.3b.   Rapid Automatic Processing

As mentioned before, a more rapid automatic processing system has been introduced recently by the Eastman Kodak Comp. in U.S.A., and its subsidiaries. The Kodak Rapid Processing X-ray system may be carried out in some types of the existing 'X-Omat processors' after suitable conversion. With this system a fully processed dried film is available in 90 seconds. This system is based on a special X-ray film and special chemicals operating at temperatures ranging from 39–41°C, i.e. approx. 103°–106°F. In a rapid 'X-Omat processor' of the smallest type 350 sheets of film can be processed and dried satisfactorily per hour.

9.3c.   Dunking System

Using this type of transport system the films must be loaded first onto hangers or in clips before being mechanically transported by means of chains and wheels, and dipped or suspended in succession into the various processing tanks. Elevated temperature of the chemicals is again used if compared with manual processing. The speed of processing using the dunking system is of a similar order to that used for the roller systems, excluding rapid-processing.

Using a dunking system films are attached to a special clip bar, which is facilitated by a special loading table, before they are passed through the processor. After loading, the clip bars with films attached are placed on the loading ramps and then automation takes over. The films are progressively moved through the various processing tanks as the two lifting arms perform upward and downward movements. Thus the films are lifted out and deposited in the various compartments. The temperature in the development tank is controlled thermostatically, and both the main developer and fixer tanks have a recirculating system around which the solutions are passed continuously. Replenishment is controlled for instance by a chicken-feeder principle. After processing, each clip bar with its attached film passes between a pair of rubber squeegee rollers which remove surplus moisture from the films before they are passed into the drying cabinet. The normal time taken to process and dry a film is under 11 minutes. Other types of automatic processor using the method of dunking include the 'Elema Schönander', 'Procomat Rapid', the 'Fischer Fax-Ray', the 'Ken X-Automatic', and the 'Refrema' Processors.

BIBLIOGRAPHY

Chesney, D. N., and Chesney, M. O. (1965). *Radiographic Photography*. Blackwell, Oxford.

Clerk, L. P. (1954). In A. Kraszna-Krausz (Ed), *Photography. Theory and Practice*. Sir Isaac Pitman, London.

Fuchs, A. W. (1958). *Principles of Radiographic Exposure and Processing*. Charles Thomas, Springfield, Illinois.

Giafkides, P. (1960). *Photographic Chemistry*. Vols. I and II. Fountain Press, London.

John, D. H. O., and Field, G. T. J. (1963). *A Textbook of Photographic Chemistry*. Chapman & Hall., London.

John, D. H. O. (1967). *Radiographic Processing in Medicine and Industry*. The Focal Press, London.

Neblette, C. B. (1962). *Photography. Its Materials and Processing*. Van Nostrand, New York.

Southworth, J., and Bentley, T. L. J. (1957). *Photographic Chemicals and Chemistry*. Sir Isaac Pitman, London.

Towers, S. W. (1957). Films and their processing. In W. J. Wiltshire (Ed.), *A Further Handbook of Industrial Radiology*, Ch. 5. E. Arnold, London.

10

Industrial Radiography

10.1 INTRODUCTION

Amongst the many nondestructive testing methods used for the detection of internal defects in industrial materials, radiography has proved to be the most universal technique. Industrial radiography using mainly high voltage X-ray machines experienced its greatest expansion during the second world war. In this period γ-rays emitted by radium and radon were also used to supplement the work with X-rays. In the postwar period research on atomic energy initiated the production of γ-ray emitting radio-isotopes which, due to their relatively low cost, contributed much to the widespread use of γ-ray radiography. Also the small size of isotope sources permits their use at sites which are inaccessible to the rather bulky X-ray tubes.

Whereas the range of kilovoltages used in medical radiography is restricted between about 40 and 150 kV_p, industrial radiography covers an unlimited range of radiation qualities, i.e. radiations generated from a few kV to megavolts.

The photographic considerations which are common to both medical and industrial radiography, such as the 'fundamentals of radiography' have been discussed in the relevant sections of Chapter 7. Many other items which have been treated before, such as radiographic material and its sensitometric characteristics, the processing of X-ray films and some considerations on image quality apply to a great extent to industrial radiography as well and in order to avoid repetition only brief reference will be made to these in the following sections. On the other hand any characteristic features of material which are typical for industrial radiography will be discussed in detail.

400

Furthermore the techniques characteristic to industrial radiography, such as the exposure techniques, the production of exposure charts, the methods of assessing the image quality and the sensitivity of the technique in use, the radiographic inspection of welds and castings, the use of radioisotopes and specialized methods will be considered in the following sections. As in the previous chapter on 'Medical radiography' it is assumed that the reader is sufficiently familiar with the basic construction and control of X-ray apparatus.

10.2   RADIOGRAPHIC MATERIAL AND ITS USE

10.2a   No-Screen X-Ray Films

In industrial radiography use is made chiefly of no-screen double coated X-ray film* (see p. 307) with or without lead intensifying screens (see p. 404). X-Ray screen films in conjunction with salt intensifying screens are used only to a small extent and the main reasons for this will be explained later.

There is a wide range of no-screen X-ray films available for industrial radiography which differ in speed, contrast and grain size. The smaller the average grain size of these, the slower in general is the film speed and the higher the inherent contrast (at higher densities) and the better the definition attainable. Hence if an industrial radiographer has to deal for instance with hair line cracks in a casting, which may be very difficult to detect, the film of his choice would be a fine grained no-screen X-ray film. Although finer grain can sometimes be obtained by the use of special developers or by shorter developing times with a conventional X-ray developer, such methods are usually applied at the expense of the inherent contrast of the emulsion and it is generally more satisfactory to select a fine grained film.

The basic characteristics of no-screen X-ray films have been described on pp. 103–119 and 307. Quantitative values of the sensitivity of commerically available no-screen X-ray films and their respective gradients are given in the next two sections.

Sensitivity.   The number of roentgens required to obtain a density of 1 above fog for the fast variety of commercially available no-screen X-ray film is of the order of 0.02 R in the region of maximum spectral sensitivity, i.e. for X-rays generated at 80 kV_p and filtered by 0.4 mm Cu (equivalent to about 40 keV). In the energy region of 60-cobalt γ-rays (i.e. between 1100–1300 keV), about 0.5 R are required to

---

* Often called 'Direct type X-ray film'.

obtain a density of 1 above fog. The data quoted above correspond to D/R values of 50 and 2 respectively within the straight line region of the density-exposure curve (see p. 107). The slowest and very fine grained variety of no-screen X-ray film requires at least about fifteen times the number of roentgens to obtain $D = 1$ for the two qualities of radiations quoted above. The sensitivity figures mentioned in the preceding notes refer to direct exposures without lead intensifying screens and they are, of course, average values, as the speeds of commercially available no-screen X-ray films differ widely between various film manufacturers.

Gradients. One of the important features characterizing no-screen X-ray films used in industrial radiography is illustrated in Figure 10.1 which shows the change of gradient $dD/d \log E$ against the net density for four different types of no-screen film. Film No. 1 in this graph corresponds to a fairly coarse grained high speed no-screen film and film No. 4 is an extremely fine grained, but very slow film, having the highest gradient in the high densities as seen from this family of curves. The properties of films Nos. 2 and 3 lie between those mentioned before. In order to utilize the high gradient which some of these films exhibit at high densities the radiographs are usually inspected on high intensity illuminators (see p. 429). It is partly this feature of achieving high contrast combined with low film graininess that makes these film types so attractive in industrial radiography. (R. L. Durant, 1948, 1951, 1966; C. Croxson, 1957; S. W. Towers, 1957.) Further it can be shown that the advantages of slower films as against faster films are based on statistical arguments about the respective number of quanta required per unit area to obtain a certain density. It will be shown that the signal to noise ratio is more favorable in the use of slower material compared with fast films. This aspect is discussed in more detail on pages 420–422.

Processing

The processing of these films is similar to that in medical radiography, using the same type of developer, temperature, and development times. Generally 'standardized' processing conditions (see p. 384) are advantageous as in medical radiography. It should also be stressed that in general the slower the variety of no-screen X-ray film, the more it will withstand longer development times without giving excessive fog. Using longer development times of up to 15 minutes at the recommended temperature, the exposure of some industrial

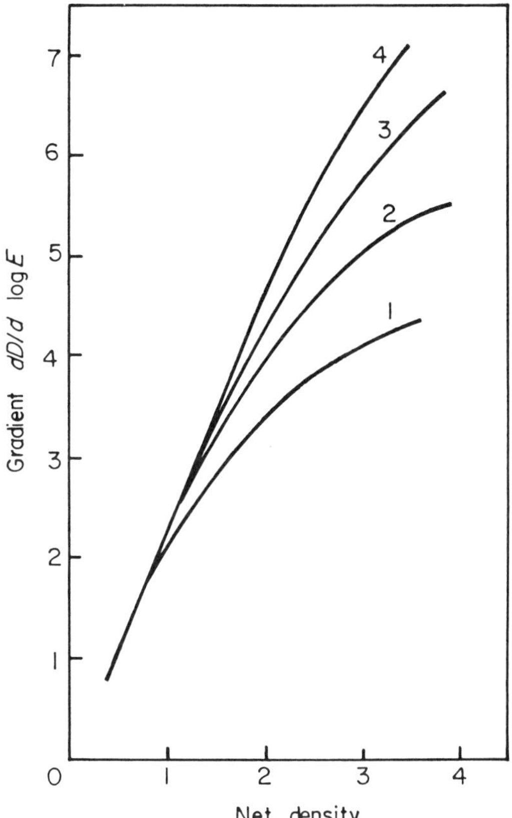

Figure 10.1. Gradient as function of net density for industrial no-screen X-ray films. 1. Very fast (coarse grained) (2) Medium fast (medium grain size) (3) Slow (fine grained) (4) Very slow (extra fine grained).

X-ray films can be reduced considerably, if compared with the exposure required for conventional development times, but not without sacrifice in image quality. (S. W. Towers, 1957; R. H. Herz, 1943). Automatic processing systems have recently been introduced for industrial X-ray films, but these usually demand a longer processing cycle than medical X-ray films, owing to their relatively heavier emulsion coatings.

Speed as Function of Radiation Quality

The dependence of the speed of no-screen X-ray films on the quality of radiation is discussed in detail on pages 121–128. Because the range of radiation qualities used in industrial radiography is practically unlimited, the effect of the speed change with radiations generated with increasing kilovoltages is important. It has been shown (p. 123) that the speed D/R to direct X-rays is on the average about 20–50 times greater at 40 keV than between 200–2000 keV. In the latter range the speed is practically constant, whereas it decreases from about 40 keV towards lower keV's. Experiments have shown that the film sensitivity, if referred to constant roentgen units, increases by about 25 per cent at energies from 2 to 8 MeV and remains fairly constant up to about 30 MeV. Because there is some uncertainty involved in measuring roentgen units of radiations exceeding energies of about 3 MeV, the determination of film speed has to be taken with some reservation.

Procedure of Sensitometry

The broad principles of sensitometry of X-ray films have been described on pages 101–103, using a slit sensitometer. It should be mentioned that with the very penetrating radiations used in industrial radiography, sensitometric exposures cannot be carried out through a lead slit, because the radiation would partly penetrate the lead. If it is desired to produce a characteristic curve of a film, say at 400 kV_p, this should be done by making separate exposures on different pieces or strips of film, which should then be processed together or the arrangement for intensity scale exposures, as described on page 102, can be used.

A method which does not provide a characteristic curve, but a quantitative relationship between density and log exposure, consists of using a metal step wedge, the relative thicknesses of each step have been calculated so as to give known transmitted quantities of radiation for a given quality of radiation (see also page 103). The method does not provide a characteristic curve in the true sense, because the quality of radiation striking the film varies from step to step owing to the different absorption taking place in each step. (H. E. Seemann and B. Roth, 1960, 1962; G. M. Corney and H. E. Seemann, 1947.)

10.2b   Lead Intensifying Screens

Commercially available lead intensifying screens consist of a pair of lead foils (usually with glossy surface) on a support, e.g. of thin cardboard. The lead foils usually contain a few per cent of antimony or bismuth which increase the hardness and resistance to surface abrasions. Abrasions may have a deteriorating effect on the image quality. Other heavy elements such as tantalum, platinum and gold have been proposed for the use of intensifying screens, but because of the high cost of these metals, lead is preferred. The intensifying action of metal screens is based on the photoelectric effect, i.e. they emit electrons, when excited by X- or γ-rays; they also emit characteristic lead radiation, but the main contribution to the intensification is due to the effect of electrons on the photographic emulsion. The characteristic K-radiation of Pb is excited at 87.6 keV, this means that the kinetic energy of the photoelectrons released in this process is equal

TABLE 10.1. Intensifying Factors of Lead Screens*
as a Function of the Quality of Radiation

kV_p and radiation sources	Intensifying factor
100	0.8
150	1.5
200	3.0
400	4.0–5.0
^{192}Ir	4.0–5.0
^{60}Co	2.5

* The intensifying factors are strongly influenced by filtration e.g. of the specimen and by the waveform of the high voltage supply.

to the X-ray energy (in keV) applied minus the excitation (or binding)-energy which is required to remove the electron from the K-orbit. Thus an electron released for instance at an X-ray energy of 120 keV has a kinetic energy of only $120 - 87.6 = 32.4$ keV. An electron of this energy has an average range in an X-ray film of 8 microns. The excitation voltage required for the L-radiation of lead is 15.8 keV.

Lead intensifying screens are used in a similar manner to salt intensifying screens, i.e. a no-screen X-ray film is sandwiched between a pair of lead screens. The intensifying factor, which is defined as the exposure required without screens over that with screens depends essentially on the thickness of the lead foils and on the quality of the radiation applied. Table 10.1 above shows the intensifying factors as a function of the peak kilovoltage of X-rays without filtration for a combination of lead screens of 0.006 inch thickness for the back screen and of 0.004 inch thickness for the front screen. The front screen is that screen through which the radiation enters first on its passage through a cassette.

The last mentioned sources in this table refer to the use of radioisotopes ^{192}Ir and [60]Co (see p. 433–436). It should be mentioned that the figures quoted in this table represent average values from various sources.

The intensifying factor of 0.8 at 100 kV_p (see Table 10.1) indicates that longer exposure times are required at this kilovoltage than without the use of lead screens. The reasons for this behavior are that the excitation of electrons at 100 kV_p is negligible, the K-electron energy is relatively small at this kilovoltage and the X-ray absorption by the front screen becomes more noticeable towards softer radiation. Even at 150 kV_p the intensifying factor is only 1.5, but increases to 5 at about 400 kV_p. The intensifying factors mentioned for ^{192}Ir and ^{60}Co refer to γ-rays emitted by these isotopes which give somewhat similar images in contrast to those obtained with X-rays generated at 400–600 kV_p and 1250 keV respectively. The drop of the intensifying factor using ^{60}Co radiation is apparently caused by the considerable reduction of the photoelectric and Compton attenuation coefficients for lead within this range of energy of 1.1–1.3 MeV (see Figure 1.11) and to some extent because less electrons are emitted in the forward direction (see Figure 6.4). The physical nature of the emission of electrons by lead screens as a function of quantum energy is very complex, as it is not only influenced by the attenuation coefficients of lead, but also by the range of electrons, their specific ionizing power, the stopping power of lead for electrons, the degree of homogeneity of the X-ray beam, the ratio of forward to backward emission of electrons (see Figure 6.4), the respective, thicknesses of screens and other factors.

The thicknesses of lead intensifying screens to be used generally depends on the quality of radiation applied and the following Table 10.2 gives some guiding data.

Although the intensification attainable with lead screens is not

TABLE 10.2. Suggested Thicknesses of Lead Intensifying Screens for Various Radiation Qualities

Radiation quality	Front screen (inch)	Back screen (inch)
X-rays between		
120–140 kV_p	0.001–0.002	0.006
140–300 kV_p	0.002–0.005	0.006
300–500 kV_p	0.005–0.007	0.006 (minimum)
[192]Ir & [60]Co	0.005–0.01	0.006 (minimum)
1–10 MeV	0.005–0.02	0.01–0.02

very great, it is certainly worth while to make use of it particularly as there are several great advantages connected with the application of lead screens.

(1) They do not impair the definition attainable, as long as there is an intimate contact between the screens and film, assuming that there are no scratches on the surface of the screens.

(2) *The most important advantage of lead screens is undoubtedly the fact that they improve the image quality by reducing scattered radiation reaching the film.* This is because the screens act as filters for scattered radiation, the latter being softer, on the average, than the primary radiation and are therefore more easily absorbed. This is more pronounced for the obliquely scattered radiation, which suffers more absorption due to its greater path in the lead.

(3) Lead screens also act as preferential filters for the softer components of the primary radiation, which is particularly important in specimens of wide variations in thickness (see also p. 456).

(4) Lead screens preferentially intensify the higher energy image forming radiation.

In the region of 200–400 kV_p lead or other metal screens are generally used for the purpose of reducing scattered radiation and less for the reduction of exposure time. (H. E. Seemann, 1937; D. Polansky and others, 1964.)

10.2c Screen X-ray Films

Double coated screen X-ray films and salt intensifying screens and their characteristics have been discussed in detail on pages 282 and 297. The shape of the characteristic curve of screen films in conjunction with salt screens is basically independent of the quality of radiation. With decreasing intensifying factor (see p. 301) the influence of the contribution of direct X-rays becomes greater and in these circumstances the shape of the characteristic curve will be influenced.

It has already been mentioned that relatively little use is made of salt intensifying screens in industrial radiography. If salt screens are applied, then it is usually the fine grain low speed variety of screens.

The intensifying factors of salt screens increase with higher quantum energies and a peak was found with filtered X-rays generated at about 250 kV_p. (Half value layer of 2.5 mm Cu) (R. H. Herz, 1956). Using short exposure times of only a few seconds and high quantum energies up to 250–300 kV_p combined with strong filtration, e.g. by the specimen to be radiographed, intensifying factors may reach very high values indeed. With X-rays generated at 180 kV_p and filtered by $\frac{1}{4}$ inch cu, using 5 seconds exposure time with screen film and salt screens (medium calcium tungstate type) an intensifying factor (I.F.) of 550 was measured.

The apparent advantage of this rather high intensifying factor (I.F.) is however upset in practice by the following facts. (1) In industrial radiography exposure times are on the average of the order of 1–10 minutes. In this range of exposure time reciprocity failure with screen light exposures occurs which tends to reduce the I.F. (2) The high I.F. must in some way be considered as an illusion, because it is based on a rather unrealistic definition, which refers to exposures using screen X-ray films with and without screens. Screen films exposed without screens have in general low speed and low gradient if compared with no-screen film of similar grain size. A more realistic approach would therefore be a measurement of the exposure required for no-screen film with lead screens over that of a screen film with salt screens and this would lead to lower I.F.'s than mentioned before. Furthermore the high gradient of a no-screen film at high densities is so advantageous that a screen film with screens cannot compete with the inherent contrast (and definition) revealed by a no-screen film with lead screens. An example may illustrate this:

(a) Assuming a 1 inch steel weld is radiographed with fine grained salt screens and screen film at 140 kV$_p$ and 10 mA at 36 inches focus to film distance (F.F.D.), the exposure time required would be 5 seconds.

(b) A fine grain no-screen film exposure with lead intensifying screens using the same weld also taken at 140 kV$_p$, 10 mA would require 10 minutes, but with considerably improved definition than in exposure (a). In practice such a low kilovoltage would not be chosen for a 1 inch weld. In order to obtain a radiograph of similar contrast to that in (a) the exposure would preferably be made with 200 kV$_p$, using the same fine grain no-screen film in conjunction with lead screens.

(c) A 200 kV$_p$, 10 mA exposure of the same weld would require 100 seconds exposure time, using again fine grain no-screen film in conjunction with lead screens.

Thus it is seen that the ratio of exposure time of case (c) to (a) would be 20 but the improved image quality of (c) would be preferable in spite of the increase in exposure time.

10.3 RADIOGRAPHIC CONTRAST AND DEFINITION

The industrial radiologist often arranges his exposure technique so that the resulting radiograph of a specimen conforms within a certain level of 'image quality sensitivity'. By sensitivity is meant the minimum thickness variation of the specimen which is visible on the radiograph and which is expressed as a percentage of the total thickness of the specimen to be radiographed. A discussion of some of the principal factors influencing this sensitivity forms the content of this section.

In order to deal with the problem described we have to study the factors which allow the smallest possible variation of the thickness of a specimen to be detected. This will depend on the contrast and the definition attainable in the radiograph. Changes in radiographic contrast and definition are influenced basically by the same factors as those described in the previous chapter on medical radiography (see p. 313). The fact however that very much higher quantum energies are involved in industrial than in medical radiography and that salt screens are rarely used, that movement is usually absent and other variations make it necessary to reconsider the problems involved. The latter arguments include for instance that short exposure times, although important for economical reasons, are not the guiding factor, as image quality has usually the first priority.

10.3a Contrast

We have seen in Chapter 7 that the ray or subject contrast (see p. 268) in radiography is chiefly influenced by the quality of radiation and the radiographic contrast by the gradient (or inherent contrast) of the film. In medical radiography the gradient of the film is generally kept constant and the most powerful tool in controlling the contrast is undoubtedly the kilovoltage at which the radiation is generated.

In industrial radiography both kilovoltage and gradient are of importance, the latter particularly at high densities. The influence of contrast due to changes in kilovoltage is less striking in the higher ranges of kilovoltage, than in the lower kV range. This is due to the relatively small changes of absorption coefficients from approximately 250 keV upwards if compared with the rather extensive changes from about 250 keV downwards. Hence one utilizes in industrial radiography the high gradients of the variety of commercially available no-screen X-ray films at high densities in connection with other advantages which the finer grained films offer.

10.3b Effect of Scatter. (Build-up Factor)

A factor which plays a very important part in influencing 'ray' or 'subject' contrast is the formation of scattered radiation and in particular the ratio of the intensity of scattered to direct radiation

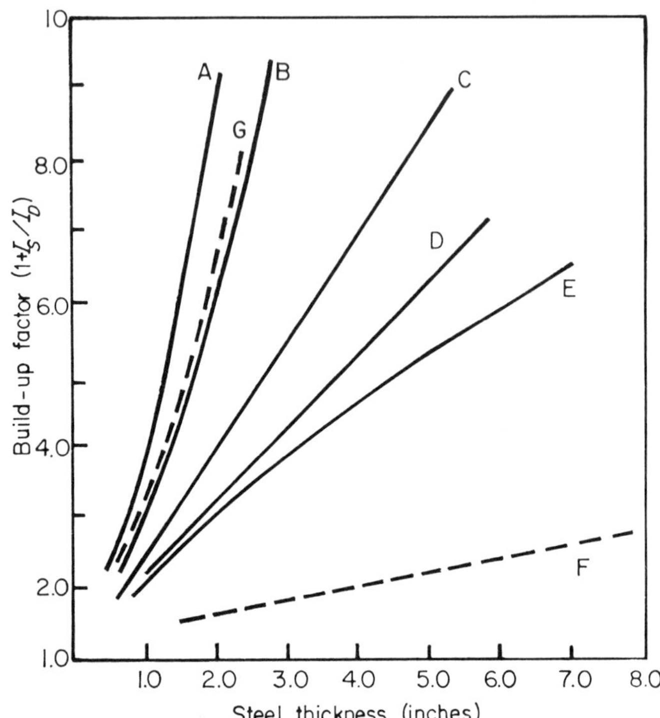

Figure 10.2. Experimental values of build-up factor for steel using different X-ray energies.
A 200 keV   B 400 keV   C 1000 keV   D 2000 keV   E 5 MeV   *F 15 MeV   *G 300 keV (* These curves 'F' and 'G' are due to H. Möller and H. Weeber, Forschungsber. Landes Nordrhein-Westfalen Report No. 1305 (1963).) (*After* R. Halmshaw, (1966). *Physics of Industrial Radiology*. Heywood Books, London, Elsevier, New York.)

which changes with radiation energy and other factors. The relative contribution of the various components of scattered radiation in different materials as a function of kilovoltage is shown in Figure 1.9–1.11 pages 25–26. Furthermore the contribution of scattered radiation depends on the width of the beam that strikes the specimen and on the thickness of the latter. Hence we distinguish between narrow beam and broad beam geometry. In the narrow beam geometry the beam is collimated through a narrow hole in a block of lead with the effect that the influence of scatter in the beam is at a minimum (see Figure 4.12). In the broad beam geometry scattered radiation from all directions within the specimen enters into the beam and the contribution of scatter will be the greater the wider the beam and the thicker the specimen, and changes in a more complicated fashion with the energy of radiation and the material concerned. A considerable amount of work has been carried out to establish the contribution of scatter with respect to the primary beam as a function of the diameter of the beam, of the type of material and its thickness. (W. Franz, 1935; R. Meakin, 1957; R. Halmshaw, 1966.)

The ratio of the total radiation intensity (I_T) i.e. the sum of the direct (I_D) and the scattered intensity (I_s) over the direct intensity (I_D) is referred to the build-up factor and this plays an important part in the formation of image contrast. The build-up factor is thus defined by

$$\frac{I_D + I_s}{I_D} = 1 + I_s/I_D \tag{10.1}$$

In Figures 10.2 and 10.3 due to R. Halmshaw (1966) the relationship between build-up factors and thickness of steel is shown for various energies of X-rays (Figure 10.2) and also for γ-rays emitted by various radioactive sources (Figure 10.3).

The very high build-up factors indicate that the radiation reaching the film is mainly composed of non-image forming scattered radiation. It is shown in Figure 10.2 that the build-up factor for a given steel thickness decreases with X-ray energy from 200 keV to 15 MeV and similar conclusions can be drawn from the results of Figure 10.3 with respect to exposures with γ-rays. (R. Meakin, 1957; R. Halmshaw, 1966.)

10.3c Definition

There are four main types of unsharpness to be considered in industrial radiography
    (1) Unsharpness caused by film U_f
    (2) Unsharpness caused by radiation U_r
    (3) Unsharpness caused by geometry U_g
    (4) Unsharpness caused by screen U_s
Unsharpness due to movement of object relative to film can be neglected, as this is very rare in industrial radiography.

(1) *Unsharpness Caused by Film (U_f)*

It is customary to consider the unsharpness U_f due to the characteristics of the film. On this basis it is commonplace to accept that the finer the grain of the film the better the image definition. In recent years H. R. Splettstosser (1967) and others have suggested that the statistical aspect of this problem is more important, since the grains can be regarded as counters of photons. This aspect will be discussed

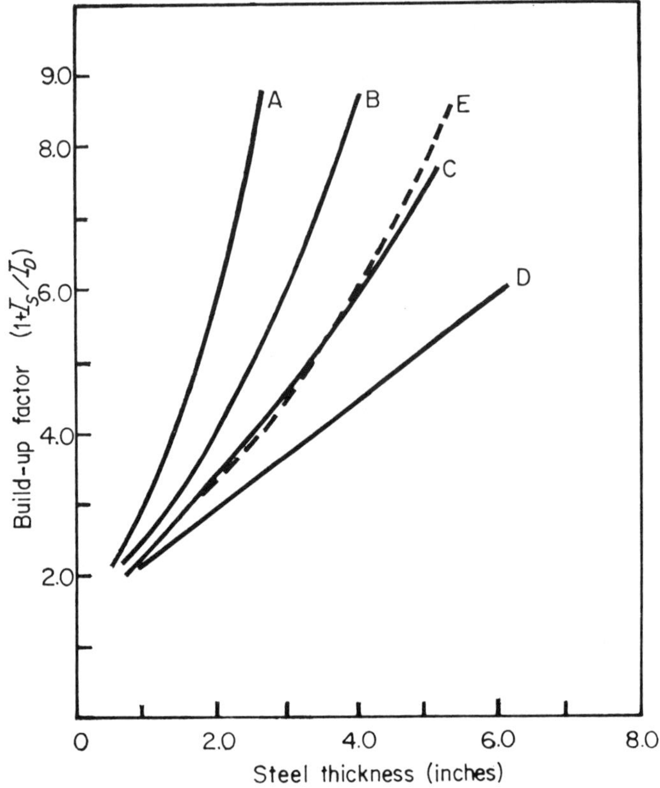

Figure 10.3. Experimental values of build-up factors for steel using γ-rays.
A. Iridium-192; B. Cesium-137; C. Cobalt-60. D. Radium (radon) radiation.
E. Cobalt-60.
(The curve E is due to H. Möller and H. Weeber. Forschungsber. Landes Nord-
rhein-Westfalen, Report No. 1305, 1963.) (*After* R. Halmshaw, (1966) *Physics of
Industrial Radiology*. Heywood, London; Elsevier, New York.)

more fully on page 420. At this point we will consider the commonly
accepted concept of film unsharpness.

Film unsharpness U_f at a given quality of radiation depends on the
graininess revealed by the film and this is a function of the amplification, i.e. on the film speed required. Considering the recording of
minute detail, such as hair-line cracks, these can undoubtedly be
detected more easily with a finer grained than with a coarse grained
film.

(2) *Unsharpness Caused by Radiation* (U_r)

The effect of change of graininess (mottle) in X-ray film with X-rays
of increasing quantum energy is a phenomenon which can be very

striking indeed. It is basically caused by quantum fluctuations and will be revealed in any film by increasing the quantum energy of the incident and absorbed radiation (see p. 138). Furthermore there is an additional effect which is superimposed and intimately related to the effect just described. This is due to the fact that with higher quantum energies electrons of increasing path lengths are released and affect adjacent grains at image boundaries, thus contributing to the unsharpness U_r. R. Halmshaw (1966) has made a detailed study of the unsharpness due to this effect which he calls film-unsharpness.* He investigated U_r in fractions of millimeters which he obtained from microdensitometer tracings made on enlarged

TABLE 10.3. Geometrical Unsharpness as Experienced under Practical Conditions

Radiation	Focal spot or source diameter in mm	Defect-film distance in mm	Focus or source-film distance in mm	U_g in mm
80 kVp X-rays	2	5	500	0.02
250 kVp X-rays	5	37	1000	0.185
400 kVp X-rays	8	75	1000	0.60
^{60}Co γ-rays	4	100	1000	0.40

images of the original radiographs of an edge taken on fine grained film. Figure 10.4 shows U_r as a function of X-ray energy for a fine grained film.

Figure 10.4(a) illustrates the growth of U_r in the MeV region and Figure 10.4(b) the initial region of curve (a) on an enlarged scale. It is seen from this graph that the blur becomes very appreciable with high energy X- and γ-rays. Precise measurements of this kind require the use of extremely small slit widths in the photometer or micro-densitometer and Halmshaw found that the equivalent slit width he used was 0.002 mm. This he calculated by referring back to the original radiographic film. (R. Halmshaw, 1955a, 1964, and 1966.)

(3) *Unsharpness Caused by Geometry* (U_g)

This is the unsharpness due to the finite size of the focal spot or size of the radioactive source and the source-object-film distances. The resulting penumbra effect is discussed in some detail on pages 261–266. A few practical examples of U_g as they arise from typical industrial radiographic exposures are quoted in Table 10.3 above.

* As the author is of the opinion that the effect measured by Halmshaw is caused by 'radiation' and not by the film, it is given the abbreviation U_r, i.e. unsharpness due to radiation.

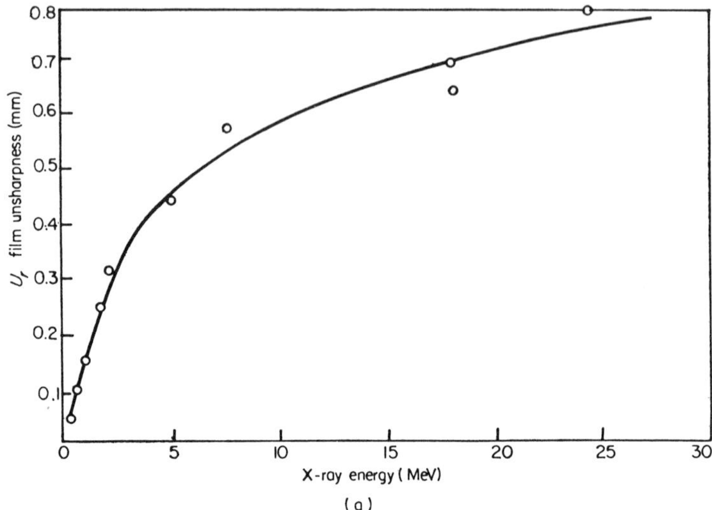

Figure 10.4. (a) Experimental values of unsharpness due to radiation as a function of X-ray energy on a fine grain X-ray film. (See footnote p. 413.)

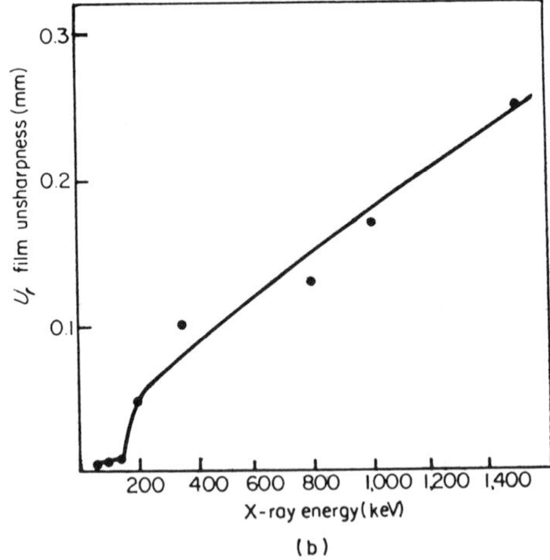

(b) Experimental values of unsharpness due to radiation as a function of X-ray energy on a fine grain X-ray film. (Lower energy region of Figure (a).) (See footnote p. 413.) (*After* R. Halmshaw (1966). *Physics of Industrial Radiology.* Heywood, London. Elsevier, New York.)

The table shows that the diameter of focal spots of X-ray tubes may be as high as 5 and 8 mm in diameter for X-ray units working in the region of 200–500 kV$_p$ and that the unsharpness values due to geometry at high energies may exceed those of the unsharpness due to radiation. The closer the defect may be brought to the film, the greater the focus- or source to film distance and the smaller the focal spot or source, the smaller will be U_g. Whenever practicable these factors should be considered in order to reduce U_g, but a compromise has often to be made, because of the long exposures required with long focus to film distances ('FFD's') and the restricted loading capacities of X-ray tubes.

(4) *Unsharpness Caused by Screen* (U_s)

Salt Intensifying Screens.   Data on the unsharpness of salt screens have been published by several authors and the average values of U_s for High Definition and High Speed screens have been quoted to lie between about 0.2 and 0.5 mm respectively. (H. S. Tasker, 1945; H. A. Klasens, 1946; A. Nemet, W. F. Cox and G. B. Walker, 1946; R. H. Morgan, 1949; R. Halmshaw, 1966.)

The screen unsharpness values quoted above include the unsharpness due to film and radiation, but as seen from Figure 10.4(b), the latter is only a very small fraction of U_s in the region of 100–200 keV. On page 345 screen resolution values in cycles per mm are given which were derived by means of modulation transfer functions. See also p. 349 on the relationship between unsharpness and resolving power.

Although theoretically it may be expected that salt screen unsharpness should increase towards higher quantum energies due to the increasing path lengths of electrons released from salt screen crystals, no evidence of this has been shown, as far as the author is aware. Within a limited range of kV, K. Rossmann and G. Lubberts (1966) have found that salt screens do not reveal any changes in resolving power with increase of kV$_p$ from 50–110 kV$_p$ using 0.25 mm Cu filtration. Their investigations are based on experiments using line spread and modulation transfer functions (see p. 343). It is possible that the effect of light spread with increasing hardness of radiation is small, because of the stopping of electrons by the heavy tungsten atoms in calcium tungstate screens.

Lead Intensifying Screens.   As has been mentioned before, the application of lead intensifying screens does not impair the definition, as long as the contact between film and screens is firm and uniform over the total area of the film. Hence U_s for lead screens can be expected to be equal to U_r. The reason that lead screens do not impair definition is due to the fact that electrons are emitted almost entirely from very near the surface of the lead foils, as those released in the interior are stopped by the lead itself. Using very high quantum energies the kinetic energies of electrons are increased so much that their ranges in the lead foil becomes considerable. Using extremely high quantum energies such as 25 MeV, a $0.040''$ front lead screen has been used to achieve maximum speed and sharpness. In this case the lead front screen was used to prevent the electrons emitted from the specimen from reaching the film.

10.3d   Quantum Mottle

The effects of quantum mottle using salt screen exposures have been discussed in some detail in Chapter 8 and the reader is referred to pages 335–342. The factors influencing this mottle in practice are mentioned on page 339. The occurrence of quantum mottle is not necessarily restricted to the use of salt screens, but may just as well be shown with or without lead screens, depending on the kilovoltage or quantum energy in use (see pp. 138 and 412).

10.4   ASSESSMENT OF IMAGE QUALITY

The considerations on contrast and definition as stated in the last section forms the basis of the calculations and control of the sensitivity of defect detection, as temporarily defined on page 409.

Measurements of the sensitivity of defect detection may be carried out by placing during exposure a thin step wedge of the same material as the specimen or some other suitable test object on top of the specimen remote from the film. In this case the sensitivity of defect detection in per cent would be defined by

$$\frac{\text{thickness of the thinnest visible layer}}{\text{total thickness of material under wedge}} \times 100\% \qquad (10.2)$$

This definition which refers to a 'thickness sensitivity' is based chiefly on the just visible contrast between the densities obtained below the specimen itself and that of the specimen plus the thinnest visible layer of the step wedge test object. If the total specimen thickness is for instance 1 inch and that of the thinnest visible superimposed step of the step wedge 0.01 inch we would speak of a 1 per cent thickness sensitivity. Thus the smaller the value of sensitivity, the better the technique applied. The terms lower or higher sensitivity should not be used, as they might be misleading. This is because a higher value of sensitivity is indicative of a worsening of the technique providing the test wedge is the same.

The step wedge placed on the specimen is frequently called a 'penetrameter' or a 'step penetrameter'. In recent years this term has been replaced by one with a wider meaning namely that of an 'image quality indicator' or abbreviated IQI and one refers also to the IQI sensitivity. An IQI should not be limited to an assessment of thickness sensitivity only, but should take care of detail sensitivity as well. Several different types of IQI's serving this purpose are available. Some of these consist of wires of various diameters or of plates or step wedges having holes of different diameters. Details of their design will be discussed in a later section.

Hence the definition of the sensitivity of defect detection as expressed in the beginning of this section should be formulated in a wider sense, e.g. by the smallest visible diameter of wire, or depth and diameter of hole in a plate as seen in a radiograph with respect to the total thickness of the specimen.

10.4a. Theoretical Approach

As the sensitivity of defect detection refers to a contrast and detail sensitivity, a theoretical approach should take care of both aspects. First however we want to discuss the thickness sensitivity or contrast sensitivity.

Thickness Sensitivity

The prediction or calculation of the 'thickness' or contrast' sensitivity is fairly straightforward providing that the visibility of the defect is little influenced by unsharpness. The radiographic perception of a thin layer of not too small an object made of the same material as the specimen and placed on top of it remote from the film can be assessed as follows.

As shown on page 341 the equation of the characteristic curve is

$$D = G \log_{10} E + C \tag{10.3}$$

where D is the density, G is the gradient which depends on E, E is the exposure (it) and C is a constant. For a very small change in density ΔD (which is here assumed to be the smallest detectable density difference)

$$\Delta D = 0.43 G \frac{\Delta E}{E} \tag{10.4}$$

and if t is constant

$$\Delta D = 0.43 G_D \frac{\Delta i}{i} \tag{10.5}$$

where G_D is the gradient at density D.

Although ΔD is assumed to be the smallest visible density difference it must be admitted that the many factors influencing the eye's visibility have not been considered here. It is however assumed that the viewing conditions are at an optimum.

We know further from page 18 equation (1.15) that

$$i = i_0 e^{-\mu x} \tag{10.6}$$

where i is the intensity, μ the linear absorption coefficient and x the thickness of the material concerned. Thus for the smallest thickness variations Δx detectable

$$\Delta x = - \frac{\Delta i}{\mu i} \tag{10.7}$$

Combining the two equations (10.5) and (10.7), we find

$$\Delta x = - \frac{2.3\,\Delta D}{\mu G_D} \tag{10.8}$$

This equation refers to a narrow beam attenuation excluding the effects of scatter. If we want to take care of scatter and if we assume that its effect produces a uniform film density we have to make use of the build-up factor $1 + I_s/I_D$. (See equation (10.1) page 411.) The value of Δx can then be expressed by

$$\Delta x = - \frac{2.3\,\Delta D}{\mu G_D} (1 + I_s/I_D) \tag{10.9}$$

If the sensitivity of defect detection (s) is defined by

$$s = \frac{\Delta x}{x} \times 100\% $$

then

$$s = - \frac{2.3\,\Delta D}{\mu G_D x} (1 + I_s/I_D) \times 100\% \tag{10.10}$$

Similar equations have been derived by R. Meakin (1957). R. Halmshaw (1966) has calculated the percentage thickness sensitivity by placing feeler gauges on steel plates using the above formula (equation (10.10)) for various radiation qualities, film densities and thickness of steel and iron and he found in general excellent agreement

between experimental and calculated values. A few of Halmshaw's
sensitivity data are quoted below. (See Table 10.4.)

Halmshaw has also used equation (10.10) to calculate the minimum
detectable thickness for a wide range of X-ray energies and he found,
in agreement with practical experience, that for 1–2 inches of steel
the sensitivity became worse, i.e. towards higher percentage sensitivity
values with increasing X-ray energy. On the other hand he found
that the reverse was the case for 4″ steel thickness using X-rays in

TABLE 10.4. Thickness Sensitivities using Feeler Gauges on Plates (Comparison of experimental and calculated values, using fine grained film (after R. Halmshaw, 1966))

Radiation	Specimen thickness	Source-film distance inches	Film density	Thickness sensitivity (percent) (observed)	(calculated)
^{60}Co γ-rays	1″ steel	10	0.9	1.4	1.45
^{60}Co γ-rays	2″ steel	20	1.7	0.6	0.6
^{192}Ir γ-rays	$\frac{1}{2}$″ steel	20	2.2	0.8	0.8
^{192}Ir γ-rays	3″ steel	20	1.9	0.6	0.66
4 MeV X-rays	2″ Fe	60	1.3	0.7	0.7
4 MeV X-rays	12″ Fe	60	1.3	0.6	0.51
210 kV X-rays	1″ Fe	36	2.2	0.2	0.22
140 kV X-rays	$\frac{1}{2}$″ Fe	36	1.9	0.3	0.33

the megavoltage range of energy where again better thickness
sensitivities were obtained.

Detail and Total Sensitivity

So far we have dealt with a theoretical and experimental approach
to the thickness sensitivity only. If the aspects of 'detail', i.e. the
definition, should be included in such an approach, the problem
becomes considerably more complex. In the total sensitivity, i.e.
the one combining thickness and detail, not only variations of thickness, but also those of the type, shape and size of detail (wires, holes
or cavities) represented by the IQI have to be considered.

Several attempts to solve this problem have been made by E. A.
Burrill (1952) and R. Halmshaw (1966). Burrill derived a more
general formula which apparently applies to any shape of detail and
is basically an extension of the equation (10.10) for contrast or thickness sensitivity making use of the Klasen's unsharpness term of

$\sqrt[3]{(U_g{}^3 + U_f{}^3)}$ (see p. 333) as seen from the following equation

$$S = \frac{\Delta x}{x} = \frac{2.3\alpha}{Gc}\left(\frac{1}{\mu x} + \frac{k}{\mu}\right) \times (\sqrt[3]{(U_g{}^3 + U_f{}^3 + U_r{}^3)})100\% \quad (10.11)$$

Film and Radiation   Unsharpness
observer factors   factors
factor

where

$$\Delta x = \text{thickness change of specimen}$$
$$x = \text{thickness of specimen}$$
$$\alpha = \text{minimum perceptible density difference } 0.01$$
$$G = \text{film gradient}$$
$$c = \text{limit of resolution of eye } (0.003 \text{ inch at } 24 \text{ inches})$$
$$\mu = \text{linear absorption coefficient (inch}^{-1})$$
$$k = \text{scattering factor (ratio scattered/direct radiation)}$$

The equation consists of three main factors as indicated below Burrill's equation.

The approach by Halmshaw and his coworkers deals mainly with a calculation of sensitivity using wires and drilled holes and the reader who is interested in this subject is referred to the detailed and excellent review by this author (R. Halmshaw, 1966).

Statistics of Radiographic Sensitivity

Radiographic sensitivity is strongly influenced by quantum fluctuations and with this consideration in mind E. T. Larson and H. F. Nitka (1961) and H. R. Splettstosser (1967) have derived IQI sensitivity by using a statistical argument together with experimental studies.

This may be considered by the following arguments. The photographic density (D) measured within a small area, for instance the area of a hole in a penetrameter (or IQI), is subject to variations arising from fluctuations in the number of X-ray photons. Assuming that we are working within the region of the density-exposure curve in which D is proportional to exposure E, we have

$$D = kE \qquad (10.12)$$

where k is a constant depending on the speed of the film (see p. 117). If the number of photons causing the image is N in the area A, then $E = N/A$ and

$$D = kN/A \qquad (10.13)$$

Now the fluctuation in N may be put equal to the standard deviation of N, namely $\sqrt{N}$. Hence for the standard deviation of the density $\sigma(D)$ arising from fluctuation in the photons we have

$$\sigma(D) = \frac{k\sigma(N)}{A} = \frac{k\sqrt{N}}{A} \tag{10.14}$$

This represents the noisy background against which we have to detect a signal.

Suppose that the number of photons per unit area in the region near the image is N_b/A, then the signal to be detected is the difference between the image density kN/A minus the background density kN_b/A, i.e.

$$\text{Signal} = \frac{k}{A}(N - N_b) \tag{10.15}$$

Hence

$$\frac{\text{Signal}}{\text{Noise}} = \frac{k(N - N_b)}{A\sigma(D)} = \frac{N - N_b}{\sqrt{N_b}} \tag{10.16}$$

We can therefore write

$$\frac{\text{Signal}}{\text{Noise}} = \sqrt{N_b}\left(\frac{N}{N_b} - 1\right) = \sqrt{AE_b}\left(\frac{E}{E_b} - 1\right) \tag{10.17}$$

This shows that for *a given subject contrast* (expressed by the factor $((N/N_b) - 1)$), which is solely dependent on the object to be radiographed, the signal/noise is larger, the larger N_b, that is the slower the film speed used in making the radiograph.

An example shows how the information content of a radiograph depends on the exposure. Consider a hole of an image quality indicator, which is penetrated by 101 photons and the surrounding area by 100 photons, i.e. a signal contrast of only 1 per cent. The signal that is to be made visible over that of the background is thus 1 photon. As the standard deviation of the background is $\sqrt{100} = 10$, it would be impossible to detect the signal in any one sampling of the two areas, because the variation in the background exposure is much greater than the signal.

The following table shows how the signal to noise ratio increases with the magnitude of exposure for constant subject contrast.

A. Rose (1946, 1948), see page 296, has shown that a ratio of signal to noise of about 5 is required for a just visible image, given reasonable viewing conditions. From Table 10.5 (p. 422) it follows that an image may just be made visible with a background (N_b) of 300,000, as the signal/noise ratio is then 5.47.

If this should occur for instance at a density of $D = 3$, a finer grained film may record a just visible image at a density of $D = 1$,

TABLE 10.5. Change of Ratio of Signal to Noise with Increasing Magnitude of Exposure

Exposure units	Background (N_b)	Signal $(N - N_b)$	Noise $(\sqrt{N_b})$	$\dfrac{\text{Signal}}{\text{Noise}}$
1	100	1	10	0.1
4	400	4	20	0.2
16	1,600	16	40	0.4
100	10,000	100	100	1.0
1,000	100,000	1,000	316.22	3.16
3,000	300,000	3,000	547.72	5.47
10,000	1,000,000	10,000	1,000	10.0

if the same number of photon interactions take place in the same area. Using this slower film an even smaller hole should be visible for a higher density. *Thus a slow film, i.e. one that counts many photons yields higher radiographic sensitivity than a fast film requiring fewer photons.*

Detailed quantitative experiments have been carried out especially by H. R. Splettstosser (1967), using circular holes of various diameters and depths in a lead sheet. Good agreement has been shown by him between theory and experiment using films of different speeds, which were exposed to low and high quantum energies, the latter producing nearly maximum graininess.

10.4b   Image Quality Indicators. (IQI's)

Design

An image quality indicator should be small enough to be placed on a region of a specimen without interfering with important parts of the specimen to be radiographed. It should be sensitive to relatively small changes in technique such as kilovoltage, focus-to-film-to-object distances, density level of the film, film type etc. and it should provide an easy and unambiguous method of reading the percentage sensitivity on the radiographic image. Furthermore it should be available for a wide range of specimen thicknesses and material.

Amongst the many constructions of IQI's proposed in the past, there are essentially three basic designs in present use. These are

1. The single thickness plaque type with holes.
2. The wire type.
3. The step/hole type.

Type (1) is almost universally used in U.S.A., where it is called a penetrameter. (The term IQI is now more generally applied in Europe.)

Types (2) and (3) are recommended by the International Institute of Welding (IIW) and are based on German and French designs respectively.

1. *The American ASTM plaque penetrameter.* The ASTM (American Society for Testing Material) penetrameter consists of a plaque of uniform thickness in which three holes are drilled of different diameters. The material of the plaque is the same as that of the specimen to be radiographed (see Figure 10.5). If the thickness of the plaque is T, the hole diameters are T, $2T$, and $4T$. Some other

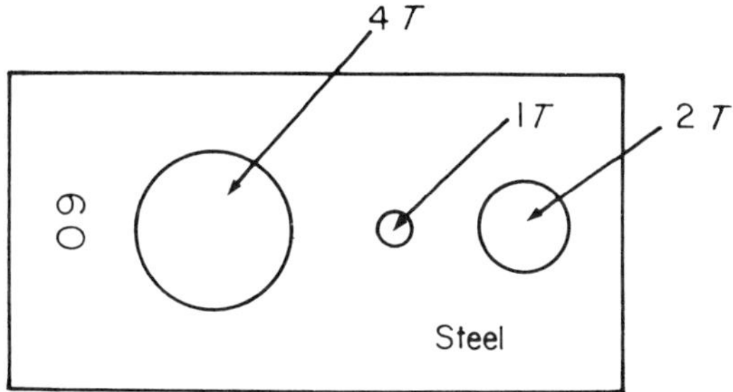

Figure 10.5. American plaque type penetrameter.

American codes specify also $3T$ and other modifications of the same general principle of penetrameter. The thinnest plaque that is specified by the ASTM code is 0.005 inch, and has holes of 0.01 inch diameter.

If for instance a 2 per cent sensitivity is required, a plaque representing 2 per cent of the specimen thickness is used and if the $2T$ hole specified is visible on the radiograph, this then represents a 2 per cent sensitivity. This method assumes however that a wide range of penetrameters is available so that a plaque representing 2 per cent of the specimen thickness can be selected. The ASTM code E 142 calls for a rather similar design of IQI which is used to indicate quality levels in radiography. If the diameters of the holes are $1T$, $2T$, and $4T$ and if all holes are visible, the sensitivity is 1.4 per cent; if only the two larger holes are visible, the sensitivity is 2 per cent and if only the largest hole is visible, the sensitivity is 2.8 per cent.

2. *Wire type.* The wire type IQI (British Standard 3971:1966) which is similar to the German Standard DIN 54109:1962 consists

TABLE 10.6. Wire Diameters

Wire No.	Diameter mm	Wire No.	Diameter mm	Wire No.	Diameter mm
1	0.032	8	0.160	15	0.80
2	0.040	9	0.200	16	1.00
3	0.050	10	0.250	17	1.25
4	0.063	11	0.320	18	1.60
5	0.080	12	0.400	19	2.00
6	0.100	13	0.500	20	2.50
7	0.125	14	0.630	21	3.20

of straight wires having a length of 30 mm and diameters as quoted in the Table 10.6 above and are spaced at intervals of 5 mm. The wires should be made of the same material as the specimen (see section on material of IQI's).

The wires are mounted between two thin transparent sheets of polyethylene and standard models of seven consecutive wires from Table 10.6 are used (see Figure 10.6). The range of wires quoted in the table above is suitable for specimens from about 3–250 mm thickness.

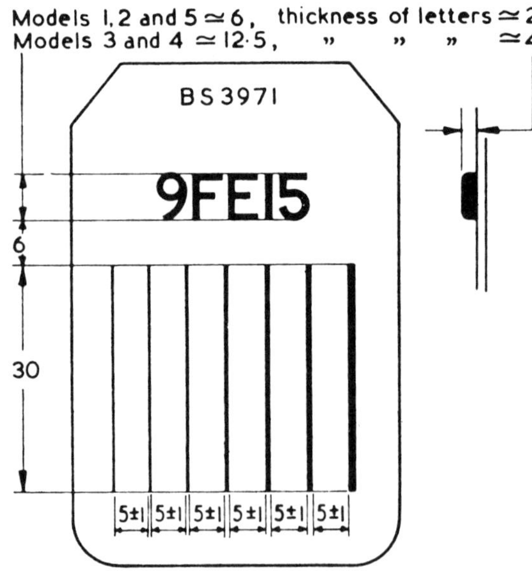

Figure 10.6. Sketch of wire type image quality indicator. (All dimensions are given in millimeters.) (*After* British Standard 3971: 1966.)

Furthermore it is essential that the wire IQI is marked with lead numbers indicating the thinnest and thickest wires and the material of which the wires are made. As a measure of image quality, sensitivity is usually expressed by

$$S_{(wire)} = \frac{\text{diameter of smallest visible wire on the radiograph}}{\text{specimen thickness under IQI}} \times 100\%$$

It should be mentioned that the image of the thinnest visible wire might not appear as a continuous line in the radiograph and only a part of the total length of the wire may be detected. Hence certain rulings with regard to the percentage length that should be detected may be adopted for a consistent judgement.

3. *Step/hole type.* The step/hole type IQI British Standard 3971:1966 which is similar to the French AFNOR (Association Française de Normalisation) NF A-04 304 pattern, consists of a series of uniform thickness metal plaques each containing one or two holes drilled through the full thickness and at right angles to the upper surface. The diameter of the hole is equal to the thickness of the plaque. The step thicknesses and diameters of the drilled holes are quoted in the Table 10.7 below.

For the steps 1–8 two holes are drilled and for the steps 9–18 one hole is drilled in the manner shown in Figure 10.7. The plaques

TABLE 10.7. Hole and Step Dimensions.*

Step No.	Diameter and step thickness mm	Step No.	Diameter and step thickness mm
1	0.125	10	1.00
2	0.160	11	1.25
3	0.200	12	1.60
4	0.250	13	2.00
5	0.320	14	2.50
6	0.400	15	3.20
7	0.500	16	4.00
8	0.630	17	5.00
9	0.800	18	6.30

* Extracts from British Standard 3971:1966 are reproduced by permission of The British Standards Institution, 2, Park Street, London, W.1 from whom copies of the complete standard may be obtained.

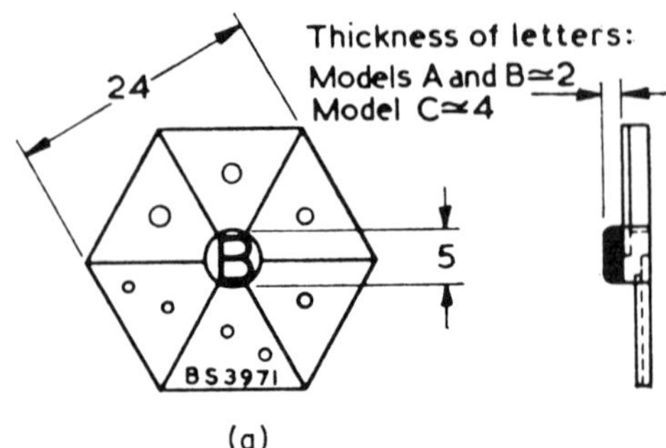

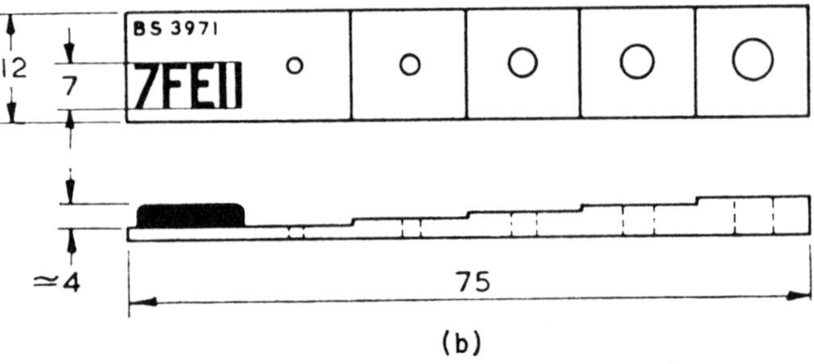

Figure 10.7. Sketches of step/hole type image quality indicators. (All dimensions
are given in millimeters.) (*After* British Standard 3971: 1966.)

should be made of the same basic material as the specimen to be
radiographed (see section on material of I.Q.I.'s). The individual
steps may be rectangular having side lengths of 12.5 mm. The plaque
may also take the shape of a hexagon having triangular shaped steps
as is shown in Figure 10.7. The step/hole IQI should be marked with
lead numbers and figures indicating the model in use.

The sensitivity is judged by the visibility of the image of the holes
and is defined by

$$S_{(step/hole)} = \frac{\text{diameter of smallest visible hole on the radiograph}}{\text{specimen thickness}} \times 100\%$$

The thicknesses quoted in the table above can be used with specimens
from about 3–250 mm thickness.

The British Standard 3971:1966 gives for 'guidance only' some
values of acceptable sensitivity for steel specimens (see Table 10.8).

Table 10.8A refers to X-ray exposures using a fine grained X-ray film and Table B refers to γ-ray exposures. It should be noted again that a smaller numerical value represents a better sensitivity. For instance if a step/hole sensitivity of 5.1 per cent is obtained with an $\frac{1}{8}$th″ thick steel specimen, this can be regarded as an adequate technique. On the other hand a 0.7 per cent sensitivity, using a wire type IQI, should be expected with a specimen of 6″ thickness. Both cases refer to X-ray exposures.

TABLE 10.8. IQI Sensitivities (From British Standard 3971: 1966)

A. Specimen thickness		IQI sensitivity per cent		B. Specimen thickness		IQI sensitivity per cent	
		Wire type	Step/hole type			Wire type	Step/hole type
mm	inch			mm	inch		
3	$\frac{1}{8}$	2.4	5.1	3	$\frac{1}{8}$	—	—
6	$\frac{1}{4}$	1.6	3.6	6	$\frac{1}{4}$	—	—
12.5	$\frac{1}{2}$	1.4	3.0	12.5	$\frac{1}{2}$	2.4	4.6
25	1	1.2	2.5	25	1	1.7	3.0
40	$1\frac{1}{2}$	1.1	2.1	40	$1\frac{1}{2}$	1.5	2.5
50	2	1.0	1.8	50	2	1.3	2.2
75	3	0.9	1.6	75	3	1.1	2.0
100	4	0.8	1.4	100	4	1.0	1.8
150	6	0.7	1.3	150	6	0.9	1.8

(Extracts from (B.S. 3971 : 1966 *Image Quality Indicator for Radiography and Recommendations for their use*) are reproduced by permission of the British Standards Institution, 2 Park Street, London W.1, from whom copies of the complete standard may be obtained.)

IQI Material

It has been mentioned that the IQI should preferably be made of the same material as the specimen. This might not be possible in all cases. With regard to ferrous materials an IQI of low carbon steel is normally adequate for use on all forms of steel and iron, except for a few very highly alloyed metals containing iron but which are not usually referred to as steels. For copper and its alloys an IQI of copper is generally regarded as adequate. For the inspection of welds in aluminum and magnesium alloys, where the composition of the weld may be different from that of the parent metal, an IQI of pure aluminum or pure magnesium respectively should be used.

Position of IQI's

The IQI should preferably be placed on top of the specimen, i.e. remote from the film, as this position leads to the poorest definition attainable owing to the greatest possible object to film distance. It is equally important to place the IQI in a position on the specimen where there is least interference with defects in the specimen. In

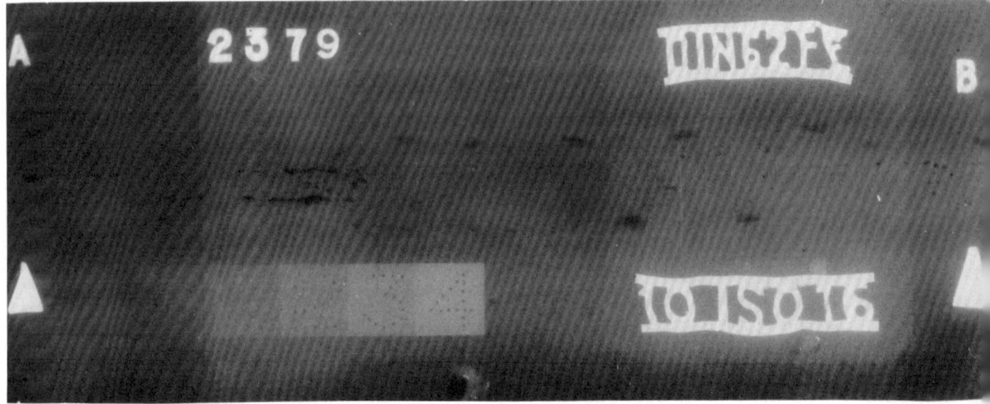

Figure 10.8. Radiograph of a $\frac{1}{2}$ inch (low carbon steel) weld showing two types of 'Image Quality Indicators' (a) wire type and (b) step/hole type.
Technique: 120 kV_p, 10 mA, 2 minutes exposure at 36 inch focus to film distance. Medium no-screen X-ray film with lead intensifying screens. (*By courtesy of* Mr. E. J. Grimwade, Kodak School of Industrial Radiography, Kodak Ltd., England.)

weld radiography the step/hole IQI should be placed close to and parallel with the weld, whereas the wire type IQI should be placed across the weld with wires situated at right angles to the axis of the weld.

Figure 10.8 shows a typical radiograph of a weld on which two different types of IQI's can be seen. In weld radiographs it is sometimes specified to place an IQI on each end of a weld; this is done because of the relative difference in projection from the focal spot of an X-ray tube towards both ends of the specimen as caused by the line type of focal spot (see p. 326).

In a specimen having wide variations of thickness it may be advantageous either to place the IQI on the thickest region of the specimen or preferably to make use of several IQI's.

IQI Sensitivity Readings

The reading of the radiographic image of IQI's is naturally subjective and the judgement of various observers in reading a radiograph showing the same IQI may differ. Furthermore the application of different types of IQI's may lead to the assessment of different sensitivity values. This is plausible as the inherent detail rendering may vary from one IQI design to another. Thus the percentage sensitivity of defect detection may become meaningless unless the type of IQI and the exact conditions of its use are specified in detail. Also the viewing conditions of the radiographs should be specified in order to obtain reproducible results. Such conditions may be for instance the brightness level of the illuminator (see p. 275), the masking of the radiograph from extraneous light and viewing in a darkened room. In general it is aimed that the radiographic image of the IQI in no-screen X-ray films is read at densities between 1.7–3, so utilizing the high gradients of this type of film.

Brightness Levels

In order to be able to read films at density levels between 1.7–3 rather high brightness levels of viewing screens of the order of 1000 candelas per square foot are required (see explanatory notes below).

If it is necessary for the eye to detect small detail at low contrast the brightness of the light reaching the eye should be of the order of about 1–10 millilamberts. Below this value the discernibility of the eye decreases. Thus at a density of 2, which transmits 1/100th of the incident light intensity, the brightness of the illuminator or viewing screen should be about 100–1000 millilamberts. At a density of 3 it should be between 1000–10,000 millilamberts which roughly corresponds to 300–3000 candelas per square foot. The British Standard 2600 : 1962 suggests for a density of 3 a brightness level of 1000 candelas per square foot.

Explanatory notes. Brightness may be expressed either in luminous intensity in candelas per square meter or by the luminous flux emitted by unit area of the surface, the unit being the lambert (L).

The lambert is the average brightness of a perfectly diffusing surface which emits or reflects 1 lumen per cm^2. (1 millilambert is the equivalent brightness of a diffusing surface emitting or reflecting 1 millilumen per cm^2.)

1 lumen is the quantity of luminous flux received on an area of 1 square foot on the surface of a sphere having a radius of 1 foot from a point source of 1 candela placed at its center.

1 candela per square foot equals 3.14 foot-lamberts.

1 millilambert equals 0.929 foot-lamberts.

1 millilambert equals 0.295 candela per square foot.

1 candela per square foot is equivalent to 10.76 candelas per square meter.

Visibility

A decisive factor in IQI readings is the ability of the eye to discern small details at low contrast. Measurements of this kind have been carried out by various authors in order to establish the minimum detectable density difference as a function of detail size, shape, brightness, and distance between eye and film. E. M. Lowry, 1931; T. Neef, 1925; A. Rose, 1946; E. S. Lamar, S. Hecht, and others, 1947; E. T. Larson and H. F. Nitka, 1961; R. Halmshaw, 1966; H. R. Splettstosser, 1967).

From the experiments of some of these authors it follows that at a brightness of about 10 millilamberts a density difference of 0.006 on a film is approximately the minimum which can be detected under the most favorable conditions. The work by Lamar and others deals with the minimum detectable density difference of images varying in total size and in their ratio of length to width. For narrow long line images it was found that the value of $D_{min} = 0.006$ still holds, whereas the value increases slightly for small spot sized images. As is to be expected the value of D_{min} increases with decreasing image area or in other words more contrast is required the smaller the image area in order to keep D_{min} at a constant value.

IQI Results and Techniques

In Table 10.8 of British Standard 3971:1966 we have seen the levels of percentage sensitivity which are regarded to be adequate for the wire and step/hole IQI's for various thicknesses of steel. It is

TABLE 10.9. Expected Percentage Sensitivity Values for Steel using Various Techniques and Wire-type IQI

K Vp	Specimen Thickness inch	mm	Type of film	Screens	Film-density	Exposure mA min.	FFD cm	Expected % sensitivity (wire type) IQI
150	½	12.5	Fine grain (medium speed)	Pb	2.0	50	100	0.7
200	1	25	Fine grain	Pb	2.0	15	100	1.0
250	1	25	Fine grain	Pb	2.5	10	100	1.2
300	2	50	Fine grain	Pb	2.5	35	100	0.6
400	3	75	Fine grain	Pb	2.0	110	80	0.9
^{60}Co	3	75	Fine grain	Pb	2.0	600*	100	0.8
^{192}Ir	2	50	Very fine grain (slow)	Pb	2.7	1200*	50	0.9

* The exposures in these cases are expressed in terms of curie-hours.

undoubtedly of interest to the reader to know the technical conditions at which such sensitivities can be achieved. A few typical values are therefore shown in Table 10.9. It should be mentioned that the technical data quoted in this table are not necessarily the only ones with which the particular sensitivities can be obtained in practice.

Summarizing this section on 'Image Quality Indicators' it may be concluded that the use of IQI's gives a good indication of the quality of the radiographic technique applied, but it does not provide any direct information on the detectability of the smallest flaws in the specimen, nor does it provide any exact information about the size of defect that can be revealed by the technique applied. An overrating of the importance of the IQI and too much reliance on its use can easily lead to misinterpretation of radiographic findings.

10.5 RADIOACTIVE SOURCES

10.5a Natural Radioactive Sources

Natural sources of γ-rays, such as radium, radon and to a lesser extent mesothorium and radiothorium were used in the very early development of industrial radiography.

Radium (^{226}Ra) having a half life of 1590 years emits a spectrum of γ-rays, the principal emission lines of which are equivalent to energies of 2.2, 1.8, 1.1, 0.6 and 0.3 MeV. Radium decays by the emission of α-particles to a heavy inert gas radon (^{222}Rn) which itself emits α-particles and decays to radium A, B, C, C'(C''), D, E, F before it becomes stable Ra G, being inactive lead.

Radon has a half life of only 3.85 days, but it provides the same γ-emission as radium. Whereas radium sources which have sufficient activity to be useful in industrial radiography are rather large in dimensions, radon sources of high activity can be concentrated into a much smaller volume. For instance a 250 millicurie (mCi) radium source being in the form of radium bromide may have dimensions of a cylinder of about 8 mm in length and 6 mm in diameter. A radon source having 6 times the activity in terms of millicuries may be concentrated in a cube of about 1 mm^3.

Although radon is a more useful source because of its high activity within a small volume, its main drawback is its short half life. The small volume makes it particularly useful from the point of view of the radiographic definition attainable.

Natural sources of radioactivity are only rarely used nowadays in industrial radiography and have been replaced by artificially made radioisotopes.

10.5b   Artificial Sources

Artificial radioisotopes are either produced by neutron-activation in an atomic reactor (atomic pile) or by extraction from fission products. Isotopes can be produced in a cyclotron as well, but the yield is usually too small to be of use for radiographic purposes. A typical reaction for the production of cobalt-60 (^{60}Co) is the (n, γ) reaction whereby a thermal neutron is captured by a nucleus such as $^{59}_{27}$Co, thus

$$^{59}_{27}\text{Co} + {}^{1}_{0}\text{n} \rightarrow {}^{60}_{27}\text{Co} + \gamma$$

$^{59}_{27}$Co is the naturally occurring most abundant cobalt metal.

A few isotopes useful in industrial radiography are extracted from fission products; the latter may be due to neutron capture in uranium-235. The most useful isotope for radiography produced in this way is cesium-137 (^{137}Cs) which forms part of used uranium fuel of a reactor.

A source which is produced by neutron capture in a reactor, such as ^{60}Co can be reactivated after its partial decay, whereas this cannot be done with a fission source.

10.5c   K-Factor

Before we review the chief physical properties of those radio-isotopes which are particularly useful in industrial radiography we must introduce the so called K-factor which is an important term referring to the specific output of various radioactive isotopes. The K-factor is the rate of exposure in roentgens per hour at 1 cm from a γ-point source of 1 millicurie activity, often abbreviated to Rh cm/mCi. The derivation of the K-factor is found from the definition of the curie (see p. 50) and from the photon flux per roentgen (see p. 613 in the Appendix).

In quoting the output of various radioisotopes it is more convenient to use a unit 10 times smaller than the K-factor, as this corresponds more to practical conditions of industrial radiography. Thus in the following we refer to roentgen per hour at 1 meter distance per curie (Rhm/Ci). The Rhm/Ci values for various radioactive isotopes are quoted in Table 10.10 of the next section.

10.5d   Properties of Sources

The requirements for a radioactive source to be useful in industrial radiography are (1) that its half life is not too short, (2) that its *specific activity*, i.e. its activity per unit volume is high, combined with a high K-factor, (3) that it emits radiation which is useful for radiography and (4) that it is readily available and that its cost is reasonable.

Some of the properties of radioactive sources are quoted in Table 10.10. Radium and radon are mentioned as well in this table for interest, although relatively little use is made of these at present. The table shows the sources in the order of decreasing energies of γ-ray emission. Of the artificial isotopes, ^{60}Co, ^{137}Cs, and ^{192}Ir are the most commonly used for industrial radiography; the use of

TABLE 10.10. Properties of Some Radioactive Sources

Source	Half life	Main energies (MeV)	K-factor ($\times 10^{-1}$) (Rhm/Ci)	Approx. range of metal thickness covered
Radium-226	1590 years	1.8, 1.1 0.6, 0.3, 0.2	0.83	2–8″ steel
Radon-222	3.85 days	1.8, 1.1 0.6, 0.3, 0.2	0.83	2–8″ steel
Cobalt-60	5.3 years	1.33, 1.17	1.3	2–8″ steel
Iridium-192	74.4 days	0.47, 0.32 0.31, 0.296	0.5	$\frac{1}{4}$–$3\frac{1}{2}$″ steel
Caesium-137	33 years	0.662	0.37	1–4″ steel
Thulium-170	127 days	0.084 0.052	0.02*	up to 1″ aluminum or $\frac{1}{2}$″ steel

* The K-factor ($\times$ 0.1) of ^{170}Tm is only approximate, as the data on this differ in literature, owing to the fact that the output is in this case a function of the dimensions of the source due to Bremsstrahlung absorption. (See p. 436.)

^{170}Tm is rare for the reasons given below. The table quotes the sources, their half lives, their main emission lines in MeV, their radioactive output (K-factor $\times$ 0.1) in Rhm/Ci and finally the approximate range of metal thicknesses which they cover radiographically.

A few specific remarks about the artificial isotopes mentioned in the table above and about their mode of decay may be of interest to the reader.

Cobalt-60. The decay scheme of ^{60}Co is shown in Figure 10.9 below. The oblique arrow indicates that the ^{60}Co nucleus emits a β-particle of 0.31 MeV energy whereby the nucleus is transformed into an excited state of ^{60}Ni, which releases two γ-quanta in cascade, as

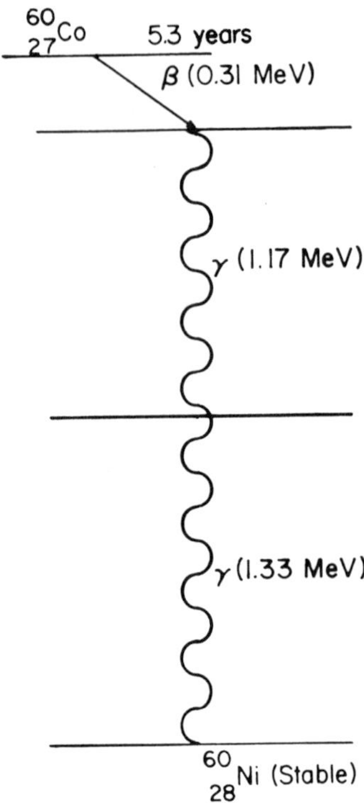

Figure 10.9. Decay Scheme of ^{60}Co.

indicated by the wavy lines. These two quanta have energies of 1.17 MeV and 1.33 MeV. On returning to the ground state the nucleus becomes the stable isotope ^{60}Ni. A typical decay curve for ^{60}Co is shown in the graph Figure 10.10 p. 435 in which the activity is plotted on the ordinate and the decay time (in months) on the abscissa axis.

Sources of a few curies of ^{60}Co are being used in many industrial laboratories. More recently ^{60}Co sources of a few thousand curies have been used for radiography as alternative sources for linear accelerators and betatrons, as ^{60}Co sources are considerably cheaper than the latter. The very high energies of radiation between 10–100 MeV emitted by some linear accelerators and betatrons cannot be reached by radioactive sources. Small sources of e.g. 1–25 curies of ^{60}Co are supplied in the United Kingdom in stainless steel capsules of small dimension such as 2×2 and 4×4 mm depending on the activity. (see p. 437).

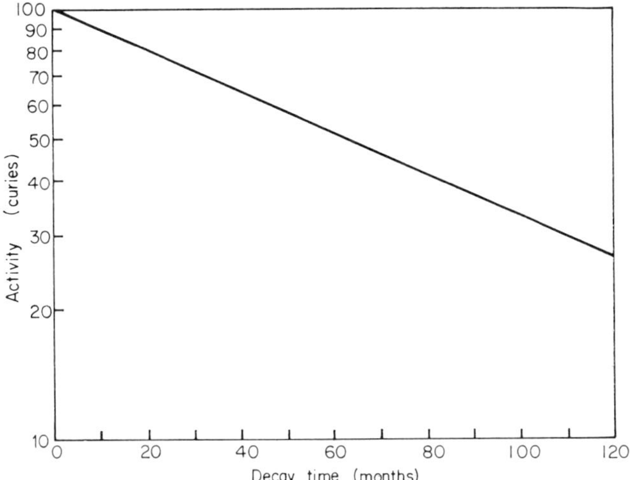

Figure 10.10. Decay curve of ^{60}Co. (Activity as a function of decay time in months.)

Cesium-137. (The British spelling of Cs is caesium, the American spelling is cesium.) As mentioned before ^{137}Cs is a fission product. Its decay scheme is shown below in Figure 10.11 and this is more complex than that of ^{60}Co. As seen in the diagram 8 per cent of the electrons emitted have energies of 1.17 MeV and 92 per cent 0.51 MeV.

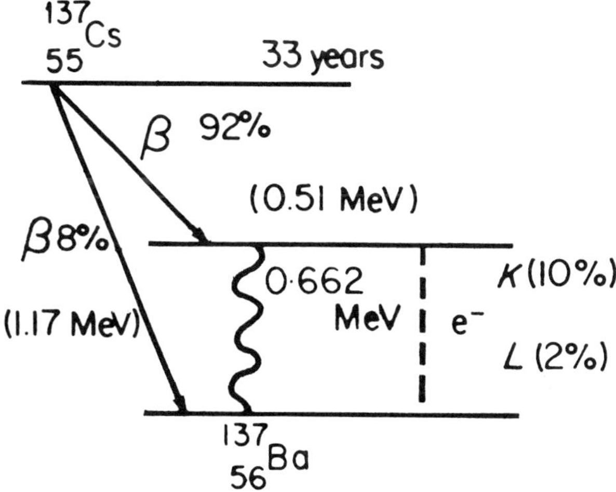

Figure 10.11. Decay Scheme of ^{137}Cs.

By the emission of the latter β-particle, an excited state of barium-137, being in an isomeric state (see p. 173) is produced. On returning to the ground state a quantum of 0.662 MeV (indicated by the wavy line) is emitted. In this case also so called conversion electrons are produced in the K and L shell (see p. 11). As indicated in the diagram in 10% of the disintegrations a K-electron of 662 keV– 37.4 keV = 624.6 keV and in 2% of the disintegrations 662 keV– 5.6 keV = 656.4 keV L-electrons are released, whereby 37.4 keV and 5.6 keV are the respective binding energies of the K and L shells. The conversion electrons are produced when a γ-ray quantum from the nucleus interacts with one of its own shell electrons.

Only the γ-ray emission is of importance for radiography and the β-emission is not utilized. The β's are usually stopped by the stainless steel capsule in which the isotope is embedded. As seen from the table on radioactive sources ^{137}Cs has a relatively small output (Rhm/Ci) when compared with that of ^{60}Co, but its half life is much longer.

Iridium-192.   The decay scheme of ^{192}Ir which is a very popular source is fairly complex and will not be discussed here. The average quantum energy of photons emitted by iridium is considerably lower than that of ^{137}Cs, but its half life is relatively short (74.4 days). A four curie source can be built into a capsule of 1×1 mm dimensions, as its specific activity is very high. (R. Halmshaw, 1954).

Thulium-170. ^{170}Tm was originally thought to be an important source of fairly soft radiation, as this is relatively rare amongst radioisotopes. The emission of radiation from ^{170}Tm is however more complex, than was first thought and the total output of γ-rays is estimated, according to R. Halmshaw (1955b) and K. Liden and N. Starfeldt (1955), to consist of 10 per cent 0.084 MeV photons, 45 per cent 0.052 MeV photons and 45 per cent of Bremsstrahlung photons, which latter have a mean energy of about 0.160 MeV. The Bremsstrahlung is caused by the emission and stoppage of 0.968 and 0.884 MeV β-rays.

It is thus seen that the fairly strong Bremsstrahlung component of about 0.16 MeV energy consists of photons of considerably higher energy than the actual γ-emission of the isotope thulium. Hence the radiation emitted by ^{170}Tm is not as soft as was originally hoped for. Due to the very low output (Rhm/Ci) and the long exposure times involved, thulium is of limited value, but it is used occasionally in cases where X-ray tubes cannot be applied owing to their bulkiness.

Other Radioactive Sources. Apart from the isotopes mentioned, many others have been used and quoted in literature, as being useful for industrial radiography. Such isotopes are e.g. xenon-133, samarium-145, europium-152, europium-155, americium-240, tantalum-182 and others. There are very few suitable isotopes which emit photons in the lower energy X-ray region, e.g. in the important range of about 60–250 keV. Either their half-lives are too short or their emission is masked by Bremsstrahlung of too high an energy or their

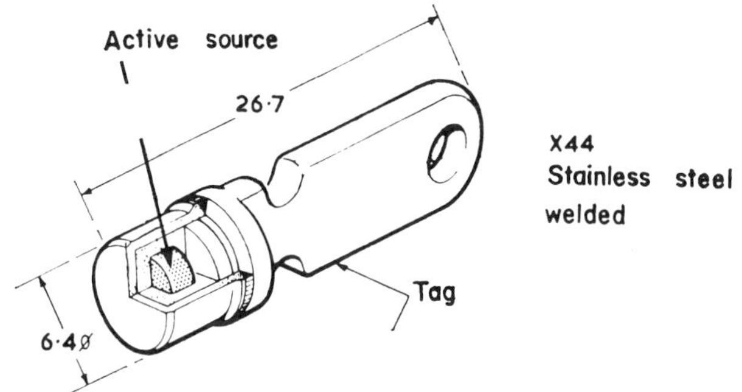

Figure 10.12. Section through a radioactive source capsule. (*By courtesy of* the Radiochemical Centre, Amersham, Bucks, England.)

output is inconveniently low. (S. Rumyantsev (year not stated), B. N. Clack, 1957; R. Halmshaw, 1955b.)

10.5e Source Containers

Source Capsules. Radioactive sources for industrial radiography are supplied in metal capsules of various designs in different countries. The typical shape of one of the standardized capsules supplied by the Atomic Energy Research Establishment, Harwell and the Radiochemical Center, Amersham, England is shown in Figure 10.12.

The actual sources are embedded in cylinders of equal height and diameter of sizes 0.5, 1, 2, 4, and 6 mm and the dimensions depend on the activity of the source they contain. The source cylinders are housed in capsules made of aluminum alloy or stainless steel. The isotope contained in the capsule is identified by engraving on the shaft of the capsule. The thickness of the capsule is great enough to stop all β's emitted by the source. Some self-absorption of the radiation emitted by the source is unavoidable, as the radiation penetrates the source itself. The sealed sources are usually held firmly in source holders or sleeves which are provided with identification marks.

The sources should only be handled by remote control, i.e. by a rod several feet in length which is either screwed or otherwise fixed into the holder of the capsule. The maximum permissible dose rate given by each source at a given distance can easily be evaluated from the known K-factor, if its activity is known.

Containers.   The chief functions of a container are (a) to house the source holder with capsule containing the radioactive source so that the dose-rate at the surface of the container is reduced to a prescribed safe level and (b) to enable the source to be removed by remote control

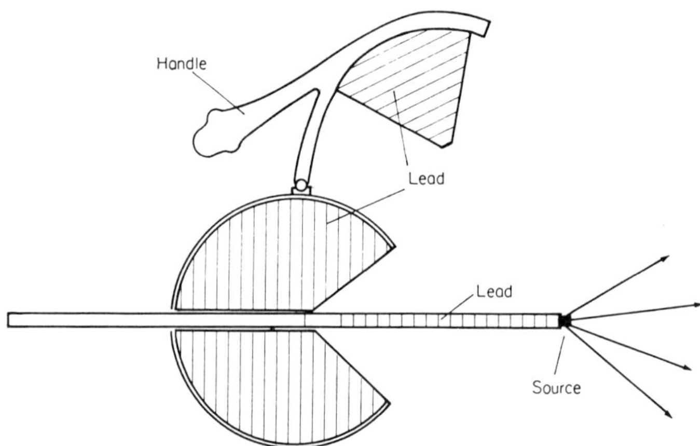

Figure 10.13. Open container with radioactive source in exposure position.

to its exposure position, so that the operator does not receive a dose exceeding that of the maximum permissible level (see p. 50). The materials used for the construction of containers are lead or lead alloys (e.g. lead, tungsten-copper) or depleted uranium.

Containers are constructed according to two different designs. One type serves as storage for the source only and when it is to be used it is removed from the container. In the other type the source remains attached to the container during exposure. In the latter type the container is opened and the source is lifted on a support rod to the exposure position, or the source may be pushed forward so that it emerges from a conical opening to a predetermined position. All these operations are carried out by remote control (British Standard 4097:1966). The diagram Figure 10.13 illustrates a design permitting the last mentioned operation. For kilocurie sources much more complex constructions of container have to be used.

10.5f γ-Ray Exposure Techniques

Frequently it is required to make use of the full radiation emitted by a source; this may occur if the source is to be placed for instance at the center of curvature of a high pressure vessel for a single exposure of a circumferential weld (see p. 461). As another example it may be intended to radiograph a larger number of castings simultaneously, as shown in Figure 10.14.

It is usual practice to complete the exposure arrangements before the source is placed in position. This gives the operator the opportunity to build up his castings, welds, cassettes at the correct distance

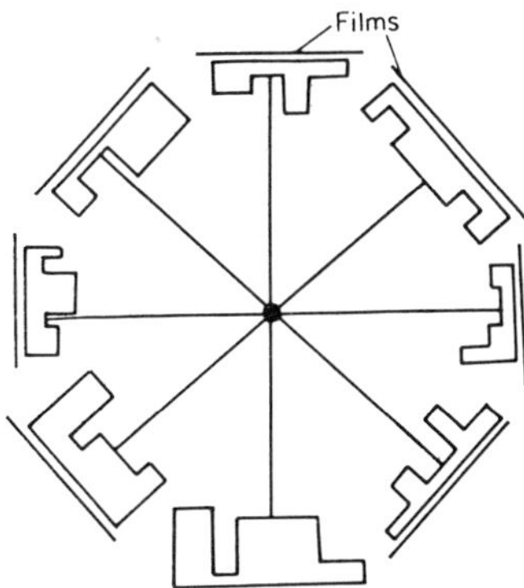

Figure 10.14. Simultaneous exposure of eight castings using a radioactive source.

between source and film in full safety. Furthermore he may use a dummy source to gain experience in handling of the source.

10.5g Safety Considerations

Because of the radiation hazards involved, radioactive isotope sources must be handled by remote control. In spite of the use of considerable thicknesses of lead shielding, relatively great distances are required for the operator in order to be at the maximum permissible dose rate of 2.5 millirads per hour (see p. 50). This will be seen by reference to the graphs of Figures 10.15 and 10.16. These show the distances for ^{60}Co and ^{137}Cs at which the maximum permissible dose rate of 2.5 millirads per hour is not exceeded, when using various activities of these sources and different shielding thicknesses of lead.

It is seen for instance that using a 2 curie source of ^{60}Co with a lead shielding of 3 inches (7.5 cm), the operator must be at a distance of 3.6 meters in order to be at the recommended safety level. Using

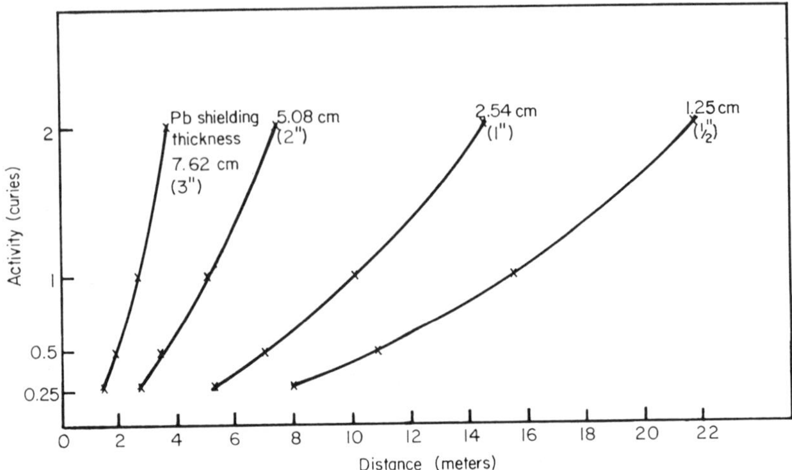

Figure 10.15. *Cobalt-60.* Distances in meters at which the dose rate does not exceed 2.5 millirads per hour for various activities in curies and for various thicknesses of lead shielding.

a 2 curie ^{137}Cs source and 2 inches (5 cm) lead shielding the appropriate distance is only 0.87 meters.

Because of the fact that the surface dose rate of a portable container usually exceeds the maximum permissible dose rate per hour, the containers must be kept inside a concrete barrier for which the distances for safe   manipulation have been calculated. Containers which

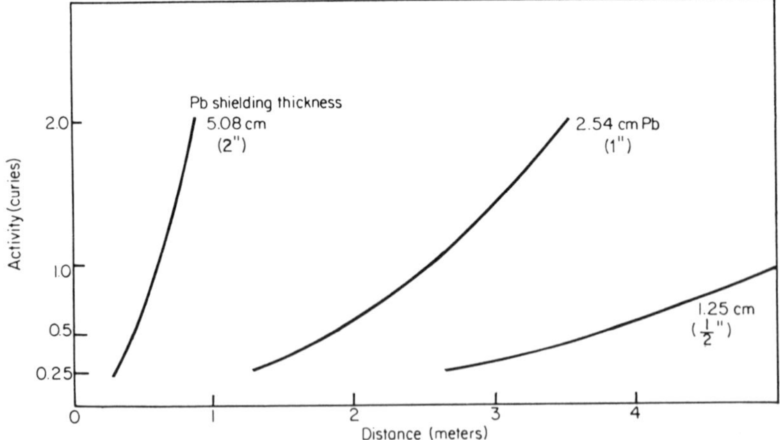

Figure 10.16. *Cesium-137.* Distances in meters at which the dose rate does not exceed 2.5 millirads per hour for various activities in curies and for various thicknesses of lead shielding.

provide maximum permissible dose rate for curie sources require so much lead protection that the container would hardly be considered as being transportable.

In many laboratories a pit is provided in the floor of the operating room for the storage of several sources in their respective containers. These are then moved in position either by mechanical means or manually within as short a time as possible. (For literature on safety considerations, see references at end of Chapter 2, pp. 51-52).

10.6 GENERAL EXPOSURE TECHNIQUES

The choice of radiographic technique to be used in industrial radiography depends on the type of work involved, i.e. on the specimen material, its thickness and partly on the IQI sensitivity required. In some cases the cost of the equipment and the economy in time play an equally important role when choosing a particular technique. Hence quite frequently compromises have to be made, which do not always lead to best sensitivity, but may lead to short exposure times. Such techniques may be quite justified if the work in hand does not warrant best IQI sensitivity.

10.6a Choice of Radiation Source

The first considerations about techniques are directed to the choice of source, whether an X-ray equipment or a radioactive source should be used in a particular type of work. It is therefore essential to discuss the merits and also the disadvantages of both radiation sources for industrial radiography.

(1) *X-ray Equipment*

Merits.

1. The quality of the radiation can be varied in wide limits.

2. The exposure times are of the order of minutes instead of hours if compared with radioactive sources, although very high activity sources permit also exposure in minutes.

3. Protection against X-rays is only required during the exposure. If the high tension of the X-ray unit has been switched off, there are no radiation hazards any more.

4. In the low energy region (50–250 kV_p) there is no useful radioactive source which can fully compete with X-ray equipment.

Disadvantages.

1. The expense involved in the purchase of an X-ray unit and in its maintenance are usually greater than with radioisotopes. This is

particularly true for X-ray units suitable for the examination of steel from 3–6 inches.

2. X-Ray units are in general bulky and only the lower energy units are transportable. They all require an electrical and usually a water supply.

(2) *Radioactive Sources*

Merits.

1. Small size which permits radioisotopes to be used at sites inaccessible to X-ray tubes.

2. In general no need for electrical or water supply.

3. Supply of high energy radiation for exposure up to 6 or even 8 inches of steel.

4. Easy transportation.

5. Relatively low purchase and running costs if compared with X-ray equipment. The container for the higher activity sources may be rather expensive in relation to the source.

6. Many specimens can be exposed simultaneously.

Disadvantages.

1. Long exposure times involved.

2. Limited use, as they are unsuitable for thin specimens and for light alloys.

3. Each type of isotope emits radiation of a given quality which cannot be changed.

4. Radioactive sources require shielding when they are not in use.

In conclusion it may be said that the alternative between X-ray equipment and radioactive isotopes depends on the type and thickness of the material to be radiographed, on the IQI sensitivity required, on the time available for carrying out the exposure and on the relative convenience of the two types of source.

10.6b   Penetration and Thickness

In order to achieve a reasonable level of sensitivity there must be an optimum relation between penetrating power of the radiation and the thickness of a given material. The following table and Figure 10.17 give a guide to the maximum thicknesses of light alloys and also of steel which can be radiographed by radiations of various qualities expressed in kV_p using different types of no-screen X-ray film in the case of light alloys.

The following Table 10.11 is taken from 'Recommendations for the X-ray examination of fusion welded joints in light alloys.' Brit. Welding J. (1958).

It is seen from this table that the faster the film used for a given thickness of light alloys the lower is the kV_p recommended. This is because of the poorer statistics involved using the faster film, when fewer quanta per unit area are required than with slower films (see p. 420). The main consideration using a faster film is of course the reduction in exposure time.

TABLE 10.11. Guide to Maximum kV_p for Light Alloys of Various Thicknesses*

Thickness in inch	Maximum kilovoltage		
	With ultra fine grain no-screen film	With fine grain no-screen film	With medium grain no-screen film
up to $\frac{1}{4}$	70	60	50
$\frac{1}{4}-\frac{1}{2}$	80	70	55
$\frac{1}{2}-\frac{3}{4}$	95	80	60
$\frac{3}{4}-1$	105	90	65
$1-1\frac{1}{4}$	115	100	70
$1\frac{1}{4}-1\frac{1}{2}$	125	110	80
$1\frac{1}{2}-2$	145	130	90
$2-2\frac{1}{2}$	165	150	110
$2\frac{1}{2}-3$	185	170	120
$3-4$	230	210	150

* From BWRA Report 66B/58

From the diagram of Figure 10.17 the recommended X-ray tube voltages for different thicknesses of steel (in the examination of fusion welded butt joints) can be derived. The curves B, C, and D refer to the use of various types of film in conjunction with lead screens. In general the finer grained the type of film used the better is the IQI sensitivity to be expected. Curve A refers to the use of High Definition salt screens in conjunction with X-ray screen film. The diagram is taken from the British Standard 2600:1962 in which it is suggested that the recommended values of kilovoltage are not exceeded. For all the exposures with lead screens densities between 2–3 should be aimed at, whereas for the salt screen exposures (curve A) a density not less than 1.3 and not greater than 2.3 is recommended. These density values are inclusive of fog density being not greater than 0.3. The British Standard 2600:1962 also gives details of the required exposure conditions.

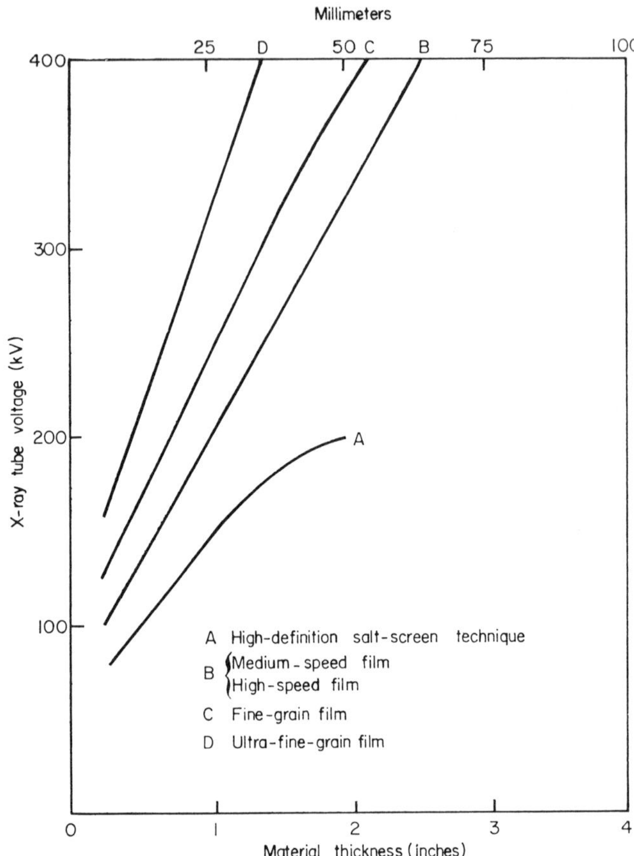

Figure 10.17. Maximum X-ray tube voltage as a function of steel thicknesses (in inches) for various types of film 'B, C, and D' no-screen X-ray films with Pb screens. 'A' for screen films with High Definition salt screens. (*After* British Standard 2600:1962.)

10.6c   Focus to Film Distances

In order to obtain a good IQI sensitivity the geometric unsharpness (U_g) must be as small as possible and this can be achieved by choosing a great focus to film distance. There are of course practical limitations to great focus to film distances, because of the long exposure times involved (see inverse square law, page 21). The British Standard 2600:1962 gives appropriate specifications for different types of photographic material. These are summarized in the graph Figure 10.18 which illustrates the relationship between the recommended

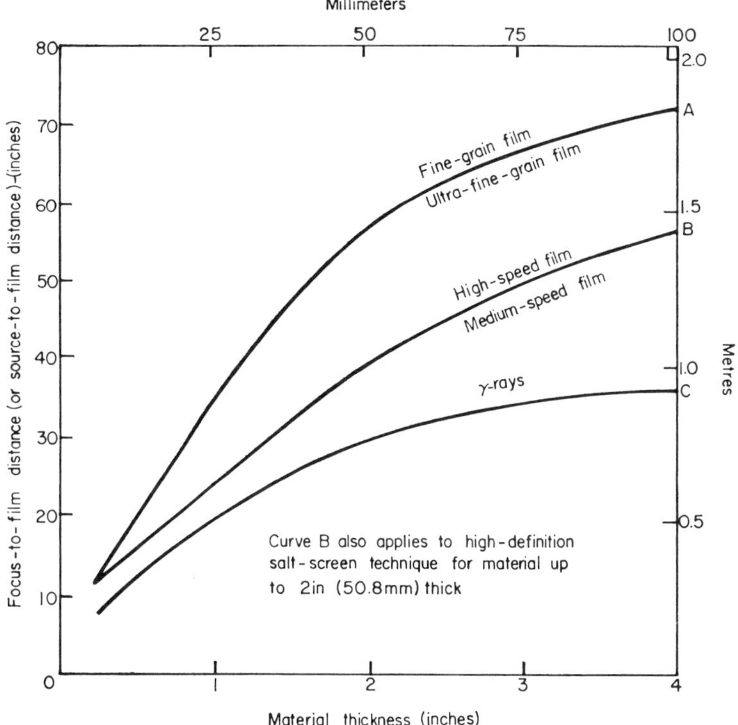

Figure 10.18. Recommended minimum values of focus to film distance (or source to film distance) for various thicknesses of steel in inches. (*After* British Standard 2600:1962.)

focus to film distance (or source to film distance) in inches and in meters against the thickness of steel in inches and millimeters. The recommendations expressed by this graph are based on good practice and on a compromise between too great focus to film distance and reasonable exposure conditions.

Figure 10.18 is based on a focal spot of maximum effective dimension of 7 mm for curves A and B and of source diameter of 4 mm for curve C. If one has to deal with other focal spot or source diameters the new focus-film distances required can be calculated from

$$F_2 = s\,\frac{F_1}{7} \tag{10.18}$$

where F_1 is the focus to film distance (FFD) for a 7 mm effective focal spot. F_2 is the new FFD and s is the maximum effective dimension (in mm) of a focal spot different from 7 mm. Similarly for radioactive sources of d mm (other than 4 mm) the equation

$$F_2 = d \, \frac{F_1}{4} \tag{10.19}$$

holds, where F_1 is the source to film distance for a 4 mm diameter source. F_2 is the new FFD and d the diameter of the source in mm if different from 4 mm.

(Extracts from (B.S. 2600:1962 *General Recommendations of the Radiographic Examination of Fusion Welded Butt Joints in Steel*) are reproduced by permission of the British Standards Institution, 2, Park Street, London, W.1, from whom copies of the complete standards may be obtained.)

10.6d   Exposure Charts

The determination of the correct radiographic exposure for a particular X-ray unit, specimen and film (including lead- or salt-intensifying screens) can either be done by trial and error, by reference to experience by others, or by means of an exposure chart. An exposure chart shows the relationship between the thickness of a particular material against the exposure, e.g. in mA minutes (or in curie hours) required in order to obtain a certain density for X-rays generated at various kilovoltages. Although exposure charts are available from publications or from apparatus and film-manufacturers, it is preferable to produce the charts for the individual X-ray unit, type of film (screens) and processing conditions in use. The calibration of X-ray apparatus is not standardized sufficiently to make reliable use of published exposure charts at any X-ray unit. In some cases it may be possible to establish a correction factor by which any value of an already existing exposure chart can be multiplied in order to obtain the correct result.

Exposure charts should be made for the material and thicknesses most likely to be used and should be based on standardized processing conditions (see p. 384). The most straightforward, although somewhat laborious method, is to make sensitometric exposures of a metal stepwedge built of the material and thicknesses of interest. The stepwedge is placed above a sheet of lead of e.g. $\frac{1}{2}''$ thickness which is provided with a slit of about $\frac{1}{2}''$ width and the wedge should cover the whole length of the slit. Below the lead sheet is the cassette loaded for instance with no-screen X-ray film. Provision should be made for the cassette to be shifted for each subsequent exposure, say at $100 \, \mathrm{kV_p}$, at a given focus to film distance, for increasing values of milliampereseconds, which turned out to be (in the following experiment) 21, 37, 70, 152, 304 mA-sec. In this way sensitometric film stepwedge images are obtained for increasing exposures after processing. After reading the densities these are plotted against thickness for various mA-seconds and a family of curves (see Figure 10.19) is obtained for one kV_p. A similar procedure may then be adopted for X-rays generated at other kilovoltages.

In order to arrive at the final exposure chart, lines are drawn parallel to the abscissa axis of Figure 10.19, e.g. through a density

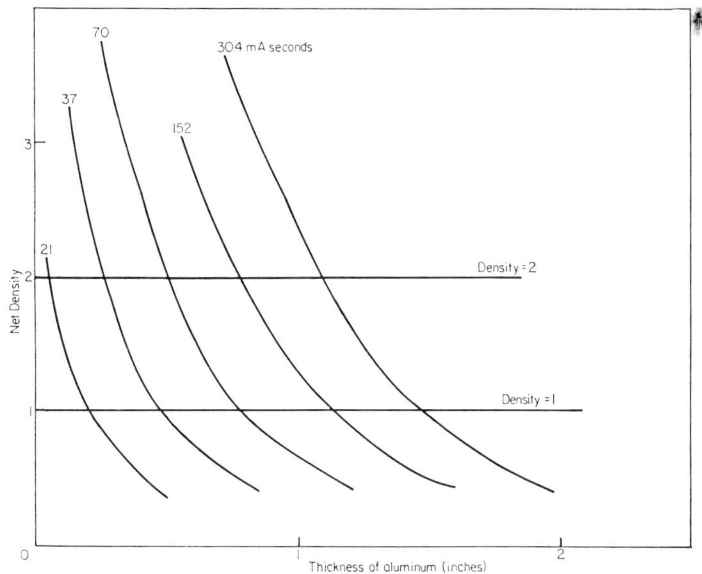

Figure 10.19. Relationship between step thickness of an aluminum step wedge and net density for exposures at various mA seconds using X-rays generated at 100 kV_p. (Medium fast no-screen X-ray film.)

of unity and a density of 2 and the respective thicknesses for each mA-second can be found. Using the appropriate charts for other kilovoltages, an exposure chart such as that in Figure 10.20 can be plotted for any desired density level.

If radiations of very high quantum energies such as above 400–500 kV_p are used, it is necessary to expose individual film strips, (for instance sandwiched between lead screens), instead of exposing through a slit because of the possible penetration of the hard radiation through the lead sheet provided with the slit. Similar procedures should be followed, if an exposure chart is to be made with a radioactive source. This precaution will be appreciated, if one considers that $\frac{1}{2}''$ lead reduces the incident intensity of ^{60}Co radiation by only one half.

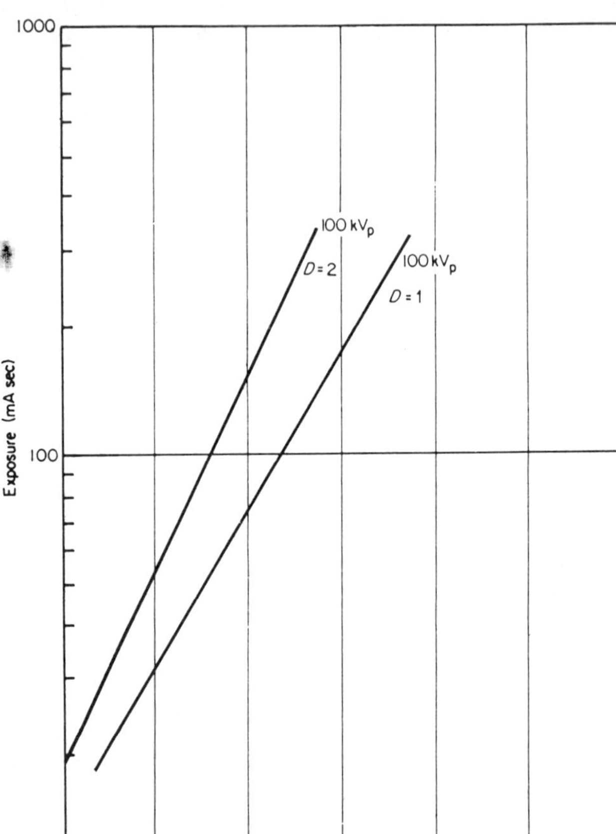

Figure 10.20. Exposure chart for aluminum based on the experimental results of Figure 10.19, i.e. 100 kV_p X-rays. The chart refers to net densities of 1 and 2 using a medium fast no-screen X-ray film and standard processing conditions.

Apart from the experimentally rather laborious method just described, there are others which essentially require less experimental data. In one of these methods only a step wedge exposure and a characteristic curve of the film with the screens to be used are required for each kilovoltage. The procedure of arriving at an exposure chart is as follows.

1. Only one exposure is made for instance of a steel step wedge at say 200 kV_p and 10 mA minutes at a given focus to film distance. The densities obtained after processing are read and tabulated against

TABLE 10.12. Data Required for Obtaining an Exposure Chart
Based on Characteristic Curve of the Film in Use

Steel thickness in inch $k* = 10,$	D (net) $E_{(D=2)} = 60$	E_r	New exposure in mA min. for $D = 2$ $k(E_{(D=2)}/E_r)$
0.8	3.1	120	5.0
0.9	2.4	80	7.5
1.0	1.9	55	11.5
1.1	1.48	35	17.2
1.2	1.14	24	25.0
1.3	0.82	16	37.5
1.4	0.62	11	54.5
1.5	0.47	7	86.0

* k = original exposure.

the thicknesses as shown as an example in the first two columns of
Table 10.12 above.

2. The relative exposures E_r corresponding to these densities are
then found from the characteristic curve (see Figure 10.21(a)) and
are tabulated in column three of Table 10.12. It is essential that the

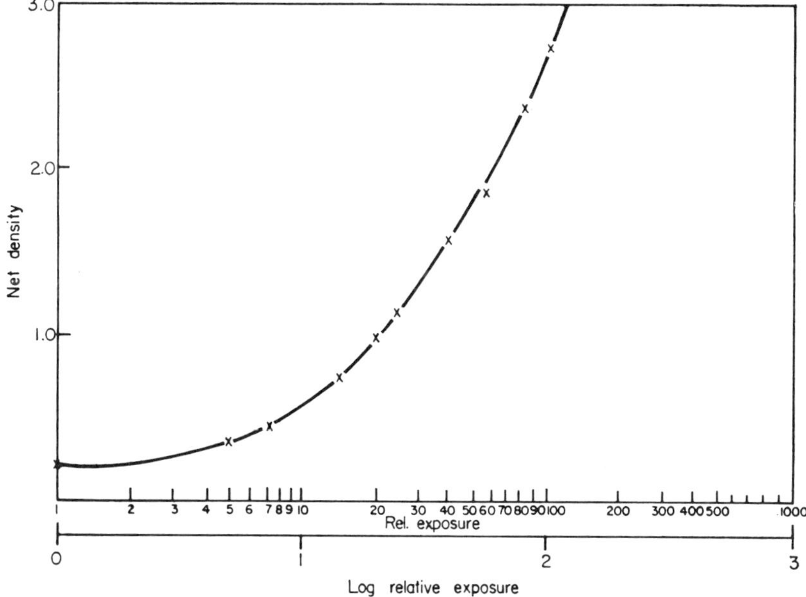

Figure 10.21. (a) Characteristic curve of a fine grain no-screen film sandwiched
between lead intensifying screens. (Standard processing conditions.)

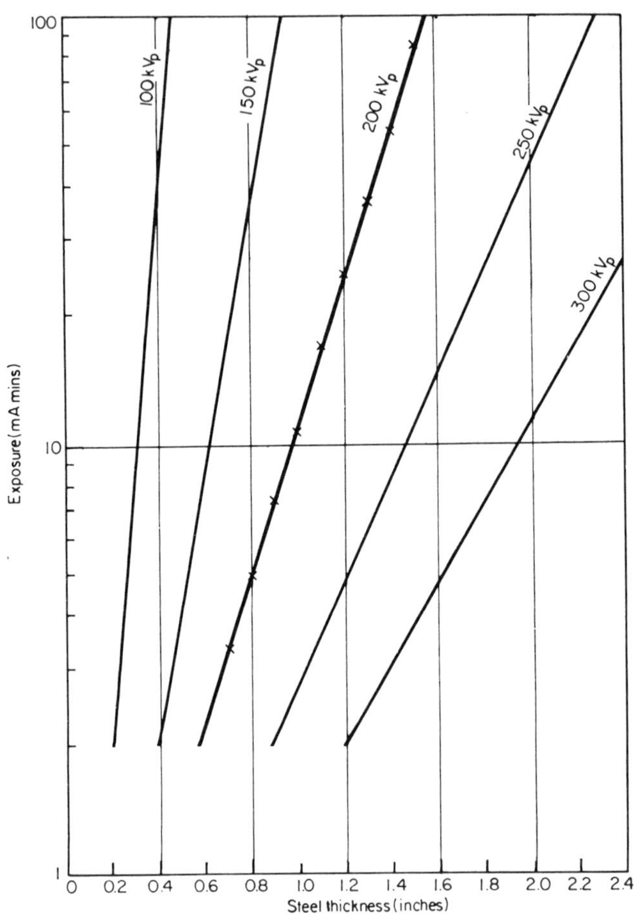

Figure 10.21. (b) Exposure chart for steel based on a net density of 2 for X-rays generated at peak kilovoltages from 100–300 kV. Focus to film distance 36 inch. Fine grain X-ray film with lead intensifying screens. (Standard processing conditions.)

film from which the characteristic curve was obtained is processed under the same conditions as the step-wedge film.

3. If an exposure chart is required for a radiographic density of 2, then we have to find the exposure corresponding to $D = 2$ above fog from the same characteristic curve, which is $E_{(D=2)} = 60$.

4. The new exposure values in mA minutes for each step of the wedge based on a density of 2 are then found from the relation

$$\frac{\text{New exposure (mA min)}}{\text{Original exposure (mA min)}} = \frac{E_{(D=2)}}{E_r} \tag{10.20}$$

If we denote the original exposure by k (being 10 mA min in our case), we find the new exposure from

$$\text{New exposure (mA min)} = k \, \frac{E_{(D=2)}}{E_r} \qquad (10.21)$$

The calculated values for the new exposure, using the equation above are tabulated in column four of the table and by plotting

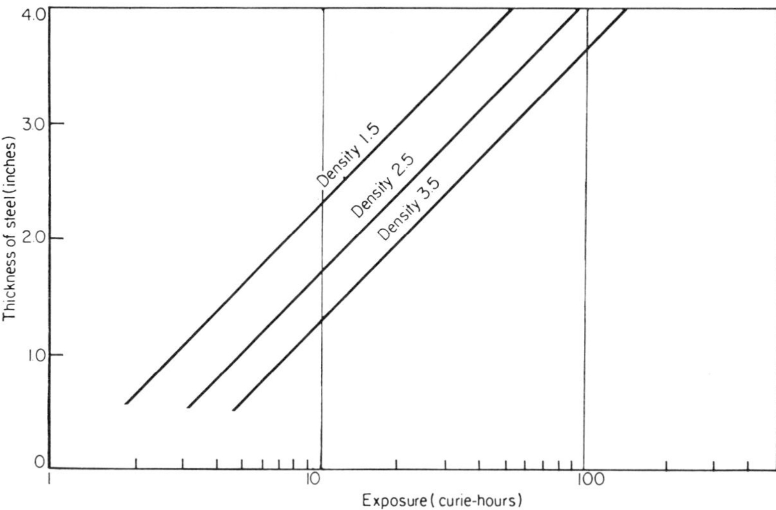

Figure 10.22. *Cobalt-60.* Typical exposure curves for steel using Kodak Industrial X-ray film (Industrex Type D Film) with lead intensifying screens. Source to film distance 36 inch (91.5 cm) Development: 4 minutes. Developer: DX 80 at 20°C (68°F). (*By courtesy of* Kodak Ltd., England.)

column one against four we obtain the required exposure chart (see Figure 10.21(b) for 200 kV$_p$). For each kilovoltage a special step wedge exposure has to be made and the same procedure has to be repeated in order to obtain an exposure chart for other kV's.

Exposure charts for steel applied to γ-rays of ^{60}Co and ^{137}Cs are shown in Figures 10.22 and 10.23.

The abscissa axis is plotted in terms of Curie hours and the ordinate in terms of thicknesses of steel. The three straight lines in each graph represent values for densities of 1.5, 2.5, and 3.5. Both graphs are taken from Kodak Ltd. (England), Data Sheet I N 16 on 'Gamma Radiography' and the charts refer to the use of medium speed no-screen X-ray films in conjunction with lead screens. It should be noted that it is irrelevant whether the exposure for example of 10

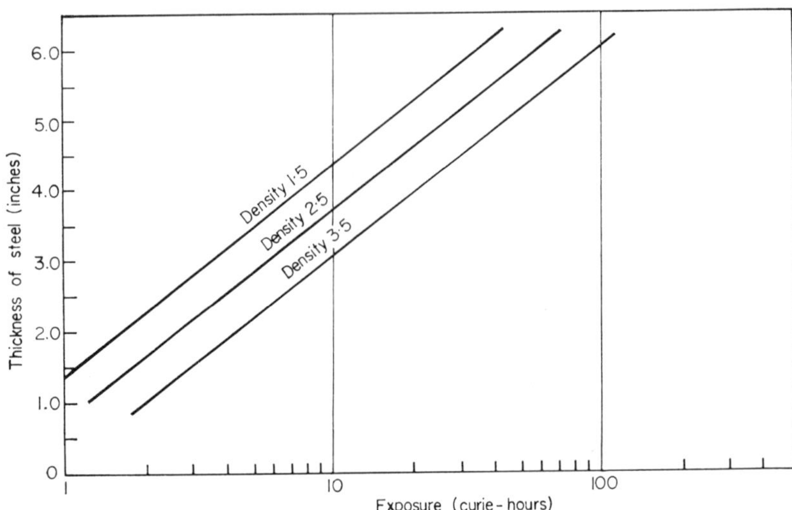

Figure 10.23. *Cesium-137.* Typical exposure curves for steel using Kodak Industrial X-ray film (Industrex Type D Film) with lead intensifying screens. Source to film distance 36 inch (91.5 cm). Development 4 minutes. Developer: DX 80 at 20°C (68°F). (*By courtesy of* Kodak Ltd., England.)

curie hours is carried out using a 1 curie source for 10 hours or using a 10 curie source for 1 hour, as long as no-screen X-ray films with or without lead screens are used (no reciprocity failure occurs).

10.6e   Reducing the Effects of Scattered Radiation

Scattering

Scattered radiation impairs the image quality of a radiograph particularly with respect to contrast and several methods may be applied to reduce the effects of scattering, although it is not possible to eliminate them entirely. In general the disturbing effects of scattering are greatest approximately in the region between 150–500 kV_p and become less troublesome towards higher energies within the region of 1 MeV X-rays and of γ-rays from radioisotopes. The effects of scatter are particularly disturbing in thick specimens as mentioned before.

In spite of the fact that the physics of scattered radiation, although complex, is well understood, it is rarely possible to predict precisely the amount of scattered radiation with respect to the primary radiation under practical conditions. This is partly due to the angular dependence of the intensity of scattering on the quality of the radiation and the effect of multiple scattering (see pp. 16, 24 and 315).

Reference has been made to the effects of scattering under the subheading 'Build-up factor' on p. 409, and in theoretical considerations on thickness sensitivity. Most of the disturbing effects of scatter originate within the specimen, but they may also be caused by the floor and walls of the room, by the film support or by any object placed in the vicinity of the specimen or cassette. Backscattering from below the cassette can be minimized by using lead sheet behind the cassette.

The more common methods of reducing scattered radiation which originates within the specimen are the use of cones, masks, diaphragms, blocking media and filtration.

Cones

Probably the most successful means of reducing scattering is the reduction of the cross-sectional area of the X-ray beam impinging on the specimen. This may be done for instance by means of a cone of lead; the thickness of the lead required depends on the maximum kV_p for which it is to be used.

Diaphragms and Masks

Other more frequently used methods consist of the use of adjustable diaphragms fixed on the tube head or masks of thick lead which are placed on top of the specimen. These reduce the cross-sectional area of the beam striking the specimen and thus prevent the formation of scatter below these regions which would otherwise reach the image.

Blocking Media

Other means of reducing scattering are the use of so called blocking media which may consist of cut sheets or strips of lead placed tightly around a specimen for reasons which will be explained later. The cutting of lead may not be economical unless a specimen of the same shape is radiographed several times. Therefore more flexible blocking media are recommended which can be reused and shaped to fit around the specimen. These can be made for instance of 'Plasticine' loaded with lead powder. In order to prevent this mixture from becoming crumbly, petroleum jelly may be added; also barium sulfate pastes and other suitable substances have been recommended for this purpose. Flexible blocking media should be moulded tightly around the specimen.

The reason for the use of blocking media is that the film is very sensitive to the softer components of the radiation scattered near the border of the specimen (undercutting radiation) and also to the direct unattenuated beam impinging on the film outside the specimen, as this contains all the soft components of the primary X-ray beam. Furthermore the radiation penetrates increasingly towards the edge of the specimen as shown in Figure 10.24 below.

All these effects combined may entirely obscure the border and even the interior of the specimen, unless blocking media are used. (See Figure 10.25).

Similar pastes as those mentioned above or pastes made of some substance of similar absorption to the specimen are often used with

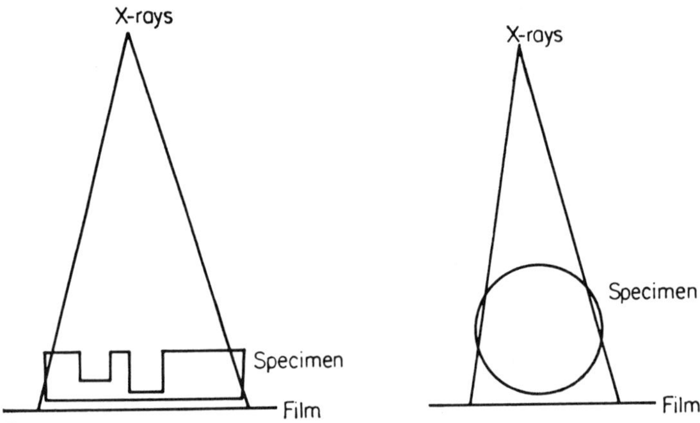

Figure 10.24. The attenuation of the primary X-ray beam decreases towards the borders of the specimen. Hence an increasing amount of the softer components of the primary beam reaches the film.

objects having wide variations of thickness. Thin regions and holes in the specimen may be completely masked by backscattering from the unattenuated beam and by scatter originating from thicker parts. Furthermore the differences in absorption of the radiation between thick and thin regions of a specimen may easily lead to strong overexposure within the region of the image of the thinner part. In such cases it is often possible to compensate for the wide variations of thickness by building up the thin region with heavy liquids (such as lead acetate), if this appears to be practicable within a particular specimen.

Filtration

A metal filter when placed between the X-ray source and the film hardens the primary X-ray beam, as the filter preferentially absorbs

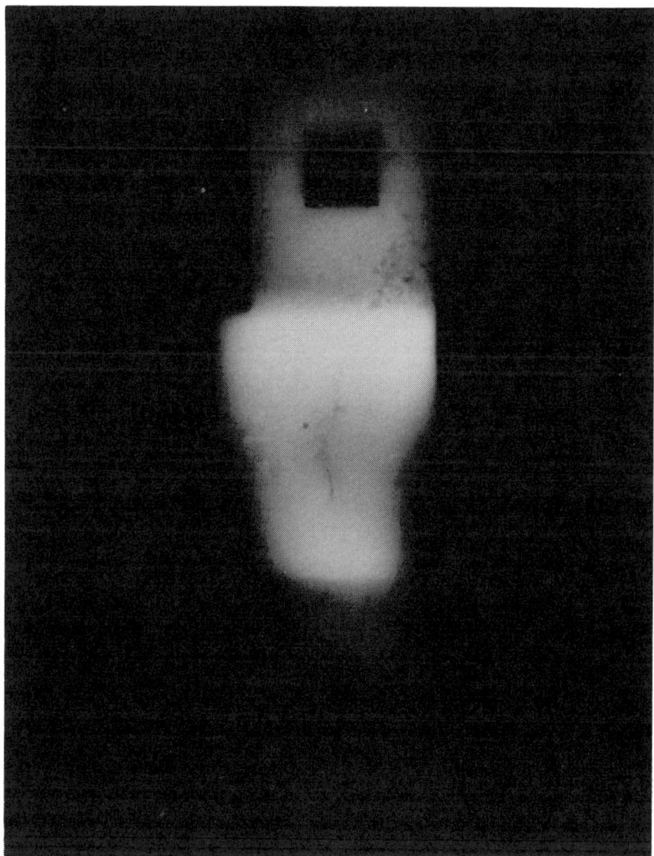

Figure 10.25. This radiograph of a symmetrical cast iron bracket (sandcasting $\frac{1}{2}$–$\frac{3}{4}$ inch thick) shows the effects of scattered radiation. One side of the casting was masked during exposure by the use of paste in the square hole and around the edges. On the other unmasked side scattered radiation and radiographic undercut have caused a drastic loss of image quality by obscuring detail. The defects revealed in this radiograph are sponginess and filamentary shrinkage. Technique: 180 kV$_p$, 10 mA, 30 secs. exposure time at 36 inch focus to film distance. Medium fine grain no-screen X-ray film. (*By courtesy of* E. J. Grimwade, Kodak School for Industrial Radiography, Kodak Ltd., England.)

the softer components of the radiation. The harder radiation reduces contrast and thus in general the sensitivity of defect detection.

Filter Material

Filters may consist of aluminum for light alloys, of a few mm copper for small thicknesses of steel, but for greater thicknesses of steel lead intensifying screens are usually used of about 0.01–0.02″ thickness in the region of 200 kV$_p$ and of 0.02–0.05″ in the region of 400 kV$_p$.

Position of Filters

Filters may be placed close to the X-ray tube, because they can then be small in size and any blemishes in the filters will be smoothed out in the radiographic image because of the large distance between filter and the film. On the other hand lead intensifying screens in the cassette act in addition to their intensifying action as filters simultaneously and reduce scattered radiation.

Filter Near the Tube. For the reasons mentioned before, the disturbing radiographic effects with an exposure of a specimen, having wide variations of thickness, can be reduced by the use of a filter near the X-ray tube. This lowers the ratio of the attenuated to the unattenuated radiation, and causes less undercut. The excessive contrast between thick and thin regions is thus smoothed out which leads to an improved radiographic sensitivity. The high densities caused at the borders of the specimen will also be reduced by the filter because of the removal of the softer components of the beam.

Filter between Specimen and Film.   Considering that the incoherent part of scattered radiation is softer than the primary radiation and because this softening effect becomes relatively greater the harder the primary radiation, lead intensifying screens in the cassette can be very efficient in removing the softer components of the scattered radiation.

Potter-Bucky Grids

Potter-Bucky grids, which are practically indispensable in medical radiography are not used in general in industrial radiography. The reason for this is chiefly that their efficiency depends largely on their grid ratio as explained on page 318. Considering the very hard radiation applied in the industrial field the grids are impracticable because the lead strips must be very thick in order to be efficient.

10.7   RADIOGRAPHIC INSPECTION

Some of the most widespread applications of industrial radiography are undoubtedly in the examination of welds and castings and it is the purpose of the following sections to review briefly some of the essential principles of inspection methods in this field.

10.7a   Weld Radiography

Weld radiography is essentially a method of flaw detection and after any flaws have been revealed, it is necessary to assess their possible influence on the serviceability of the weld within a particular structure assembly. Weld radiography serves another purpose by providing the welder with a guide to the soundness of his welding technique. (L. Mullins, 1957; E. Fuchs, L. Mullins, and S. H. Smith, 1947).

Codes and Standards

Probably the most important step in the history of welding occurred in 1930 when the United States Navy agreed to accept welded steam vessels only if these have been subjected to radiographic weld inspection. A few years later the American Petroleum Institute, the American Society of Mechanical Engineers and Lloyd's Register of Shipping in the United Kingdom published codes of practice for welded pressure vessels.

Now similar codes on the radiographic approval of welded structures, class I* pressure vessels such as for nuclear power plants, storage tanks, pipe lines and others have been prepared in various countries. The use of these codes implies high standards of radiographic examination. Some of the relevant British and American standards and codes are referred to in the list of references at the end of this chapter.

Weld Defects

A thorough discussion of weld and casting defects is beyond the scope of this book, but a brief review of the subject will familiarize the reader with some of the basic facts.

The appearance of defects in weld radiographs may be caused by
1. Surface marks on the weld or parent plate
2. Internal defects in the weld
3. Adventitious images due to faulty handling or processing of films.

Let us consider first items 1 and 3. The surface defects such as undercut, surface pitting, chipping marks and others typical of welding techniques can be detected by visual inspection of the surface of the weld. Blemishes due to film processing and other artifacts will be quickly recognized by the more experienced radiologist.

---

* Class I pressure vessels are of special design for high duty performance.

The internal defects should be identified in terms of their radiographic appearance such as shape, uniformity of photographic density, their surrounding, contrast, sharpness, size, and position.

A technical knowledge of the preparation of welds will help considerably in the interpretation of the radiograph. The drawings in Figure 10.26 show typical butt welds diagrammatically. A few typical internal defects in fusion butt welds are listed below. For more details the reader is referred to the British Standard 499:1965, Part 3. 'Welding Terms and Symbols', on which the following definitions are based.

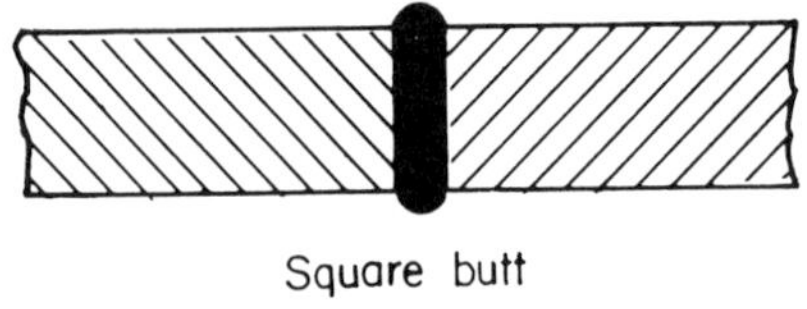

Figure 10.26. Examples of butt welds.

Gas pore.   This is a cavity, generally under 1/16th inch in diameter, formed by entrapped gas during the solidification of molten metal. When the cavity is over 1/16th inch in diameter it may be referred to as a blow hole. The radiographic appearance of a gas pore is a sharply defined dark shadow of circular contour (good sensitivity is required to detect individual gas pores).

Wormhole.   An elongated or tubular cavity caused by entrapped gas. It appears in the radiograph as a dark shadow the shape of which depends on the orientation.

Inclusions.   Slag or foreign matter entrapped during welding. The image of the defect is usually more irregular in shape than that of a gas pore. The relative density of the image of an inclusion depends on the relative radiation absorption by the inclusion and the parent metal. For example tungsten inclusions in aluminum welds give light images and oxide inclusions give dark images.

Lack of fusion. i.e. lack of union in a weld:
(a) between weld metal and parent metal
(b) between parent metal and parent metal
(c) between weld metal and weld metal.

It may occur as a lack of side fusion or root fusion or inter-run fusion. Its detection depends on its orientation with respect to the beam direction.

Incomplete root penetration. This is a failure of the weld metal to extend into the root of a joint. It appears in the radiograph as a dark continuous or intermittent linear band the edges of which may be straight or irregular.

Cracks. This is a linear discontinuity produced by fracture. Cracks may be defined as longitudinal, transverse, edge, crater center line according to their position and form. They are generally revealed in radiographs as fine dark lines, but these may also appear diffuse and often discontinuous. The detection of a crack depends on its orientation with respect to the direction of the radiation.

Although the author is quite aware that a reader who wants to take up weld radiography requires a much more detailed study of defects than that given in this short review, the main purpose of having quoted these definitions is to give the reader an idea of the type of defects involved.

Beam Direction

An important aspect in weld radiography is the choice of beam direction which depends not only on the position and type of the weld, but also on the defect to be detected. In some cases the choice of beam direction is straightforward, as shown in Figure 10.27(a), but in other cases oblique directions have to be taken (see Figure 10.27(b)). In pipe weld radiography single or double wall techniques may have to be used, as indicated in the Figures 10.28(a), (b), and (c), 10.29, and 10.30.

Using a small diameter and thick walled pipe, a radioactive source may be placed inside (but off-center) at the inner wall of the pipe and the film is placed around the external circumference (see Figure

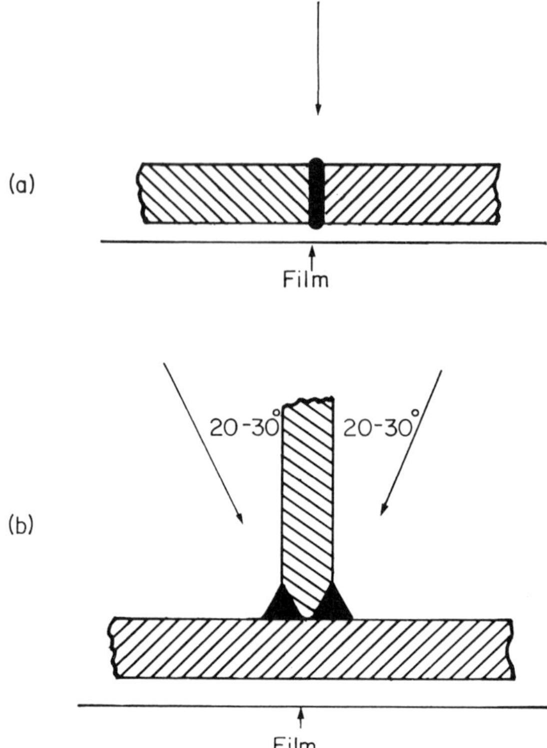

Figure 10.27. (a) Normal beam direction.
(b) Oblique beam direction.

10.28(a). Alternatively a sensitive method of radiographing a pipe weld is to place the film inside the wall, using a large source or focus to film distance (as shown in Figure 10.28b). If a rod anode X-ray tube or a radioactive source can be placed in the center of the pipe as indicated in Figure 10.28c, the whole of a circumferential weld can be radiographed in one exposure either on one single strip of film or on several shorter film strips which overlap at the ends.

If there is no access to the interior of the pipe a double wall technique may be used (see Figure 10.29) by applying a great source to film distance, so that not only the weld adjacent to the film, but also that remote from the film appear sharp in the radiograph. In order to avoid overlapping of the two weld radiographs the beam direction should be offset by $\frac{1}{5}$th of the source to film distance, as

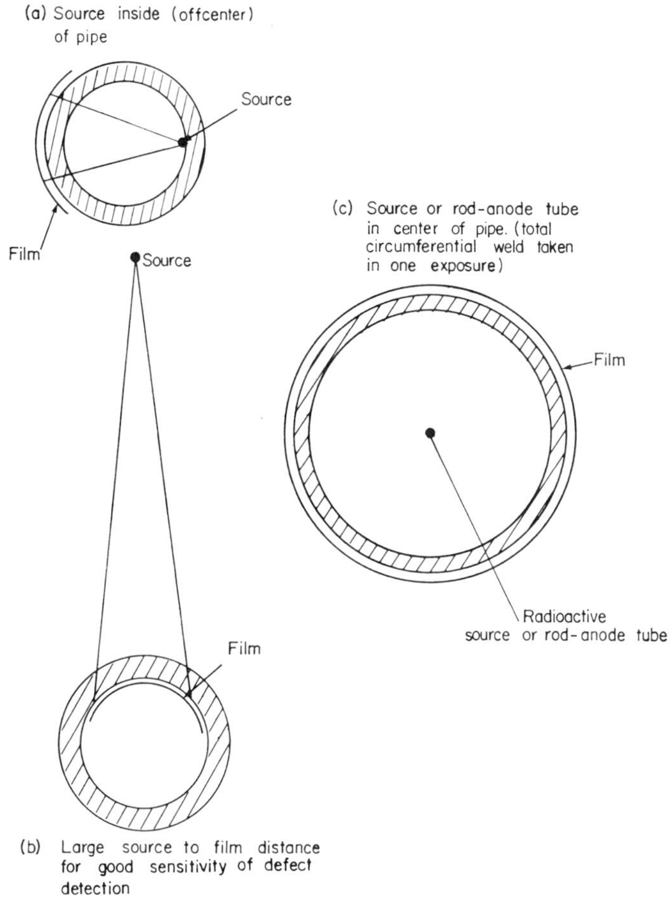

Figure 10.28. Pipe weld radiography techniques.

indicated on the right in Figure 10.29. This is called the double wall–double image technique.

If the pipe has a large diameter, a short source to film distance may be chosen by placing the source near the outside of the pipe (see Figure 10.30). Hence only one weld at a time is obtained in good definition, whereas the one adjacent to the source is blurred. This is the double wall–single image technique (see also B.S. 2910:1965).

For the radiography of welds envelope wrapped films in roll form of relatively small width are available. These can be cut at any length and sealed at the ends.

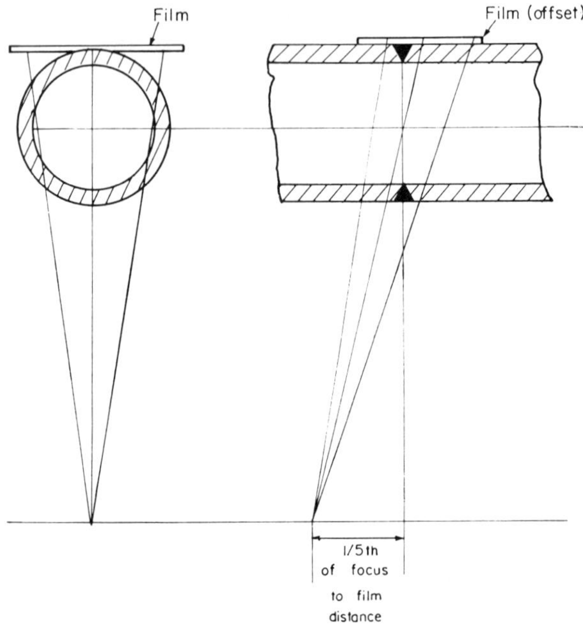

Figure 10.29. Double wall–double image technique. Tube position shifted by $\frac{1}{5}$th of focus to film distance.

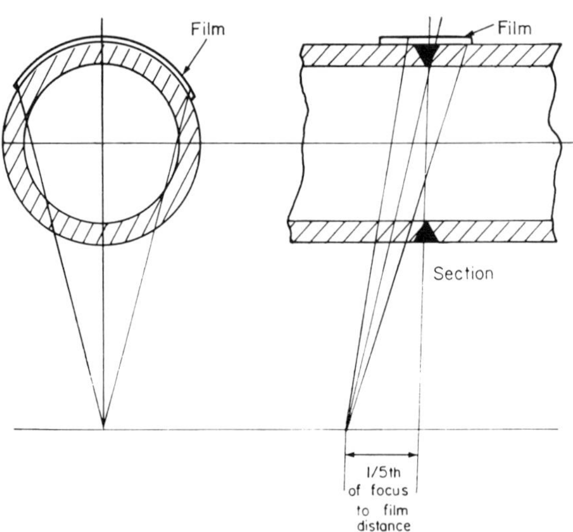

Figure 10.30. Double wall–single image technique.

Identification and Marking of Welds

When preparing longitudinal or circumferential weld radiographs of pressure vessels and other weld structures, it is essential that the section of the weld that has been radiographed can be correlated with the radiograph. To be sure that every inch of the weld is radiographed, a film slightly larger than the length of the weld is to be used, so that overlapping occurs. The length of the weld is then marked off for instance in one foot sections each provided appropriately by an identification number made of lead letters. The lead numbers, together with the IQI's are then fixed on the specimen, in order to obtain a record on the radiograph. (Details of this identification method are quoted in B.S. 2600 and B.S. 2910.)

10.7b Casting Radiography

A casting is a block of metal that has been formed in a mould. Radiography covers a wide variety of castings differing in size, shape and composition and only a brief reference will be made to some of the problems involved in their inspection. Castings may be simple or very complex in their construction. They may be so large that they can only be radiographed by using megavoltage techniques. Wide variations of thickness in a casting may present difficulties in the radiographic technique as discussed before.

Since the radiographic examination of castings is a non-destructive technique, the method is ideal for eliminating faulty castings before machining and defects may be detected, which reveal that a casting is unsuitable for service. Such defects may not necessarily be discovered on machining.

Whether a casting should be subjected to radiographic examination depends on the purpose for which it is to be used and in many large castings only an essential part is to be examined. The term 'essential' depends on the serviceability of the particular part of the specimen.

Radiography may play an important role in foundry control, thus providing a useful guide to the inspection of pilot castings. In cases of insufficient experience in interpreting radiographs, sectioning of the casting may help the beginner in checking the interpretation.

Casting Defects

Radiographic examination can disclose inhomogeneities in metals which may be caused by contraction of the metal, by gas inclusions and impurities, by interactions between the mould material and the liquid metal, by gas entrapped during the solidification process and many other causes. Some of the typical defects which have been established in castings have been described in detail in B.S. 2737 : 1956. In this standard typical defects are also illustrated and they are divided into the following main sections.

Voids, i.e. cavitation due to gas or shrinkage or both.

Cold shut, i.e. discontinuity caused by failure of stream of molten metal to unite with confluent stream or solid metal.

Cracks, i.e. a discontinuity due to the fracture of the metal during or after solidification. The defect appears in the radiograph as one or more dark lines whose width and form depend upon the type of crack, and particularly upon the radiographic technique used. The detection of a crack, or part of it, is dependent on its orientation relative to the incident radiation and consequently it may not be equally well revealed along its entire length. As the orientation diverges from the optimum direction for detection, the radiographic shadow becomes broadened and increasingly difficult to recognize.

Segregation, i.e. local concentration of any of the constituents of an alloy.

Inclusions, i.e. foreign matter entrapped in casting, and others.

For further information on the radiography of castings the reader is referred to the relevant literature at the end of this chapter.

10.7c   Non-Metallic Materials and Assemblies

Radiography is not limited to the examination of welds and castings and periodic inspection of aircraft, but has expanded to non-metallic materials, such as the detection of defects in graphite for nuclear reactors, in plastics, rubber, glass, ceramics, and other materials. Industrial radiography also plays an important part in the inspection of assemblies, such as engines, valves, fuses, cables etc. In some of this work it may be merely of importance to check the position of components, for instance it may be necessary to check the flow of metal in soldered joints or to discover hidden breaks in electrical cables. If no great demands are made on fine detail or latitude, the use of X-ray paper instead of film may be adequate. Xeroradiography has also been applied successfully in this type of work. (R. L. Durant, 1957; R. L. Durant and C. G. Pollitt, 1966).

10.8   SPECIALIZED TECHNIQUES

There are many additional techniques used in industrial radiography, some of them are frequently used in certain laboratories, others are less conventional, but they all play a part in the detection of defects.

10.8a Fluoroscopy

The viewing of a positive X-ray image on a fluorescent screen (see p. 293) called 'fluoroscopy' is mainly applied to mass inspection and it is limited in scope, owing to poorer sensitivity of defect detection, when compared with radiography. Fluoroscopy is also restricted because of the visual fatigue of the operator and because of the difficulties in providing adequate protection to him. Apart from this there is the necessity of viewing with dark adapted eyes. Sufficient protection against radiation below the maximum permissible dose rate can be achieved by adequate lead shielding up to about $200\ kV_p$ and by the provision that the eye views the fluorescent image after reflection from an inclined mirror, which is covered by lead glass. Thus the observer receives no direct radiation.

Fluoroscopy being cheaper than film radiography, may be applied in cases where less critical inspection can be accepted. Although it is frequently used in checking mass produced light alloy castings, plastic mouldings, ceramics, rubber and other materials, the main application of fluoroscopy deals with the routine inspection of manufactured articles. Here it may be used to check e.g. the correctness of position of parts in an assembly, the correct spacing of components in ordnance work for the inspection of fuses, or broken or misplaced wires in electrical appliances. In the food industry it may be used for the detection of foreign particles in cans, for frost damage in oranges, grapefruits etc. In fact there exists an almost unlimited number of applications, where it has been or might be used in future. (L. Mullins, 1949; J. C. Rockley, 1957, 1965; R. C. McMaster, 1959).

Fluoroscopy also offers the opportunity of inspecting moving parts in specimens and thus the possibility of cine-radiography from the screen (see p. 362). S. C. Fry (1945) after studying fluoroscopic images of molten metal has taken cine-radiographs of the motion of a stream of metal during the initial period of pouring metal into a mould. This type of work can now be carried out by means of image amplifiers (see p. 365), which facilitate the problem of radiation hazards particularly in connection with closed circuit television systems. (R. L. Durant and B. J. Vincent, 1964).

10.8b Localization of Defects

Industrial radiography, being a method of projecting a three-dimensional object on to a two-dimensional plane, prevents a simultaneous record of a third dimension. Often, however, it is of importance to locate a defect with respect to its space coordinates in the specimen, in order to find whether its position may be critical in regard to the serviceability of the specimen.

Stereoradiographic methods (see p. 353) which are used for three-dimensional viewing in medical radiography are rarely applied in the industrial field. The reasons for this are chiefly because there is a variety of other very simple localization techniques which can easily be applied with little special apparatus and can provide precise measurement in space. Furthermore measurements in space by stereoradiography requires expensive precision apparatus. It is also not easy for many observers to achieve a three-dimensional view of a specimen, particularly if it is thin, unless there are certain reference

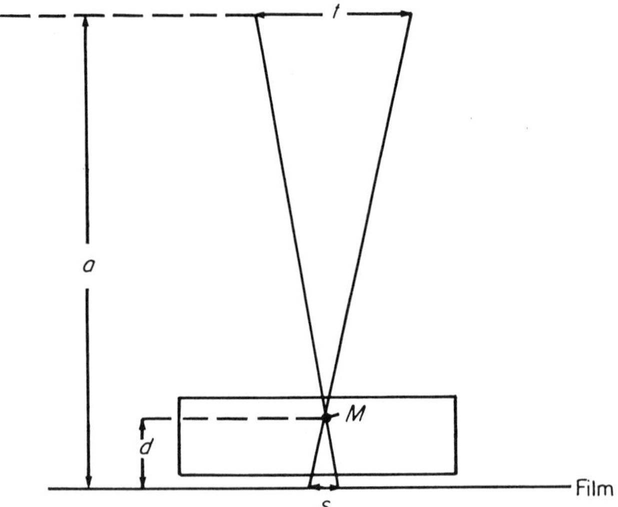

Figure 10.31. Diagram illustrating the tube shift method. a = vertical distance between focal spot and film, d = required distance between defect and film, t = tube shift distance, s = distance between image displacements, M = position of defect.

points in the specimen which can act as a guide for the viewer. (W. Watson, 1957).

Two of the mentioned localization methods will now be discussed, one is called the tube-shift method, the other the lead marker method.

Tube-Shift Method

Using the tube-shift method two exposures of the specimen are made subsequently but on the same film by shifting the X-ray tube or the radioactive source by a known distance between the two exposures, as indicated in Figure 10.31.

Suitable values for a, in the figure, which is the vertical distance between tube position and film and for t, which is the tube shift distance are for instance 60 cm and 10 cm (i.e. 24 and 4″) respectively, but as these values depend on individual circumstances a suitable ratio of $a/t = 6$ is suggested. From the geometry of similar triangles, it can be seen from Figure 10.31 that

$$\frac{a - d}{t} = \frac{d}{s} \qquad (10.22)$$

$$sa - sd = td$$

or

$$d = \frac{sa}{t + s} \qquad (10.23)$$

Hence the required distance d between the defect and the film can be found from the known values of a, t, and s which latter is the radiographic displacement of the defect M. It is important that the image shift is measured in a direction parallel to the tube shift. Furthermore if the defect is linear, i.e. considerably longer in one than in the other direction, the tube shift should be arranged in a direction perpendicular to the length of the defect. This means that the specimen should be turned in to the required position before the radiographs are taken. The measurement is based on the defect to film distance, hence an allowance should be made for distance between the underside of the specimen and the film in its cassette.

Lead Marker Method

This method has the advantage that only the thickness of the specimen must be accurately known for the localization of the defect. The depth of the latter is determined from a measurement of image displacements on the radiograph.

Two lead markers, e.g. two lead wires of about 1 inch in length may be used, one in contact with the upper surface of the specimen, the other one should be fixed on the bottom of the specimen, which itself need not be in contact with the cassette. The position of the lead wires which serve as reference markers should be parallel to the main axis of the defect. Care must be taken that the radiographic images of the markers do not coincide or superimpose on the film, when two consecutive exposures are taken on the same film at two positions of the X-ray tube. The direction of the tube shift should be perpendicular to the main axis of the defect, if this is predominantly linear. The two subsequent exposures will reveal double images of the markers and all the defects in the specimen, as indicated (for one

defect only) in Figure 10.32 which shows a radiograph (diagrammatically) of the various image shifts. The image shifts between the markers
and between the defects will be proportional to their respective
distances from the film. Hence the image shift of the bottom surface
marker is very small, depending on its distance from the film. After
the image shifts have been measured on the radiograph, a straight line
graph can be drawn indicating the relationship between the known
image shifts and the corresponding distances of the markers from the
lower surface of the specimen. Knowing the image shift of the defect

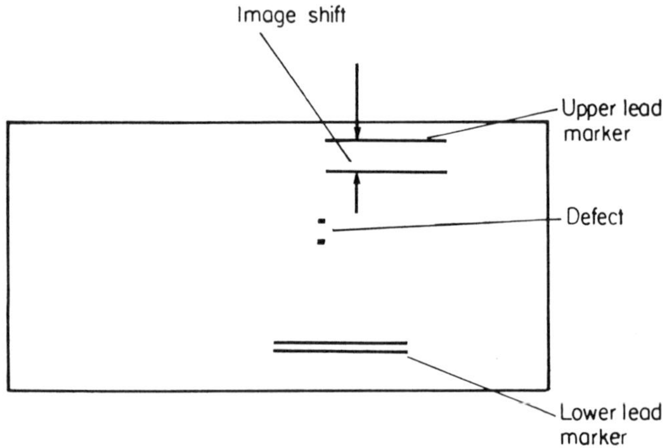

Figure 10.32. Lead marker method. Radiograph (diagrammatically) showing
image shifts of upper and lower lead markers and of defect after two consecutive
exposures on same film.

a derivation of its depth within the specimen can be made using the
graph (Figure 10.33).

A typical example may help the understanding of the principle of
the method. Below are given (a) the essential data, (b) the non-
essential data and (c) the data as evaluated from the exposed radiograph.

(a) *Essential data:*

Thickness of specimen   1.5 inches

(b) *Non-essential data:*

Tube shift distance   6 inches
Distance between focal spot and upper surface of
   specimen   20 inches
Distance between bottom of specimen and film   0.5 inch

(c) *Data evaluated from radiograph:*
Image shift of upper lead marker   0.65 inch
Image shift of lower lead marker   0.16 inch
Image shift of defect   0.4 inch

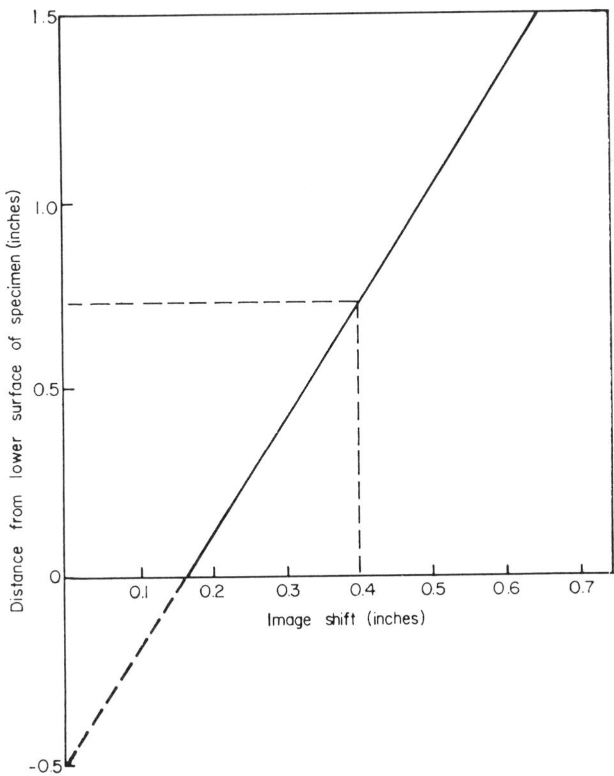

Figure 10.33. Relationship between distance of markers and defect from lower surface of specimen (in inches) and respective image shift (in inches) in the radiograph.

By plotting the distances of the markers from the bottom of the specimen on the ordinate and the respective image shifts of the lower and upper lead markers on the abscissa axis, we obtain two points in the graph through which a straight line is drawn, see Figure 10.33.

As the image shift of the defect was found from the radiograph to be 0.4 inches, the defect is 0.73 inches from the bottom of the specimen.

By extrapolating the straight line towards the negative part of the ordinate, we find that the vertical distance between the lower surface of the specimen and film is 0.5 inches which corresponds to the actual value taken in our case.

Thus it is seen that apart from the data evaluated from the radiograph only the thickness of the specimen need be known in order to determine the position of the defect.

10.8c   Film Sandwich Method

Specimens varying widely in thickness often present the difficulty that the whole specimen cannot be presented adequately in one radiograph, as has been mentioned on several occasions. Although the use of filters or harder radiation has been recommended in a previous section (p. 454) these methods show only a limited success. Another suitable method to overcome the difficulty is to cover the variations in thickness by the use of several films and screens exposed simultaneously. This is usually done by exposing through a variety of slower films placed in the front of the cassette and then through a faster type of film placed behind the slow ones. Several examples of such combinations of films and screens are quoted below and the key to this table is given by the following abbreviations.

$$
\begin{array}{ll}
A & \text{slow speed film} \\
B & \text{medium speed film} \\
C & \text{high speed film} \\
S_F & \text{front lead intensifying screen} \\
S_B & \text{back lead intensifying screen}
\end{array}
$$

In the following Table 10.13 combinations of film and screen are listed in vertical columns; this means that the first mentioned film or screen in each vertical column is struck first by radiation.

A variety of combinations is shown and the reader has to find which of these is particularly suitable for the specimen in question and the kilovoltage in use. As a further variable the kilovoltage may be

TABLE 10.13. Suitable Combinations of Film and Screen in Sandwich Method

A	A	S_F	S_F	S_F	S_F
S_F	S_F	A	A	A	A
B	B	B	S_B	S_B	S_B
	S_B		B	S_F	C
				B	

reduced or raised accordingly. It will be appreciated that the basic idea of the method is to reveal the thick portions of the specimen in the fast films and the thin portions in the slow variety.

Another sandwich method has been suggested whereby several films and screens are registered during exposure in such a way that they can, after processing, be viewed when superimposed on each other. This technique may provide increased contrast by the super-position of the various under-exposed films when viewed on a high intensity illuminator and a wider variety of thicknesses showing lower contrast can be seen when only one film is viewed, using a conventional viewing box.

10.8d Flash Radiography

Flash radiography sometimes called 'high speed radiography' is a technique used for radiographing rapidly moving objects, such as projectiles in flight, explosive phenomena, moving pistons and other objects in motion. The problem involved in flash radiography is chiefly concerned with the production of exposures of very short duration and extremely high output. This means that very specialized X-ray equipment is needed to carry out this work. One of the chief radiographic considerations in flash radiography is to obtain a reasonable image sharpness in spite of the high velocity of the motion of the object to be radiographed. A simple calculation of an arbitrary example shows that for a velocity of an object of say 50 meters per second, an exposure time of 10 microseconds is required, in order to obtain a movement unsharpness of 0.5 mm. Assuming that the thickness of the moving object is $\frac{1}{2}$ inch steel and that $200\,\mathrm{kV_p}$ should be used in conjunction with screen X-ray film and salt screen, about 1–2 mA seconds would be required at 1 meter distance from the focal spot in order to obtain a density of unity. In general direct X-ray film cannot be used, as the output of X-ray machines is usually not great enough to obtain sufficent density. In radiography of projectiles even shorter exposure times than those mentioned above, namely of the order of 1 microsecond are required, thus 1 mA second would lead to a momentary tube current of 1000 amperes. Hence it is not surprising that special X-ray equipment is needed to carry out the short exposures at extremely high tube currents.

Furthermore special trigger circuits are required to provide the synchronization of the X-ray flash with the motion of the object. Much research has been devoted to the development of X-ray flash equipment and X-ray flash tubes and a brief indication of this development will be given.

The function of some of the X-ray units used in the past was based on the discharge of a capacitor charged to high voltage and discharged to the X-ray tube. The trigger mechanism was built into the flash tubes. With some of these units X-rays were generated between 180 to 300 kV using very high tube currents at 10 microseconds duration and shorter. Special types of X-ray flash units were developed more recently by the Field Emission Corporation, U.S.A. which make use of a cold cathode X-ray tube (Fexitron type). Some of these tubes work at 300–600 kV_p using maximum currents of 1000 amperes at fractions of 1 microsecond. A hot cathode tube developed by the same corporation is available for maximum ratings up to 600 kV_p, 2000 amperes and 0.2 microsecond exposure time. The effective focal spot of this tube is 2.5×7 mm.

Linear accelerators and betatrons have also been used successfully in flash radiography at megavoltage energies with pulse lengths of 3–6 microseconds.

For the development work on X-ray equipment and flash X-ray tubes the reader is referred to papers by R. H. Kingdon and H. E. Tanis (1938), M. Steenbeck (1938), R. Schall (1956), R. F. Thumwood (1959), and on the Fexitron type flash X-ray equipment see references to W. P. Dyke and W. W. Dolan (1956), E. E. Martin, J. K. Trolan and W. P. Dyke (1960), W. P. Dyke, and F. M. Charbonnier (1960).

If the X-ray flash exposure is synchronized with a particular phase of a cyclic motion a radiograph can be taken of a machine, e.g. of a motor whilst in running condition (stroboscopic radiography.) The exposure may be made using a series of X-ray pulses, but only one pulse per cycle is utilized. This method allows the study of the effects of vibrations on components of a machine and may be recorded or viewed on a television equipment connected to an image amplifier. C. Mitchell and R. A. Pulk (1956), B. J. Vincent (1960).

A review article on the subject of flash radiography is given by R. Meakin (1962).

10.8e   Color Radiography

The idea of using color in radiography is not new and several proposals have been made in the past. Advantages of color compared with conventional radiography have been claimed by M. B. Bryce (1955), J. M. Blais, and A. K. Schwerin (1955). It is probably true to say that the human eye is in general more sensitive to small differences in color than to small differences in grey. Successful attempts to use color in industrial radiography have been made notably by N. S. Beyer (1961) and by R. W. Parish and D. A. W. Pullen (1964, 1966).

Color radiography is still in a stage of development in a few laboratories only. From the work done so far, it appears that its main advantage if compared with conventional radiography is an improvement of latitude due to large differences of color gradation. The technique makes use of commercially available color films in combination with salt and also with lead intensifying screens. The arrangement used by Beyer is shown in Figure 10.34.

The diagram shows a section through color film sandwiched between salt and lead intensifying screens. The arrangement illustrated in

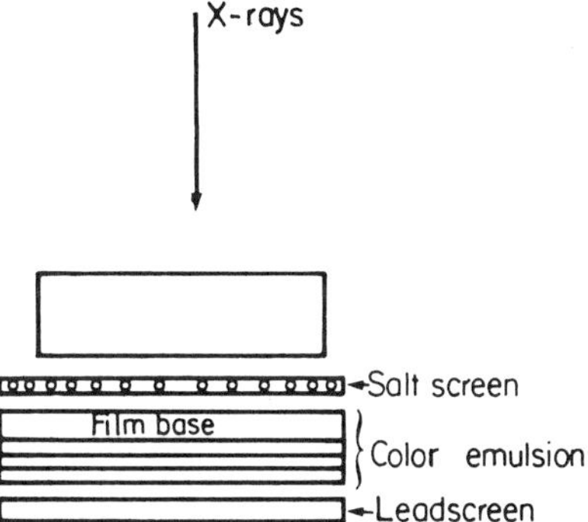

Figure 10.34. Arrangement for color radiography; diagram illustrating section through color film sandwiched between salt and lead intensifying screens. (*After* N. S. Beyer (1961).)

Figure 10.34 can also be reversed or otherwise modified to suit individual conditions. After the exposure to X-rays or γ-rays is given, the color film is developed conventionally but an additional exposure during development has been found advantageous. This additional flash exposure is carried out by colored light, e.g. yellow, red or green during an interruption of the development procedure after partial development has taken place. The flash exposure, according to an example given by Beyer, takes $2\frac{1}{2}$ seconds using light from a high intensity lamp, filtered by a Kodak Series 7 filter, the lamp being at 36″ from the film. After the flash exposure normal processing is completed. The purpose of the flash exposure is that the image of the top emulsion layer is printed on the two other layers of the color film. It has not been found necessary to adopt the flashing technique in all cases. The exposure times required for color radiography are of the same order as those used for fine grain low speed no-screen film with lead screens and any radiation qualities can be used with this sytem. Some quite remarkable color radiographs have been shown by N. S. Beyer and by R. W. Parish and D. A. W.

Pullen using Ektachrome and Ektacolor films. The latter authors of the Atomic Energy Establishment at Harwell, England use color radiography as a routine method.

The color radiographs by Beyer show partly superimposed samples of uranium foils of 0.001″ and 0.002″ thickness and reveal for each thickness distinct color changes from yellow-brown-orange to red, when the film was exposed to a red flash light; a change to green appeared in the radiograph of the thickest foil, after another exposure using green flash light.

Beyer took some color radiographs in the field of crack detection, using fluorescent solutions which filled the cracks of the specimen. The solution fluoresced in the red and the radiographs show after processing fine 'red' crack lines against a dark greenish background.

There are so many modifications possible in the use of color film techniques that much more experimental work is required in this field, before a final judgement can be given as to the usefulness of the method. From the evidence given so far, it appears that in certain cases differences in a specimen can be revealed in one color radiograph for which two conventional radiographs would be required to give the same information.

10.8f   Electron Radiography

Electron radiography is a specialized method which plays a role in forensic applications, in the identification of water marks in documents, of fingerprints, textures, handwritings etc. It is also used in philatelic studies and in some metal investigations. Its potential advantage in other fields has probably not been fully realized yet. There are various modifications of electron radiographic methods, some of which are based on transmission, others on surface emission of electrons.

Transmission Electron Radiography

A thin sample such as a forged document, a postage stamp or a banknote having water marks is sandwiched between a lead foil and a photographic film. The combination of the lead foil-specimen-film system is placed in an X-ray cassette to ensure firm contact, the lead screen facing the X-ray tube. During exposure the X-ray tube is generated at 200–250 kV_p and the beam is heavily filtered by about 5–10 mm of copper to eliminate the softer components of the radiation to which the film is more sensitive. An additional aluminum filter of about 2 mm thickness below the copper filter provides some improvement in reducing the fog due to characteristic Cu-radiation. The mechanism of exposure is based on the emission of electrons from the lead foil and the electrons are differently absorbed by the various areas of the thin specimen. In this way electrons are more easily transmitted through the less dense regions and those composed of elements of low atomic number, than through the denser regions and those made up of elements of higher atomic number. Hence the water mark, being due to a separation of the paper fibres, may be revealed as dark lines on the film. High contrast images can be obtained by this method. It is advisable to use a lead intensifying screen as it is applied in industrial radiography. As film material Process film or a fine grained X-ray film may be used.

It should further be mentioned that the thickness of the specimen paper which can be penetrated by the electrons emitted by the lead is limited to about 200 microns. Some surface emission of electrons and characteristic radiation excited by X-rays from the elements composing the print or ink on the document may produce an image on their own but not of the water mark (see next section). Although the use of Grenz rays, i.e. very soft X-rays can also be used to show watermarks, the method described is in general more easily applied in practice. (H. S. Tasker and S. W. Towers, 1945; H. C. Pollack, C. F. Bridgman, and H. R. Splettstosser, 1955; W. H. S. Cheavin, 1957, and 1958; D. Graham and H. C. Gray, 1964).

Surface Emission Electron Radiography

Study of Metal Surface.   This method which is primarily due to J. J. Trillat (1948) is also based on the use of hard X-rays of the order of $200\text{--}250\ \mathrm{kV_p}$ which are heavily filtered to eliminate the softer components of the radiation. The X-ray beam passes first through a fine grained photographic emulsion which is in intimate contact with a polished surface of the metal sheet or block to be examined. The hard radiation excites electrons and characteristic radiation from the polished surface of the specimen. As the response of the photographic emulsion to electrons is so much greater than that to the hard primary X-rays passing through the emulsion, the latter will have little effect on the film. If the specimen surface contains elements of different atomic number a differentiation between the elements may be revealed, as the electron emission differs between elements of various atomic number in quantity as well as in range. If the atomic numbers are very close, differentiation may be less striking or may be difficult to obtain. The advantage of the method lies in the fact that thick specimens can be used, although back scatter from the specimen may cause some fog. The density distribution achieved in the photographic material can be measured by means of a microdensitometer and can be related quantitatively if a photographic calibration of the constituents has been carried out. Enlargement of the electron radiograph thus obtained may be advantageous.

Surface Emission From Documents and Stamps.   It has already been pointed out that an analogous method to that just described can be applied with documents and stamps, when only electron emission and characteristic radiation excited by hard X-rays are responsible for the exposure on the film. The ink used in the stamp manufacture is often composed of mercuric oxide which is a good electron emitter.

Study of Fingerprints.   More recently further electron radiographic methods have been suggested by D. Graham and H. C. Gray (1965), which permit electron radiographic records of fingerprints to be taken. This study is carried out by spreading a fine powder of lead particles (200–400 meshes per inch) on the finger print which may practically be on any material, metal, wood, paper or even human skin. The lead dust adheres to the fingerprint and is less likely to adhere to areas outside it. Excess lead dust and also lead dust surrounding the fingerprint have to be removed by tapping before exposure. On exposure the photographic image is produced by the electrons emitted from the lead dust, when excited by the hard X-ray beam. The exposure technique can be carried out either by placing the document with fingerprint together with the photographic material into a cassette in an arrangement as shown in Figure 10.35(a) or if no cassette can be used, as the fingerprint may be on a wall or on a door, by an arrangement as illustrated in Figure 10.35b. In this figure it is assumed that the exposure is carried out in a darkened room. (For exposures in daylight a special cassette is required, as mentioned towards the end of this section.)

X-Rays of 200 kV_p are used with about 1 cm Cu filtration. A fine grained no-screen X-ray film or a Process type film are recommended as photographic material. As seen from Figure 10.35(a) and (b) in addition an aluminum filter and film base are introduced according to the suggestion of Graham and Gray (1965). The aluminum serves to absorb the characteristic radiation of Cu and the film base serves to absorb the soft characteristic Al radiation. As long as the lead powdered fingerprint can be brought to the X-ray unit, there is no particular difficulty, but this is of course not always possible. The present author found that the use of radioisotopes may overcome the difficulty of transport of a bulky X-ray unit. Successful, fingerprint electron-radiographs have been taken with ^{137}Cs and ^{192}Ir sources, without Cu-filtration.

For exposures to be made in daylight, a special cassette is required which permits the film to be pressed in firm contact with the lead powdered fingerprint after a sliding cassette lid has been removed. If there is a spacing between the specimen and the recording film, the image will be blurred.

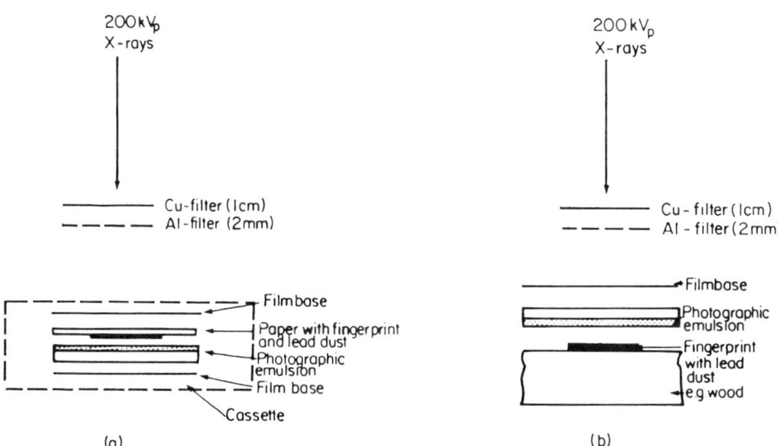

Figure 10.35. Arrangements for electron-radiographic fingerprint exposures, using 200 kV$_p$ X-rays.
(a) With cassette. (b) Without cassette, but in the dark. (*After* D. Graham and H. C. Gray (1965).)

The advantage of this type of fingerprint electron radiography is that only the fingerprint and no other background will be recorded on the film. It further offers the possibility of recording the fingerprint on a cylindrical surface as the flexible film may be fixed around this.

BIBLIOGRAPHY

Clauser, H. R., *Practical Radiography in Industry*. Reinhold, New York.

Crowther, J. A. (1944) (Ed.) *Handbook of Industrial Radiology*, Edward Arnold, London.

Eggert, J., and Gajewski, H., *Einführung in die technishe Röntgenphotographie*, S. Hirzel, Leipzig.

Halmshaw, R. (1966) (Ed.) *Physics of Industrial Radiology*, Heywood, London. Elsevier, New York.

Hinsley, J. F. (1959). *Non-Destructive Testing*, MacDonald and Evans, London.

Kodak Limited England, *Industrial Radiography* and Eastman Kodak Company U.S.A. *Radiography in Modern Industry*.

Low, K. S., (1940). *Metallurgical and Industrial Radiology.*, Pitman, New York.

McMaster, R. C. (1959). *Non-Destructive Testing Handbook.* Vol. I and Vol. II, Ronald Press, New York.

Rockley, J. C. (1965). *An Introduction to Industrial Radiology.* Butterworths, London.

Sproull, W. T., *X-rays in Practice.* McGraw-Hill, New York.

Wiltshire, W. J. (1957) (Editor) *A Further Handbook of Industrial Radiology,* Edward Arnold, London.

REFERENCES

American Standards Association. Inc., (1964). *Method for The Sensitometry of Industrial X-Ray Films for Energies up to 3 million Electron Volts.* PH2, 8.

American Society of Mechanical Engineers. A.S.M.E. Sec. 1. *Code for Power Boilers* Sec. 8. *Code for Unfired Pressure Vessels.*

American Society for Testing Materials. A.S.T.M./E94-52T. *Recommended Practice for Radiographic Tests.*

American Society for Testing Materials. A.S.T.M./E155-60T. *Reference Radiographs for the Inspection of Aluminum and Magnesium Castings.* Series 2.

American Society for Testing Materials. A.S.T.M./E99-55T. *Reference Radiographs for Steel Welds.*

American Society for Testing Materials. A.S.T.M./E98-53T. *Reference Radiographs for the Inspection of Aluminum and Magnesium Castings.*

American Society for Testing Materials. A.S.T.M./E71-52T. *Radiographic Standards for Steel Castings.*

American Petroleum Institute. A.P.I. 650, *Welded Oil Storage Tanks.*

American Standards Association. A.S.A. B.31. *Refinery Piping.*

Beyer N. S. (1961). Color Radiography. *Second Symposium on Physics and Nondestructive Testing,* held at Argonne National Laboratory, October. ANL-6515, p. 203.

Blais, J. M., and Schwerin, A. K. (1955). A new color radiographic process. *Radiography.* **21.** 254.

British Standard 499 Part 3: (1965). *Welding Terms and Symbols. Terminology of and Abbreviations for Fusion Weld Imperfections as Revealed by Radiography.*

British Standard 2600: (1962). *General Recommendation for the Radiographic Examination of Fusion Welded Butt Joints in Steel.*

British Standard 2654 Part 2. *Vertical Mild Steel Welded Storage Tanks with Butt Welded Shells for the Petroleum Industry.*

British Standard 2737: (1956). *Terminology of Internal Defects in Castings as Revealed by Radiography.*

British Standard 2910: (1965). *General Recommendations for the Radiographic Examination of Fusion Welded Circumferential Butt Joints in Steel Pipes.*

British Standard 3451: (1962). *Methods of Testing Fusion Welds in Aluminium and Aluminium Alloys.*

British Standard 3971: (1966). *Image Quality Indicators for Radiography and Recommendations for their use.*

British Standard 4080: (1966). *Methods for Nondestructive Testing of Steel Castings (Metric units).*

British Standard 4097: (1966). *Gamma Radiography Exposure Containers for Industrial Purposes and their Source Holders (Metric units).*

B.W.R.A. Report. (1958). Recommendations for the X-ray examination of fusion welded joints in light alloys. *Brit. Welding J.* **5,** 489.

Bryce, M. B. (1955). Experimental color radiography. Brit. J. Radiol. **28,** 552.

Burril, E. A. (1952). Modern techniques in high-voltage radiography. *Nondestructive Testing* **11**(2) 23.

Cheavin, W. H. S. (1957, 1958). Watermarks and their detection. *Sanders' Philat. Journal.* **12,** 187 and **2,** 225.

Clack, B. N. (1957). Natural and artificial sources for gamma radiography. In W. J. Wiltshire (Ed.), *A Further Handbook of Industrial Radiology.* Ch. 2. Edward Arnold, London.

Corney, G. M., and Seemann, H. E. (1947). A logarithmic step tablet for X-rays. *Nondestructive Testing* **6,** 227.

Croxson, C. (1957). Radiographic quality and sensitivity. In W. J. Wiltshire (Ed.), *A Further Handbook of Industrial Radiology.* Ch. 4. Edward Arnold, London.

Durant, R. L. (1948). Characteristics of some non-screen films used in industrial radiology. *J. Sci. Instr.* **25,** 105.

Durant, R. L. (1951). Characteristics of some further films suitable for use in industrial radiography. *Brit. J. Appl. Phys.* **2,** 32.

Durant, R. L. (1957). Industrial xeroradiography. In W. J. Wiltshire (Ed.) *A Further Handbook of Industrial Radiology.* Ch. 14. Edward Arnold, London.

Durant, R. L. (1966). The response of recording and screening media. In R. Halmshaw (Ed.), *Physics of Industrial Radiology.* Heywood Books, London. Elsevier.

Durant, R. L., and Pollit, C. G. (1966). Xeroradiography and ionography. In R. H. Halmshaw (Ed.). *Physics of Industrial Radiology.* Ch. 15. Heywood Books London, Elsevier, New York.

Durant, R. L., and Vincent, B. J. (1964). Fluoroscopy with high energy X-rays. *Proceedings. 4th International Conference on Nondestructive Testing,* Butterworths, London, p. 61.

Dyke, W. P. and Charbonnier, F. M. (1960). Electrical stability and life of the heated field emission cathode. *J. Appl. Phys.* **31,** 790.

Dyke, W. P., and Dolan, W. W. (1956). *Advances in Electronics and Electron Physics,* Vol. 8. Academic Press, New York., p. 90.

Franz, W. (1935). Rayleigh scattering of hard radiation by heavy atoms. *Zt. f. Phys.* **98,** 314.

Fry, S. L. (1945). Cineradiography. Application to the investigation of metal pouring. *Metal Industry*, 67. Nos. 1 & 2 p. 2 and 25.

Fuchs, E., Mullins, L., and Smith, S. H. (1947). Inspector's approach to radiographs of mild steel butt welds. *Trans. Inst. Weldg.*, **10**, 19.

Graham, D., and Gray, H. C. (1964). The Investigation of suspect receipts by electron radiography. *X-Ray Focus*. **5**, 3.

Graham, D., and Gray, H. C. (1965). The application of X-ray techniques in forensic investigations. *X-Ray Focus*, **6**, 2.

Halmshaw, R. (1954). Use and scope of iridium-192 for the radiography of steel. *Brit. J. Appl. Phys.*, **5**, 238.

Halmshaw, R. (1955a). The factors involved in an assessment of radiographic definition, *J. Phot. Sci.*, **3**, 161.

Halmshaw, R. (1955b). Thulium-170 for industrial radiography. *Brit. J. Appl. Phys.* **6**, 8.

Halmshaw, R. (1964). Photographic aspects of the definition attainable in radiographs. *J. Phot. Sci.*, **12**, 2, 110.

Halmshaw, R. (Ed) (1966). *Physics of Industrial Radiology*. Heywood Books, London, American Elsevier, New York.

Herz, R. H. (1943). Some fundamental aspects in the processing of X-ray films. *Phot. J.*, **83**, 343.

Herz, R. H. (1956). The spectral quantum and energy efficiency of calcium-tungstate X-ray intensifying screens. *Brit. J. Appl. Phys.*, **7**, 182.

Hinsley, J. F. (1957). Castings and their radiography. In W. J. Wiltshire (Ed.), *A Further Handbook of Industrial Radiography*. Ch. 10. Edward Arnold, London.

International Institute of Welding, I.I. W. T. C5, *Collection of Reference Radiographs of Welds*.

Kingdon, R. D., and Tanis, H. E. (1938). Experiments with a condenser discharge X-ray tube. *Phys. Rev.*, **53**, 128.

Klasens, H. A. (1946). Measurement and calculation of unsharpness combinations in X-ray photography. *Philips Res.* **1**, 241.

Knipe, G. F. G. (1943). The sensitometric estimation of radiographic exposures. *Phot. J.*, 338.

Larson, E. T., and Nitka, H. F. (1961). The meaning of characteristic curves. *Second Symposium on Physics and Nondestructive Testing*. Argonne National Laboratory A.N.L.-6515 p. 87.

Lamar, E. S., Hecht, S., Shlaer, S. and Hendley, C. (1947). Size, shape and contrast in detection of targets by daylight vision. *J. Opt. Soc. Am.*, **37**, 531.

Liden, K., and Starfeldt, N. (1955). Thulium-170 for industrial radiography. *Brit. J. Appl. Phys.* **6**, 262.

Lowry, E. M. (1931). The photometric sensibility of the eye and the precision of photometric observations. *J. Opt. Soc. Am.*, **21**, 132.

Martin, E. E., Trolan, J. K., and Dyke, W. P. (1960). Stable high density field emission cold cathode. *J. Appl. Phys.*, **31**, 782.

Meakin, R. (1957). Scattered radiation. In W. J. Wiltshire (Ed.), *A Further Handbook of Industrial Radiology*. Ch. 3. Edward Arnold, London.

Meakin, R. (1962). Flash radiography. In *Progress in Applied Material Research*, Vol. 4. Heywood Books, London. pp. 43–79.

Mitchell, C., and Pulk, R. A. (1956). Applications and problems in stroboradiography. *Nondestructive Testing*, **14**, (5)24.

Morgan, R. H. (1949). An analysis of the physical factors controlling the diagnostic quality of roentgen images. Part 5. Unsharpness. *Am. J. Roentgenol.* **62,** 870.

Mullins, L. (1949). Some uncommon applications of industrial radiography. In J. A. Crowther. (Ed.), *Handbook of Industrial Radiology*, Ch. 8. Edward Arnold, London.

Mullins, L. (1957). Weld radiography. In W. J. Wiltshire (Ed.), *A Further Handbook of Industrial Radiology*. Ch. 9. Edward Arnold, London.

Neeff, Th. (1925). Metall Untersuchung mittels Röntgenstrahlen. *Zt. f. Techn. Phys.* **6,** 208.

Nemet, A., Cox, W. F., and Walker, G. B. (1946). Blurring in radiography. *Brit. J. Radiol.*, **19,** 257.

Nutting, P. G. (1916). Effects of brightness and contrast in vision. *Trans. Illum. Eng. Soc.*, **11.** 939.

Parish, R. W., and Pullen, D. A. W. (1964, 1966). The development of colour radiography into a routine service at A.E.R.E. Harwell. *Br. J. Nondestructive Testing*, **6,** 103. and A.E.R.E.-R 4515.

Polansky, D., Holloway, J. A., and Criscuolo, E. L. (1964). Characteristics of metallic screens in the megavolt X-ray region. *Proc. 4th International Conference on Non-Destructive Testing*. Butterworth, London. p. 80.

Pollack, H. C., Bridgeman, C. F., and Splettstosser, H. R. (1955). The Investigation of postage stamps. *Med. Radiography and Phot.*, **31,** 2.

Rockley, J. C. (1957). Fluoroscopy. In W. J. Wiltshire (Ed.), *A Further Handbook of Industrial Radiology*. Ch. 7. Edward Arnold, London.

Rockley, J. C. (1965). *An Introduction to Industrial Radiology*. Butterworths, London.

Rose, A. (1946). A unified approach to the performance of photographic film, television pick-up tubes and the human eye. *A.J.S.M.P.E.*, **47,** 273.

Rose, A. (1948). Sensitivity performance of the human eye on an absolute scale. *J. Opt. Soc. Am.* **38,** 196.

Rossman, K., and Lubberts, G. (1966). Some characteristics of the line spread function and modulation transfer function of medical radiographic films and screen-film systems, *Radiology*, **86,** 2, 235.

Rumyantsev, S., *Industrial Radiology. Foreign Languages Publishing House*. Moscow.

Schall, R. (1956). *Proceedings of 3rd Congress on High Speed Photography* (1956). Butterworth, London, p. 228.

Seeman, H. E. (1937). Some physical and radiographic properties of metallic intensifying screens. *J. Appl. Phys.*, **8,** 836.

Seemann, H. E., and Roth, B. (1960). New stepped wedges for radiography, *Acta. Radiol.*, **53,** 215.

Seemann, H. E., and Roth, B. (1962). A new stepped wedge design for industrial radiography. *Nondestructive Testing,* **20,** 37.

Splettstosser, H. R. (1967). The visibility of detail obtainable with industrial X-ray film. *Mater. Eval.,* **25,** 245.

Steenbeck, M. (1938). *Wiss. Veröff.* Siemens Werke, Berlin, **17, 363.**

Tasker, H. S. (1944). The response of photographic materials to X-rays. In J. A. Crowther, *Handbook of Industrial Radiology.* Ch. 4. Edward Arnold, London.

Tasker, H. S. (1945). Some properties and uses of X-ray intensifying screens. *Phot. J.* Section B, Vol. 85B., 4, 75.

Tasker, H. S., and Towers, S. W. (1945). Electron radiography, using secondary β-radiation from lead intensifying screens. *Nature* **156,** 50.

Thumwood, R. F. (1959). Production of X-rays during a low-pressure gas discharge. *Brit. J. Appl. Phys.,* **10,** 147.

Towers, S. W. (1957). Films and their processing. In W. J. Wiltshire (Ed). *A Further Handbook of Industrial Radiology.* Ch. 5. Edward Arnold, London.

Trillat, J. J., (1948). Electronic radiography and microradiography. *J. Appl. Phys.* **19,** 844.

Vincent, B. J. (1960). Stroboscopic radiography with a linear accelerator. *Brit. J. Appl. Phys.,* **11,** 132.

Watson, W. (1957). Stereo-methods. In W. J. Wiltshire (Ed.), *A Further Handbook of Industrial Radiology.* Ch. 13. Edward Arnold, London.

11

Neutron-Radiography

11.1 INTRODUCTION

Neutron radiography is complementary to X- and γ-ray radiography and its significance is based on the fact that the absorption by materials exposed to neutrons differs entirely from that exposed to X- and γ-rays. The absorption of thermal neutrons in light and heavy material is practically the reverse of that of X-rays. This difference of absorption leads to radiographic discrimination of certain elements, which is not possible with conventional X- and γ-ray radiography. Because of the low absorption of thermal neutrons in some heavy and thick specimens, neutron radiography can lead to shorter exposure times than γ-radiography. Apart from this the method is useful in the inspection of radioactive materials, such as radioactive fuel element containers, which can be carried out without fogging the photographic material. Certain light elements can easily be distinguished radiographically.

Neutron radiography is essentially used in non-destructive testing of industrial and engineering assemblies. A considerable amount of research has been carried out with regard to the techniques and potentialities of neutron radiography within the last ten years, indicating a growing interest in this method.

Neutron radiographic techniques originate from investigations by H. Kallmann and E. Kuhn, Germany in 1935, O. Peter (1946), H. Kallmann (1948) and were taken up again in England by J. Thewlis (1956). Most of the detailed work in the last decade was reported by H. Berger (1960–1968), Metallurgy Division, Argonne National Laboratory, Argonne, Illinois, U.S.A. The following sections are mainly based on his work.

483

This chapter will deal chiefly with the neutron absorption characteristics of the elements of the periodic system, with the sources of neutrons and photographic detection techniques and with some applications of neutron radiography.

11.2   THE ABSORPTION OF NEUTRONS

Whatever radiation is used in radiography, the resulting image is essentially controlled by the differential absorption of the radiation

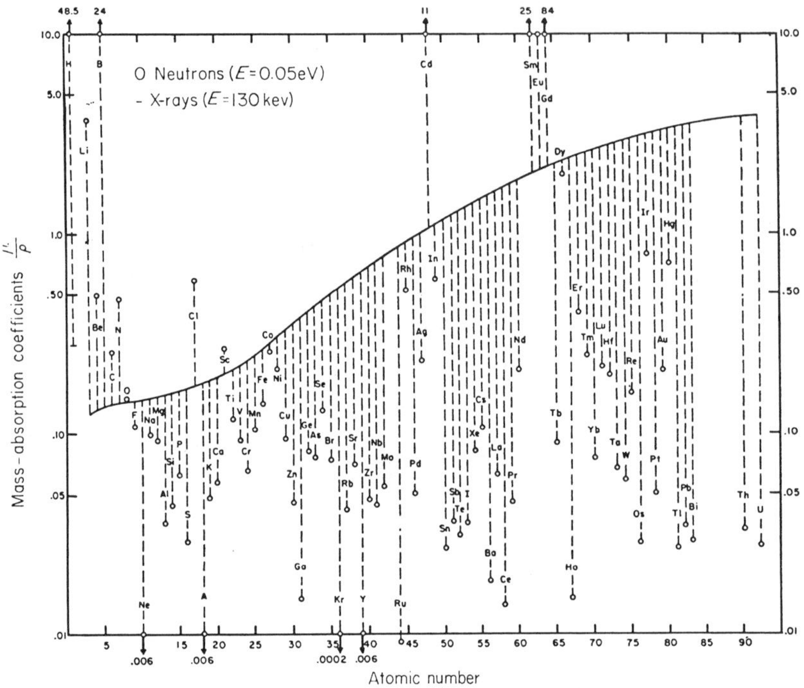

Figure 11.1. Mass-absorption coefficients of the elements for neutrons and X-rays. (*After* J. Thewlis, *Brit. J. Appl. Phys.* **7**. 345 (1956).)

by the specimen to be radiographed. As the interest in neutron radiography stems from the difference in the relative absorption of neutrons versus X-rays, the potentialities of neutron radiography can best be appreciated by comparing the relative absorption by elements of increasing atomic number using two particular wavelengths of the two radiations. In Figure 11.1, which is due to J. Thewlis (1956), the total mass-absorption coefficients (μ/ρ) are shown as a function of the atomic number of elements for X-rays ($\lambda = 0.096$ Å, which is equivalent to 130 keV) and thermal neutrons of wavelength 1.08 Å equivalent to 0.07 eV.* The full curve refers to the mass-absorption coefficients for X-rays and the small circles indicate the position of the mass-absorption coefficients for thermal neutrons. It should be noted that the mass-absorption coefficients are drawn on a logarithmic scale. Whereas the X-ray absorption increases steadily with the atomic number of the elements, the neutron absorption coefficients appear to be randomly distributed amongst the elements. It is further seen in the diagram (Figure 11.1) that some of the light elements show very high absorption and most heavier elements, relatively low absorption of neutrons. Some of the light elements which do not differ very much in atomic number, such as beryllium and boron, or boron and carbon, reveal considerable differences in neutron absorption. This indicates the possibility of radiographic discrimination between neighboring elements using neutrons. This is not feasible with X-rays, as the X-ray absorption coefficients of the elements mentioned above are very similar. Furthermore the fact that the neutron absorption coefficients for most heavier elements are relatively small suggests the possibility of neutron radiography of thick and heavy materials and a possible reduction of exposure time if compared with that for X- and γ-rays.

An example, often being quoted in connection with the thermal neutron absorption, is the feasibility of assessing the height of the water level in a vessel built of lead, using a neutron radiograph.

The graph (Figure 11.1) may be useful as a guide to the reader in deciding whether a discrimination between certain elements, in which he is interested, is possible. In comparing the possibilities with those of X-rays, it must be kept in mind that the absorption coefficients for X-rays shown in this graph refer only to one quality of radiation i.e. to about 130 keV.

11.3 NEUTRON SOURCES

Since fast neutrons in general do not show such large differences of absorption between elements, as indicated in Figure 11.1 for thermal neutrons, fast neutrons have to be slowed down, i.e. moderated to thermal neutrons, in order to be useful for radiography. This moderation however causes a loss of intensity so that a relatively

---

* The monochromatized neutron beam originated from one having a mean energy of 0.05 eV. The relationship between the neutron wavelength λ (Å) and the energy (E) in eV is expressed by the de Broglie equation $\lambda = \dfrac{2.87 \times 10^{-9}}{\sqrt{E}}$ cm.

high yield of the order of 10^{11} fast neutrons per second is required in order to produce sufficient intensity of a collimated thermal neutron beam for radiography. H. Berger (1963a, 1963b) has shown that an intensity of about 10^5 thermal neutrons per cm^2 sec is adequate for many radiographic applications using reasonable exposure times.

Neutron sources suitable for radiography are accelerators, e.g. of the Van der Graaf type, isotope sources and nuclear reactors.

11.3a   Neutron Accelerator Sources

The operation of accelerators is based on the bombardment of various targets by high energy deuterons or protons, using some of the following reactions

$$^9Be(d,n)^{10}B$$
$$^3H(d,n)^4He$$
$$^2H(d,n)^3He$$
$$^7Li(p,n)^7Be$$

The tritium (^3H) target reaction appeared, at first, to be particularly useful, as it was found that it requires only a relatively low acceleration voltage of the order of 300–400 kV in order to produce about 10^8 neutrons per second per microampere, whereas most of the other reactions mentioned above require voltages of several MeV. On the other hand, disadvantages in the practical use of tritium targets have been found due to relatively short target lives and the fact that rather a high loss occurs when high energy neutrons produced in this reaction are moderated to slow neutrons.

Yields of 10^{10}–10^{11} neutrons per second in all directions may be achieved with accelerators providing a useful thermal yield of 10^5–10^6 n per cm^2 sec. The reduction to a useful yield is caused by the moderation of fast to thermal neutrons and by the collimation of the thermal neutron beam, which is required for resolution in radiography. The yield of the neutron production of a given reaction depends on the flux and kinetic energy of the bombarding particles and also on the configuration of the target.

Photoneutron (γ, n) reactions have also been used for the production of neutrons in which certain targets are bombarded with high energy X-rays or γ-rays. These neutron sources suffer however from the disadvantage of having a fairly high γ-ray background. (See pps. 14 and 245.)

A more detailed survey of accelerator neutron sources and the required accelerating voltages and relative thermal neutron yields have been given recently by E. A. Burrill and M. M. MacGregor (1960), by E. A. Burrill and J. Hirschfield (1961) and in the book on 'Neutron Radiography' by H. Berger (1965).

11.3b Isotope Neutron Sources

There are chiefly two types of reactions used for isotope neutron sources, namely the (α, n) and the (γ, n) reactions. Some useful commercially available neutron sources of this kind, using the (α, n) reactions are listed below (Table 11.1) together with the respective half lives, neutron yields per curie sec and the γ-ray dose in mR per hour at one meter from a one curie source.

TABLE 11.1. Isotope Neutron Sources Based on (α, n) Reaction

Source	Half-life	Neutron yield per curie sec.	γ-ray dose in mR per hour per Ci at 1 m
^{226}Ra-Be	1650 y	1.2×10^7	830
^{241}Am-Be	458 y	2.7×10^6	1
^{228}Th-Be	1.9 y	2.0×10^7	600
^{210}Po-Be	138 d	2.5×10^6	0.1

The sources consist of a mixture of a light element target with an α-emitting isotope sealed in a platinum or steel capsule.

Of the many available isotope neutron sources using photoneutron (γ, n) reactions, it is particularly the ^{124}Sb-Be reaction which is useful for neutron radiography and this is for three distinct reasons.

1. In this reaction an isotope is involved with a reasonably long half-life of 60 days;

2. The maximum neutron energy produced is only 0.024 MeV. This means that less will be lost by moderation to thermal neutrons, than by using higher energy neutrons initially;

3. It has a fairly high neutron yield of 3.2×10^6 neutrons per second per curie. E. J. Hennelly (1961) reported on a ^{124}Sb-Be photoneutron source having a total yield of 10^{10} fast neutrons per second. E. A. Warman (1962, 1965) even proposed a portable ^{124}Sb-Be source for which the major neutron shielding is only required when the source is in use. All the mentioned sources apart from the Po-Be and Am-Be sources, emit a strong γ-ray background.

11.3c Nuclear Reactor Sources

One of the most intense thermal neutron sources suitable for radiography is undoubtedly a nuclear reactor, but because of its exceptionally high cost and its large size only few laboratories are able to afford the purchase, apart from those laboratories using a reactor essentially for other purposes. Many applications of neutron radiography are related to problems occurring in the nuclear energy industry and hence the pioneer work in this field was carried out using nuclear reactors. In general these are capable of providing useful collimated thermal neutron beams of 10^6–10^8 neutrons per cm^2 sec.

H. Berger (1962a, 1964a) makes use of a γ-ray free monoenergetic neutron source which is supplied by the Argonne CP-5 reactor by

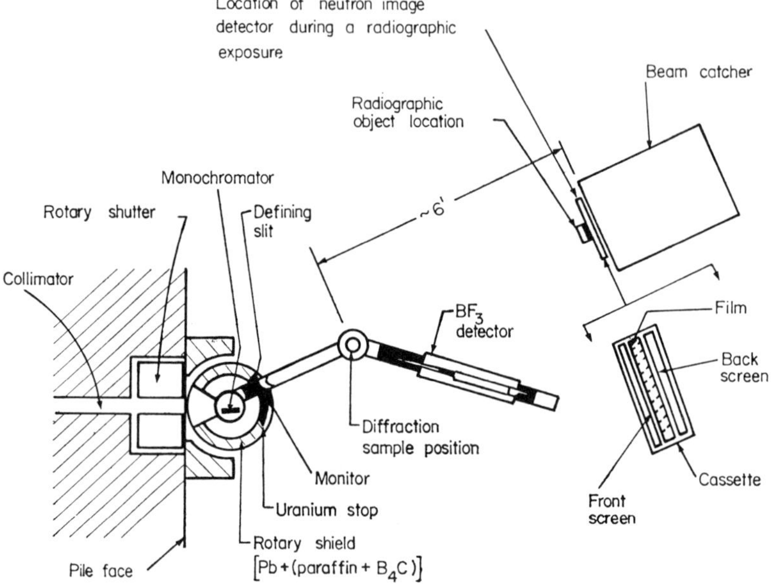

Figure 11.2. Schematic drawing of neutron spectrometer showing arrangement used for neutron radiography. (*After* H. Berger, *Nondestructive Testing*. May–June (1962).)

applying a neutron spectrometer. This was set up originally for neutron diffraction experiments. The arrangement for Berger's method is shown in Figure 11.2.

A collimated beam of neutrons from the nuclear reactor is directed towards a monochromatizing crystal. The emerging monoenergetic neutron beam travels through a slit system at the end of which it can be detected or measured by a boronfluoride (BF_3) counter. The part of the thermal beam utilized for radiography is the one falling on the beam catcher. The approximate cross-section of this beam is $2\frac{1}{2} \times 3\frac{1}{2}$ inches with a central portion of about 0.5×1 inches, which contains the maximum beam intensity of 3×10^5 neutrons per cm^2 sec, using a neutron wavelength of 1.56 Å (0.034 eV).

11.3d Neutron Spectra and Collimation

The neutron spectra produced by most sources are continuous spectra with some peak intensities occurring at certain energy levels. A typical spectrum of a Po-Be source is shown in Figure 11.3.

The sources are surrounded by several inches of moderating material such as paraffin wax, water or other hydrogenous substances in order to obtain thermal neutrons.

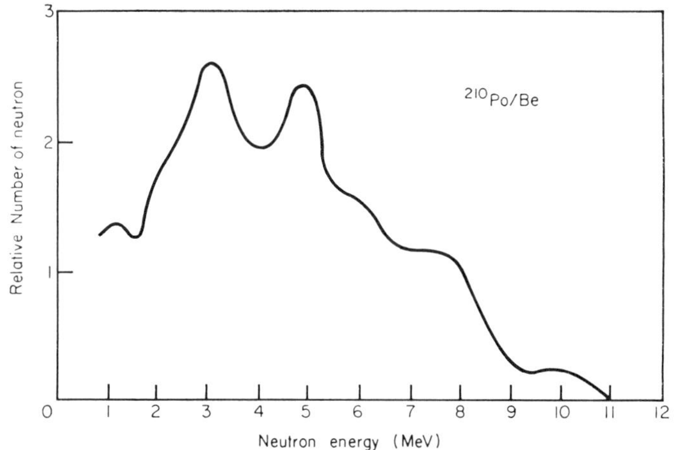

Figure 11.3. Neutron spectrum of a polonium-beryllium source. (*After* B. G. Whitmore and W. B. Baker, *Phys. Rev.* **78**. 799 (1950).)

Precautions have to be taken to reduce the γ-ray background of most sources, as this tends to obscure the neutron radiograph. This may be done to some extent by means of filters or to a more limited extent by an increase of the source to film distance. The most efficient method of eliminating the effect of γ-rays on the photographic emulsion is by means of image detection systems which are insensitive to γ-rays (transfer method). This aspect will be discussed in the next section.

Neutron beams do not originate from a point but form reasonably good parallel beams after collimation. The collimators may consist for instance of an aperture in cadmium, which has a very high cross-section for thermal neutrons. Owing to the fact that neutron beams do not diverge like X-rays from a point source, it is not possible to obtain enlarged images. The propagation of these beams does not obey the inverse square law of distances.

In this section on neutron sources it is of importance to draw the reader's attention to the need for proper protection against neutrons. Although neutrons are not directly ionizing particles, they produce ionization in the human body by collision and by nuclear reactions, as outlined on page 34. The proper protection material against neutrons consists of moderating material, such as paraffin wax and other hydrogenous substances. Boron is often included in shielding material because of its high capture cross-section for thermal neutrons. In most cases additional shielding against γ-rays is obviously necessary.

11.4   PHOTOGRAPHIC DETECTION TECHNIQUES

11.4a   Direct Exposure and Transfer Techniques

Although photographic emulsions respond directly to thermal neutrons, the photographic sensitivity is too low for practical use in neutron radiography (see p. 172). An efficient use of the photographic process is only possible by applying so called conversion reactions. Without these, neutron radiography would hardly be possible at convenient exposure times. Such conversion reactions are utilized in screens consisting of foils of certain elements or some isotopes, which emit α, β or γ-rays when exposed to thermal neutrons. These screens are therefore placed in direct contact with photographic emulsions. A list of suitable converter screens is given in Table 11.2.

In the second column the isotope involved in the reaction is quoted, in the third column the relative percentage abundance is given. The fourth column shows the cross-sections for thermal neutrons, the fifth column the reaction involved and the last column the respective half-lives. The optimum thicknesses of converter screens to be used will be given later (see Table 11.3).

Basically one distinguishes two different techniques, one in which the screens are placed in contact with photographic emulsions during exposure to neutrons (*direct exposure technique*) and one in which the screens are first activated during neutron exposure to become radioactive and are then placed in contact with a photographic emulsion, (*transfer technique*). In this latter method the radiation emitted during radioactive decay is utilized in the photographic process.

Screens of lithium, boron, cadmium, and gadolinium (see Table 11.2) show little tendency to become radioactive whilst bombarded with thermal neutrons and only emit immediate radiation on exposure. These screens are useful in the direct exposure technique. Whereas all the other materials mentioned in Table 11.2 are suitable for use in the transfer technique, they can also be used in the direct exposure technique, as they all also emit prompt radiation, i.e. they emit radiation whilst being bombarded by neutrons.

TABLE 11.2. Characteristics of Several Neutron Photographic Image Intensifier Materials
(By courtesy of H. Berger, *Neutron Radiography*, Elsevier, Amsterdam, London and New York (1965).)

Material	Isotope involved in reaction	Relative natural abundance (%)	Cross-section for reaction (thermal neutrons, velocity = 2,200 meters per sec) (barns)	Reaction[†]	Half-life s = sec m = min h = hours d = days
Lithium	Lithium-6	7.52	910	$^6\mathrm{Li}(n, \alpha)^3\mathrm{H}$	
Boron	Boron-10	18.8	3,770	$^{10}\mathrm{B}(n, \alpha)^7\mathrm{Li}$	
Rhodium	Rhodium-103	100	12	$^{103}\mathrm{Rh}(n)^{104m}\mathrm{Rh}$	4.5 m
			140	$^{103}\mathrm{Rh}(n)^{104}\mathrm{Rh}$	44 s
Silver	Silver-107	51.35	44	$^{107}\mathrm{Ag}(n)^{108}\mathrm{Ag}$	2.3 m
	Silver-109	48.65	2.8	$^{109}\mathrm{Ag}(n)^{110m}\mathrm{Ag}$	270 d
			110	$^{109}\mathrm{Ag}(n)^{110}\mathrm{Ag}$	24.2 s
Cadmium	Cadmium-113	12.26	20,000	$^{113}\mathrm{Cd}(n, \gamma)^{114}\mathrm{Cd}$	
Indium	Indium-115	95.77	155	$^{115}\mathrm{In}(n)^{116m}\mathrm{In}$	54.1 m
			52	$^{115}\mathrm{In}(n)^{116}\mathrm{In}$	13 s
Samarium	Samarium-149	13.8	40,800	$^{149}\mathrm{Sm}(n, \gamma)^{150}\mathrm{Sm}$	
	Samarium-152	26.8	140	$^{152}\mathrm{Sm}(n)^{153}\mathrm{Sm}$	47 h
Gadolinium	Gadolinium-155	14.73	61,000	$^{155}\mathrm{Gd}(n, \gamma)^{156}\mathrm{Gd}$	
	Gadolinium-157	15.68	240,000	$^{157}\mathrm{Gd}(n, \gamma)^{158}\mathrm{Gd}$	
Dysprosium	Dysprosium-164	28.1	500	$^{164}\mathrm{Dy}(n)^{165m}\mathrm{Dy}$	1.25 m
			2,000	$^{164}\mathrm{Dy}(n)^{165}\mathrm{Dy}$	140 m
Gold	Gold-197	100	96	$^{197}\mathrm{Au}(n)^{198}\mathrm{Au}$	2.7 d

* The data for this table were obtained from:

B. T. Feld, *Experimental Nuclear Physics*. Vol. 2, Edited by E. Segré, Wiley, New York.

D. J. Hughes and J. A. Harvey, Report BNL-325, Brookhaven Nat. Lab., Upton, New York.

D. J. Hughes and R. B. Schwartz, Report BNL-325, Supplement 1, Brookhaven Nat. Lab., Upton, New York (1957) and BNL-325, 2nd Edition (1958).

D. J. Strominger, J. M. Hollander and G. T. Seaborg. *Rev. Modern Physics*, **30**, 585 (1958).

† Metastable excited states are indicated by the *m* designation as in STROMINGER *et al.* (1958).

TABLE 11.3. Neutron Image Resolution Summary

After H. Berger (1963c) (By courtesy of *J. Appl. Phys*, **34**. 4. 914 (1963).)

Converter screen material thickness[a] and orientation[b]	Resolution[c]		Screen thickness investigated, mils	Neutron[d] exposure time for AA film density of 1.5, min.	Remarks
	Cadmium test piece	Gadolinium test piece			
0.5 Gd, back	1.2 mils	0.4 mils	0.25 to 2	11.1	Single-metal screen exposure methods
3 Cd, back	1.2	—	1 to 30	15.1	
3 Rh, back	2	—	3 to 10	17.1	
5 In, front	2	—	2 to 30	23	
10 Dy, back	2	—	5 to 10	8.1	
5 Ag, front	2 to 3.6	—	5 to 30	21.75	
Li–6F powder, back	1.2	1.0	60	25	Compressed Li–6F powder, 95.6% enriched
B-10, ZnS (Ag)[e]	1.2	3	<1	8 (type F X-ray film)	Vapor deposited phosphor on B-10 layer[f]
Li-6, ZnS (Ag)[e]	2	—	8	0.1 (Tri-X film)	Modified Stedman-type scintillator, 96% Li-6 enriched[g]
Li-6, ZnS (Ag)[e]	2	—	60	0.1 (Tri-X film)	1:4 powder mixture[h] using 96% enriched Li-6F and ZnS (Ag)
Li-6 loaded glass[e]	2	—	75	1	Cerium activated silicate glass[i] containing 2.5% Li, 96% enriched with Li-6
B-10, ZnS (Ag)[e]	2	—	12	1	Sun-type scintillator,[j] 92% enriched with B-10
10–10 Rh	2 to 3.6	—		4.3	Previously reported double-screen methods
20–30 In	3.6	—		5.5	

18–18 Ag	3.6	—		7.5	
10–20 Cd	20	—		6	
0.25–2 Gd	1.2	0.4 to 0.7		4.8	Fastest double Gd technique
10 Rh-2 Gd	1.2	1 to 3		3.85	Fastest metal screen method
3 Au transfer	1.2	1.0	3 to 10	240	Transfer-exposure methods
2 In transfer	2	—	2 to 30	163	
10 Dy transfer	2	—	5 to 10	14.6	

[a] All numbers given are in mils (thousandths of an inch).

[b] The word front or back refers to the location of the screen with respect to the film. A front screen would be used between the film and neutron source. For double screens the film was sandwiched between the front screen (first number indicated) and the back screen. The transfer techniques in the thicknesses indicated showed a slight preference for both speed and resolution for placing the film on the neutron source side of the exposed screen.

[c] The number indicated in each column is the minimum separation, in mils, between holes through each test piece which could be resolved on a neutron radiograph taken by the indicated method. The cadmium test piece was 0.020 in. thick and had 0.020 in.-diam. holes. The minimum separation was 0.0012 in. The gadolinium test piece was 0.002 in. thick and had 0.005 in.-diam. holes. The minimum separation was 0.0004 in.

[d] The thermal neutron intensity was 3×10^5/cm^2-sec. The radioactive materials were permitted a 3 half-life decay on the film in addition to the neutron exposure shown. Radioactive transfer foils were placed in a film loaded cassette within 30 sec after completion of the neutron exposure.

[e] The scintillators were all used as back screens only.

[f] A vapor deposited layer of ZnS(Ag), 1 mg/cm^2 thick was applied on a thin B-10 layer on Pyrex glass.

[g] This was a modified NE421 scintillator from Nuclear Enterprises, Ltd. See R. Stedman, *Rev. Sci. Instr.* **31**, 1156 (1960).

[h] See S. P. Wang, C. G. Shull and W. C. Phillips, *Rev. Sci. Instr.* **32**, 126. (1962).

[i] Scintillating glass available from Nuclear Enterprises, Ltd.

[j] This was a modified Nuclear Enterprises, Ltd, type NE402 scintillator. See K. H. Sun, P. R. Malmberg, and F. A. Pecjack, (1956), Nucleonics **14**, No. 7, 46, for a general description of the scintillator.

In general there is little to be gained by permitting the screens to decay after neutron exposure when using the 'direct exposure' technique, but the activity of certain screens such as that of indium, with a half life of 54 minutes, will add appreciably to the initial exposure. The 'direct exposure' technique is the faster of the two techniques.

One of the main advantages of the 'transfer technique' is the fact that the converter screen is only affected by neutrons and not by any other radiation, such as γ-rays present in the neutron beam. Thus during the subsequent exposure of the photographic emulsion the transfer radiograph is not exposed to any other radiation than that originally excited by neutrons. This indicates the possibility of taking radiographs of radioactive specimens without the photographic layer being affected by γ-rays emitted from the radioactive specimen. A further advantage of the transfer technique is that the photographic exposure can take place remote from the location of the neutron source.

11.4b   Photographic Material and Resolution

Photographic material used in neutron radiography is predominantly a double coated X-ray film. α-emitters, such as ^{10}B or ^{6}Li converter screens (see Table 11.2) are generally used in conjunction with silver activated ZnS fluorescent screens. These emit light when excited by α-particles which may be recorded by light sensitive emulsions which are relatively insensitive to γ-rays.

Both the direct exposure and the transfer technique have been studied extensively by H. Berger (1962a, 1962b, 1962c, 1963a, 1963c) with regard to the relative speed of the various converter systems, their optimum thicknesses, as single and double screens and with regard to image sharpness. The results of Berger's investigations are summarized in Table 11.3.

Column 1 of this table shows the converter screen materials used together with their relative orientation, i.e. whether as single front or back screen or double screens. The figures quoted before the materials refer to thousandths of inch thickness. Hence 3 Cd, back means the use of a back screen, i.e. remote from the neutron source and the thickness of the screen is 0.003 inches. Reference to a front screen means the use between film and neutron source. If two thicknesses are quoted, the first refers to the front, the second to the back screen. Only the three last mentioned converters refer to the 'transfer' method.

The resolution tests mentioned in the second column were carried out by evaluating neutron radiographs taken from cadmium and gadolinium test objects containing a number of drilled holes with decreasing spaces between them. In the cadmium test object the spacings between the 0.02″ diameter holes in the 0.02″ thick test object varied from 0.03 to 0.0012 inches. The gadolinium test object was made using gadolinium as the base material. The hole size was 0.005″ and the hole separations were 0.0004, 0.0007, 0.003 and 0.001″. The resolution values quoted are the result of various observers and the figures quoted in Table 11.3 present the minimum separation in thousandths of an inch between holes which could be resolved by the eye on a neutron radiograph.

With regard to resolution the gadolinium screens rank as the best of all the systems in the 'direct exposure' technique and that of gold as the best in the 'transfer technique'. The advantage of gadolinium with regard to resolution is believed to be due to the soft radiation which originates from the first 0.00025″ of the screen thickness. Although the scintillation (phosphor) screens permit fairly good resolution the neutron radiographs are severely influenced by a grainy (mottled) background and the exposures suffer obviously from reciprocity failure.

11.4c   Exposure Times

The Table 11.3 also shows the approximate exposure times in minutes required for each converter screen using Eastman Kodak Industrial X-ray Film Type AA. For those converters which are connected with phosphor exposures, the respective film materials used are given in the table. All the exposures were carried out with the neutron spectrometer referred to before, using a neutron intensity of $3 \times 10^5/cm^2$ sec. The radioactive screen materials were given a 3 half-life decay on the film in addition to the neutron exposure quoted. The transfer screens were placed in a film loaded cassette within 30 seconds after completion of the neutron exposure. The scintillators were all used as back screens. The relative speeds of the various converter systems can be estimated from the column on neutron exposures in minutes which all refer to a density of 1.5. By far the fastest system is that of ^{6}LiZnS and the slowest is that obtained with the transfer methods. It was further found that the exposure times could be reduced to one fifth of those quoted in the table, when using Eastman Kodak X-ray Film Type KK instead of the Eastman Kodak Type AA Industrial X-ray film used for the values of the table, but the latter film reveals less graininess.

A comparison of the relative exposure times required for increasing

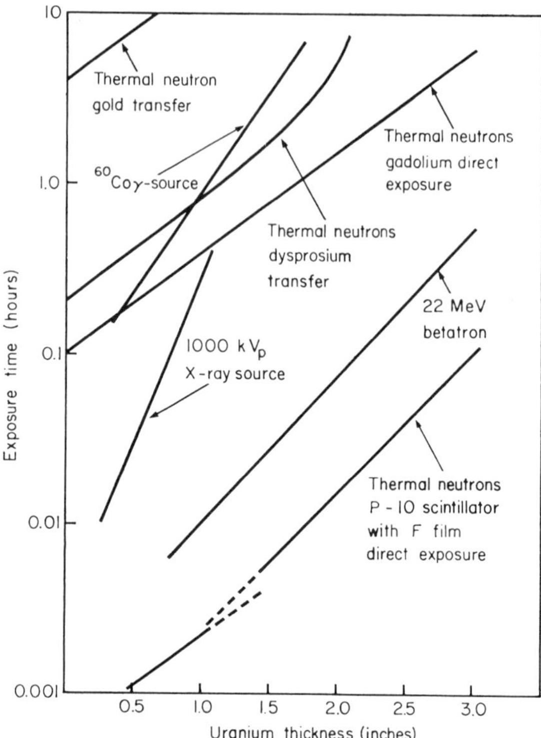

Figure 11.4. Exposure times as a function of natural uranium thicknesses for a number of radiation sources and for several neutron techniques. All data shown are for Eastman Kodak Industrial X-ray Film Type AA except the scintillator technique, where Type F film is indicated on the curve. The ^{60}Co source was 25 curies used at a distance of 30 inches. A GE X-ray 1000 kVp unit was used with the transmitted beam 3 mA, 72″ distant. The betatron was operated at an output of 100 R/min at a meter and at a distance of 72″. The thermal neutron intensity used for all the neutron data was 3×10^5 n/cm^2 sec. (*After* H. Berger, *Nondestructive Testing* Nov.–Dec. (1963).)

uranium thicknesses for various neutron converter systems, using 3×10^5 thermal neutrons per cm^2 sec, a 1000 kV$_p$ X-ray source and a 25 curie ^{60}Co source is shown in the graph of Figure 11.4 also due to H. Berger.

It is seen that in spite of the relatively small neutron intensity used in this investigation the neutron exposure times are in general shorter than those with X-rays and γ-rays of ^{60}Co. The thickness sensitivity for neutrons was found to be about 1–2 per cent, apart from the exposures taken with scintillator screens.

11.5 APPLICATIONS

Although the techniques of neutron radiography are in a fairly advanced stage, the potentialities of this new tool have not been fully exploited yet. This is probably due to the fact that only relatively few laboratories have made use of the technique and that most of these are connected with engineering problems occurring in the field of nuclear reactors and atomic energy. This is not surprising considering that most of the strong neutron sources are available in the large nuclear research centers. With the greater availability of suitable neutron sources to other laboratories, a fairly rapid progress in the field of neutron radiography can be expected.

The obvious advantages of neutron radiography over conventional X-ray and γ- radiographic methods are in the field of thick and heavy metals, in the examination of radioactive specimens and in the discrimination of light elements in the presence of slightly heavier materials and in other more specialized inspection methods. A few examples of these applications will be given in this section.

11.5a Heavy Metals

Laboratories that are engaged in the inspection of very heavy materials, such as tungsten, lead, bismuth, uranium, and others, will find advantages in the use of neutron radiography particularly if great thicknesses of these materials are involved. This will be evident from the graphs shown in Figure 11.5(a) and (b) which are due to H. Berger and I. R. Kraska (1964).

Figures (a) and (b) show the exposure times in hours required to give a density of 1.5 for various thicknesses of tungsten, pressed steel, uranium, and lead, using the 'transfer technique' (Figure 11.5(a)) and the 'direct exposure' technique (Figure 11.5(b)). The left graph Figure 11.5(a) shows the conditions for the transfer method using 0.010″ thick dysprosium metal screen in combination with Eastman Kodak Industrial X-ray Film Type AA for 3 half-lives or more. The upward trend of the curves at the longer exposures is caused by the approach of several half-lives of dysprosium (half-life of ^{165}Dy is 2.3 hours.) The thermal neutron intensity used in all the exposures was 10^7 neutrons per cm² sec. A fairly strong γ-ray beam and higher energy neutrons were also present. Figure 11.5(b) is based on the direct exposure technique using 0.0005″ thick gadolinium metal screen used as a back screen in conjunction with Type AA film. The dashed curve is for steel using a rhodium-gadolinium screen combination also with Type AA film. For exposures with ^{10}B loaded phosphor

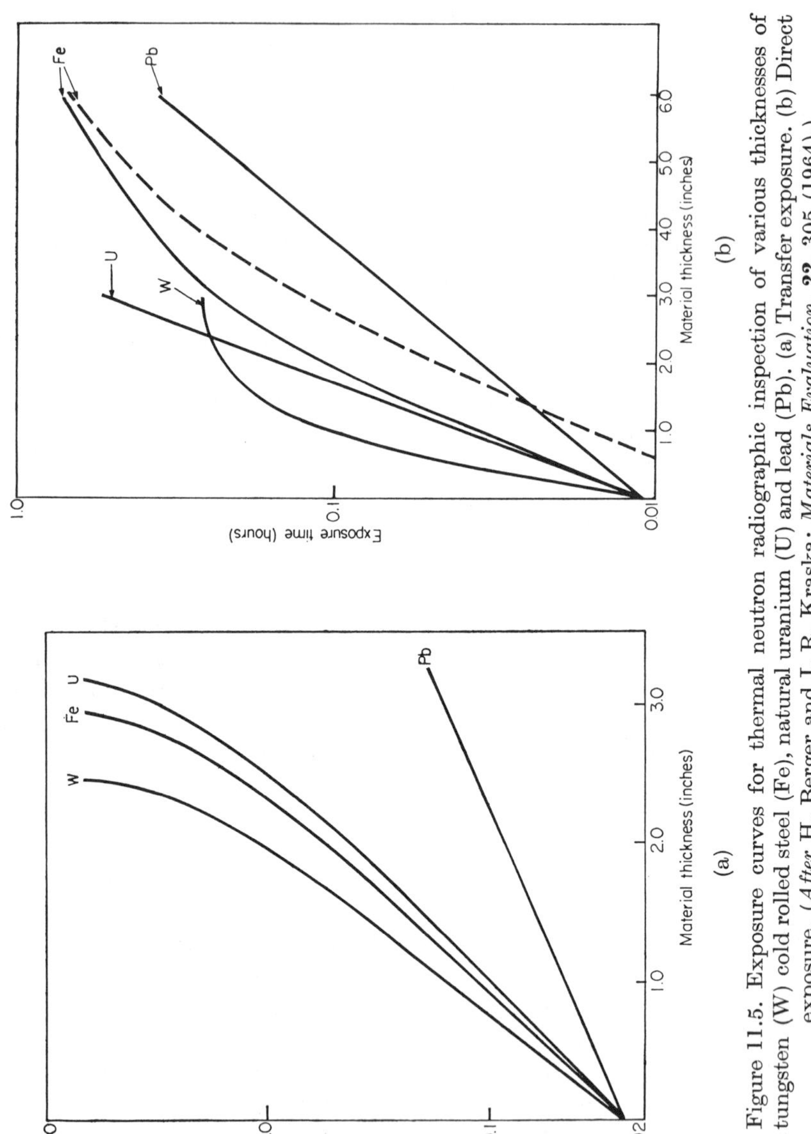

Figure 11.5. Exposure curves for thermal neutron radiographic inspection of various thicknesses of tungsten (W) cold rolled steel (Fe), natural uranium (U) and lead (Pb). (a) Transfer exposure. (b) Direct exposure. (*After* H. Berger and I. R. Kraska; *Materials Evaluation*, **22**. 305 (1964).)

in combination with Eastman Kodak Film Type F in a direct exposure technique, exposure times of only fractions of a second to a few seconds would be required. These short exposure times would be achieved however at the cost of having a fairly bad image quality indicator sensitivity of about 6–10 per cent, whereas the sensitivity is of the order of 2 per cent using the other techniques.

In Figure 11.6 a neutron radiograph of a pressed uranium oxide cylinder containing a wax filled central cavity is shown. The cylinder

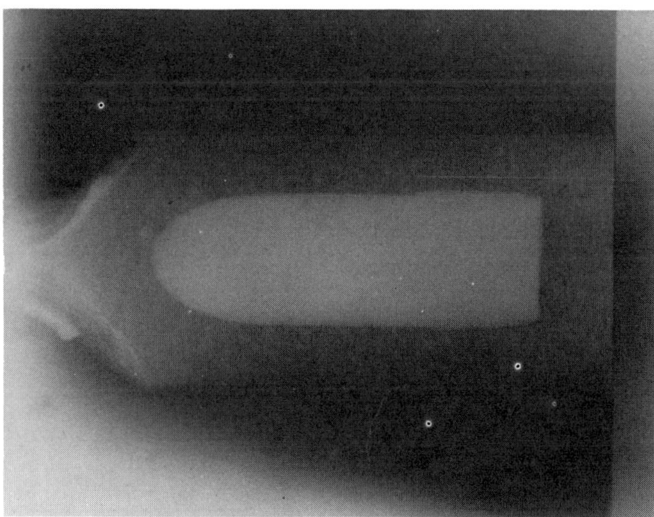

Figure 11.6. Neutron radiograph showing a pressed uranium oxide cylinder containing a wax filled central cavity. Note the high neutron absorption of the hydrogen containing wax and the relatively low neutron absorption of the 1 inch diameter uranium oxide. The non-uniformities of the uranium oxide mixture are also apparent. Exposure data: 16 min. Direct exposure technique, gadolinium screens, Eastman Kodak Industrial X-ray Film Type AA. (*After* H. Berger, *Nondestructive Testing*, Nov.–Dec. (1963).)

was prepared for use as a pycnometer. The purpose of the investigation was to locate the internal cavity so that a hole could be accurately drilled through the upper sloping wall to the top of the cavity. The high contrast between the wax and uranium made the inspection relatively simple.

11.5b   Examination of Radioactive Specimens

The nondestructive testing of radioactive fuel specimens for revealing dimensional changes and for the detection of flaws plays an important part in the inspection of nuclear reactors. Radiographic examination of highly active fuel containers using X-rays has been successfully applied, for instance by using heavy lead slit systems which are moved into position for exposure in order to prevent the photographic material from being fogged before and during exposure. Special X-ray films exhibiting an exceptionally high ratio of X-ray to γ-ray speed have also been used for this purpose.

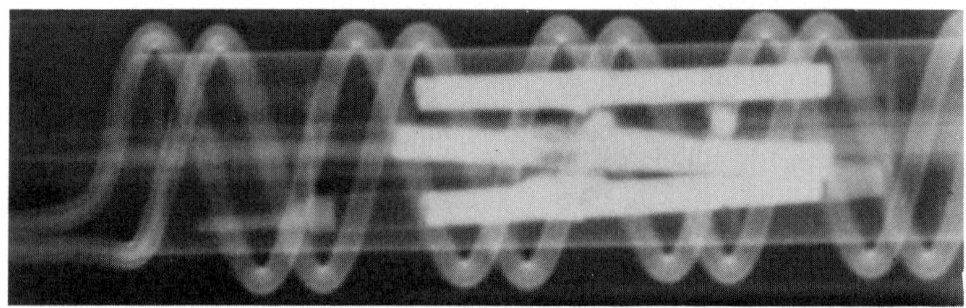

Figure 11.7 Neutron radiograph of the upper portion of an irradiated fuel capsule. This is a negative print, i.e. dark areas are areas of high neutron intensity and the three white bar shaped areas are the images of the upper tier fuel specimens. Cladding rupture can be observed. The images of the encircling heating coil and the inner tantalum cup are also readily detectable. The sodium level within the assembly near the left of this print can also be detected on the original negative. (*After* H. Berger, *Proceedings of the 4th Int. Conf. on Non-Destructive Testing* (1963) p. 117.)

The neutron radiographic transfer method which was suggested by J. Thewlis (1958) appears to be an ideal method of preventing photographic materials from being fogged by the emission of radiation from radioactive specimens. H. Berger and W. N. Beck (1963) reported on the neutron radiography of radioactive fuel specimen assemblies emitting γ-ray doses of up to 20 R/hr through 8 inches of lead. It is obvious that all precautions had to be taken with regard to the safety of the operators, i.e. remote manipulation through protective shields with regard to the γ-radiation from the specimen and to the neutrons from the source. The transfer techniques can be carried out with the screens mentioned before, such as dysprosium, indium, gold, and others. In the transfer exposures about two minutes elapsed between the end of the neutron exposure and the loading of the cassette with screen and film. A neutron radiograph of the upper position of an irradiated fuel capsule is shown in Figure 11.7.

The dark areas are the regions of high neutron intensity and the three white bar shaped areas are the images of the upper tier fuel specimens. Cladding rupture can be observed on the radiograph. The images of the encircling heating coil and the inner tantalum cup are also visible. Furthermore particles of fuel which have separated from the specimen can be seen.

11.5c Other Applications

Many other applications in reactor, rocket and missile technology are quoted by H. V. Watts (1961).

In general neutron radiography has been applied successfully in diffusion problems of boron, silicon, and germanium, in studies of moisture content of wood, in the inspection of plastic material, paper, and rubber, since small differences in the hydrogen content or in thickness of hydrogenous ingredients can be detected. Thin biological specimens such as plant and animal tissues have also been examined by neutron radiography. (H. L. Atkins, 1965).

The first thermal neutron image intensifier has recently been installed at the Argonne National Laboratory. (H. Berger, W. F. Niklas and A. Schmidt (1965)), using a silver activated cadmium-zinc-sulfide screen as the input screen. This screen contains ^{6}Li enriched LiF in a weight ratio of four parts phosphor to one part LiF. The input screen diameter is 22 cm and the useful image diameter on the output screen is 2.5 cm. With a television system the minimum detectable neutron intensity was about 10^5 thermal neutrons/cm^2 sec, which could be very much improved when using an image orthicon system. Due to the electron acceleration in the image intensifier and the reduction of image size at the output screen a much brighter image can be obtained than by conventional fluoroscopic methods. One of the additional attractions of a neutron image intensifier is the possibility of observing objects in motion.

REFERENCES

Atkins, H. L. (1965). Biological applications of neutron radiography. *Mat. Eval.* **23,** 453–8, 9.

Berger, H. (1962a). A discussion of neutron radiography. *Nondestructive Testing*, May–June 1962.

Berger, H., (1962b). Comparison of several methods for the photographic detection of thermal neutron images. *J. Appl. Phys.* **33,** 1.48.

Berger, H. (1962c). Photographic detectors for neutron diffraction. *Rev. Sci. Instr.* **33,** 8.844.

Berger, H., (1963a). Neutron radiography as an inspection technique. Conference Brit. Nat. Com. for non-destructive testing. *Proceedings of the Fourth Internat. Conf. on "Non-Destructive Testing"*. Butterworths, London, p. 113.

Berger, H. (1963c). Resolution study of photographic thermal neutron image detectors. *J. Appl. Phys.* **34,** 4.914.

Berger, H. (1963b). Neutron Radiography. A new dimension in radiography. Nov. Dec. 1963. *Nondestructive Testing.*

Berger, H. (1964a). Characteristics of a thermal neutron beam for neutron radiography. *Int. J. Appl. Radiation & Isotopes.* **15,** 407.

Berger, H. (1964b). Radiography as a tool of nondestructive testing. *Forest Prod. J.* 190.

Berger, H. (1965). *"Neutron Radiography"*. Elsevier, New York.

Berger, H., and Beck, W. N., (1963). Neutron radiographic inspection of radioactive irradiated reactor fuel specimens. *Nuclear Science & Eng.,* **15,** 411.

Berger, H., and Kraska, I. R. (1963). An evaluation of a fast scintillator-Polaroid film camera for neutron image detection, *Nondestructive Testing*, May–June, 1963.

Berger, H., and Kraska, I. R. (1964). Neutron radiographic inspection of heavy metals and hydrogenous materials. *Mat. Eval.* **22**.305.

Berger, H., Kraska, I. R., and Dickens, R. E. (1964). A Polaroid film method for transfer neutron radiography. *Nuclear Science & Eng.* **18,** 137. (Letter to Editor).

Berger, H., Niklas, W. F., and Schmidt, A. (1965). An operational thermal neutron image intensifier. *J. Appl. Phys.* **36.** 6.2093.

Burrill, E. A., and Hirschfield, J. (1961). Activation analysis for materials testing and research. *Second Sympos. Phys. & Non Destructive Testing* (Argonne National Laboratory) ANL-6515.

Burrill, E. A. and MacGregor, M. H. (1960). Using accelerator neutrons. *Nucleonics* **18**(12) 64.

Hennelly, E. J. (1961). Intense Sb-Be sources make 10^{10} neutrons/sec. *Nucleonics* **19,** 3.124.

Kallmann, H. (1948) Neutron radiography. *Research* **1,** 254.

Peter, O. (1946). Neutronen-Durchleuchtung. *Zt. Naturf.* **1,** 557.

Thewlis, J. (1956). Neutron Radiography. *J. Appl. Phys.* **7,** 345.

Thewlis, J. (1958). Neutron Radiography. *Progr. in Nondestructive Testing.* I. 112. 1958. Heywood, London.

Warman, E. A. (1962, 1965). 22nd. Nat. Conv. Soc. for Nondestructive Testing. N.Y. Neutron radiography in field use. *Mat. Eval.,* **23,** 543.

Watts, H. V. (1961). Investigations in neutron imaging. *Second Symp. on Phys. & Non Destructive Testing.* Argonne National Laboratory. ANL-6515. 53–80.

12

Autoradiography

12.1 INTRODUCTION

Autoradiography is a photographic method for the recording of the distribution and location of radioactive material within a specimen. This method cannot be replaced by Geiger or scintillation counters as these do not normally discriminate between radioactive sources, which are very close together and may be separated by distances as small as a few microns only; nor are these latter detectors capable of giving a pictorial survey of the distribution of radioactive material unless scanning methods are introduced.

The basis of the autoradiographic technique is the placing of a specimen, such as a leaf of a plant or a section through a bone, containing a radioactive substance in contact with a suitable photographic emulsion. After exposure to the radiation emitted by the radioactive material and processing, the photographic layer reveals the location of the radioactive material within the specimen. The image obtained is called an autoradiograph. As far as the technique refers only to gross sections, i.e. to macro-autoradiography, the method has been known for a considerable time.

An entirely fresh stimulus was however given to autoradiography with the greater availability of so many species of artificial radio-isotopes, together with the introduction of new types of nuclear particle emulsions. The latter make it possible to take autoradiographs of microscope sections containing radioactive substances. By further improvements in the resolution attainable, the method applied, for instance, to biology, even permits its use on a cellular level within the limits of light microscopy. More recently the resolution was further improved using electron microscope autoradiography.

It is particularly in the hands of biologists that autoradiography has developed into a very important and successful tool for studying cells, nuclei, and chromosomes of which the uptake and location of tracer substances in the various stages of cell division can be recorded and, if desired, even quantitatively evaluated.

One of the fields in which autoradiography plays such an important part at present is for instance in the study of DNA synthesis (DNA = deoxyribonucleic acid). Very intensive research is being done on DNA, being one of the chemical substances on which life and its reproduction is based. Essentially, a radioactive tracer is injected into the tissues of an animal. After the death of the animal a histological section of the animal's tissues is brought in contact with a photographic layer. During exposure the radiation emitted by the tracer will affect the silver bromide grains only at the location of the tracer. The tracer used in this case may be radioactive hydrogen (^{3}H or tritium), which is introduced into the molecule of thymidine. Thus the electrons emitted by tritiated thymidine, a substance used for DNA-synthesis, produce developed grains in the processed autoradiograph, by impact on the emulsion only where there are cells incorporating precursor into DNA. Hence the percentage of cells synthesizing DNA at a given time and sometimes the rate of cell division can be measured under varied conditions.

As will be shown later the histological section remains in intimate contact with the photographic emulsion during the whole operation of exposure, processing and drying and this permits an accurate alignment between tissue and autoradiographic record when inspected under the microscope. Thus a precise correlation between the structural details of the specimen and the autoradiographic record can be made. This type of autoradiograph is usually produced as a record made up of a few developed grains only, instead of the conventional distribution of optical densities.

It should be mentioned that autoradiography is not restricted to biological applications, but is also used in metallurgy and in industrial and other applications of which a few examples will be given.

The following sections deal separately with macro- and with micro-autoradiography, as the techniques of the two methods differ. The various techniques can roughly be divided as shown in the family tree on p. 505.

Finally, it should be mentioned that there is no special section in this chapter dealing exclusively with the processing of autoradiographic emulsions. The reason for this is that each type of emulsion and technique used in this field requires its individual processing

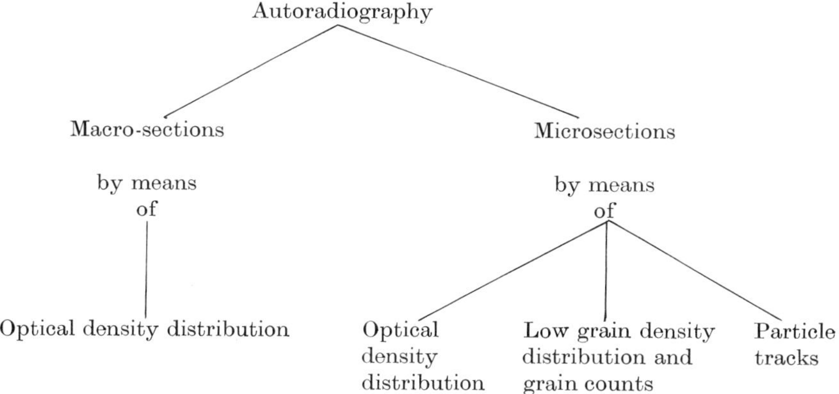

conditions which are usually given in the instruction leaflets supplied by the manufacturers of the particular emulsion. Brief references to processing will however be given in connection with stripping films and electron-microscopic techniques.

12.2   MACRO-AUTORADIOGRAPHY

12.2a.   Principle

Macro-autoradiography entails the recording of large areas or macro-sections of plants, minerals, skins of fish or sheets of wood containing radioactive materials. In general the technique is straight-forward and does not require such familiarity with the characteristic properties of photographic emulsions as is necessary when dealing with micro-autoradiography. As mentioned before, during exposure the photographic material, usually fine-grained medium or coarse-grained fast no-screen X-ray film, is placed in direct contact with the dry specimen sheet containing the radioactive substance. Good contact between specimen and film is usually achieved in an X-ray cassette or in a printing frame. After exposure the X-ray film is processed in a conventional manner, as described on pages 376–385.

A very simple application is shown as an example in Figure 12.1 which illustrates an autoradiograph of an ashtree seedling. The root of the seedling was first placed for a few hours in a weak aqueous solution containing radioactive phosphorus (^{32}P). After the plant, nourished by this solution, had taken up some of the radioactive phosphorus, it was placed in contact with a fast double coated X-ray film for about a one day exposure. Preferably the exposure is carried

Figure 12.1. Autoradiograph of an ashtree seedling after take-up of phosphorus—32. (The young leaves are shown darker than the older ones, because the
former have taken up more radioactive phosphorus.) (*By courtesy of* Kodak
Ltd., England.)

out in a vacuum cassette (H. F. Sherwood, 1947). The radioactive
phosphorus decays with the emission of high speed electrons. Many
of the electrons passing through the emulsion are stopped by the
silver-bromide grains which are rendered developable during processing. The local distribution of density in the film will be proportional
to the local distribution of electrons impinging on the film. This
proportionality between density and exposure holds up to a certain
density (see p. 105) and this depends on the type of film in use.
Figure 12.1 of the plant autoradiograph shows that higher densities
are obtained with the younger than with the older leaves, indicating
that the younger have concentrated more ^{32}P than the older
ones.

12.2b. Photographic Requirements

The main photographic questions arising in the autoradiography
of macrosections are the choice of the most suitable emulsion for
achieving the resolution, contrast, and exposure time required. Since
all these factors are related to one another, only a compromise ca
be achieved between one or two of these factors.

The problem of exposure time is usually of secondary importance
in autoradiography of gross specimens, unless the specimen contains
so little radioactive material that extremely long times (weeks or
months) might be involved. In such cases the fastest available X-ray
film would be a suitable choice. Sufficient contrast is generally achieved
depending on the distribution of the radioactive material. If this
should be a problem, one of the finer grained X-ray films may be
used at high density levels, where the inherent contrast is very high.
In this case high intensity viewing boxes or illuminators would have
to be used for the inspection of films at a density level between 2–3
above fog.

Resolution in autoradiography, as in other fields of radiography,
depends on geometrical and photographic considerations. The geometrical arrangement is entirely different from that in conventional
radiography. In the latter, radiation travels from a distant quasi-point
source through the specimen on to the photographic layer and the
greater the source to film distance, the better the resolution. In autoradiography the sources are distributed throughout the specimen and
the closer the contact, i.e. the smaller the distance between radiation
source and film, the better the resolution. In addition the geometry
is strongly influenced by the thicknesses of the specimen and of the
photographic emulsion and by the range of the particle radiation
emitted in the process. (See range relations of charged particles on
p. 603.) The thinner the specimen and the thinner the photographic
layer, the better is the resolution attainable. Furthermore, in general
the shorter the range of the particle the better the resolution to be
expected, and this is particularly so, when the specimen is relatively
thick. The reason for this is that electrons originating from regions
remote from the photographic layer will have less chance to penetrate
the specimen the shorter their range.

For instance, in the case of the leaves of the ashtree seedling in
Figure 12.1 where ^{32}P was used, relatively bad resolution is to be
expected, as very high energy electrons (max. energy 1.7 MeV) are
involved which penetrate and diverge not only through the specimen
but also through the emulsion.

The resolution attainable using macrosections is rather limited, but in general, sufficient for the purpose for which the autoradiographs are intended, even if relatively thick specimens from about 50–300 microns thickness are used. Autoradiographs of macro-sections are only rarely enlarged, because a general survey of the distribution of a radioactive substance is all that is required. If, however, enlargements of about 5 times should be required, the use of the finer grained no-screen X-ray films (see pp. 308 and 403) is recommended for the autoradiograph.

The isotopes used in autoradiography are generally electron- or positron- and sometimes α-emitters but γ-ray emitters are more rarely used. In many cases however the electron emission is accompanied by that of γ-rays. Since the photographic response to electrons and α's is so much greater than that to hard γ-rays (on the average about 50–100 times greater) only a low background from γ-rays would be expected. This however depends largely on the relative percentages of electrons and photons involved in the exposure.

If the energy of the participating electrons is not higher than about 100 keV (see p. 164) only one coating side of the double coated X-ray film is utilized. Electrons of higher energy affect both emulsion layers. If only one coating side is exposed, the unexposed silver bromide grains of the opposite emulsion layer will be fixed out during processing, so that hardly any disadvantage occurs due to the presence of two emulsions. Obviously an improvement in speed and contrast is gained by the full utilization of both emulsions; this is however offset by the better resolution attainable, if only one emulsion layer is used.* Inspecting the exposed, processed and dry double coated X-ray film by reflected light from both sides, it can be checked whether essentially both, or only one coating side, has been affected. If the latter occurs, the unexposed layer has a more glossy appearance than the exposed layer.

How the resolution is affected by the thickness of emulsion and specimen and by any separation between the two layers, will be discussed in more detail in one of the following sections on 'micro-auto-radiography.'

12.2c. Exposure to Electrons

As the autoradiography of gross sections is in general carried out by means of no-screen X-ray films, it is possible to give the reader an approximate guide to the quantitative response of these emulsions to β-particles and high energy γ-rays. For most of the fastest types

---

* One exposed layer can be removed, if necessary, by wiping with a piece of cotton wool, soaked in a solution of 10% sodium hypochlorite.

of no-screen X-ray film about 10^7 incident electrons per cm² are required to obtain a density of unity above fog. Hence, if the number of electrons emitted per unit area or of the radioactivity, for instance in terms of microcuries per unit area, of the specimen is known or can be calculated, the exposure time can be derived (see formulae 12.7–12.15 on pp. 528–529). The response to γ-rays of 1 MeV energy of a fast no-screen X-ray emulsion is of the order of 10^9 photons per cm² in order to obtain a density of unity; in other words it is 100 times slower than the response to electrons.

12.2d.   Exposure to α-Particles

Autoradiography of gross sections emitting α-particles does not present any particular difficulties and also fast no-screen X-ray films

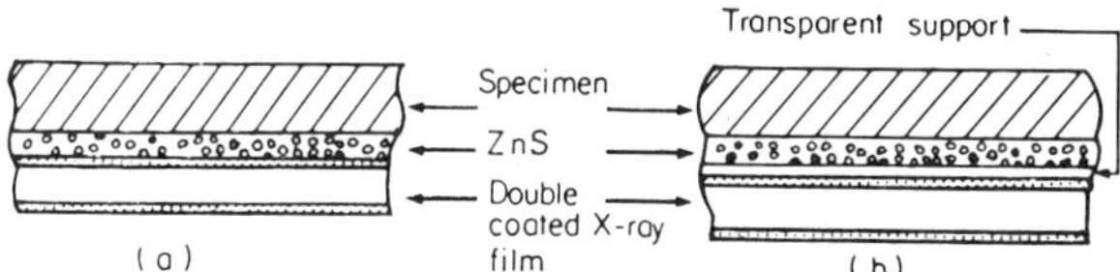

Figure 12.2. Arrangement for macro-autoradiography with α-particles when using fluorescent screens for intensification.

can be used. As each α-particle travels through the emulsion, it renders a row of grains developable on subsequent processing depending on its range in the emulsion (see pp. 603–604).

In order to reduce exposure time, a thin layer of fluorescent material may be sandwiched between specimen and photographic layer (see Figure 12.2(a)). This system of film-fluorescent screen and specimen is only advantageous if the spectral emission of the fluorescent material is adjusted to the maximum spectral sensitivity of the photographic layer. Hence use may be made of a screen type X-ray film which is most sensitive in the blue light region (max. 4200–4400 Å) in conjunction with a blue emitting, silver activated, zinc sulfide screen. Although the intensification obtainable may be very considerable, this advantage is partly offset by a loss of definition due to the separation between specimen and film and the diffusing light from the fluorescent layer. The fluorescent layer must be extremely thin and should have no support, as the α-particles with their extremely short range (see pp. 32 and 603) would otherwise be unable to penetrate it. If a support of the fluorescent layer cannot be avoided this should also be extremely thin and transparent and be placed in contact with the emulsion as illustrated diagrammatically in Figure 12.2(b).

J. J. C. Hsieh, F. P. Hungate and S. A. Wilson (1965) have described the preparation of the phosphor which they used for the above purpose and which they obtained from the U.S. Radium Corporation. The phosphor is ZnS:Ag P7-920B. About 25 gram of it was added to 100 ml of 2 per cent gelatin solution. The coarse particles were allowed to settle out from the warmed solution for 1–2 minutes. The suspension was then decanted and coated directly onto the mounted specimen or deposited by sedimentation on a double sided aluminum coated Mylar film (1 mg per cm^2). A uniform deposition of 12–14 mg per cm^2 was found to be optimum. The authors mention that, although the aluminum coated Mylar film absorbs more α-particle energy than one without backing, the loss due to absorption was compensated by the reflection of light from the aluminum coating.

The above mentioned authors used Polaroid Polascope film 410 and Polaroid Land Type 57 film in their experiments instead of screen type X-ray film. They arranged the ZnS-Ag and photographic film on aluminum coated Mylar film in a light-tight unit so that the exposure could be carried out in full light.

If the length of exposure time is immaterial, direct exposures with α-particles are preferable, as the resolution attainable with the specimen in direct contact with the emulsion is far superior to that with an interposed fluorescent screen.

12.2e.   Applications

Strictly speaking, macro-autoradiography deals with gross layers, i.e. the film in contact with the specimen produces an image which is practically of the same size as the specimen. This however does not prevent the use of enlargements, as has been mentioned before. Some of the applications to be mentioned in the following require enlargements of about 5–10 times or more, in order to reveal the information required. If the necessary resolution cannot be achieved by these methods micro-autoradiographic techniques such as those mentioned later have to be applied. A few typical examples of macro-autoradiography will now be discussed; some of these may well be refined by the use of micro-autoradiography.

Distribution of Radioactivity in Radium Needles

One of the most straightforward examples of macro-autoradiography is the examination of radium needles as used in radiotherapy. These needles are cell loaded, i.e. the radium salt, being mixed with a filler, is loaded into individual cells of about 1 cm length and 1 mm diameter; the whole needle may be as long as 6 cm. The sealed cells are loaded into platinum capsules which are sealed as well. As needles may have been bent or even broken during use, periodic tests have to be made by means of a combined radiographic-autoradiographic technique. A conventional radiograph is taken, using a few seconds exposure with soft X-rays in order to show the contour of the needle and the radiation emitted by the radium blackens the film at the same time, thus showing the distribution of the radium in the needles and any leakage that may have occurred.

Relative Amounts of Thorium in Tungsten Wires

A manufacturer producing thoriated tungsten wire for the production of lamps or valves wants to check the thorium content of the wires. Thorium is a natural radioactive material emitting α-particles (electrons and γ-rays). By placing lengths of wire having varying content of thorium on a sheet of fast no-screen X-ray film, the α-particles of the thorium will blacken the emulsion below and, after processing the film the relative densities formed along, i.e., below the wires, will give a relative measure of the respective thorium content of the wires. A much more sensitive method is to use a fast nuclear emulsion. Then the relative number of α-particle tracks per unit area can be counted under the microscope and hence a relative measure of the thorium content and its uniformity of distribution can be made. A calibration can be made to permit a quantitative or absolute evaluation.

Study of Friction

It it is of interest to study the efficiency of various lubricating materials, a radioactive tracer may be incorporated in one of the components of a machine causing frictional damage. It is assumed here that the damage caused by friction is so minute that extremely sensitive methods are required for their detection. After the operation of the machine in which frictional damage may have occurred on an adjacent component, the latter can then be brought into contact with a photographic emulsion for autoradiographic examination. The autoradiograph may reveal how much of the radioactive material has been transferred from the component containing the radioactive substance during the operation of the machine and, in many cases, the location of these transfers.

In another application it was of interest to find out how much pressure can be applied by a stylus of polished chromed steel, which is in constant contact with a moving band of plastic, without causing transfer of the chromium to the plastic. The chromium plated stainless steel stylus was sent to a reactor for making the chromium radioactive. After various pressures were exerted on the plastic it could be ascertained that a pressure of 250 grams per cm² did not produce friction, whereas a pressure of 500 grams per cm² caused a distinct transfer of radioactive chromium on to the plastic. In each case the plastic sheet was placed in contact with a no-screen X-ray film to reveal blackening from the radioactive chromium (see Figure 12.3).

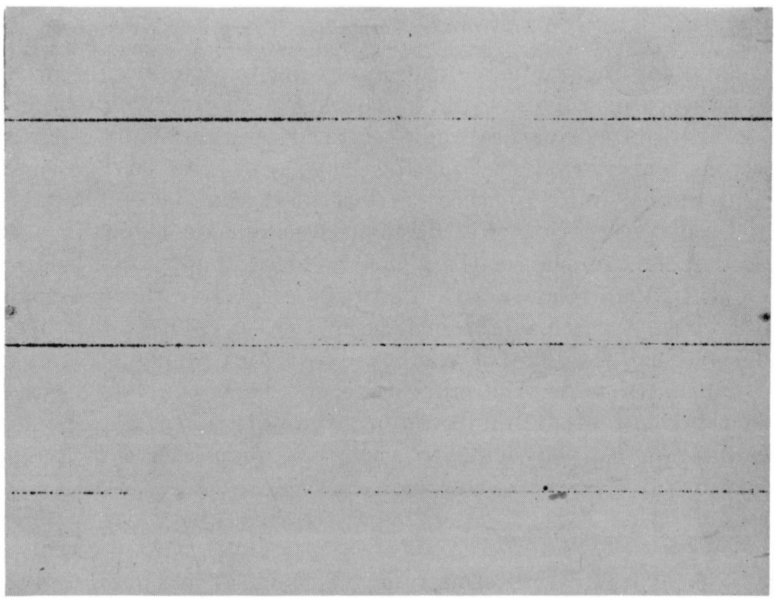

Figure 12.3. Autoradiographs taken on X-ray film, showing the transfer of radioactive chromium from a plated stylus onto a plastic sheet by friction, using various pressures (0.5, 1 and 2 kg per cm²) (*After* R. H. Herz, Research Laboratories, Kodak Ltd., England.)

Restoration of Faded Photographic Prints by Neutrons

A rather unusual example of an application of autoradiography has been shown by E. Ostroff (1965). He intended to restore some very old and faded photographic prints of great historical interest. As the prints are some of the oldest in existence and produced by the great British pioneer of photography W. H. F. Talbot, Ostroff did not want to subject them to chemical treatment, which might have damaged the pictures. Hence he irradiated them with neutrons, thus forming radioactive silver isotopes. After neutron irradiation the faded prints were placed in contact with fine grained X-ray films for exposure to electrons emitted by the radioactive silver. The processed

X-ray films then revealed autoradiographic copies, i.e. restored reproductions of the original prints.

Study of Metals

Radioactive tracer techniques offer great potentialities for the study of alloys, particularly in those cases where the various metal constituents are not uniformly distributed. In a series of papers A. Kohn (1951a, 1951b, 1953a, 1953b) has studied the homogenization of phosphorus and arsenic dendritic segregations in steel. Either a

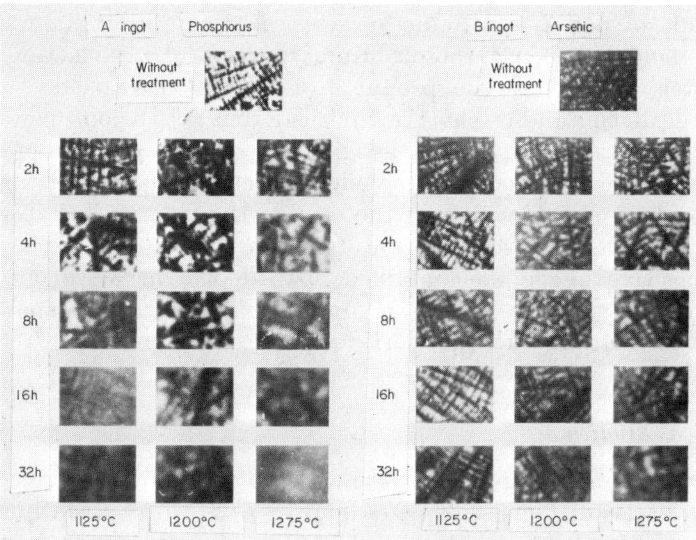

Figure 12.4. Dendritic segregation of radioactive phosphorus and radioactive arsenic in steel at various temperatures and different times of treatment. The autoradiograph shows that the homogeneity is almost reached after 32 hours at 1275°C with phosphorus, but not with arsenic. (*After* A. Kohn, l'Institut de Recherches de la Sidérurgie, Saint Germain-En-Laye, France.)

specific radioactive element is introduced into the liquid steel during melting or a sample taken from an ingot is irradiated with slow neutrons in a nuclear reactor, thus activating some of the components, namely phosphorus and arsenic. The latter method is particularly useful if elements having a high neutron activation cross-section are employed, for iron has a very low activation cross-section. After increasing hours of annealing up to 1275°C, auto-radiographs of the metal samples showed that the phosphorus distribution was almost completely homogeneous, while the arsenic segregation though lowered, was still noticeable (see Figure 12.4).

In other studies on cooling and solidification in casting metals small quantities of radioactive gold were examined. It was then found that one part per million was sufficient to produce an autoradiograph from a polished section of an ingot. (See references on Investigation on corrosion inhibitors (L. Matteoli and P. Logi, 1966) and on Friction studies (E. Rabinowicz, 1950).)

Study of Surface Potential

Rather unusual experiments were carried out by T. Westermark and L. G. Erwall (1952) on the autoradiographic study of contact potentials of polished metallic surfaces. By maintaining a uniform electric field between a radiothorium source and a collecting disc, the latter will collect the products of thoron emanation. With a disc of lead containing short cylindrical rods of various metals, it could be shown that the metal regions received a larger deposit of thorium emanation than the surrounding lead. Evidence of this was given by an autoradiograph of the disc. If the polarity between disc and source was reversed, the autoradiograph was reversed as well, i.e. a positive image was obtained instead of a negative one. The photographic densities corresponding to various metals showed a certain correlation with the work function of the metal with regard to lead.

Biological Studies

Macro-autoradiography is successfully applied in certain studies of the uptake of bone-seeking radioisotopes such as strontium and cesium. (J. Jowsey and others, 1953.) It is also used in the study of the metabolism of plants and during the time of nuclear bomb tests, it was used in connection with the uptake of fission products. The majority of autoradiographic studies in biological experiments are, however, carried out by means of micro-autoradiography.

12.3 MICRO-AUTORADIOGRAPHY

When dealing with autoradiography of microsections the problem of high resolution becomes of paramount importance. This is because high precision in the localization of radioactive tracers is essential if the distribution between or within biological cells is to be studied. Since the particles emitted by a tracer are randomly emitted in all directions they will affect a photographic layer over an area greater than the area from which they originate. The resolution, therefore

depends on the distance between each element of the source and the
emulsion, on the thickness of both specimen and emulsion, on the
grain size and on the range of the particles concerned. This problem
has been studied by theoretical and by experimental approaches.
In the following a review of the theoretical treatment will be given
and the experimental procedure will be described later (see p. 549).

12.3a. Resolution (Theoretical)

Doniach and Pelc's Method

I. Doniach and S. R. Pelc (1950) considered the distribution of
developed grains in an emulsion due to particles emitted from radioactive material in the form of a rod of length d having a very small
diameter and being perpendicular to the plane of the emulsion of
thickness a, at a distance δ from the specimen surface (see Figure 12.5).
In the following calculations absorption and scattering effects have
not been considered.

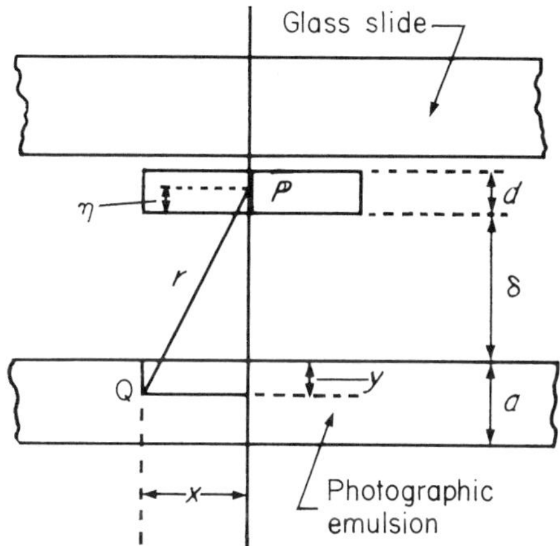

d = specimen thickness
δ = distance between surface of specimen and emulsion
a = thickness of photographic emulsion
P = point in rod containing radioactive material
η = distance between P and surface of specimen
r = distance between P and Q
Q = point in photographic emulsion
y = depth of point Q in emulsion
x = length of perpendicular from Q to extension of rod.

Figure 12.5. Geometry of exposure for considerations on resolution. (*After* I.
Doniach and S. R. Pelc, *Brit. J. Radiol.* **XXIII**, 267.184. 1950).

Let E be the exposure, N the total number of grains, n the number of developed grains and k a factor depending on the characteristics of the photographic emulsion, such as grain size, grain sensitivity and on the type and energy of particles. Considering only moderate densities (high densities are undesirable generally for various reasons to be considered later, see p. 554) one may express the relationship between the various parameters by

$$n = kNE \tag{12.1}$$

For a rod which can be regarded as a disc source of zero diameter, the effect of an element P (see Figure 12.5) on the point Q in the emulsion can be expressed by

$$dn = \frac{kNE\, dy\, d\eta}{4\pi r^2} \tag{12.2}$$

On integration over the total length d of the rod and the thickness of the emulsion a this expression leads to

$$n = \int_{\eta=0}^{d} d\eta \int_{y=0}^{a} \frac{kEN}{4\pi^2 r}\, dy \tag{12.3}$$

$$= \frac{kEN}{4\pi} \int_{\eta=0}^{d} d\eta \int_{y=0}^{a} \frac{dy}{(\eta + \delta + y)^2 + x^2} \tag{12.4}$$

The solution of this equation leads to a fairly complex expression, but the graphical results of the computation of the expression for different values of the rod length d, separation between surface of specimen and emulsion δ and thickness of the photographic emulsion a are shown in Figure 12.6(a) and (b).

The curves in Figure 12.6(a) indicate the relative spread of the number of grains for various distances between the photographic emulsion and specimen surface, based on a length of rod of 5 microns containing the radioactive substance for an emulsion thickness of 15 microns. From the graph (a) it can be seen that the resolving power (in terms of half-width of the curves) for gaps δ from 3–0.1 microns is improved from 17–3 microns respectively. Figure (b) shows similar conditions for a rod length d of 2 microns and for an emulsion thickness a of 2 microns. If the gap δ is reduced from 1–0.1 microns, the resolving power will be improved from 5–2 microns. A further reduction of δ to 0.01 micron would bring an improvement to 1.5 microns.

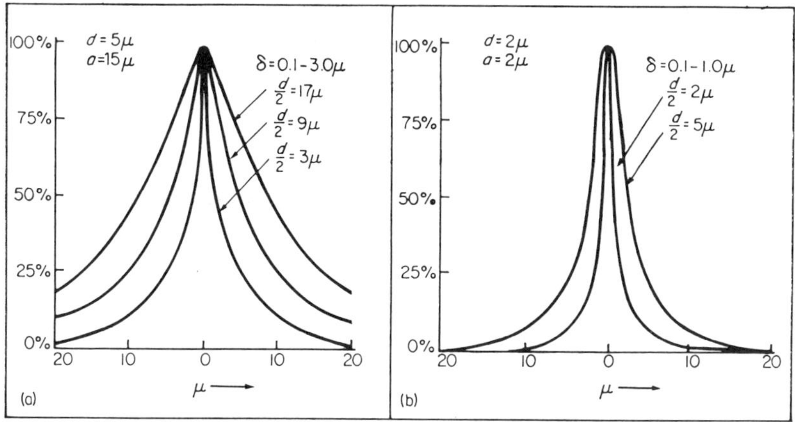

Figure 12.6. Results of Doniach and Pelc's computation illustrating the percentage spread of grains in terms of microns. The resolving power is evaluated from the half width $d/2$ of the curves under the conditions quoted below. Figure (a) shows the spread for gaps δ (distance between specimen and emulsion) of 0.1 to 3 microns for a rod length $d = 5$ microns and an emulsion thickness $a = 15$ microns. Figure (b) shows the spread for a gap δ of 0.1 to 1.0 micron for $d = 2$ microns and $a = 2$ microns.

Doniach and Pelc conclude from these results that the distance between photographic emulsion and specimen is of the greatest importance and that optimum results may be obtained with a reasonably fine grained emulsion of between 2–4 microns thickness, using a technique that ensures a contact better than 0.1 microns. In order to give the reader a survey of the calculated results by Doniach and Pelc, the main figures of interest are summarized in Table 12.1.

It can be concluded from Pelc's and Doniach's calculations, that

TABLE 12.1. Minimum Resolvable Separation Values for Various Specimen and Emulsion Thicknesses

Specimen thickness	Emulsion thickness	Minimum resolvable (δ) separation for gaps of		
		$0.1\ \mu$	$1.0\ \mu$	$3\ \mu$
$5\ \mu$	$15\ \mu$	$3\ \mu$	$9\ \mu$	$17\ \mu$
$2\ \mu$	$5\ \mu$	$2\ \mu$	$5.5\ \mu$	$10\ \mu$
$2\ \mu$	$2\ \mu$	$2\ \mu$	$5\ \mu$	—

if specimen and emulsion thickness are equal and the gap is about 0.1 micron, the resolving power due to geometrical reasons is approximately equal to one of the thicknesses.

Lamerton and Harriss's Method

L. E. Lamerton and E. B. Harriss (1954) considered a different approach to the problem of resolution in micro-autoradiography.

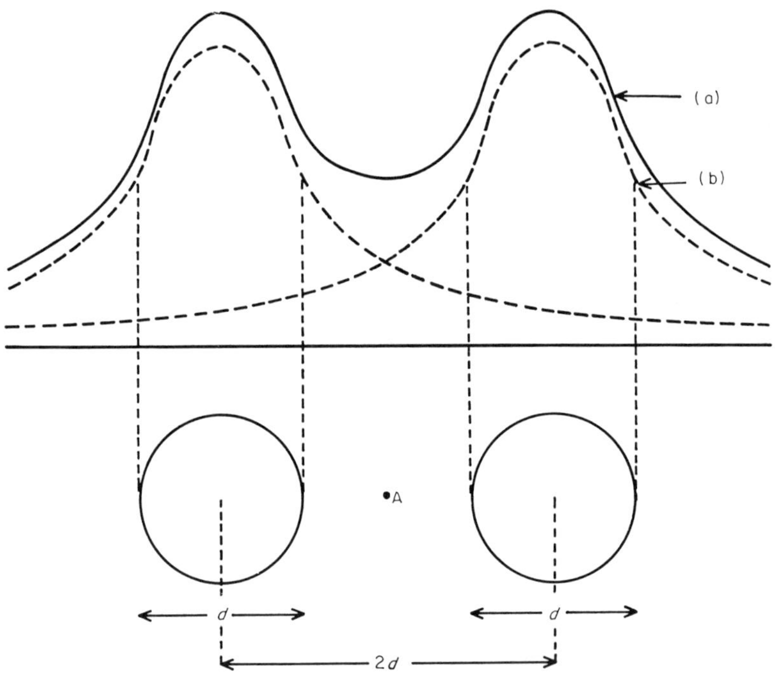

Figure 12.7. The spread of the autoradiographic image. The assumption is made that two autoradiographic images can just be resolved when the blackening from one cylinder at the midpoint A is half that at the edge of the cylinder. (a) Combined blackening variation. (b) Blackening variation from single source. (*After* L. F. Lamerton and E. B. Harriss, *J. Phot. Sci.* **2.** 135.) (1954)

These authors suggested that it would be more satisfactory to base the arguments on sources of finite dimensions. In connection with their theoretical treatment on cylindrical radioactive sources they suggested the following definition of resolution (see Figure 12.7).

'The resolution of a given autoradiographic technique shall be defined as the distance d, if the images of two uniformly active cylindrical sources of diameter d can just be resolved, when the centres are separated by a distance $2d$.'

The criterion assumed here is that two autoradiographic images can just be resolved when the blackening from one cylinder at the midpoint A is half that at the edge of the cylinder. Although this may be regarded as a somewhat arbitrary assumption, it simulates practical conditions and provides a basis for calculations. If this criterion is fulfilled, the blackening variation will be as shown in Figure 12.7.

Lamerton and Harriss's calculations on cylindrical sources are

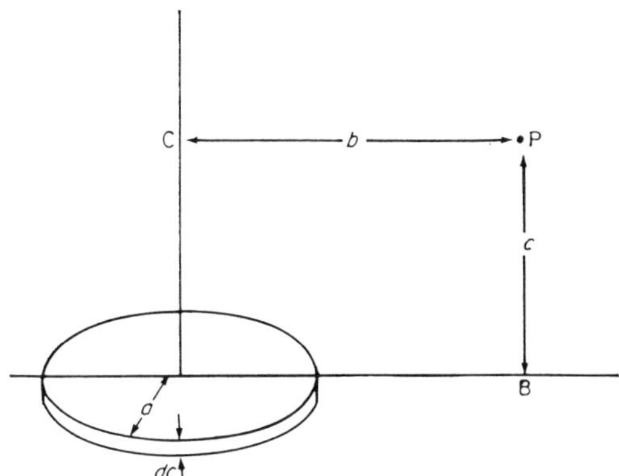

Figure 12.8. Geometry for the calculation of intensity values for cylindrical sources using a disc source. (*After* R. Sievert, *Acta Radiol.* **1.** 89. (1921) and L. F. Lamerton and E. B. Harriss, *J. Phot. Sci.* **2.** 135. 1954).

based on formulae originally developed by R. Sievert (1921) for disc sources, as illustrated in Figure 12.8. The intensity at any point P near a disc source is given by

$$I_p = \pi \, \rho \, k \, dc \, \log_e \frac{a^2 - b^2 + c^2 + \sqrt{(a^2 - b^2 + c^2)^2 + 4c^2b^2}}{2c^2}$$

$$(12.5)$$

where ρ is the activity per unit volume of the source and k is the intensity at unit distance from unit point source. Using this equation it is found that iso-intensity contours around the disc sources lie on ellipses. (For further derivations the reader is referred to the original paper by Lamerton and Harriss (1954).)

Calculations of this kind were used for different values of the thickness of emulsion and specimen and of the separation between the latter and for various radii of cylindrical sources. Finally, the distance from the cylinder for which the mean intensity through the thickness of the emulsion falls to one half of that of the edge of the cylinder, was computed. The conversion from grain density to intensity was carried out, assuming that grain density is proportional to radiation intensity. The results of the calculations of the distances $d_{\frac{1}{2}}$ outside a cylinder at which density is one half of that at the edge, for cylinders of various radii, are shown in Table 12.2.

TABLE 12.2. Distances $d_{\frac{1}{2}}$ for Cylinders of Various Radii. (After Lamerton and Harris, 1954)

Conditions	Values of $d_{\frac{1}{2}}$ (distance to half mean intensity at edge)		
	1 μ radius cylinder	5 μ radius cylinder	10 μ radius cylinder
2 μ thick specimen 2 μ thick emulsion	1.02 μ	1.94 μ	2.6 μ
5 μ thick specimen 5 μ thick emulsion	1.92 μ	3.28 μ	4.4 μ
2 μ thick specimen 5 μ thick emulsion (or vice versa)	1.36 μ	2.52 μ	3.6 μ

It is seen from this table that the values of $d_{\frac{1}{2}}$ for a given specimen and emulsion thickness increase with the diameter of the cylinder. Hence the size of the cylinder should be considered in the calculations on resolution as shown by Lamerton and Harriss. It will be shown later, however, that these values will be strongly influenced by radiation from the neighborhood of the cylinder (crossfire effect).

From the calculated values of $d_{\frac{1}{2}}$ the following resolution figures were computed as shown in Table 12.3.

Although the results by Lamerton and Harriss are of similar order to those obtained by Doniach and Pelc they are not strictly comparable, because the emulsion and specimen thicknesses, as well as the gaps chosen by the two groups of authors, differ. All the values shown in these tables indicate the fundamental importance of keeping the

TABLE 12.3. Theoretical Values of Resolution (after Lamerton and Harriss (1954))

Emulsion thickness (microns)	Specimen thickness (microns)	Gap between specimen and emulsion (microns)	Resolution (microns)
2	2	0	2.1
2	2	0.5	3.4
5	5	0	5.1
5	5	0.5	6.4
5	2	0	3.3
5	2	0.5	5.0
20	5	0	9.3
20	5	4	20.6

separation between specimen and emulsion and the respective thicknesses of both at a minimum in order to achieve good resolution.

Crossfire-Effect

Lamerton and Harriss investigated also the grain density below a given structure of the specimen when influenced by the radiation from other regions of the specimen. This latter effect, which they called 'crossfire effect' was examined by calculating the contribution to the main intensity in the emulsion overlying the cylindrical structure of given dimensions from the more distant parts of the specimen, assuming that for small distances of r, the change of ionization from a point source of a β-ray emitter follows reasonably well either an $e^{-\mu r}/r$ or an $e^{-\mu r}/r^2$ law. μ is an experimentally determined linear absorption coefficient for a thin extended source.

It was found that the recognition of autoradiographic detail is strongly dependent on crossfire-effects. It could be shown for isotopes such as ^{32}P and ^{35}S that (a) the contribution by crossfire is relatively high for very small structures, (b) the use of thin specimens and thin emulsions tends to reduce the crossfire-effects and (c) low energy electrons, such as those from ^{35}S, reveal less effect from crossfire than high energy electrons such as those from ^{32}P.

12.3b. Autoradiographic Emulsions

After having discussed on a theoretical basis the essential conditions for achieving good resolution, it is now necessary to describe the types of emulsion, their sensitivity and the techniques with which this resolution can be achieved in practice.

Micro-autoradiography makes use chiefly of the nuclear type of emulsion (see p. 60) which characterized by fine grain size, not greater than 0.1–0.4 microns in diameter, narrow grain size distribution and very close packing of grains, i.e. high concentration of silver halide in gelatin. The average physical density of the dry emulsion is 3.6–4 g per cm². These features combined with very thin emulsion layers are the prerequisites for good resolution in autoradiography. Whether the final aim of the chosen technique is to inspect a low grain density distribution, where the number of grains per unit area may have to be counted, or an optical density, or even particle tracks (see the family tree on p. 505), nuclear particle type emulsions have been found to be the most suitable emulsions for the purpose.

Special autoradiographic emulsions (of the nuclear type) for high resolution work are available from several manufacturers: some are supplied on glass, some on film base, some in strippable form or sometimes even in liquid form. These emulsions differ in average grain size and in grain sensitivity. For recording particle tracks thickly coated emulsions are usually required. The choice between these emulsions and the form in which they are supplied depends largely on the requirements of speed and resolution and on the techniques to be used. A survey of some of the available varieties is given in Table 12.4.

The significance of the various types will be better appreciated after the reader has familiarized himself with the techniques of using these emulsions and also with the concept of deriving their relative sensitivities. The latter aspect will be discussed first and later sections will deal with the various techniques.

Sensitivity of Nuclear Emulsions

It is important at this stage that the reader has sufficient understanding of the meaning of sensitivity as applied to the nuclear type emulsions when exposed to charged particles of various energies. The response of such emulsions depends, like that of any other emulsion, on the number of ions required per grain to render this developable. Although the basic ideas have been discussed on pages 64 and 159, they will be presented in the following from the point of view of their significance to autoradiography.

Rate of Energy Loss Per Unit Length of Path

The sensitivity of the nuclear type of emulsion is often given in terms of the energy of the particle at which its passage through the

TABLE 12.4. Autoradiographic Emulsions

Product	Applications and special properties	Nominal emulsion thickness (before processing) (microns)	Mean diameter of undeveloped grains (microns)
Gevaert-Agfa N.V. **Germany and Belgium** 'Scienta' NUC 3.07 Plate	Emulsion for recording particles of high ionizing power; α-particles, electrons up to 100 keV applied in high resolution including Electron-microscope autoradiography. (Also available as liquid emulsion and as sheets without support)	25, 50, 100	0.07
'Scienta' NUC 7.15 Plate	Emulsion sensitive to charged particles of all energies. (Also available as liquid emulsion and as sheets without support)	25, 50, 100, 200	0.15
Ilford Ltd. **United Kingdom** Nuclear Research Plate G 5[1]	Emulsions sensitive to all charged particles of any energy.	50	0.27
Nuclear Research Plate K 5[1]		50	0.20
Nuclear Research Plate L 4[1]		50	0.14
Nuclear Research Plate K 2	Emulsion sensitive to very slow electrons. (Also available as liquid emulsion and as stripping plates)	50	0.20
Eastman Kodak Comp. **U.S.A.** Autoradiographic Plate, Type No-Screen	For recording β- and α-rays, when low radiation flux dictates sacrifice of grain size for sensitivity.	25	1.25
Autoradiographic Plate Type A	For high contrast recording of β- and α-emitters.	25	0.65
Nuclear Track Emulsion, Type NTB3[2]	Liquid emulsion of highest sensitivity; records all charged particles of any energy.		0.34

TABLE 12.4 (Continued)

Product	Applications and special properties	Nominal emulsion thickness (before processing) (microns)	Mean diameter undeveloped grains (microns)
Eastman Kodak Comp. U.S.A.			
Nuclear Track Emulsion, Type NTB2[2]	Liquid emulsion. Records electrons up to 0.2 MeV.		0.26
Nuclear Track Emulsion, Type NTB	Liquid emulsion for recording electrons up to 30 keV and high velocity α-particles.		0.29
Nuclear Track Emulsion, Type NTA[2]	Liquid emulsion for recording moderate energy α-particles.		0.22
Nuclear Track Emulsion Type NTE	Liquid emulsion for very high resolution, including electron-microscope autoradiography		0.06
SWR Plate	suitable for ^{3}H chromatography		
Kodak Ltd., United Kingdom			
Fine-grain Autoradiographic Stripping Plate AR10	Fine grain strippable emulsion of nuclear track type for recording β particles, but not tracks. α-particles can also be recorded.	5μ supported by 10μ gelatin	0.15–0.2
Experimental Scientific Plate V. 1042	Stripping plate coated with same AR10 emulsion. Gives higher resolution with thin specimens.	2μ supported by 10μ gelatin	0.15–0.2
Experimental Scientific Plate V. 1055	Pellicle for reconstituting as liquid emulsion (AR 10 type).		0.15–0.2
Experimental Scientific Plate V. 1056	Non-stripping AR10 Plate for use in apposition technique with water soluble tracers.	5	0.15–0.2
Experimental Scientific Plate V. 1060	Same as V. 1056, but with thinner emulsion for higher resolution.	2	0.15–0.2
Experimental Scientific Plate V. 1062*	Unsupported AR 10 stripping emulsion for use in differential autoradiography.	4	0.15–0.2
Fast Autoradiographic Stripping Plate AR 50	Fast stripping emulsion. (X-ray emulsion type) for coarse autoradiography)	12μ supported by 10μ gelatin	1.2

* See E. O. Field, K. B. Dawson and J. E. Gibbs, *Autoradiographic differentiation of tritium and another beta emitter by a combined color-coupling and double-stripping film technique.* Stain Technology, Vol. 40. No. 5. pp. 295–300. 1965.

From all the manufacturers mentioned above X-ray films of various speeds and grain size are available, which are suitable for macro-autoradiography.

[1] All three emulsions also available as liquid emulsions and, as stripping plates with 5 micron thick emulsion supported by 10 micron thick gelatin.

[2] Also available as plates.

emulsion leads to a developable grain. As an example in Figure 12.9 we see the sensitivity levels of four types of Nuclear Track Emulsion (NTA, NTB, NTB2, and NTB3) as supplied by the Eastman Kodak Co., U.S.A. for electrons of various energies. This graph shows the rate of energy loss dE/dX in eV per micron of silver bromide on the ordinate and the kinetic energy of the particle (electron) in terms of MeV on the abscissa axis. The curve leads to a minimum of ionization at about 1 MeV, as discussed on pages 96 and 159.

The sensitivity levels for each type of emulsion are indicated on the left side of the curve of Figure 12.9 and each type of emulsion

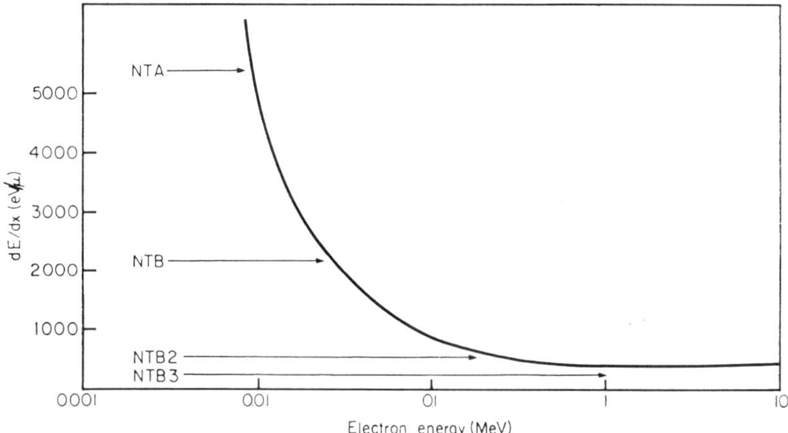

Figure 12.9. Rate of energy loss (eV per micron) for electrons in Eastman Kodak Nuclear Track Plates as a function of particle energy in MeV (theoretical). The approximate sensitivity levels for several types of plates NTA, NTB, NTB2 and NTB3 are indicated by arrows. Each level represents a characteristic rate of energy loss above which grains become developable for particle track recording. In the case of NTA, no electron track can be formed, as the range of electrons is too short at the given energy.

is represented by a characteristic rate of energy loss at which grains become developable to form a track, i.e. a row of grains in the emulsion (see page 159). The lower the rate of energy loss per micron at which a certain type of emulsion responds to an electron of given energy, the more sensitive is the emulsion. If we consider, for instance, a fast electron of 1 MeV initial energy travelling through an emulsion layer, it will not produce developed grains in NTA, NTB or NTB2 emulsion, unless it slows down in the emulsion to energies of less than 10^{-2} MeV in NTA, to 2.8×10^{-2} MeV in NTB and 4×10^{-1} MeV in NTB2 plates. These values can be read off directly from Figure 12.9. On the other hand a 1 MeV electron will render a grain developable at its initial energy in the NTB3 emulsion which is sensitive to all charged particles of any energy. The reason that electrons of maximum energy of 1 MeV emitted by a radioisotope nevertheless render grains developable in all the emulsions referred to in Figure 12.9, is that the emission of most isotopes covers a whole spectrum of electron energies. Hence there will be a sufficient number which slow down in the emulsion to the energy values quoted above. This means that the grain yield, i.e. the number of grains rendered developable per incident electron of 1 MeV, would be highest in the NTB3 and lowest in the NTA emulsion.

Particle Tracks

Apart from α-tracks which are straight line rows of grains, the technique of recording particle tracks is not used much in autoradiography. Nevertheless the theoretical concept of the formation of electron tracks and its connection with sensitivity of nuclear track emulsions should be understood by workers in the field of autoradiography. This is important not only for fully utilizing the potentialities of the photographic methods, but also for interpreting the results. In order to record a longer part of the range of a particle, thick emulsion layers have to be used. Referring to the example given before, an electron of 1 MeV would be fully recorded in an NTB3 emulsion only if the path of the electron (i.e. about 1000 microns) was entirely in the emulsion. As shown on page 97 the number of developed grains per unit length of track would be greatest towards the end of the track, when the electron slows down, as its ionizing power increases towards lower energy. An NTB2 emulsion would record the track of a 1 MeV electront only from about 400 KeV downwards to lower energies and the oher emulsions at the corresponding values quoted before. The NTA emulsion, for instance, would be incapable of recording a track, as it would only respond from less than 10^{-2} MeV, which is less than 10 keV. As can be seen from Figure 14.2 (see p. 603) which shows the ranges of electrons and α-particles as function of the energy of the particle, one cannot speak of a 10 KeV electron track, as the range corresponds to less than 1 micron. The graph Figure 14.2 is useful in any considerations on the path length of electrons or α-particles passing through nuclear emulsions.

Density and Number of Incident Electrons

Our previous considerations on sensitivity are useful for the distinction of various types of nuclear emulsions, but the worker in the field of autoradiography needs a more practical approach. He wants for instance to know the actual number of electrons per unit area required to obtain either a certain number of grains per unit area or a certain density level. Furthermore he is interested in the exposure time required to obtain this level of density for any radioisotope he might be using.

Grain Yield

Such an approach to sensitivity has been made for nuclear track stripping emulsions by determining experimentally the grain yield (η) for electrons emitted by various radioisotopes. The following Table 12.5 shows the experimental grain yield, i.e. the number of

TABLE 12.5. Grain Yield
(for Kodak AR10 Autoradiographic Stripping Plate)

Isotope	Max. energy in MeV	Grain Yield				
		Lamerton and Harris (1954)	Cormack (1955)	Levi (1957)	Herz (1951)	Barnard and Marbrook (1961)
^{14}C	0.155			2		
^{35}S	0.167	1.8				
^{59}Fe	0.46		1.6			
^{131}I	0.81		1.8		1.9	
^{32}P	1.7		0.75		0.8	
^{3}H	0.018					0.85

grains rendered developable per incident electron for the Kodak AR10 Autoradiographic Stripping Plate (For details of this plate, see list on autoradiographic emulsions pp. 523–524.)

For grain yield determinations it is important to use isotope sources of known activity which are uniformly distributed over the area and placed in direct contact with the emulsion layer in order to simulate autoradiographic exposure conditions. Lamerton and Harriss (1954) have shown that grain yields determined by external sources lead to slightly different results if compared with those obtained in direct contact. A further precaution must be considered when determining grain yields: this is that stripping emulsions are stretched after been wetted and the stretching may be non-uniform.

From the table on grain yields it can be seen that the ratio of the maximum to minimum yield is not greater than 2.6, in spite of the wide range of energies of electrons referred to in Table 12.5. As would be expected from the previous discussions, the table shows that the grain yield decreases with higher average energy of electrons. The slight variations of grain yields are due to errors of measurement.

Knowing the grain yield (η) for a particular radioisotope

$$\eta = \frac{N}{n_e} \tag{12.6}$$

where N is the number of developed grains per cm^2 and n_e the number of incident electrons per cm^2, we can determine the optical density, by using Nutting's formula (equation 4.30, see p. 151).

$$D = 0.4343 N\bar{a} \cdot 2.6 \tag{12.7}$$

where $\bar{a}$ is the average grain area presented to the incident electrons. Since the grains of Kodak Autoradiographic Stripping Emulsion AR10 have a cubic shape, having a mean side length of 0.175 microns, $\bar{a} = 3 \times 10^{-10}\,cm^2\,(1\,\mu^2 = 10^{-8}\,cm^2)$. If we substitute N from equation (12.6) we have

$$D = 0.4343 \eta n_e \bar{a} \times 2.6 \tag{12.8}$$

and from this equation the density for a given number of incident electrons per cm^2 can be predicted with a fair approximation, if the grain yield is known for the particular isotope.

Initial Radioactivity Required

If it is desired to know the initial activity of a given isotope to obtain a certain number of electrons per cm^2 for a given exposure time, this can be derived by the following considerations (see p. 9).

Let P_0 be the number of atoms originally present and P_t the number of atoms at time t. Since the rate of decay is proportional to the number of atoms present

$$-\frac{dP}{dt} = \lambda P \tag{12.9}$$

i.e.

$$P_t = P_0 e^{-\lambda t} \tag{12.10}$$

where λ is the decay constant which can be derived from its relationship to the half life H of the particular isotope (see equation 1.8, p. 9), namely

$$\lambda = \frac{0.693}{H} \tag{12.11}$$

In an interval t the number of electrons emitted is equal to the number of atoms which have decayed which is $(P_0 - P_t)$.

The initial number P_0 is related to the initial rate of decay by the expression

$$\left|\frac{dP}{dt}\right|_0 = \lambda P_0 \tag{12.12}$$

Hence if $n_e = (P_0 - P_t)$ is the number of electrons emitted

$$n_e = \left|\frac{dP}{dt}\right|_0 \frac{1}{\lambda}\,(1 - e^{-\lambda t}) \tag{12.13}$$

or

$$n_e = (\mu\text{Ci})_0 \frac{3.7 \times 10^4}{\lambda}\,(1 - e^{-\lambda t}) \tag{12.14}$$

since 1 microcurie (μCi) is equivalent to 3.7×10^4 disintegrations per second (see p. 50). The initial activity A_0 in μCi is therefore

$$A_0 = \frac{2n_e\lambda}{3.7 \times 10^4(1 - e^{-\lambda t})} \tag{12.15}$$

The factor 2 has been inserted because only half of the number of electrons can be utilized photographically, the other half is emitted in a direction opposite to the photographic emulsion layer. Hence double the number of microcuries are required to obtain the necessary number of electrons incident on the emulsion.

The results of such calculations and two experimental results are given in the Table 12.6 below. This shows the number of electrons and the initial activity in microcuries per cm² to obtain a density of 0.2 in 85 days exposure time for the Kodak AR10 autoradiographic stripping emulsion.

TABLE 12.6. Number of Electrons per cm² and Activity in microcuries per cm² to Obtain a Density of 0.2 in 85 Days (For Kodak Autoradiographic Stripping Plate AR10)

Isotope	Max. energy of electrons MeV	Number of electrons per cm²		Initial activity $(\mu\text{Ci})_0$ per cm²
		Exper. $\times 10^9$	Theoret. $\times 10^9$	
^{14}C	0.155	0.75	0.77	5.94 10^{-3}
^{36}Cl	0.71	1.0	0.9	6.48 10^{-3}
^{32}P	1.7		2.04[1]	6.30 10^{-2}

[1] In view of the short half life of ^{32}P (14 days), an exposure of approximately 28 days would be sufficient.

Sensitization

If it is desired to improve the sensitivity of stripping emulsions this can be done for instance by bathing the emulsion before exposure in a 2 per cent solution of triethanolamine for 10 minutes and drying the emulsion before use. In this way short tracks of 10–20 grains per electron (see Figure 12.10) have been recorded, using ^{131}I with an increase of background fog of only twice that without treatment.

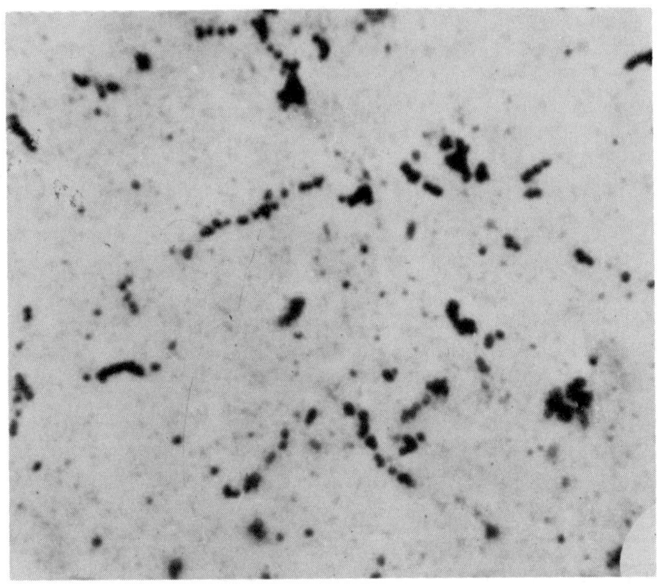

Figure 12.10. Electron tracks in Kodak AR 10 Stripping Plates after treatment in triethanolamine. (*After* R. H. Herz, *Lab. Invest*, **3**.1.71 (1959).)

(T. H. James, W. Vaselow and R. F. Quirk, 1948; P. Demers, 1958; K. S. Bogomolov, I. A. Ruditskaya, and A. A. Siroiskaya, 1958; R. H. Herz, 1959).

12.3c. Autoradiographic Techniques

Many different techniques have been suggested to suit various requirements and some of these will be discussed in the following.

L. F. Bélanger and C. P. Leblond (1946) detached photographic emulsions from lantern plates, melted the emulsion and painted a thin layer of it on a stained histological section. The emulsion attached to the section was left in a light-tight container for drying and exposure. The whole system was then placed in the processing solution

and the specimen remained in permanent contact with the emulsion
for microscopic inspection.

The permanent superimposition of the photographic emulsion and
specimen and their intimate contact with one another is of great
significance as it assists the location of the autoradiographic image
with regard to the spatial distribution of the radioactive material
within the specimen (see also L. F. Bélanger, 1950).

Stripping Techniques

Probably the most convenient method suggested involves the use
of stripping film or stripping plate as described by S. R. Pelc (1947);
G. A. Boyd, and A. I. Williams (1948); A. M. MacDonald, J. Cobb,
A. K. Solomon and D. Steinberg (1949); H. Yagoda (1949); R. W.
Berriman, R. H. Herz, and G. W. W. Stevens (1950); G. A. Boyd
(1955). The basic idea is that an emulsion layer is fastened to a very thin
support which is in turn mounted on a temporary support. Both the
emulsion and its thin support can be stripped and transferred to the
specimen. The thin support is needed for protection of the emulsion
layer which is only about 5 microns thick and therefore easily
damaged and very delicate to handle. The advantage of the thin
support, depending on the material used, is that it may provide an
impermeable layer between the emulsion and the radioactive specimen, avoiding any artifacts by chemical action or abrasion. This
advantage is gained at the expense of resolution, as can be expected
from the various calculated values (see Tables 12.1 to 12.3 and
Figure 12.22, p. 552).

In the Kodak Autoradiographic Stripping Plate AR10, as supplied
by Kodak Ltd, England, the emulsion is reinforced by a gelatin
layer of 8 microns thickness. With this material the emulsion can be
placed in direct contact with the specimen and processed and stained
through the gelatin. The procedure involved in the use of the AR10
Stripping Plate is illustrated in Figure 12.11.

An area of the emulsion sufficiently large to cover the entire
specimen is cut out with a margin at least a quarter to half an inch
all round. The microscope slide bearing the specimen containing the
radioactive substance is 'subbed' before use, i.e. coated with a layer
that ensures good wet adhesion of the emulsion when processed. A
method for subbing microscope slides is given in the Appendix p. 616.
The microscope slide with its fixed specimen is then placed on the
bottom of a glass dish filled with distilled water. The emulsion and
its gelatin support are stripped from the glass plate and placed on the
surface of the water with the emulsion side downwards, facing the

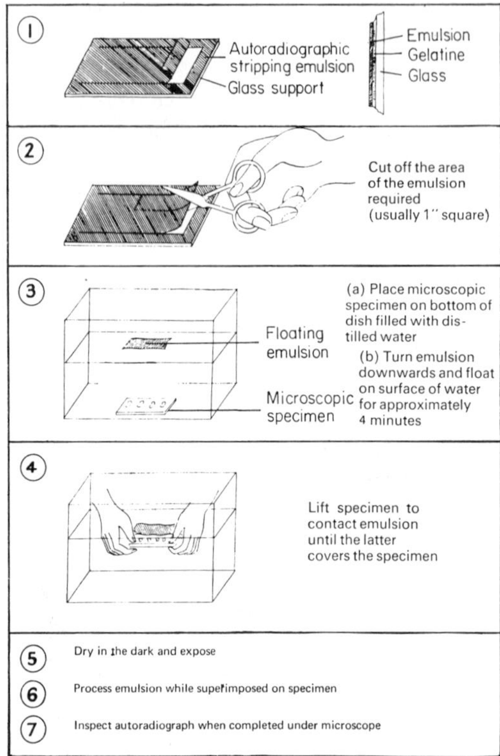

Figure 12.11. Procedure of autoradiographic technique as recommended for Kodak AR10 Autoradiographic Stripping Plate. (*By courtesy* of Kodak Ltd., England. Data Sheet SC-10.)

specimen, as indicated in Figure 12.11. As it swells, the emulsion-gelatin composite first crumples and then stretches out flat. It should be permitted to swell for approximately 4 minutes and may then be lifted from the water by raising the slide underneath it.

The specimen with the superimposed emulsion layer should next be dried in a stream of cold air and placed in a light-tight box for exposure. This may last from a few days to several weeks depending on the activity of the radioactive isotope and its half life.

Processing is then carried out with the emulsion layer in permanent contact with the specimen, using one of the conventional X-ray developers.

I. Doniach and S. R. Pelc (1950) found a satisfactory technique of staining sections through the gelatin and emulsion after processing, provided the emulsion was thin. The common nuclear stains used in cytology, such as the various haematoxylins and methyline blue work well. Particularly good results were obtained using celestin blue and Mayers haemalum. Typical examples of autoradiographs using stripping emulsions are shown in Figures 12.12(A–C) and 12.13.

Study of Soluble Labelled Compounds

A difficulty of fundamental importance in autoradiography is encountered if water-soluble tracers have to be employed in conjunction with the wet stripping film technique which has been described above.

R. S. Russell, F. K. Sanders and O. N. Bishop (1949) described a method of locating water-soluble ^{32}P in plant tissue. The tissue is dehydrated at $-70\,°$C in alcoholic basic lead acetate, which precipitates the phosphate *in situ*. Before it is sectioned or placed in contact with the stripping emulsion, the tissue is embedded in wax. Other freeze-drying techniques are described by F. P. W. Winteringham, A. Harrison, and J. H. Hammond (1950) and by J. E. Harris, J. F. Sloane, and D. T. King (1950).

A very successful and simple technique for the study of soluble labelled compounds was recently introduced by T. C. Appleton (1964). In this method a stripping emulsion is placed (emulsion side facing upwards) on a cover slide glass, the latter having been smeared with glycerin-albumin. The arrangement is shown in Figure 12.14.

The film cover-slips are then dried in a stream of air at room temperature for 2–3 hours and stored in light-tight boxes at 2–4°C. Immediately before use the film cover-slips are cooled to -5 to -10°C. Tissue sections of 5 to 6 microns in thickness are cut in a cryostat at -20 to -30°C. After sectioning each tissue section is then placed against the emulsion surface where it remains during exposure at -20°C and maintains its position during processing.

The fact that perfect adhesion bonds are formed at such a low temperature is not fully understood, but experience has shown that a perfect contact exists not only between emulsion and specimen, but that the section sticks securely to the film during exposure and subsequent processing. The exposure is carried out in a light-tight plastic box provided with a desiccant at a temperature below -20°C. No appreciable loss of sensitivity of the emulsion was observed by exposing at this low temperature.

After exposure the attached tissue section was fixed in Wohlmann's fixative (5 ml glacial acetic acid, 95 ml absolute ethanol) at 17°C for

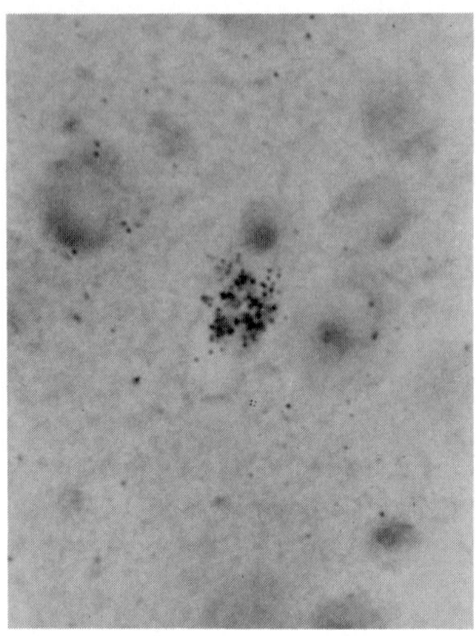

12.12 A

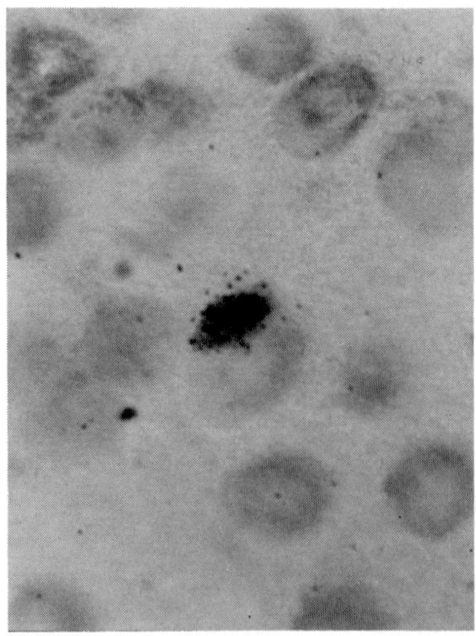

12.12 B

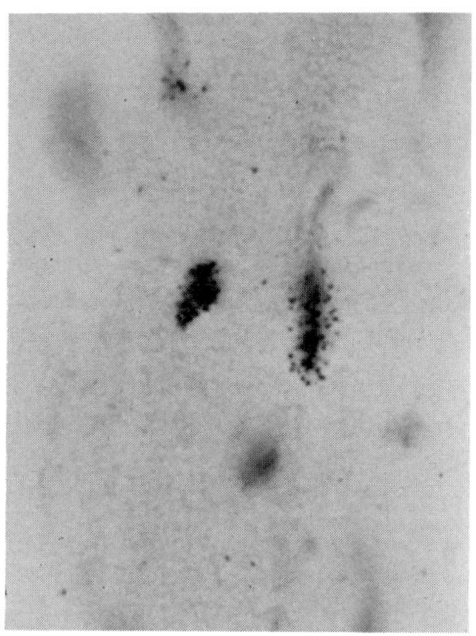

12.12 C

Figure 12.12. Autoradiographs of labelled DNA (stripping film technique). Adult mice were injected with ^{3}H-thymidine, a specific precursor for DNA, and killed 24 hours later. After 250 days exposure the autoradiographs were processed and stained. No divided cells are found in the brain or muscle of adult animals. The labelling shown therefore indicates renewal of DNA. A. A labelled neuron from the brain of an adult mouse. B. A labelled glial cell from the brain of an adult mouse. C. Three labelled nuclei in the heart muscle of an adult mouse. Note the weak labelling of one nucleus. (*By courtesy of* S. R. Pelc, Medical Research Council, Biophysics Research Unit. Dept. of Biophysics, King's College, London.)

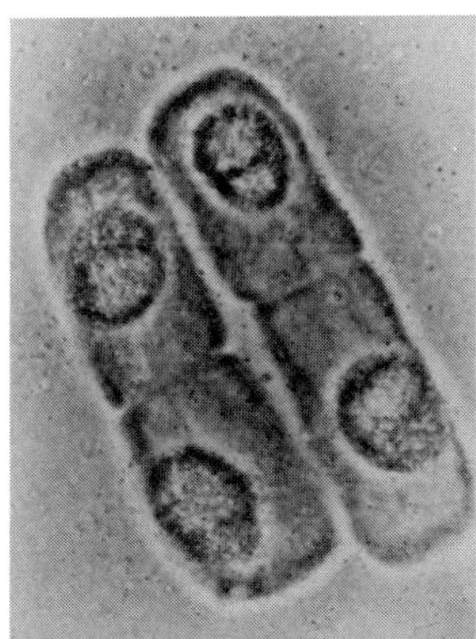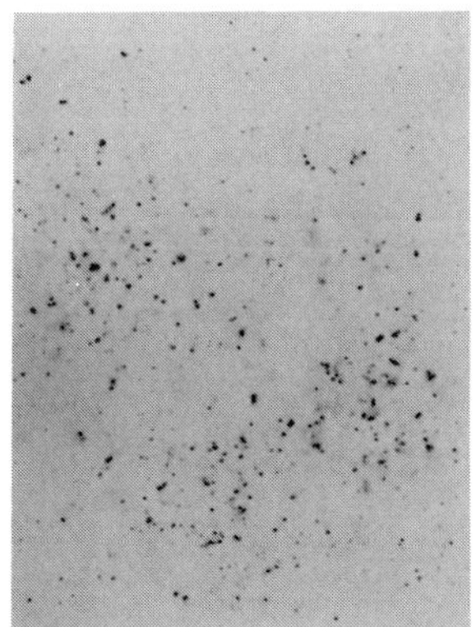

Figure 12.13. Low grain density autoradiograph of squashes of bean root meristems. Left: Photomicrograph of four resting (non-dividing) bean root cells. ($\times$2000) Right: Same area showing autoradiograph over nuclei. One cell did not synthesize. ($\times$2000) (*After* S. R. Pelc, London.)

535

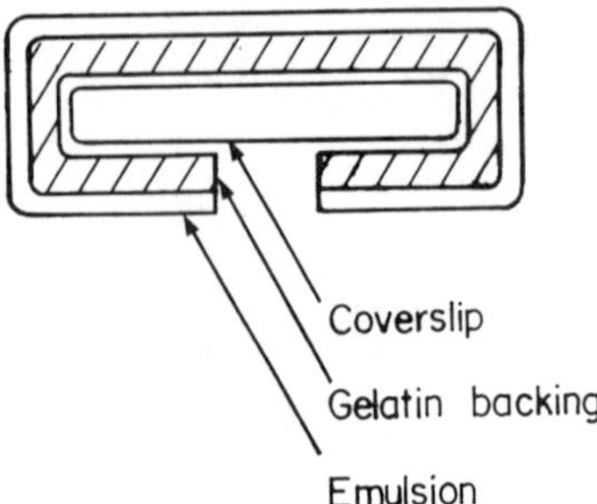

Figure 12.14. Coverslip covered with stripping film emulsion upward. The gelatin backing of the film is in close apposition to the cover slip. (*After* T. C. Appleton. 1964).

1 minute and washed for 1 minute each in four changes of tap water (pH > 7). The system of cover-slip–emulsion–specimen is then processed conventionally and the tissue section, being on the outside, may be stained after processing is completed.

After drying the whole composite of cover-slip–emulsion–tissue is inverted and mounted on a microscope slide with the tissue section in contact with the glass side. Thus the completed autoradiograph is finally built up exactly as in the conventional stripping film method, but with the advantage that any soluble tracer substances are kept in their original position. (S. R. Pelc and T. C. Appleton, 1965). (See Figures 12.15A and B.)

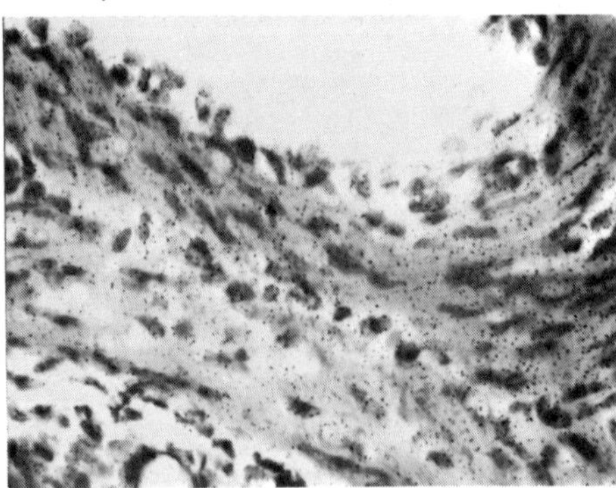

Figure 12.15. (A) Water soluble autoradiograph of the distribution of tritiated thymidine in smooth muscle of the colon of a mouse 2 minutes after intravenous injection. Note the even labelling throughout the muscle. (Haematoxylin and eosin, $\times 400$). (The autoradiographs A and B were prepared by a technique which allows the retension of the soluble labelled compound. *After* T. C. Appleton (1964) *J. Roy. Microscop. Soc.* **83**.3.277. (1964).)

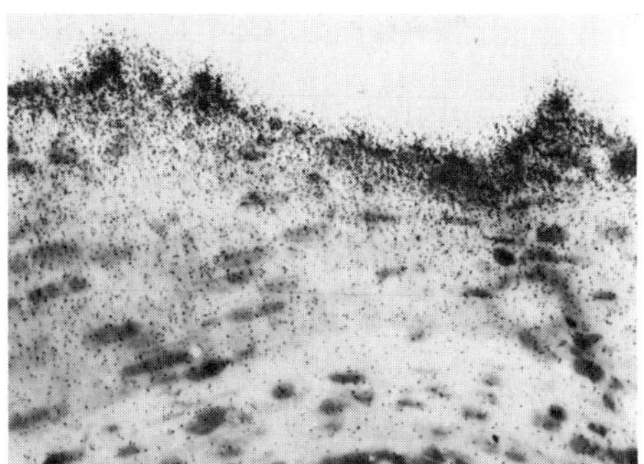

Figure 12.15. (B) Water soluble autoradiograph of the distribution of tritiated thymidine in smooth muscle, 2 minutes after intraperitoneal injection. The thymidine can be seen diffusing in from the peritoneal fluid bathing the outside of the gut; the diffusion appears to be unaffected by nuclei or muscle fibres. (Haematoxylin and eosin $\times 400$) (*After* S. R. Pelc and T. C. Appleton, *Nature* **205**.4978.1287 (1965).)

Dipping Method. (Liquid Emulsions)

Techniques of dipping microscope slides with attached histological specimens in liquid emulsions have become quite popular in recent years chiefly for two reasons. (1) Several types of nuclear emulsion of different sensitivity, which are otherwise not available in strippable form, can be used and (2) some workers consider the dipping method as less time consuming than the stripping method. (J. S. Arnold, 1954).

After histological slides have been prepared by conventional methods, the liquid emulsion container which was stored in a refrigerator, is heated up to a temperature of between 40–50°C. The emulsion is then stirred in the dark by means of a glass rod before the microscope slide is dipped into the liquid emulsion, so that the specimen attached to the slide is fully immersed. It is then immediately withdrawn and allowed to drain vertically onto a gauze pad for a few seconds. The drying is carried out horizontally in a light-tight box preferably containing a tray for several other slides. D. L. Joftes (1959) uses a drying agent in the box, and an open vessel containing a saturated solution of KNO_2 providing a relative humidity of 47 per cent. Other workers recommend a lower relative humidity. After sufficient exposure has been given, the slides may be processed in a conventional manner, keeping permanent contact between specimen and emulsion, just as with stripping emulsions. After drying in a clean air stream cover slips or a thick protective coating e.g. of acrylic spray plastic may be given. Special equipment for liquid emulsion techniques for autoradiography of biological specimens is now available commercially, which is known as the Con-Rad/Joftes Fluid emulsion system.

The disadvantage of the dipping method if compared with the stripping method is that it is neither possible to control the thickness of the coatings exactly, nor to achieve uniform or reproducible thicknesses in coating. The technique is now frequently used in connection with ^{3}H, the important isotope of hydrogen (called tritium), which emits electrons of a maximum energy of 18 keV; the mean energy is of the order of 6 keV. Using tritium, thickness considerations are less critical, as the majority of ^{3}H electrons do not penetrate deeper into the emulsion than the top layers of grains. The average range of tritium electrons in silver-bromide has been found to be about 0.9 microns. E. A. Evans (1966); R. Baserga and K. Nemeroff (1962, 1963); E. A. Barnard and I. Marbrook (1961).

Amongst the many modifications of specialized high resolution techniques the following two methods are considered to be of impor-. tance in the author's view. One of these is of interest for autoradiographic procedures in which chemically reactive surfaces are to be considered, specifically in connection with metallurgical problems. The other one permits the simultaneous distinction between ^{3}H electrons and those of other β-emitters.

Study of Chemically Reactive Surfaces

G. T. Rogers and J. D. H. Hughes (1963) have described a method permitting high resolution which is suitable for metallurgical research applications. In this field the resolution was often restricted due to chemically reactive solids causing fogging, particularly during the wet processing method.

In most stripping and dipping methods the autoradiographs must be developed *in situ*, and a protective film of about 1 micron thickness has to be used for separating specimen and emulsion to prevent chemical reaction between specimen and silver bromide grains. Using the method of Rogers and Hughes, a film consisting of a single layer of silver bromide crystals and backed by a Formvar layer is applied to a microscope slide. The silver bromide/Formvar layer is then stripped from the slide, softened with methyl ethyl ketone and is applied to a separate 0.3 micron layer of polyisobutylene coated on another glass slide. In use, the layer thus supercoated with polyisobutylene is again softened with methyl ethyl ketone and then applied to the specimen. As the solvent evaporates the silver bromide layer conforms to the surface topography. After exposure, which is done in a desiccator containing silica gel, the film is stripped from the specimen and processed.

The single grain layer of silver bromide referred to earlier, is obtained by coating the microscope slide with silver by vacuum evaporation, which is then converted to silver bromide in bromine vapour. The silver bromide grain size varies with the surface density of the evaporated silver and with the conditions of bromination. Under favorable conditions the authors claim that a resolution of 1–2 microns can be obtained with grains smaller than 0.3 microns.

The claimed advantages of this method are that use of a supercoat and emulsion softened in a non-aqueous solvent prevents chemical interaction when the emulsion is applied even to very reactive metals. In addition, stripping *before* processing prevents interaction during development and replication of the shape of the specimen surface by the plastic layer facilitates re-registration with the sample after processing.

Differentiation Between Tritium and Other β-ray Emitters

In this technique a distinction is made possible between cells labelled with ^{3}H, with ^{14}C and with both isotopes together in the same histological preparation. The method which is due to K. B. Dawson, E. O. Field, and G. W. W. Stevens (1962) is based on the difference between the mean range of ^{3}H electrons ($<1\,\mu$) and ^{14}C electrons ($<60\,\mu$) and the use of two stripping film layers, which are separated by a thin layer of celloidin. A special unsupported stripping film (Experimental Scientific Plate No. V 1062, supplied by Kodak Ltd. England) is applied first to the thoroughly washed specimen containing cells labelled with ^{3}H and [14]C. This film is then allowed to dry for 10–15 minutes in the dark, then exposed for 24 hours and processed after which the emulsion thickness is about 2.5 microns. The silver grains due to the exposure by ^{3}H electrons are situated predominantly on the surface of the emulsion. The developed grains are then bleached for 1 minute in a solution containing 5 per cent potassium ferricyanide and 2.5 per cent potassium bromide. After washing in water the silver grains are colored by immersion for 5 minutes in a dye freshly prepared as a 9:1 mixture of two solutions A and B respectively, having the following composition

Solution A:	Sodium sulfite (anhydrous)	2 grams
	Genochrome (May & Baker, London)	2 grams
	Sodium carbonate (anhydrous)	20 grams
	Potassium bromide	1 gram
	Water	900 ml
Solution B:	2,4-dichloro-1-naphthol	1 gram
	Industrial spirit	100 ml

These solutions can be stored in dark bottles and mixed on the day of use.

The staining by this solution results in blue colored grains which contrast well with the pink of Feulgen stained cell nuclei.

After washing the film is dried and dipped in a 0.5 per cent solution of celloidin in 1:1 ether alcohol. Then a standard stripping film is applied over the celloidin and exposed for 10 days and processed in a conventional manner.

The first layer which is in contact with the tissues records predominantly ^{3}H electrons and the second layer exclusively ^{14}C electrons. The coloring of the silver grains in one layer permits the grains in the two layers to be differentiated without the need for separate focusing. The inert celloidin layer applied between the two layers is about 0.1 μ thick and its purpose is to insulate the first autoradiograph from the second layer. If no coloring is used, a distinction between the effects of ^{3}H and ^{14}C electrons can of course be made by focussing with an oil immersion objective. As the exposure times with ^{3}H and ^{14}C can be chosen separately as required, any relation between the number of electrons from each isotope can be achieved. (R. Baserga and K. Nemeroff, 1962; E. O. Field, K. B. Dawson, and J. E. Gibbs, 1965).

Autoradiography with the Electron Microscope

The potentialities of micro-autoradiography have still not been fully utilized, as is apparent from recent developments in which the electron microscope has been applied to a further improvement of resolution. Increase of magnification as it is attainable with the electron microscope can be utilized only if (a) smaller grain sizes are used than in conventional autoradiography and (b) if the thickness of emulsion layers and histological sections is further reduced. Obviously the resolution is also greatly influenced by the range of the exposing particles involved.

The smaller the developed grain size and the thinner the layers of emulsion and specimen section, the better is the alignment between the location of labelled structural detail in the specimen and the corresponding position of the developed silver grains.

The desired changes are however confronted with new inherent difficulties. A reduction of grain size limits the intrinsic sensitivity, i.e. the average length of passage of electrons through a grain is reduced, with the result that the energy loss an electron suffers per grain may be too small to render it developable. Hence several hits may be required for developability. On the other hand the chance of hitting the same grain twice or three times decreases with further reduction of grain size.

For a full utilization of the resolution of the electron microscope the average diameter of the developed grain should be of the order of $\frac{1}{10}$ of the smallest grain diameter of about 0.2 microns as used in conventional micro-autoradiography. This would then be of the order of 0.02 microns = 200 Å. The energy loss of a 10 keV electron in an undeveloped silver bromide grain of a diameter of 200 Å is about 100 eV. This energy is far below the minimum which is required for a single grain hit developability of grains having about 10 times the diameter. Thus, as expected, several hits will be necessary in a grain of diameter of 0.02 microns to render it developable. As will be shown in this section photographic manufacturers have produced suitable emulsions having an average grain size of almost the desired dimension.

The other requirements mentioned above referring to the thickness of emulsion and specimen section are of even greater importance than the grain size. The emulsion thickness should be a monolayer of grains and that of the section should be an ultramicrotome section. All these changes tend to reduce the chance of obtaining a statistically adequate yield of silver grains which must be guaranteed before certainty in the interpretation of EM autoradiographs is reached. Hence there is a tendency to compromise between these various competing factors by using intermediate grain sizes, and it can be said that the development of EM autoradiography in recent years has already achieved very noticeable results.

M. M. Salpeter and L. Bachmann (1964, 1965) and L. Bachmann and M. M. Salpeter (1965) have described detailed methods for obtaining autoradiographs of increased resolution with the electron microscope of specimens labelled with ^{3}H-thymidine, using a magnification of approximately 10–40,000. Because of the future importance of this technique a few details will be given in the next sections, as they are reported by M. M. Salpeter and L. Bachmann (1964, 1965) and G. C. Budd and S. R. Pelc (1964).

(1) *Emulsions.* The available liquid emulsions suitable for electron microscope autoradiography are for instance Agfa-Gevaert 'Scienta' Nuc. 3.07 (average grain diameter 0.07 microns), Ilford L4 (average grain diameter 0.14 microns) and Eastman Kodak NTE (average grain diameter 0.06 microns).

(2) *Preparation of Emulsion.* In order to obtain very thin emulsion layers, which is of fundamental importance in this technique, the

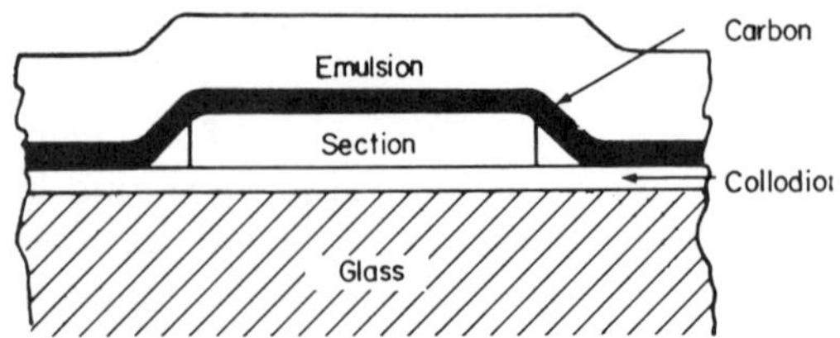

Figure 12.16. Build-up of specimen sandwich for electron microscope auto-radiography. (*After* M. M. Salpeter and L. Bachmann. *J. Cell Biol.* **22**.2.469 (1964).)

authors had to remove some of the gelatin from the emulsion. The NTE emulsion was therefore diluted by using 10 c.c. of water per gram of emulsion. This was carried out in a water bath at 60–70°C. The emulsion was then centrifuged in a preheated centrifuge until the supernatant was clear. Then the supernatant was discarded and the remaining concentrated emulsion was resuspended in warm water (1–4 c.c. per gram of the original emulsion). Very uniform emulsion layers are made by giving a few drops of rediluted centrifuged emulsion (kept at 60°C) on to horizontally held slides. The emulsion is then drained off and air dried in vertical position. The emulsion thickness is checked by interference colors, using reflected white light. An examination of the centrifuged NTE emulsion, revealing silver interference color has been shown to be a monolayer of 600 Å thickness.

(3) *Specimen Preparation.* Tritium labelled tissue sections are transferred onto a glass slide coated with a very thin layer of collodion on which the histological section is placed. On this section a carbon layer of 50 Å thickness is vacuum coated. The emulsion prepared as described before is then coated on the carbon layer, as shown in Figure 12.16.

The purpose of the collodion layer is to allow the stripping of the system specimen-carbon-emulsion from the glass. The purpose of the carbon layer is (1) to avoid chemical influence of tissues on the emulsion layer; (2) to protect stained sections from destaining during processing, and (3) to permit the emulsion to be coated more evenly than it would be directly on the specimen.

(4) *Gold Latensification.* Before the actual processing procedure the emulsion can be sensitized by gold latensification as described by T. H. James, W. Vanselow, and R. F. Quirk (1948), in order to increase the sensitivity of the emulsion. This is carried out by dipping the

exposed slide first in water and then soaking it for 30 seconds in a solution of 1:20 dilution of gold thiocyanate for 30 seconds and it is then washed. The gold latensification bath consists of 0.5 c.c. of a 2 per cent stock solution of gold chloride ($AuCl_3HCl \cdot 3H_2O$) which is diluted to 10 c.c. and 0.125 grams of potassium thiocyanate is dissolved in it. Then 0.15 grams potassium bromide is added to the solution after which it is diluted to 250 c.c. This solution is unstable and should be used within the same day.

(5) *Development.* The developer consists of 0.045 per cent Elon, 0.3 per cent ascorbic acid, 0.5 per cent borax and 0.1 per cent potassium bromide. Development is carried out at 24°C. The developer is unstable but when used within the same day after preparation an 8 minute development time results in developed grains of about 0.05 μ (500 Å). The developed grain size depends mainly on development time.

In another experiment the NTE emulsion was developed in Dektol developer (diluted 1:2) for 1 minute at 24°C and L4 emulsion was developed in Microdol developer for 3 minutes. Both emulsion grains showed filamentary silver (see p. 81). It was found that one developed grain was obtained for 12 electrons incident on a monolayer of grains. A three to fourfold increase of sensitivity was found when gold latensification was used in conjunction with Dektol development; one developed grain was thus obtained for every 3 electrons hitting the monolayer of grains.

After completion of processing the combined system specimen-carbon and emulsion layer was transferred from the glass to the electron microscope grid.

Because of the complexity of the procedures involved, the reader is referred to the original paper by Salpeter and Bachmann (1964).

Figure 12.17 illustrates an electron-microscope autoradiograph, taken by M. M. Salpeter.

(6) *Budd and Pelc's Membrane Technique.* In EM-autoradiography the system photographic emulsion plus labelled section is usually mounted on copper or stainless steel grids. In the past some techniques have been used in which the EM-grid is in contact with the labelled sections and emulsion during exposure and subsequent processing. The specimen grid influences the spreading of fluid emulsion and causes an increase in thickness near the grid bars and in the center of the square. A further useful method in which the grid is not applied until the autoradiographs have been processed has been

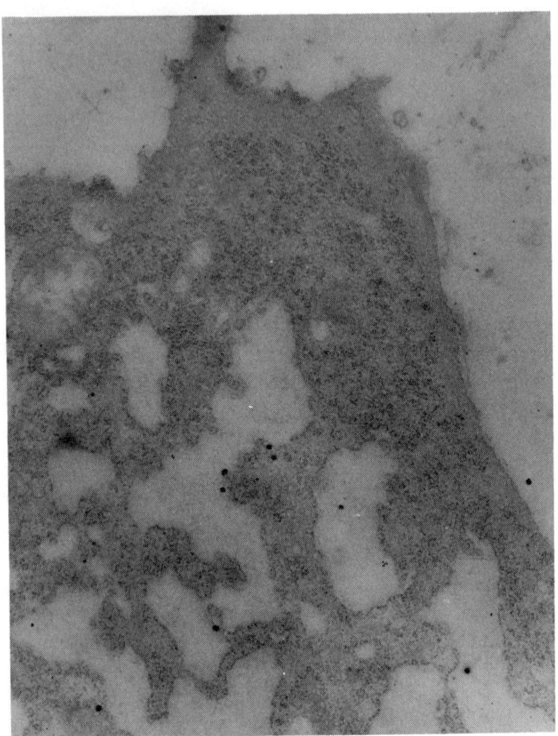

Figure 12.17. Electron-microscope autoradiograph, ×37,000. A silver section of 500Å had been stained with uranylacetate, coated with Eastman Kodak NTE emulsion. Developed by gold latensification-Olon ascorbic acid. The autoradiograph shows developing chondrocyte (cartilage cell) in the regenerating limb of the newt, triturus, labelled with ^{3}H proline (a precursor of collagen). The label is associated with the limiting membranes of the endoplasmic reticulum. (*By courtesy of* M. M. Salpeter Dpt. of Engineering Physics, College of Engineering, Cornell University, Ithaca, New York.)

described by G. C. Budd and S. R. Pelc (1964). These authors make use of a plastic membrane to support the specimen and emulsion, thus avoiding the undesirable surface effects due to the bars of a supporting grid.

The supporting membranes are prepared by dipping glass slides into 0.3–0.4 per cent polyvinyl formvar dissolved in ethylene dichloride. After scoring a line about 2 mm around the edge of the microscope slide, the membranes are floated onto distilled water and are then taken up on thin Perspex (Lucite) slides which are provided with holes. Hence each hole is bridged by a thin membrane which has been shown to be strong enough to support sections and photographic emulsion during all processes including that of examination in the electron microscope. (See Figure 12.8).

The Perspex slides are of the same dimensions as conventional glass slides, i.e. either 76 × 25 mm or 76 × 38 mm. The optimum diameter of the holes was found to be approximately 7 mm. Two holes of this diameter are usually pierced in to the smaller and six each into the larger sized slides.

Sections of labelled tissue may be cut from material, embedded for instance in epoxy resin, such as 'Araldite' on an ultramicrotome and are transferred onto each Formvar membrane using a fine brush or a wire loop. A suspension of photographic emulsion is then prepared in a water bath at 45°C. The following emulsions were used: Ilford G5 diluted 1 in 5; Ilford L4 diluted 1 in 5; and Agfa-Gevaert Scienta NUC 3.07 diluted 1 in 10. A drop of dilute emulsion is then placed on each membrane by means of a pipette. After 10 seconds the excess emulsion is poured off and the slide is left to dry for exposure in a light-tight box.

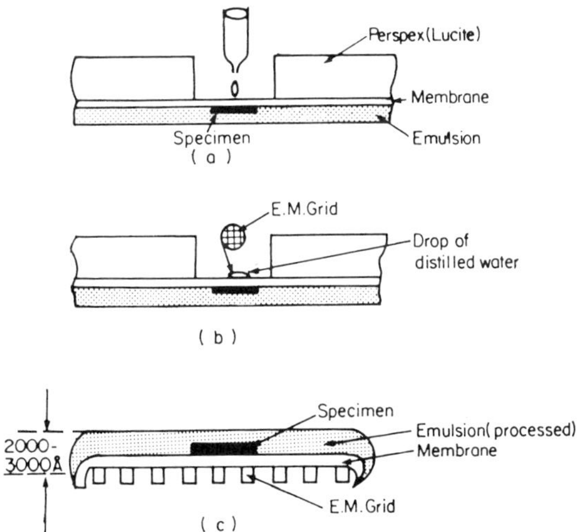

Figure 12.18. Preparation of autoradiographs for electron microscope. (a) Invert autoradiograph and drop on distilled water. (b) Place electron microscope grid through Perspex hole on membrane. (c) Carefully perforate membrane around grid, remove the grid + membrane + specimen + processed emulsion, with watchmakers forceps. (*After* G. C. Budd and G. Rowden, Biology Department, Royal Free Hospital, School of Medicine, London and S. R. Pelc, Medical Research Council, Biophysics Research Unit, King's College, London.)

After conventional processing (and staining if desired) the system slide-emulsion-section is prepared for the electron microscope (see Figure 12.18). The E.M. grids are then attached by dipping a grid into water and placing this in contact with the underside of an autoradiograph on a membrane. For removal the latter is then perforated around the grid before the water has dried out. Finally the grid plus the whole system of membrane-autoradiograph and section is removed with fine forceps and examined under the electron microscope.

Figures 12.19 and 12.20 show electron-microscope autoradiographs taken by G. C. Budd and R. Rowden, Biology Department, Royal Free Hospital, School of Medicine, London.

Further information on autoradiography with the electron microscope is found in the following papers: R. P. van Tubergen (1961); S. R. Pelc, J. D. Coombes, and G. C. Budd (1961); J. C. Hampton and H. Quastler (1961); L. G. Caro (1962); L. G. Caro, and R. P. van

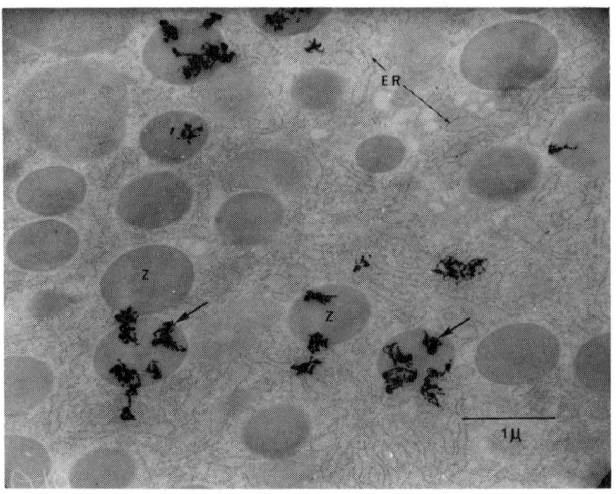

Z = zymogen granules.
E.R. = Rough surfaced endoplasmic reticulum.
→ Silver grain.

Figure 12.19. An electron microscope autoradiograph of part of a pancreatic acinar cell from a mouse killed one hour after injection of ^{3}H-glycine(1 mCi). This amino acid is incorporated into the proteins in the zymogen granules produced in these cells. There is a clear association between silver grains and these secretion granules. Membrane method. Ilford L4 emulsion, 2 months exposure. Section stained with lead citrate. Specimen embedded in Araldite Resin. ×36,000. (*By courtesy of* G. C. Budd and G. Rowden, Biology Department, Royal Free Hospital, School of Medicine, London.)

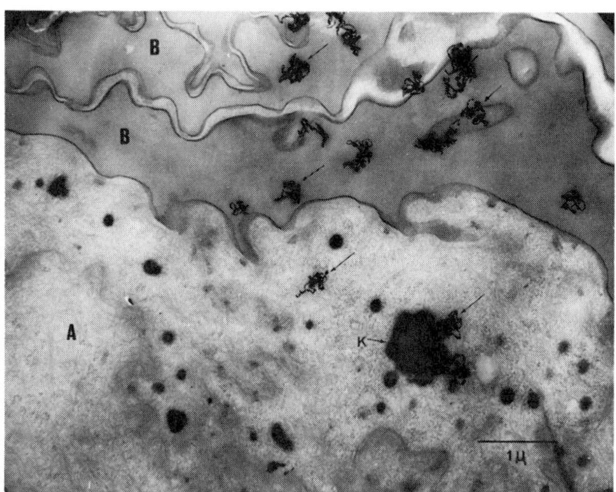

→ Silver grain. K = keratohyalin granule.

Figure 12.20. E.M. autoradiograph. A section through cells in the upper layers of keratinized(cornifies) tissues of a mouse. Twenty-four hours after injection of 0.5 mCi of ^{3}H-glycine. This amino-acid has been incorporated into proteins in cells in the upper layers of the epithelium (in this case oesophagus). The cells have then migrated so that most of the label appears in the virtually dead outer cornified region. Membrane method, Ilford L4 emulsion, two months exposure. Lower cell A is still viable and will change into the form of cells B. Upper cells B etc. can be considered as dead. Thus the labelled proteins indicated by the silver grains in the dead cells B must have arrived by earlier labelling of cells in position A which then change during twenty-four hours into cell type B and take the labelled proteins with them. This shows cell movement in the constant replacement required in keratinized epithelia such as skin. ×33,000. Details as for Figure 12.19. (*By courtesy of* G. C. Budd and G. Rowden. Biology Department, Royal Free Hospital, School of Medicine, London.)

Tubergen (1962); L. G. Caro, R. P. van Tubergen and J. Kolb (1962); G. A. Meek and M. J. Moses (1963); S. R. Pelc (1963); L. Bachmann and M. M. Salpeter (1964a, 1964b); M. M. Salpeter and L. Bachmann (1964); P. Granboulan (1965). Papers dealing specifically with resolution in this field are those by L. G. Carol (1962), M. M. Salpeter and L. Bachmann (1965), and L. Bachmann and M. M. Salpeter (1965).

Artifacts and Interlayers

The possibilities of artifacts in conventional autoradiographic records are almost unlimited and this is not surprising if one considers the rather sophisticated methods which are sometimes involved. These artifacts may be due to unwanted radiation, chemical or safelight fogging, they may be caused by air bubbles between specimen and photographic material, reticulation due to different bath temperatures during processing, latent image fading, some may appear as surface marks, spottiness, drying marks, stain etc. One of the remedies for these is, of course, careful and proper handling of the delicate photographic material and its processing procedures. It must be admitted that occasionally artifacts occur, the origin of which is very difficult to trace and only by checking each stage of the various operations may the cause be discovered and thus avoided in future records.

In some autoradiographic techniques the interposition of an impermeable layer between the emulsion and specimen has been suggested for the following reasons.

1. Ghost autoradiographs may appear due to possible chemical reactions of the substances contained in the specimen with the emulsion and this may lead to the formation of undesired latent images. Such a result will lead to illusions in the interpretation of the autoradiograph. For this reason *it is a necessary to carry out control exposures with and without a radioisotope.* Only if this control is negative, i.e. if no photographic action can be detected after a non-active specimen has been placed in contact with the emulsion for the same exposure time as required for the genuine autoradiograph, is interpretation of the actual autoradiograph permissible. (G. A. Boyd and F. A. Board, 1949; G. A. Boyd, 1955; H. Yagoda, 1949.)

2. Interlayers may be used to prevent leaching of the radioactive compound when a wet mounting technique is applied (otherwise see pp. 533 and 538).

3. There are cases where any contact of the developer with the specimen should be prevented, since the latter might undergo changes due to the reaction with processing solutions.

Whatever the material of the interposed impermeable layer, its thickness should be limited to about 1 micron and preferably less in order to ensure still acceptable resolution. C. Chapman-Andresen (1953b) has carried out a thorough investigation on this subject in which she examined materials such as celloidin, formvar, nitro-cellulose varnish and nylon in wet and dry applications particularly with respect to restripping the emulsion from the specimen. Nylon was found to be most suitable with regard to restripping properties. Dupont Nylon was dissolved by heating in isobutyl alcohol and the interlayer was made by dropping the solution onto a water surface. The interlayers of all the materials mentioned could be made as thin as 0.25 to 0.5 microns.

12.4 EXPERIMENTAL RESOLUTION TESTS ON STRIPPING EMULSIONS

A description of the experimental resolution tests has been delayed until the reader has acquired sufficient familiarity with the techniques applied to autoradiography and in particular with stripping emulsions. As mentioned in the earlier sections on the theoretical aspects of resolution as a function of specimen and emulsion thickness and their respective separations, it is of great interest to compare the theoretical with the experimental results.

12.4a. Resolution Charts (Stevens' Method)

A most elegant method for testing resolution was first described by G. W. W. Stevens (1948, 1950). He makes use of a test chart, such as that used for determining the resolution of photographic lenses. This test chart is reduced photographically onto a Kodak Maximum Resolution Plate which permits a resolution of more than 1000 lines per millimeter. The test chart shown in Figure 12.21(a) is an actual enlargement of a silver image 1.3 $\times$ 1 mm in size from such a plate. The silver image is first reconverted into a silver halide image. The latter is then toned with radioactive iodine (^{131}I). Figure 12.21(b) shows an enlargement of an autoradiograph of the small iodine test chart. The autoradiograph was obtained on a Kodak AR10 Stripping Plate in the conventional manner as indicated before.

From the known separation of the lines, the autoradiographic resolution attainable has been found to be better than about 2.3 microns which is in fair agreement with the values theoretically derived by Doniach and Pelc (1950) and by Lamerton and Harriss (1954). (See pp. 515–521). Resolution of this order was obtainable only with a thickness of specimen not exceeding 3–5 microns, since the divergence of particles produces at increasing distances greater unsharpness.

Figure 12.22 due to Stevens illustrates how the resolution is affected if the distance between specimen and film is increased.

As seen from this figure a separation of 3 microns produces a very marked loss of resolution. This effect has been studied systematically by Stevens by overcoating the test charts with gelatin of various thicknesses. The effect of resolution has been stated by Stevens in terms of lines per mm which can be detected by different separations between radioactive material and emulsion using stripped experimental nuclear track and process emulsions of 4 microns coating thickness. His results are quoted in Table 12.7.

TABLE 12.7. Experimental Resolution Tests (after Stevens)

Emulsion type	Thickness of separating layer	Resolution (lines per mm)	Line width (microns)
Process	$0\,\mu$	95	5.3
Process	$3\,\mu$	45	11.0
Process	$10\,\mu$	24	20.8
Nuclear track	$0\,\mu$	220	2.3
Nuclear track	$3\,\mu$	61	8.2
Nuclear track	$10\,\mu$	35	14.3

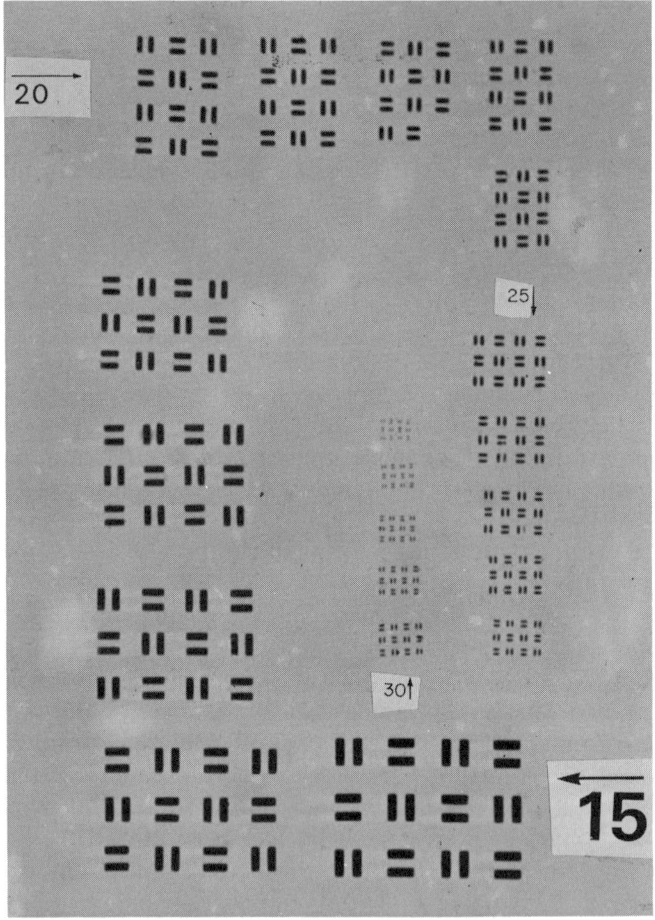

Figure 12.21 (a)

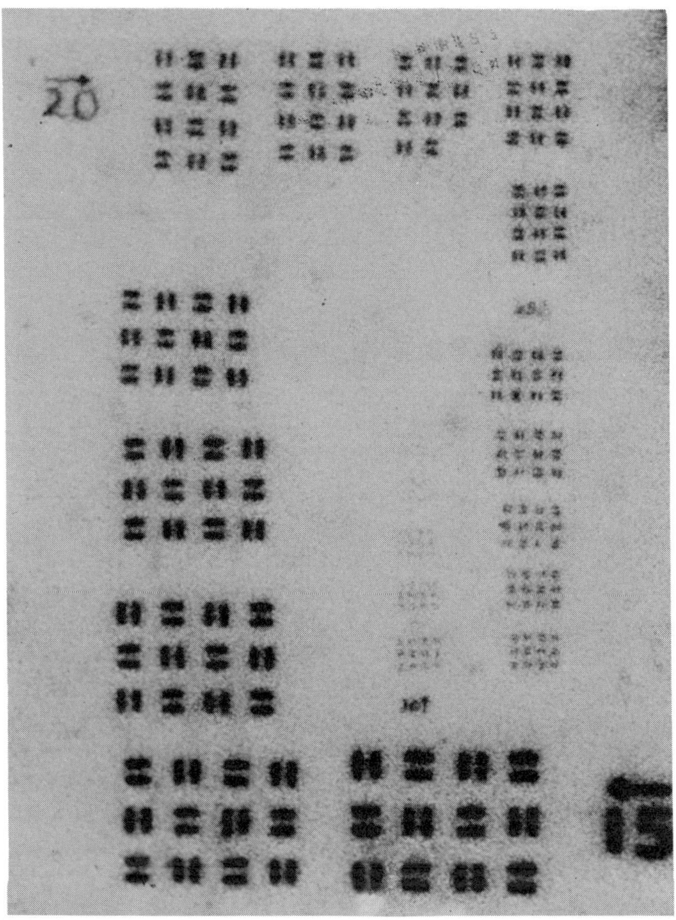

(b)

Figure 12.21. Radioactive test chart. (a) Test chart enlarged from Kodak Maximum Resolution Plate. (b) Autoradiograph of radioactive test chart containing ^{131}I. (*After* G. W. W. Stevens *Brit. J. Radiol.* **23**.723. (1950).)

If electrons of shorter range are used than those emitted by ^{131}I, the resolution will be considerably improved, as has been shown by C. Chapman-Andresen (1953a) in autoradiographs of algae containing tritium and ^{14}C, see Figure 12.23.

12.4b. Contact Versus Stripping Method

L. F. Lamerton and E. B. Harriss (1954) applied the experimental test method of Stevens to no-screen X-ray film and to experimental

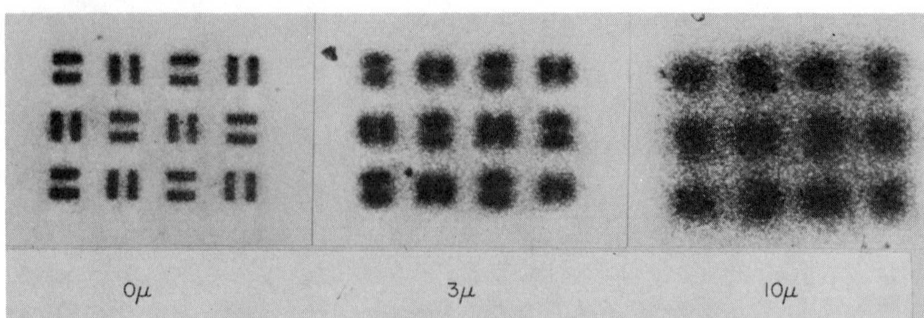

Figure 12.22 Autoradiographs of radioactive test chart taken in contact with the iodine-131 test chart, (i.e. at 3 and at 10 microns distance from it. Note the change in resolution with increasing distance. (After G. W. W. Stevens, *Brit. J. Rad.* **23**. 723. 1950).

X-ray stripping films of 24 and 12 microns thickness, in order to compare the stripping-method with the conventional contact-method. The values for resolution determined in this way were as follows

Resolution

No-screen X-ray emulsion in contact   30 μ approx.
Experimental X-ray emulsion (stripping film) (24 μ thick) 12 μ approx.
Experimental X-ray emulsion (stripping film) (12 μ thick)   9 μ approx.

This comparison gives a good demonstration of the advantages of the stripping as against the contact-method. Using the intimate contact between the stripped film and the specimen it can be assumed

 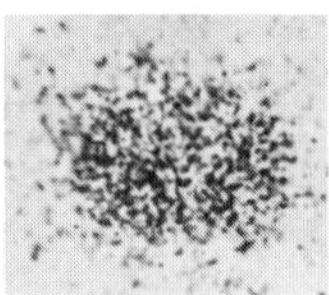 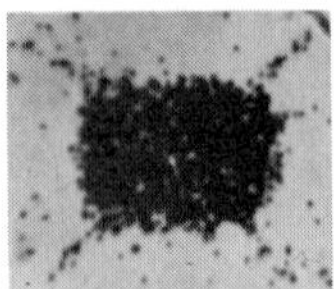

Figure 12.23. Resolution as a function of range of electrons. *Left:* Photomicrograph of a living Scenedesmus (a small alga) composed of four cells. ($\times 1800$.) *Middle:* Autoradiograph of Scenedesmus after 48 hours in a solution containing ^{14}C as carbonate. Max. Energy of ^{14}C electrons 0.155 MeV. Activity of solution $2 \times 10^{-3}\,\mu Ci$ per μl. Exposure twenty-four hours. ($\times 1800$.) *Right:* Autoradiograph of Scenedesmus after 48 hours in tritiated water containing 33 μCi per μl ($\times 1800$) Max. Energy of ^{3}H electrons is 0.018 MeV. (*By courtesy of* Mrs. C. Chapman-Andresen, Carlsberg Laboratory, Copenhagen, Denmark. *Exp. Cell Res.* **4**.1. (1953).)

that the emulsion follows the grooves or uneveness of the specimen on a microscopic scale, which is impossible with the contact method.

Using autoradiographic 'X-ray' stripping emulsions, which are commercially available, intermediate resolution between that demanded by macro- and micro-autoradiography can be obtained. Owing to their relatively high speed they permit a rapid autoradiographic test to be carried out in order to check whether an exposure on a fine grained emulsion is worth while.

12.5   QUANTITATIVE EVALUATION

In some applications of micro-autoradiography it is important to assess quantitatively the amount of radioactive substance present within a given volume of a section of a specimen. There exist basically three different techniques for which a quantitative assessment can be made. First the photographic emulsion may be strongly exposed, so that a fairly high density results; secondly the exposure can be carried out so that only a few, say about 10 grains per 100 square microns, are rendered developable and thirdly a technique may be chosen whereby particle tracks are seen under the microscope. According to the appearance of the autoradiographic record the three methods may be referred to as

(a) Optical density method,
(b) Low grain density method, and
(c) Particle track method.

12.5a.   Optical Density Method

In this case it is assumed that we have to deal with a fairly dense autoradiograph on which optical density measurements can be made, if this is found necessary. Since conventional densitometers measure relatively large uniform areas, a microdensitometer may be used.

When quantitative measurements are made, it may be necessary to know either the relative or the absolute quantities of radioisotope present within various regions of a specimen.

Relative Quantities

In the region of low exposures it is well known that the photographic density is proportional to the number of electrons incident on the emulsion. Hence by measuring the densities at various regions of the photographic emulsion, the relative activities of the specimen in these regions can be assessed. The densities up to which proportionality with exposure exists can often be found from exposures using radioactive standard sources of a given isotope. Such standards are available for some well known radioisotopes, sometimes as point sources or as uniformly coated areas of low, but known activity; although the latter is not necessary for the assessment of relative quantities.

If a wide range of densities should be covered, which is beyond that for which proportionality exists between density and exposure, the photographic material should be calibrated by means of a standard radioactive source. This can be done by plotting a density-exposure curve.

Absolute Quantities

If the activity of a standard source which is uniformly distributed over a certain area is known either in terms of microcuries per cm^2 or in number of particles emitted per cm^2 per unit time, the relationship between density and activity can easily be plotted for any emulsion from experimental results. In this and the other methods described, corrections for background must always be applied.

An alternative approach to the establishment of absolute quantities is to determine from the measured density in the autoradiograph the approximate number of developed grains per unit area, using Nutting's formula (equation 12.7, p. 528) and the grain yield, (see p. 527); finally if required equation 12.15, p. 529 may be used in order to obtain quantities in terms of microcuries per unit area. This is valid only within the straight line region of the density-exposure curve, as the grain yield holds only within this part of the curve.

The use of optical densities causes three distinct disadvantages with regard to a satisfactory evaluation of autoradiographs. These are

1. The high density may obscure important histological details in the autoradiograph and does not permit full utilization of the available resolution.

2. The exposure time involved for producing an autoradiograph of optical densities may be unduly long.

3. In order to achieve high density in a reasonably short time, the limit of radioactive material for safe incorporation into an animal may have to be exceeded, thus causing radiation damage.

For these reasons the application of the low grain density method is preferred in many applications.

12.5b.   Low Grain Density Method

Relative Quantities

The relative quantities of tracer in a specimen can be derived in a similar way as with optical densities. The average number of grains per unit area in the nuclear track stripping emulsion (AR10) has been found to be proportional to the exposure up to a density of about unity. This does not hold for tritium, as in this case chiefly the top grain layers of the emulsion are affected and self-absorption of the electrons from ^{3}H becomes significant even in a very thin specimen. (J. H. Taylor, 1953.) The relative assessment of the amount of radioactive isotope present in a cell can be made from grain counts only.

Absolute Quantities

In the low grain density method, the number of grains rendered developable per unit area can be counted under the microscope and a quantitative estimate of the activity can be approached from the grain yield, and equations 12.7 and 12.15, pp. 528–529 (R. H. Herz, 1951, 1960. S. R. Pelc, 1958). Care should be taken in these calculations when referring to the extremely fine grained emulsions, as used in conjunction with the electron microscope (see p. 540), as the grain yield of these is much less than unity. (M. M. Salpeter and L. Bachmann, 1965; L. Bachmann and M. M. Salpeter, 1965.)

Typical examples of conventional low grain density autoradiographs are those illustrated in Figures 12.12 and 12.13.

12.5c. Particle Track Method

Theoretically it can be assumed that each charged particle hitting a nuclear track emulsion sensitive at minimum ionizing power (see p. 526) will produce a recognizable track if the emulsion is sufficiently thick to allow enough of the path of the particle to remain within the emulsion. The quantitative estimate (equation 12.15, p. 529) of the amount of tracer producing these tracks could therefore be based on a simple relation between the number of tracks per unit area and the number of atoms decayed in this area. Such estimates can be made fairly easily with α-tracks only and if their number per unit area is limited. α-Tracks appear as short straight rows of grains, but even if the tracks are fairly close together, short tracks can be recognized individually under the microscope and be counted (see Figure 12.24).

If the investigation deals with β-particles (or positrons) however, many disturbing effects occur, which render the track counting most difficult. These tracks, particularly those of low energy electrons, are highly curved due to scatter in the emulsion and appear as a tangled mass of silver grains, as soon as a small number of tracks are present within a small area of the emulsion. A further complication is the

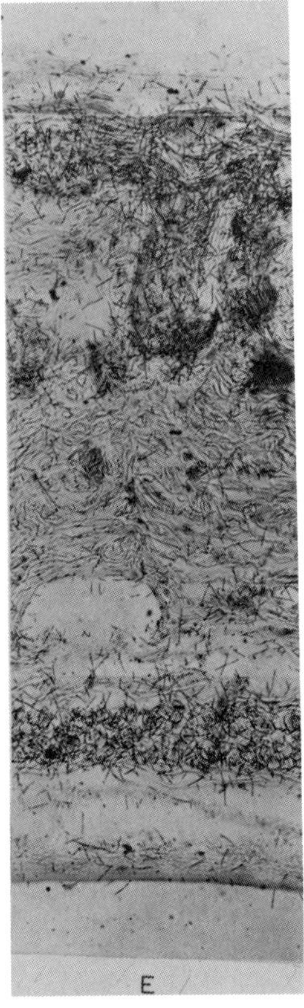

Figure 12.24. α-Particle track autoradiograph of the skin of a mouse containing polonium.

chance that some of the tracks occurring may have been caused by cosmic rays and natural radioactivity in the surrounding of the emulsion. It is chiefly for these reasons that particle track autoradiography is seldom used. (J. E. Harris, J. F. Sloane, and D. T. King (1950); H. Levi (1954); H. Levi and A. S. Hogben (1955); H. Levi and A. Nielsen (1959).)

A typical electron track autoradiograph is shown in Figure 12.25.

12.5d.   Grain Counting and Background

Counting

Grain counting is usually done visually by using an oil immersion objective ($\times 60$ or $\times 90$). It is necessary to count grains by focusing throughout the depth of the emulsion. The use of phase microscopy may facilitate the separate viewing of silver grains and tissues.

Automatic electronic grain counting has also been used for this purpose. (R. A. Dudley and S. R. Pelc, 1953). Frequently it is not the number of grains per microscopic field that is counted, but the number of grains e.g. across the nuclei or cytoplasm of a cell. Workers have to arrange their own individual techniques of grain counting.

Background

The recognition and interpretation of the effects of tracers in photographic emulsions depends largely on the amount of background present. This means that the number of developed silver grains due to the radiation emitted by the specimen must be statistically significant above the number of background grains, before the former can be regarded as being due to the effect of a tracer. Considering nuclear type stripping emulsions, the background should preferably be not greater than four grains per 100 square microns but a background of less than 1 grain per 100 square microns has been achieved. This lower level of background grains per unit area can be obtained if the darkroom illumination is kept at low intensity and for a limited time, if the emulsion is well protected from undesirable ionizing radiation during storage and exposure and if it is processed strictly according

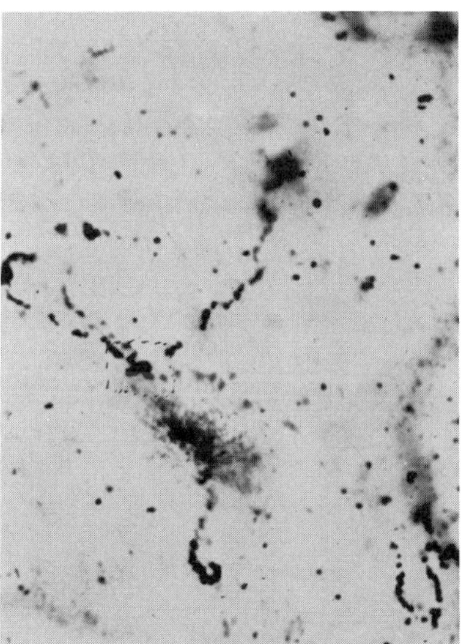

Figure 12.25. Five electron tracks emitted by ^{35}S from a group of staphylococci having taken up radiosulfamide. The bacteria are not visible on the photograph, because they are not in focus in the microscope. (*By courtesy of* F. Zajdela and A. Lacassagne. Institut du Radium, Paris.)

to the manufacturer's recommendations. Also the age and storage condition of the emulsion play an important part with respect to background fog.

As a rough guide it may be said that at least 20–50 grains per 100 square microns should be obtained if a grain count is to be regarded as significant, but this figure depends largely on individual condition and particularly on the standard deviation of the background grains per unit area.

12.5e.   Latent Image Fading

The taking of autoradiographic records may require exposure times of many days or weeks and during this period the emulsion may suffer from latent image fading (see p. 144). This can be of serious consequence particularly in the case of quantitative evaluation. Although latent image fading is beneficial during storage of unexposed film material, as it reduces background effects from cosmic rays and natural radioactivity, it should be reduced to a minimum during exposure. If the tendency to latent image fading has been observed in certain experiments this can be very much reduced by storing the emulsion and the attached specimen during exposure in a dry oxygen free inert gas, such as nitrogen or by storing it in an atmosphere of carbon dioxide. (C. R. Ray and G. W. Stevens, 1953.) The conditioning of an emulsion and the attached specimen in carbon dioxide in combination with a drying agent such as lithium chloride was found to be advantageous. (R. H. Herz, 1959.) His arrangement for conditioning is shown in Figure 12.26. The main advantage of using CO_2 as against

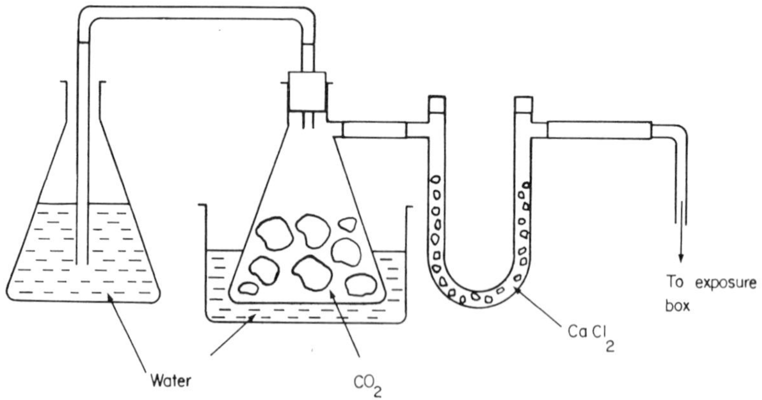

Figure 12.26. Use of CO_2 for conditioning of autoradiographic emulsions.
(*After* R. H. Herz, *Lab. Invest.* (1959).)

nitrogen for instance is that it is a heavy gas that can be retained easily in a container for a considerable time and also that it is usually available as dry ice in most research institutions. No latent image fading has been observed in the Kodak Autoradiographic Stripping Plate AR10, as long as the emulsion was in direct contact with the section, probably due to lack of free oxygen. (Miss M. G. E. Welton and S. R. Pelc, private communication.)

12.5f. Concentration of Radioactive Material

The concentration of tracer required in a specimen is a compromise between the minimum quantity needed to produce an autoradiograph in a reasonable exposure time and the amount which may affect the metabolic process in the organism either chemically or by the radiation emitted by the tracer. Although the quantity of tracer to be applied in any given case is usually determined by trial, attempts have been made by S. R. Pelc (1951) for instance to lay down certain quantitative recommendations which may be helpful to the inexperienced worker.

The rough guide given by Pelc distinguishes between minimum concentration of an isotope in an animal for short and for long lived isotopes. In the following equations originally derived by Pelc, certain assumptions have been made with particular reference to two aspects; (1) to the required average exposure times for short and long lived radioisotopes and (2) to the fact that only a part of the specimen is usually labelled. The following equations refer to the use of Kodak Autoradiographic AR10 Stripping plates.

(1) *Short Lived Isotopes*

Assuming that the thicknesses of section and emulsion are 5 microns, the minimum concentration (C_m)

$$C_m = \frac{11.5f}{H\eta} \text{ microcuries per ml}$$

where H is the half-life (in days), f is the ratio of labelled to unlabelled tissue, η is the grain yield which is the number of grains developable per incident electron.

(2) *Long Lived Isotopes*

In this case the minimum concentration C_m

$$C_m = \frac{12.5f}{t\eta} \text{ microcuries per ml}$$

where t is the exposure time in days.

Obviously these equations can only give an approximate estimate, since results depend on so many unpredictable features, such as the fate of the tracer compound in the organism, on secretion and excretion and others. As a rough rule it can be assumed that about 30–50 radioactive atoms must decay during the time of exposure in order to obtain an autoradiograph of an average cell, using the low grain density method. A further practical but very rough guide is that about 1 microcurie of activity of an isotope is required per gram of tissue to produce an autoradiograph with an exposure time of 20 days.

12.5g.   Autoradiography Versus Counters

It is a generally accepted view that electronic instruments, such as Geiger- or scintillation counters are the obvious detectors for radio-activity. It has been shown however by A. N. Davenport and G. W. W. Stevens (1955) and by G. W. W. Stevens and D. Spracklen (1962), that photographic recording of radioactivity may be useful or even preferable in certain circumstances. These may be, for instance, when electronic instruments are not available or cannot be used for certain reasons, also when permanent records of relative activities are required or when many rather weak sources of short half lives have to be assessed simultaneously. In the latter case corrections for decay might not be required, as the samples are measured all at once.

Using the photographic method, the radioactive specimens may be mounted on a sheet of plastic or glass and then be placed in direct contact with a photographic emulsion. Since image details would most probably not be required in this case, resolution is of no impor-tance and the fastest double coated no-screen X-ray film should be used. After exposure and processing the densities of the X-ray film should be measured and the relative or even absolute values of activity may be determined by reference to a density-exposure curve on the same type of film, obtained from exposures to the same radio-isotope. Obviously the same processing conditions should be used for the film to be evaluated, as for the density-exposure curve. Within the straight line region of this curve, a relative assessment can be made without reference to a calibration curve.

More favorable geometry can be attained in general by using the photographic method rather than counter measurements, as the specimens are in direct contact with the photographic material. It must be admitted that the disadvantage of the photographic method is that it usually requires specimens of uniform activity over a certain area which cannot always be achieved. If specimens of non-uniform activity are to be used greater distances between source, specimens, and film should be chosen in order to smooth out the lack of uniformity. This will undoubtedly impair the accuracy of the experiment. Furthermore 'cross-fire' effects such as the influence of the radiation from areas adjacent to the regions to be measured can be partly prevented, for instance by the use of metal frames around the specimens in contact with the film. This influence can be estimated from the range of particles involved and from the distances between the specimens on the film.

BIBLIOGRAPHY

Boyd, G. A. (1955). Autoradiography in Biology and Medicine. Academic Press, New York.

Rogers, A. (1967). Techniques of Autoradiography. Elsevier Publ. Co. Amsterdam–London–New York.

REFERENCES

Appleton, T. C. (1965). Resolving power, sensitivity and latent image fading of soluble compound autoradiographs. *J. Histochem. and Cytochem.* **14,** No. 5 p. 414, 1965.

Appleton, T. C. (1964). Autoradiography of soluble labelled compounds. *J. Roy. Microscop. Soc.*, **83,** 3, 277.

Arnold, J. S. (1954). An improved technique for liquid emulsion autoradiography. *Proc. Soc. Exp. Biol. Med.*, **85,** 1, 113.

Bachmann, L., and Salpeter, M. M. (1964a). A quantitative approach to high resolution autoradiography. *"The Third European Conference on Electron Microscopy,"* Prague.

Bachmann, L., and Salpeter, M. M. (1964b). Zur Autoradiographie im Elektronenmicroskopischen Bereich. *Naturwissenschaften*, **51,** 237.

Bachmann, L., and Salpeter, M. M. (1965). Autoradiography with the electron microscope. *Lab. Invest.*, **14,** 6, Pt. 2. 303.

Barnard, E. A., and Marbrook, I. (1961). Quantitative cytochemistry using directly applied radioactive reagents. *Nature*, **189,** 412.

Baserga, R., and Nemeroff, K. (1962). Factors which effect efficiency of autoradiography with tritiated thymidine, *Stain Technol.*, **37,** 1, 21.

Baserga, R., and Nemeroff, K. (1963). The use of standard slides in semi-quantitative radio-autography with tritiated compounds. *Stain Technol.*, **38,** 2, 111.

Bélanger, L. F. (1950). Method for routine detection of radiophosphates and other radioactive compounds in tissues: inverted autograph. *Anat. Rev.*, **107,** 149.

Bélanger, L. F., and Leblond, C. P. (1946). A method for locating radioactive elements in tissues by covering histological sections with photographic emulsions. *Endocrinology* **39,** 8.

Berriman, R. W., Herz, R. H., and Stevens, G. W. W. (1950). A new photographic material and high resolution emulsion for autoradiography. *Brit. J. Radiol.* **23,** 472.

Bogomolov, K. S., Ruditskaya, I. A., and Siroiskaya, A. A. (1958). Hypersensitization of nuclear emulsions by triethanolamine. *Zhur. Nauch. i. Priklad Fotografii i Kinematografii,* **3,** 52.

Boyd, G. A. (1955). *Autoradiography in Biology and Medicine.* Academic Press, New York.

Boyd, G. A., and Board, F. A. (1949). A preliminary report on histochemography. *Science,* **110,** 586.

Boyd, G. A., and Williams, A. I. (1948). Stripping film techniques for histological autoradiographs, *Proc. Soc. Exp. Biol. Med.,* **69,** 225.

Budd, G. C., and Pelc, S. R. (1964). The membrane method of electron-microscope autoradiography. *Stain Technol.,* **39,** 295.

Caro, L. G. (1962). High resolution autoradiography II. The problem of resolution. *J. Cell Biology,* **15,** 189.

Caro, L. G., van Tubergen, P. P., and Kolb J. (1962). High resolution autoradiography. I. Methods. *J. Cell Biology* Vol. **15(2)** 173.

Chapman-Andresen, C. (1953a). Autoradiographs of algae and ciliates exposed to tritiated water. *Exp. Cell. Res.,* **4,** 1, 239.

Chapman-Andresen, C. (1953b). Thin films for application between emulsion and isotope-containing specimen in autoradiography. *Compt. Rend. Trav. Lab., Carlsberg.* Serie Chimique, **28,** 22, 529.

Cormack, D. V. (1955). The beta-ray sensitivity of autoradiographic stripping film. *Brit. J. Radiol.,* **28,** 450.

Davenport, A. N., and Stevens, G. W. W. (1955). Comparison of radioactivities by the use of X-ray films. *Brit. J. Appl. Phys.,* **6,** 31.

Dawson, K. B., Field, E. O., and Stevens, G. W. W. (1962). Differential autoradiography of tritium and another emitter by a double stripping film technique. *Nature,* **195,** 4840, 510.

Demers, P. (1958). *Ionography, Les émulsions nucléaires, Principes et applications.* Montreal. Les Presses Universitaires de Montreal.

Doniach, I., and Pelc, S. R. (1950). Autoradiograph technique. *Brit. J. Radiol.,* **23,** 267, 184.

Dudley, R. A., and Pelc, S. R. (1953). Automatic grain counter for assessing quantitatively high resolution autoradiographs. *Nature,* **172,** 992.

Evans, E. A. (1966). *Tritium and its Compounds.* Butterworths, London.

Field, E. O., Dawson, K. B., and Gibbs, J. E. (1965). Autoradiographic differentiation of tritium and another β-emitter by a combined colour-coupling and double-stripping film technique. *Stain Technol.,* **40,** 5, 295.

Granboulan, P. (1965). Comparison of emulsions and techniques in electron-microscope radioautography. In *The Use of Radioautography in Investigating Protein Synthesis,* C. P. Leblond and K.W. Warren (Eds.) Academic Press, New York.

Hampton, J. C., and Quastler, H. (1961). Combined autoradiography and electron microscopy of thin sections of the intestinal epithelial cells of the mouse labelled with ^{3}H-Thymidine. *J. Biophys. Biochem. Cytol.*, **10**, 140.

Harris, J. E., Sloane, J. F., and King, T. D. (1950). A new technique in autoradiography, *Nature*, **166**, 25.

Herz, R. H. (1951). Photographic fundamentals of autoradiography. *Nucleonics*, **9**, 24.

Herz, R. H. (1959). Methods to improve the performance of stripping emulsions. *Lab. Invest.*, **8**, No. 71.

Herz, R. H. (1960). Autoradiography. Physics and technique of autoradiography. *IXth International Congress of Radiology* (1959) Georg Thieme Verlag Stuttgart. Urban and Schwarzenberg, München-Berlin.

Hsieh, J. J. C., Hungate, F. P., and Wilson, S. A. (1965). Autoradiography: Technique for drastic reduction of exposure time to alpha particles. *Science*, **15a**, 1821.

James, T. H., Vanselow, W., and Quirk, R. F. (1948). Gold and mercury latensification and hypersensitization for direct and physical development. *Phot. Soc. Am. J.*, **14**, 349.

Joftes, D. L. (1959). Liquid Emulsion Autoradiography with Tritium. *Lab. Invest.*, **8**, 131.

Jowsey, J., Rayner, B., Tutt, M., and Vaughan, J. (1953). The deposition of ^{90}Sr in rabbit bones following intravenous injection. *Brit. J. Exp. Pathol.*, **XXXIV**, 4. 384.

Kohn, A. (1951a). Emploi des éléments radioactifs pour l'étude des aciers et des processus industriels en métallurgie. Applications au cas de l'étude des segregations. *Rev. Met.*, **48**, No. 3, 219.

Kohn, A. (1951b). The utilization of the autoradiographic method for studying the homogeneization of phosphorus and arsenic dendritic ségrégations in steels. *The World Metallurgical Congress*, Detroit.

Kohn, A. (1953a). Étude de l'homogénéisation des Ségrégations Dendritiques de phosphore et d'Arsenic dans les aciers par la méthode autoradiographique. *Rev. Met.*, **50**, No. 2, 139.

Kohn, A. (1953b). Les Applications des Radioéléments en Métallurgie, *La Documentation Métallurgique*, **13**, 9. Serie B. No. 20.

Lacassagne, A., and Lattes, J. (1924). *Compt. Rend. Soc. Biol.*, **90**, 351.

Lamerton, L. F., and Harriss, E. B. (1954). Resolution and sensitivity Considerations in Autoradiography. *J. Phot. Sci.*, **2**, No. 4., 135.

Leblond, C. P., and Warren, K. B. (1965). In (C. P. Leblond and B. Warren.) The Use of Radioautography in Investigating Protein Synthesis. Vol. **4**. *Symposia of the International Society for Cell Biology*. Academic Press, New York.

Levi, H. (1954). Quantitative β-track autoradiography of single cells. *Exp. Cell Res.*, **7**, 44.

Levi, H. (1957). A discussion of recent advances towards quantitative autoradiography. *Exp. Cell Res. (Suppl.)* **4**, 207.

Levi, H., and Hogben, A. S. (1955). Quantitative β-track autoradiography with nuclear emulsions. *Dan. Mat. Fys. Medd.*, **30**, No. 9.

Levi, H., and Nielsen, A. (1959). Quantitative evaluation of autoradiograms on the basis of track or grain counting. *Lab. Invest.*, **8,** No. 71, 82.

Matteoli, L., and Logi, P. (1966). An improved method of quantitative autoradiography for industrial purposes. *European Atomic Energy Community, Euratom, Eur 3173.e.* Report by Instituto di Recerche Breda, Milan.

McDonald, A. M., Cobb, J., Solomon, A. K., and Steinberg, D. (1949). Stripping film technique for autoradiographs. *Proc. Soc., Exp. Biol. Med.*, **72,** 117.

Meek, G. A., and Moses, M. J. (1963). Localization of tritiated thymidine in hela cells by electron autoradiography. *J. Roy. Microscop. Soc.*, **81,** 187.

Ostroff, E. (1965). Early Fox Talbot photographs and restoration by neutron irradiation. *J. Phot. Sc.*, **13.** 5. 213.

Pelc, S. R. (1947). Autoradiographic techniques, *Nature*, **160,** 749.

Pelc, S. R. (1951). Radiation dose in tracer experiments involving autoradiography. *Ciba Found. Conf. on Isotopes in Biochemistry*, J. and A. Churchill, London, p. 122.

Pelc, S. R. (1958). Autoradiography as a cytochemical method with special reference to ^{14}C and ^{35}S. In *Cytochem. Methods.*, Vol. **1,** Academic Press, New York.

Pelc, S. R. (1963). Theory of electron-autoradiography. *J. Roy. Microscop. Soc.* **81,** pts 3 and 4, 131.

Pelc, S. R., Coombes, J. D., and Budd, G. C. (1961). On the adaptation of autoradiographic techniques for use with the electronmicroscope. *Exp. Cell Res.*, **24,** 192.

Pelc, S. R., and Appleton, T. C. (1965). Distribution of tritiated thymidine in various tissues. *Nature*, **205,** 4978, 1287.

Pelc, S. R., Appleton, T. C., and Welton, M. E. (1965). State of light autoradiography. In C. P. Leblond and K. W. Warren. *The Use of Radioautography in Investigating Protein Synthesis*. Academic Press, New York, pp. 9–22.

Rabinowicz, E. (1950). An investigation of surface damage using radioactive metals. Physics of Lubrication, Supplement No. 1. *Brit. J. Appl. Phys.*

Ray, R. C., and Stevens, G. W. W. (1953). Reciprocity failure and latent image fading in autoradiography. *Brit. J. Radiol.*, **26,** 362.

Rogers, G. T., and Hughes, J. D. H. (1963). High Resolution Autoradiography of Chemically reactive Surfaces. *Nature*, **199,** 4893, 566.

Russel, R. S., Sanders, F. K., and Bishop, O. N. (1949). Preparation of radioautographs to show contribution of ^{32}P in plant tissue. *Nature*, **163,** 639.

Salpeter, M. M., and Bachmann, L. (1964). Autoradiography with the electron microscope. *J. Cell Biol.*, **22,** No. 2, 469.

Salpeter, M. M., and Bachmann, L. (1965). Assessment of technical steps in electron microscope autoradiography. (In C. P. Leblond and K. B.

Warren) *"The Use of Radioautography in Investigating Protein Synthesis"* Academic Press, New York. pp. 23–41.

Sherwood, H. F. (1947). Vacuum exposure holder for microradiography, *Rev. Sci. Instr.*, **18,** 80–83.

Sievert, R. (1921). *Acta. Radiol,* **1,** 89.

Stevens, G. W. W. (1948). Resolution testing in autoradiography. *Nature,* **169,** 432.

Stevens, G. W. W. (1950). Radioactive microphotographs for resolution testing in autoradiography. *Brit. J. Radiol.,* **23,** 723.

Stevens, G. W. W., and Spracklen, D. (1962). Potentialities of autoradiography for comparison of radioactive samples. *"Radioisotopes in the Physical Sciences and Industry."* I.A.E.A. Vienna.

Taylor, J. H. (1956). In (G. Oster and A. W. Pollister) Autoradiography at the Cellular Level. *Physical Techniques in Biol. Research.* Vol. III Cells and Tissues. Academic Press, New York.

van Tubergen, R. P. (1961). The use of radioautography and electron microscopy for the localisation of tritium label in bacteria. *J. Biophys. Biochem. Cytol.* **9,** 219.

Welton, M. G. E., and Pelc, S. R. Private Communication.

Westermark, T., and Erwall, L. G. (1952). Study of the distribution of surface potential by means of radioactive deposits. *Nature,* **169,** 703.

Winteringham, F. P. W., Harrison, A., and Hammond, J. H. (1950). Autoradiography of water-soluble tracers in histological sections. *Nature,* **165,** 149.

Yagoda, H. (1949). *Radioactive Measurements with Nuclear Emulsions.* John Wiley, New York.

13

Microradiography

13.1 INTRODUCTION

X-Ray microradiography may be regarded as a supplementary method to optical and electron microscopy. It is attractive because it offers the possibility of revealing the interior micro-structure of the object under examination and of achieving a resolution which is at least as good as that of the optical microscope. The advantages of microradiography are due to the penetration of soft X-rays and their specific absorption properties. The mass-absorption coefficients in the range of wavelengths applied are approximately proportional both to the third power of the X-ray wavelength and atomic number of the absorber, neglecting the effects of absorption edges. The latter may be particularly useful in certain circumstances.

The term X-ray microscopy which is frequently used covers not only microradiography, but the complete field of applying X-rays in microscopic investigations, such as in quantitative and qualitative absorption, emission and micro-diffraction analysis. Scanning microscopy, using counter detection and fluorescent methods, is now widely used in the electron probe microanalyser in biological, metallurgical and industrial applications.

In this chapter we will deal with microradiography only, which covers essentially three different methods of recording photographically microscopic details in a thin specimen by means of soft X-rays. These are

(a) the contact method
(b) the point projection method and
(c) the reflection method.

The principles of these three methods are illustrated diagrammatically in Figure 13.1.

(a) *The contact method.*   This is similar to conventional radiography. The chief differences are that soft X-rays are used, because the specimen to be radiographed is thin and that high resolution has to be attained as subsequent magnification depends essentially on the resolving power of the photographic emulsion and on the optical properties of the enlarging system. The essential feature of the contact method is that the specimen is in intimate contact with the photographic material as shown in Figure 13.1.

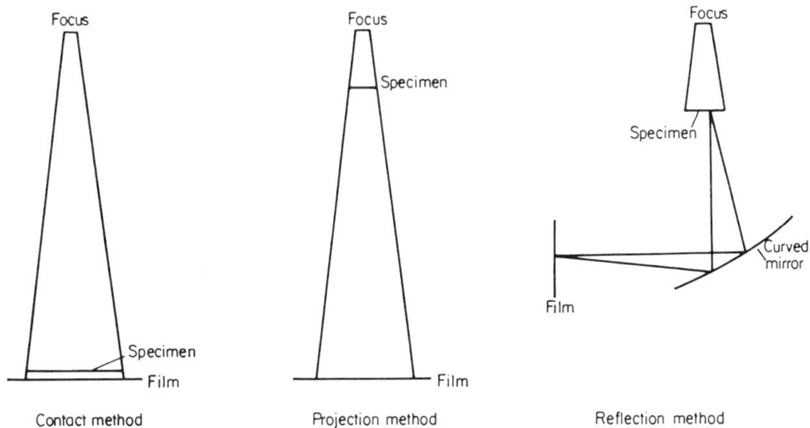

Figure 13.1. Microradiographic methods.

(b) *The point projection method.*   The point projection method makes use of electron focusing methods to obtain an extremely fine focal spot of the order of 1 micron diameter ($= 1/1000$th millimeter) or less. The enlargement is obtained by placing the specimen close to the focal spot and by projecting the details of the specimen on a 'medium grained' photographic emulsion at such a distance from the source that a high magnification, i.e. a high ratio of focus-film to focus-object distance is obtained.

(c) *The reflection method.*   In the reflection method X-rays, having penetrated the specimen, are totally reflected by curved polished mirrors and brought to focus on a photographic emulsion of high resolving power. Although a considerable amount of research has been carried out in developing this attractive method, it has not reached the stage at which it could be used for routine examination.

With all the methods mentioned a resolution of 0.1–1 micron has been attained, which is very close to and slightly better than that of the best optical microscope. Microradiography is mainly applied in the examination of biological and metallurgical specimens and also in some industrial fields. Some of these applications are briefly discussed on pages 589–596. We shall now deal with the physical, technical, and photographic principles of contact and projection microradiography. The reflection method will only be briefly mentioned, in order to discuss its fundamental features. With respect to the constructional details of the X-ray apparatus used in microradiography the reader is referred to the relevant literature and in particular to V. E. Cosslett, A. Engström and H. H. Pattee (1957), V. E. Cosslett and W. C. Nixon (1960), A. Engström, V. E. Cosslett, and H. H. Pattee (1960). Bibliography on X-ray microscopy: R. V. Ely, Micro-X-radiography and analysis 1913–1963.

13.2   CONTACT-MICRORADIOGRAPHY

13.2a   The Geometry of Projection

As already noted in the introduction, contact-microradiography is fundamentally a conventional radiographic method with the exception that the image is obtained on a very fine grained emulsion which is subsequently enlarged. The geometry of the method, in which the object is placed in direct contact with the photographic material, is discussed in connection with Figure 7.3 on page 262 from which it was concluded that the penumbra $p = bF/a$, where a is the focus to object, b the object to film distance and F is the diameter of the focal spot.

In Figure 13.2 we assume that a very thin specimen of thickness t_s is placed in contact with a photographic emulsion of thickness t_p. We assume further that our object to film or plate distance reaches from the top of the specimen to the back of the emulsion, hence

$$p = \frac{F(t_s + t_p)}{a} \tag{13.1}$$

If the focal spot diameter is 2 mm, the specimen thickness (t_s) is 50 microns, the emulsion thickness (t_p) 10 microns and the focus to object distance is 400 mm, then the geometrical unsharpness, i.e. the width of the penumbra is

$$p = 2 \times 0.06/400 = 0.0003 \text{ mm} = 0.3 \text{ microns}$$

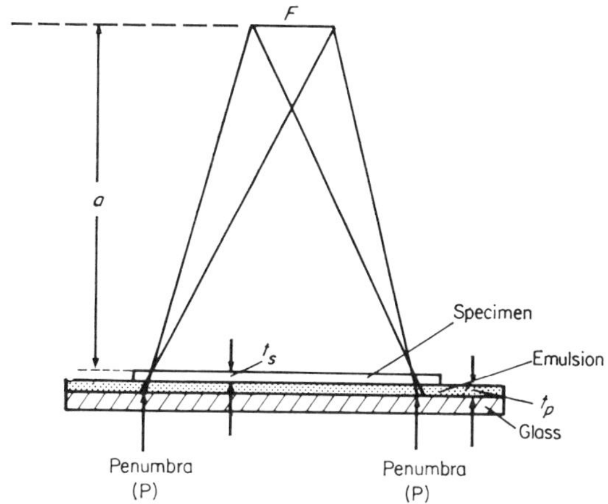

Figure 13.2. Geometry of contact-microradiography.

In Figure 13.3 we see the geometrical unsharpness in microns as a function of the specimen thickness in millimeters for focal spots of 0.1, 0.2, 0.5, 1, and 2 mm diameter.

This graph is drawn for a constant focus to object distance, i.e. $a = 200$ mm (Figure 13.2), assuming that the thickness of the emulsion is 10 microns. The penumbra values refer to the back of the emulsion, as has been mentioned before. The reason that the curves bend from about 0.1 mm specimen thickness is due to the fact that the emulsion thickness of 10 microns becomes more and more appreciable relative to the specimen thickness. The graph shows that a geometrical unsharpness of less than 1 micron can easily be achieved even with a fairly large focal spot when using specimen thicknesses of below 0.1 mm, i.e. 100 microns. In many metallurgical applications specimens are of this order of thickness or lower.

The effect of a focal spot smaller than that available in the X-ray tube can be achieved indirectly by placing a deep lead pinhole of appropriate diameter near the focal spot below the X-ray tube. This reduces the larger focal spot to an optically effective focal spot of reduced size.

The image thus obtained by the method described should be recorded on a photographic material of very high resolving power and furthermore, this image should be enlarged by an optical system of very high resolving power. Considering both these requirements some of the high geometrical resolution obtained may be sacrificed unless the greatest care is taken from the point of view of the properties of the

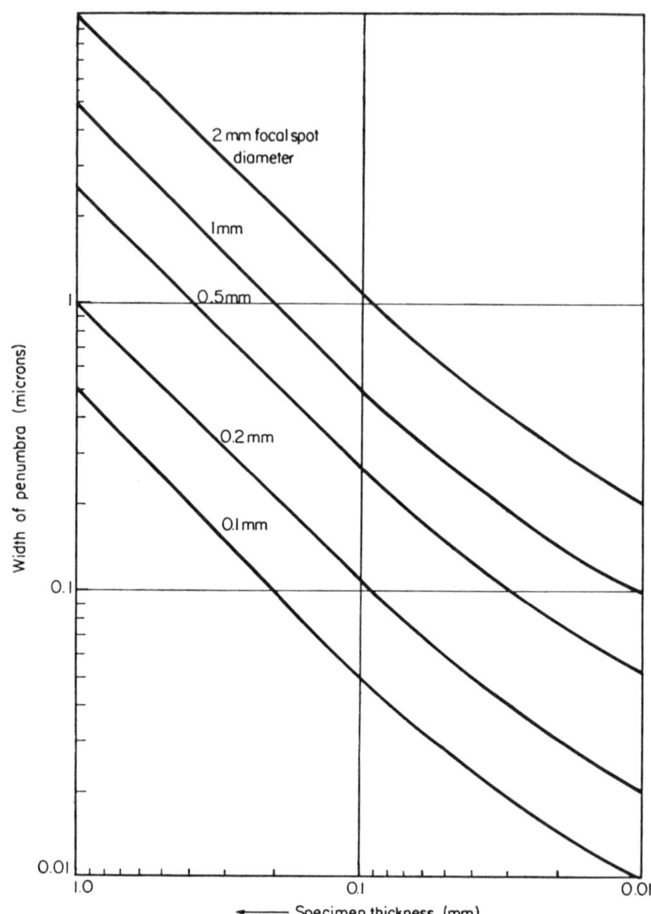

Figure 13.3. Geometrical unsharpness (microns) as a function of specimen thickness (millimeters) for various focal spot diameters. (Constant focus to object distance is 20 cm and constant emulsion thickness is 10 microns.)

emulsion and the enlarging system. There are some other factors which may influence the final resolution attainable, such as the range of electrons, scatter, and Fresnel diffraction effects, which will be discussed in a later section.

13.2b   Emulsions of High Resolving Power

Emulsions

The resolution attainable in photographic emulsions depends chiefly on the average grain size and distribution of silver halide.

Furthermore the thickness of the emulsion coating can have a considerable effect on resolution as well, as was shown by W. Ehrenberg and M. White (1957, 1959) (see Figure 13.8). Certainly at high magnifications the emulsion coating thickness should not be much greater than the depth of focus of the high aperture microscope objective to be used for the magnification of the microradiographic image. In order to achieve this, a dry emulsion layer of 1–2 micron thickness may be used and the loss in speed due to the reduced thickness may partly be compensated by using an emulsion of a high silver to gelatin ratio, i.e. a concentrated emulsion. It is to be expected that high concentration of silver halide should contribute to resolution in microradiography, as the grains are more closely packed in an emulsion of high, compared with one of lower concentration. In practice its usefulness in enhancing speed has been confirmed. (B. Combeé and A. Recourt 1957/1958.) The properties mentioned above together with suitable inherent emulsion contrast, low fog value, freedom from pinholes and abrasions, and high uniformity of coating are other important characteristics.

There are several emulsions available from photographic manufacturers which are of high resolving power, such as for instance Agfa-Gevaert Lippman, Ilford High Resolution, Kodak Ltd. Maximum Resolution, and Eastman Kodak High Resolution and others. Some of these have a resolving power of 1000 lines per mm to X-rays and even better with grain sizes of the order of 0.05 microns. Their X-ray speed is however of the order of 1000–10,000 times lower than that of a fast no-screen X-ray film. Other emulsions of intermediate resolving power and higher speed are recommended for use, if extreme resolution is not required, as for instance with thicker specimens. The relation between sensitivity of emulsions and grain size is in general reciprocal and it is therefore unnecessary to sacrifice speed by using an emulsion of extreme resolving power, if only that of a few microns is required. (A. Engström and B. Lindström, 1951.)

Processing, and particularly drying of these emulsions, has to be carried out with special care in order to avoid any impurities or dust particles sticking to the surface of the coating which, when enlarged, can be most disturbing. Hence removing of any particles from the surface of the wet emulsion by gentle wiping and drying in a dust free atmosphere is important. It is advisable to use developers and processing conditions as recommended by the manufacturers for the particular emulsion in use. Most of the photographic material for high resolving power is supplied on glass support instead of on film base. This has the advantage of rigidity and of eliminating any warping of the emulsion during exposure. Information on the behavior of various types of commercially available emulsions to ultra soft X-rays is given by B. L. Henke (1963).

In recent years attempts have been made to image the emulsion grains of a contact microradiograph by means of an electron microscope in order to overcome the limitations in resolution of the optical microscope. (A. Recourt, 1957; P. Granboulan, 1963.) This may be done by using extremely thin emulsions having grain sizes, as used for instance in electron microscoscope autoradiography (see p. 540) without support and by placing the stripped or self coated emulsion, containing the microradiographic image on an electromicroscope grid, or by using systems related to those discussed on pages 537 and 543.

Attempts have also been made to find X-ray sensitive substances other than silver bromide or silver-chloride, which can be used as grainless recording media for X-rays, when used in the electron microscope. (H. H. Pattee and R. Warnes, 1957.) Certain plastics which are degraded by irradiation with X-rays, such as polyethylene glycol methacrylate, were found suitable. In order to be used for magnification in an electron microscope these plastics can be treated by solvents or selective stains, for instance, which cause local changes of mass in the layer in proportion to the intensities of X-rays absorbed. Some success with such material has been achieved and this led to a resolution of better than 1000 Å ($= 0.1$ microns), but with the disadvantage of very long exposure times. (S. K. Asunmaa, 1960, 1963.)

Development phenomena (*Adjacency effects*)

Several phenomena influencing the accuracy of image reproduction are based on the interaction of the developer between adjacent photographic images. The results of these effects, which are generally described under the title of 'adjacency effects', may, in some cases, seriously impair the microradiographic image interpretation.

Let us consider a very sharp boundary between two adjacent images of widely different density. During development, the developer has little action in the light area and thus remains relatively fresh locally. The developer will, however, act strongly in the dark area where the solution becomes almost exhausted locally. The interaction between the developer solution and the two areas leads to a diffusion of the fresh developer into the dark area with the result that the image near the boundary will be of increased density. On the other hand, development will be retarded in the region of low density due to the diffusion of reaction products, chiefly bromide. The darker line on the high density side of the boundary is usually called a border effect, whereas the faint line within the low density is referred to as a fringe effect. The lines have also been called 'Mackie Lines'.

Another effect which is basically due to the same phenomenon as that described above occurs if two small images of different size have been produced by the same exposure. After processing these may show a higher density in the smaller area than in the larger area. The smaller the image size the less oxidation products are formed and so the bromide action is low, leading to a higher density than in the larger areas. The effect increases with the thickness of the emulsion layer. This phenomenon, which is called the "Eberhard effect", may cause errors in the quantitative evaluation of densities in small areas.

13.2c   Optical Enlargement

If a microradiograph taken on a high resolution emulsion is viewed on an optical microscope, or has to be enlarged photographically, the final resolution attainable is limited by the inherent optical properties of the microscope or other enlarging device.

The limiting resolution of the optical microscope is given by $0.5\lambda/NA$, according to Abbe, where λ is the wavelength of light and NA is the numerical aperture. If we assume for instance a numerical aperture $NA = 1.4$ and a wavelength $\lambda = 5000$ Å, the limit of the optical resolution is 0.18 microns. In order to achieve such an exceptionally high resolution, an excellent optical microscope is required together with the experience of utilizing it at its optimum performance. Hence the conditions of optical magnification are not only very stringent, but they also set certain limitations on the final resolution attainable in contact-microradiography. It is because of these limitations that it has been suggested to make use of an electron-microscope for the enlargement of the microradiograph, as was pointed out in the previous section.

13.2d   Contrast Conditions

The conditions for achieving high contrast in microradiography are essentially based on the same general principles as in radiography which have been discussed in Chapter 7 page 269. There are however several additional features, such as the position of absorption edges, the excitation of characteristic radiation etc., which may be utilized to advantage.

We consider here a thin object of a material having a linear absorption coefficient μ_2 and a thickness t_2 embedded in a matrix, which has a linear absorption coefficient μ_1 and a thickness t_1, as illustrated in Figure 13.4.

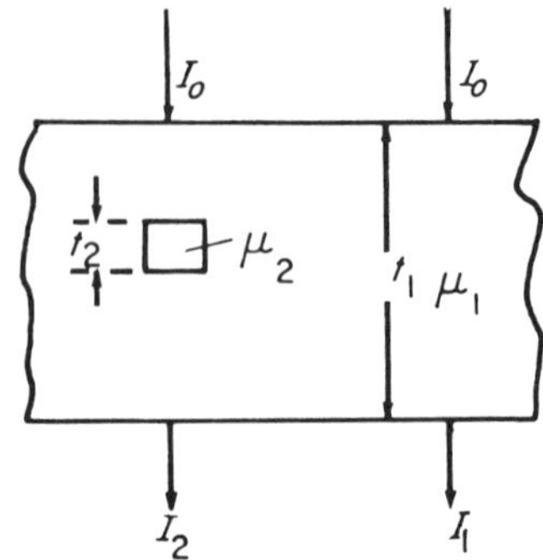

Figure 13.4. A substance of thickness t_2 embedded in a matrix of thickness t_1 of a material different from that of the substance. μ_2 and μ_1 are the respective absorptions coefficients.

The respective transmitted radiations I_1 and I_2 are then

$$I_1 = I_0 \exp\left(-\mu_1 t_1\right) \tag{13.2}$$

$$I_2 = I_0 \exp\left[-\mu_1(t_1 - t_2) - \mu_2 t_2\right] \tag{13.3}$$

and

$$I_2 - I_1 = \Delta I = I_0 \exp\left(-\mu_1 t_1\right) \times \left[\exp\left(\mu_1 - \mu_2\right)t_2 - 1\right] \tag{13.4}$$

By approximation we find that

$$\Delta I = I_0 \times \left(\mu_1 - \mu_2\right) \times t_2 \exp\left(-\mu_1 t_1\right) \tag{13.5}$$

It has been shown in equation (10.5) that a small difference in optical density, i.e. contrast in the film can be expressed by

$$\Delta D = 0.43 G_D \frac{\Delta I}{I_1} \tag{13.6}$$

where G_D is the gradient of the photographic emulsion at a density D and further

$$\frac{\Delta I}{I_1} = \left(\mu_1 - \mu_2\right)t_2 \tag{13.7}$$

By substituting $\Delta I / I_1$ in (13.6) we obtain

$$\Delta D = 0.43 G_D(\mu_1 - \mu_2)t_2 \tag{13.8}$$

Hence ΔD is proportional to the gradient of the photographic emulsion at a given density and to the difference between the linear absorption coefficients and the thickness of the substance embedded in the matrix. It is independent of the thickness of the matrix because of the fact that both the transmitted intensities I_1 and I_2 (see Figure 13.4) are equally affected by an increase or decrease in the thickness of the matrix.

For a given thickness of the embedded substance and film gradient we have to find a maximum value for $(\mu_1 - \mu_2)$ in order to obtain as high a contrast as possible. This in turn depends on the atomic number of the respective elements to be distinguished radiographically, on the X-ray wavelength or tube voltage to be used and on the contribution of secondary radiation. The use of absorption edges to enhance contrast, as mentioned in the beginning of this section, will be discussed under the section on 'Choice of radiation' page 576.

13.2e Further Factors Influencing Contrast and Resolution

We have treated the conditions of contrast so far by neglecting any influence from secondary and scattered radiation. Scattering in general is low within the range of wavelengths applied in micro-radiography, particularly if very thin specimens are used. The scattering coefficient becomes, however, of greater importance with higher kilovoltages, with elements of low atomic number, and with greater thicknesses of specimen. R. S. Sharpe (1961) drew attention to the serious drawback of secondary radiation in the microradiography of metals. The effect of the emission of characteristic radiation can spoil the contrast otherwise attainable in absorption radiography, if the incident radiation lies on the shorter wavelength side of the absorption edge of the matrix element (see also p. 578). Sharpe expressed the influence of secondary radiation by referring to the build-up factor $B = (I_p + I_s)/I_p$, (see page 411), where I_p is the intensity of the primary radiation and I_s that of the scattered and secondary radiation. He was using an iron specimen of 40 microns thickness (a) with a cobalt target and (b) with a copper target both excited at kilovoltages from 17–48 kV_p and he found values of the build-up factor B in case (a) of 3 and in case (b) of 10^4. This result shows that the image contrast of metallurgical specimens may almost disappear when using a radiation on the high energy side of the K absorption edge of the matrix element.

Another effect which may seriously reduce resolution and also contrast in the image, is the spread of photoelectrons released in the specimen and in the emulsion, due to the impact of X-rays. If an exposure is taken at 6–7 keV, the range of photoelectrons in an average photographic emulsion is of the order of one micron and this is appreciably greater than the average distance between grains in a fine grained emulsion. Using about 2 keV, the range of electrons is only about 0.1 microns which is just of the order of the grain separation in an emulsion of very high resolving power. Hence the range of photoelectrons may set a limit to the resolving power under certain circumstances.

A further effect which may influence resolution is that of Fresnel diffraction fringes caused by X-rays. This influence is generally negligible in contact microradiography unless a fairly thick specimen is used together with X-rays of long wavelength. The two features, thick specimen and very long wavelength, are however contradictory in practice, as generally shorter wavelengths are required the thicker the specimen. The subject of Fresnel diffraction effects is also discussed in the section on point projection microradiography, where its influence is more severe, but an example of the order of the width of the first fringe will be given here, using two different geometrical conditions.

If a specimen of 500 microns thickness is used at a wavelength of 5 Å, the width of the first fringe will be about 0.5 microns, whereas it is only about 0.16 microns if the specimen thickness is reduced to 50 microns, using the same wavelength as above. It is thus seen that under the practical conditions prevailing in the contact method, Fresnel diffraction effects are usually negligible with regard to image resolution (see also p. 584).

13.2f Choice of Radiation

The radiation to be used for a particular specimen should be selected so that high contrast is obtained between the substances which are to be detected and those of the matrix in which they are embedded. In order to achieve this, one of the first considerations is the choice of X-rays of a wavelength capable of penetrating the specimen. The second consideration is to find a radiation which tends to give a maximum value for the difference of absorption coefficients of the elements concerned. In many applications the use of the continuous X-ray spectrum is quite adequate. An improved discrimination between two elements is often possible by using monochromatic radiation and this is particularly so in quantitative evaluation.

In general it can be said that in the range of biological studies of thin specimens up to about 50 microns, a tube voltage from about $1\text{--}1.5\,\mathrm{kV_p}$ is required and the exposure should be carried out in a vacuum in order to eliminate the absorption of the softer components of the radiation, as can be seen from Figure 13.5 which shows the X-ray transmission through 5 cm of air as a function of wavelength. From this

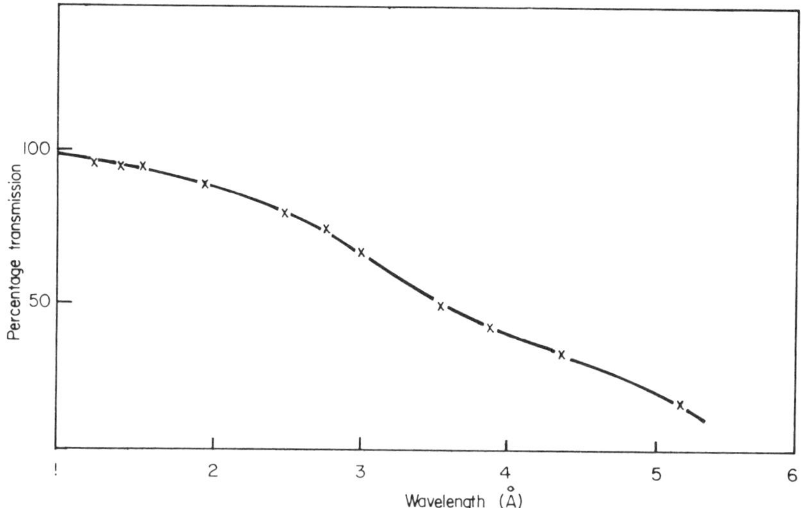

Figure 13.5. Percentage transmission of X-rays through 5 cm thickness of air as
a function of wavelength. (Taken from table in X-ray Microscopy, by V. E.
Cosslett and W. C. Nixon, Cambridge University Press, Cambridge, 1960.)

figure we conclude that at a wavelength of 5 Å (equivalent to 2.5 keV)
only about 20 per cent of the beam flux will be transmitted. At 8 Å
the transmission is only of the order of 0.1 per cent at 5 cm distance
from the source.

For the microradiography of bone sections and calcified tissues the
tube voltage may be raised to about 4–8 kV_p.

Metallurgical work is generally covered by tube voltages within
the range from about 10–40 kV_p and the position of the absorption
edges (see p. 21) of the elements under examination can play a
decisive role in the rendering of contrast. It is often possible to
utilize the increase of the absorption coefficient near the absorption
edges of two elements in order to obtain a differentiation between
these elements. One may even distinguish elements of adjacent atomic
numbers, as is shown in Figure 13.6 which illustrates the linear
absorption coefficients for Cu and Ni as a function of the wavelength.
The absorption edges of Cu and Ni in Figure 13.6 lie at wavelengths
of 1.380 Å and 1.487 Å respectively and it is seen that the characteristic K_{α_1} radiation of Zn which is 1.435 Å lies just between the
two absorption edges. Hence for instance by using K_{α_1} radiation of Zn
the relatively large difference between μ_{Ni} at the upper part of the Ni-
absorption edge and μ_{Cu} at the lower part of the Cu-edge (i.e.

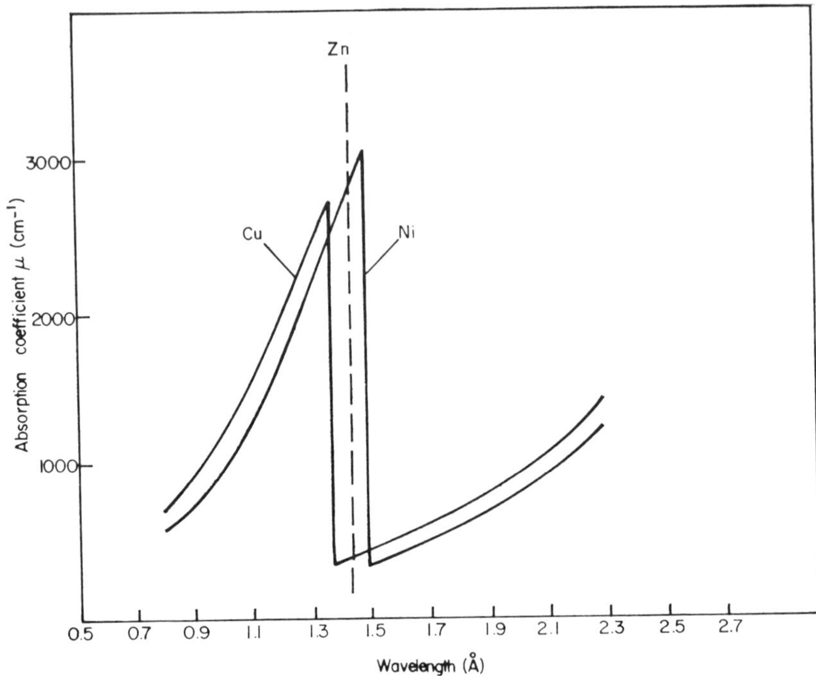

Figure 13.6. Physical basis for the microradiographic discrimination between elements of consecutive atomic numbers, such as $Cu(Z = 29)$ and $Ni(Z = 28)$ using ZnK_1 radiation. The graph shows the absorption coefficients $(\mu(cm^{-1}))$ as a function of wavelength in Å.

$\mu_{Ni} - \mu_{Cu} = 2700 - 400 = 2300$) permits high contrast to be achieved between these elements of neighboring atomic number. Similarly, differentiation can be obtained between other elements using various K_{α_1} radiations as shown in Table 13.1 on page 579. The respective K-absorption edges are also given in this table.

It should also be noted that the difference between the absorption coefficients at both sides of the K-absorption edges increases with decreasing atomic number; in other words the jump between top and bottom of the edges increases with lower atomic numbers, which is a useful feature for differentiation between elements of low atomic numbers. (R. Glocker and W. Frohnmayer (1925).)

Different characteristic radiations of sufficient intensity may be obtained by using X-ray units with demountable tube targets of the respective elements. The characteristic radiations are however superimposed on the continuous or white X-ray spectra. The intensity

TABLE 13.1. Wavelengths to be Used for Discriminating Elements of Close Atomic Numbers

Elements	Atomic Number Z	K Absorption edges Å	$K^*_{\alpha_1}$ radiation of	Wavelength (Å)
Ni	28	1.487	Cu	1.541
Co	27	1.607		
Co	27	1.607	Ni	1.658
Fe	26	1.743		
Fe	26	1.743	Co	1.789
Mn	25	1.895		
Mn	25	1.895	Fe	1.936
Cr	24	2.070		
Cr	24	2.070	Mn	2.102
V	23	2.268		

* The K_{α_2} emission lines are very close to the wavelengths of those of K_{α_1}, but have only half the intensity of those of K_{α_1}.

I of the characteristic radiation increases with the voltage across the X-ray tube in a first approximation according to

$$I \propto (V - V_0)^n$$

where V is the tube voltage, V_0 is the exciting voltage of the characteristic radiation and n is between 1–2. More refined formulae are given by V. E. Cosslett and W. C. Nixon (1960). The formulae show that the intensity of the characteristic radiation increases with tube voltage. On the other hand, care has to be taken not to exceed certain limits of tube voltage, as otherwise the radiographic contrast may be reduced by the high penetration of radiation from the continuous spectrum. Using a crystal monochromator, preferably with a curved crystal, as often applied in X-ray diffraction work, monochromatic radiation can be produced without or at least with very little background of the continuous spectrum.

13.2g Practical Considerations

As a general rule it can be said that the thinner the specimen, the better are the prospects for good resolution, and thinner specimens require softer X-rays to obtain sufficient contrast. Certain applications of contact microradiography can be made with X-ray tubes having focal spot diameters preferably not larger than 1–2 mm using X-rays generated at kilovoltages of the order of 10–40 kV_p. Medical X-ray units are not always suitable, as they usually do not permit the generation of X-rays soft enough for this particular application. In general, X-ray diffraction apparatus are preferable for the purpose, (a) because they permit the production of soft radiation and (b) because the X-ray tubes for X-ray crystallography have, in general, small focal spots. Furthermore a change of tube target is possible in certain X-ray diffraction units which are continuously evacuated. This has a particular advantage, as mentioned before, in metallurgical work, for the distinction of certain elements.

We have seen in Figure 13.3 that a geometrical resolution of 1 micron can theoretically be obtained with a focal spot of 1 mm diameter for a specimen thickness of about 0.2 mm and a focus-to-image plane distance of 20 cm. Metallurgical specimens of less than 200 microns thickness and biological specimens of similar thickness can thus be examined. Several specialized X-ray units for contact microradiography are now commercially available. These are provided with X-ray tubes having small focal spots, thin beryllium windows, adjustments for very low voltages (down to 1 kV_p) and facilities for exposing in a vacuum.

If such a specialized microradiographic unit is not available and if radiations of wavelengths shorter than about 2 Å are used no evacuation is necessary and good results can be obtained by using the foreshortened, i.e. square focal spot of a sealed X-ray tube in combination with a light tight box, in which the specimen is placed in closest contact with the photographic emulsion. The box should be provided with a low X-ray absorbing window for instance of beryllium or very thin aluminum. An infrared filter of sufficient opacity has been found useful, if other window materials are not available. The metal filters allow exposures to be carried out in daylight, assuming that the box is light tight. When using an infrared filter a dimly lit room is preferred. The use of a black paper window should be avoided, as this might superimpose its own structural details on the required microradiograph, unless it is placed at a sufficient distance from the photographic emulsion. A typical set up for contact microradiography is illustrated in Figure 13.7.

Specimen Preparation and Mounting

Only a few hints will be given on the preparation and mounting of specimens, as this depends so much on the type of specimen used.

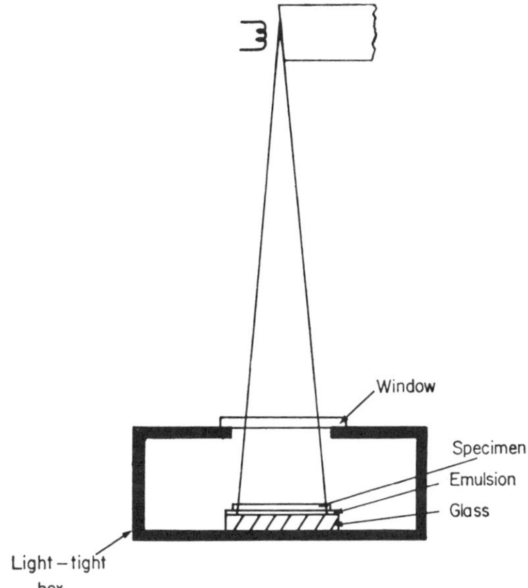

Figure 13.7. Typical arrangement for contact microradiography.

With biological specimens the freeze-drying method is certainly one of the best methods, firstly because water has to be removed from the specimen in order to improve contrast and secondly, little distortion takes place using this method. Larger specimens such as organs and bones have to be sectioned by means of a microtome, after the sample has been embedded for example in plastic or in paraffin wax, which has to be removed for exposure. Metallurgical samples have to be sectioned, and in most cases ground down, if not thin enough.

In mounting the specimen it is most important to ensure intimate contact between sample and photographic emulsion. If it is likely that the specimen may interact chemically with the emulsion or it may stick to the emulsion layer, so that removal from the layer becomes difficult, an interlayer of an extremely thin plastic material of 0.5–2 microns is advisable (see p. 548). The greater the thickness of the separating interlayer the more the geometrical resolution will be impaired. The tolerable thickness depends entirely on the resolution required and the permissible thickness of the interlayer can be assessed from geometrical considerations, based on equation (13.1). Photographic plates are not usually provided with a superlayer, as films are, and the plate may have to be coated with a layer of gelatin in order to reduce the chance of chemical interference from the specimen and abrasion of the emulsion.

Check on Resolution

Before a final exposure is taken, it is advisable to check the resolution attainable under given conditions and this can be done for example by making a microradiographic image of an electron microscope grid or one of similar design. Grids are available with bars as small as 2 microns in width. Minute diatoms, i.e. planctonic unicellular algae with silicified skeletons, (see Figure 13.8) can also be used as a test object for resolution as they have an extremely

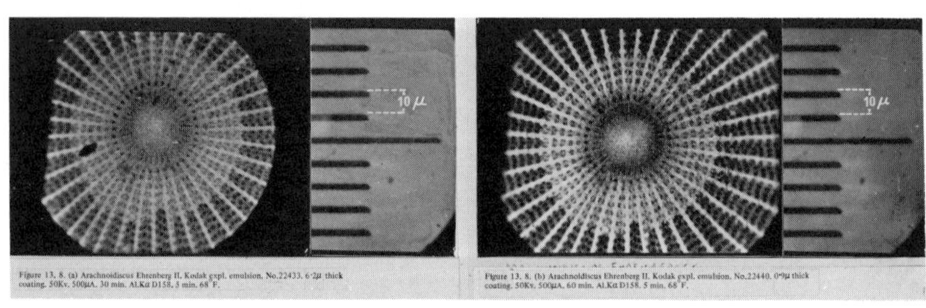

Figure 13.8. Microradiographs of Diatom Arachnoidiscos Ehrenbergii, showing the effect of emulsion thickness on the resolution in microradiography. (a) 6 micron thick emulsion (×800), (b) 1 micron thick emulsion. *By courtesy of* Prof. W. Ehrenberg and M. White and *Brit. J. Radiol* 32.380.558.1959.

fine structure. (W. Ehrenberg and M. White, 1959.) The structural dimensions of grids and diatoms should be assessed with an optical microscope before using them as test objects.

The main features influencing resolution in contact-microradiography are summarized in the diagram on page 583.

13.3　POINT-PROJECTION MICRORADIOGRAPHY

In this method an image of the object close to a point source of X-rays is projected on to a distant plane. The magnification is given by the ratio of the source-image to source-object distance. The focal spot

Chief Factors Influencing Resolution in Contact Microradiography

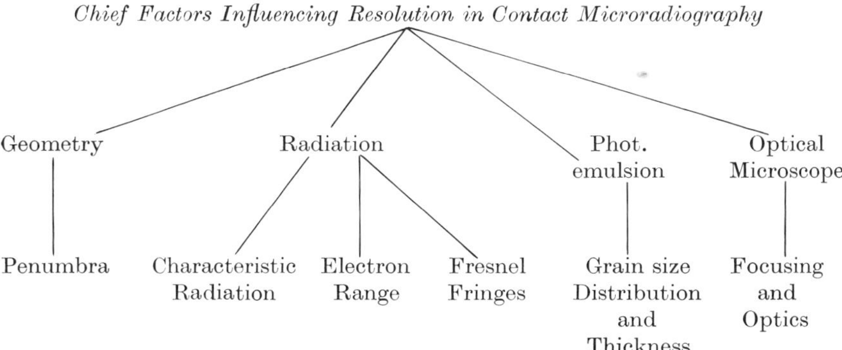

of the source must be very small indeed in order to achieve high resolution at great magnification. X-ray sources of 1 and even 0.1 micron focal spot diameter have been produced by means of electron-focusing methods by V. E. Cosslett and W. C. Nixon (1951, 1952, 1960) and X-ray sources using focal spots of 1 micron and less in diameter are commercially available.

Geometry of Projection

The geometry involved in projection microradiography thus differs from that of contact microradiography by the fact that the object is placed very near to the source and the photographic material is distant from the specimen. In Figure 13.9 these conditions are shown. F is the diameter of the focal spot, x is the width of the object, p is the penumbra, and u is the umbra. We can see from this figure that a magnified image of the object x is obtained, the magnification being given by $\dfrac{a}{a-b}$. We can also see that the width of the penumbra is

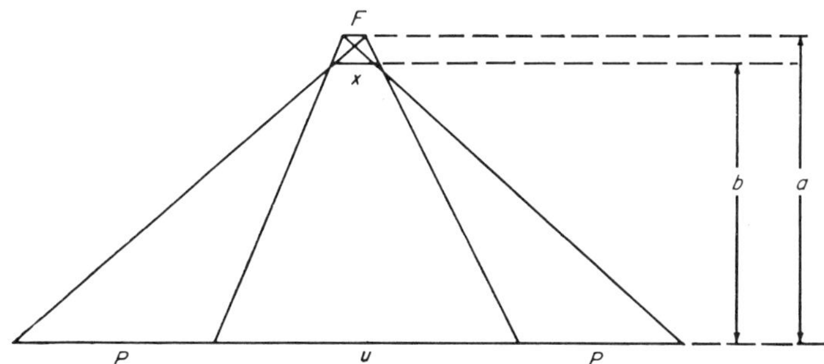

Figure 13.9. Geometry of point projection microradiography.

generally not negligible and affects the contrast of the image and the resolution attainable.

If x is smaller than $\dfrac{Fb}{a}$, the umbra vanishes (see Figure 13.9) and, as in the projection method, the object-image distance b is very nearly equal to the focus-image distance a, no umbra is formed when the object is smaller than the focal spot. However objects remain visible even if the image is formed by penumbra only. In particular it can be shown (V. E. Cosslett and W. C. Nixon, 1960) that two adjacent objects which do not form umbras can still be resolved if the separation of the centers of the objects equals the width of the focal spot. This is an analogous consideration to that given on page 518 for the case of resolution attainable in autoradiography.

Fresnel Diffraction

The high resolution attainable in point-projection microradiography is limited in certain circumstances by Fresnel diffraction fringes, as has been mentioned before. Although this effect plays a part in contact microradiography as well, its influence is more critical in the projection method.

Adjacent points in the image plane can be resolved only if their separation is greater than the width of a diffraction fringe. Interference from such diffraction fringes may occur when very high resolution is required and very soft X-rays are used, as for instance in biological work. In order to overcome this difficulty very short focus to object distances are required which may be difficult to apply in extreme cases as will be seen from the table below. This table gives data on the maximum focus-object distances that are permitted in order to avoid interference from Fresnel diffraction fringes for various X-ray wavelengths and for various resolution values in microns.

TABLE 13.2. Maximum Focus-to-Object Distances at which No Interference from Fresnel Diffraction Occurs (*After* V. E. Cosslett and W. C. Nixon (1960))

Wavelength λÅ	Resolution (microns)	Maximum Focus-to-object distance
1	1	$\leqslant 1$ cm
1	0.1	$\leqslant 100\ \mu$
10	0.1	$\leqslant 10\ \mu$

From this table we see (first row) that for a resolution of 1 micron, using a wavelength of 1 Å, the focus to object distance at which no interference from diffraction effects occurs is equal or smaller than 1 cm. This is a sufficiently large distance for most practical purposes. In the second row we see that for a resolution of 0.1 micron and $\lambda = 1$ Å the focus-object distance should be less than 100 microns in order to avoid the influence from fringe effects. Using the conditions of the last row, we find that a resolution of 0.1 microns with a wavelength of 10 Å is only possible if the focus-object distance (or the specimen distance) is less than 10 microns. This means a more serious limitation to projection microradiography. Such long wavelengths of 10 Å (equivalent to 1.24 keV) and even softer radiations are required in biological work in order to obtain sufficient contrast of tissues.

In conclusion it may be said that Fresnel diffraction interferes with projection microradiography only if extremely high resolution is required and very soft radiation has to be used.

Photographic Requirements

Considering that an enlarged image is projected on the photographic material and that usually no further magnification of this enlarged image is required, no very stringent demands have to be made on the grain size properties of the photographic material. Photographic emulsions of medium grain size and moderate speed are usually quite suitable for the purpose, as long as the minimum separation of two adjacent points in the image exceeds the average grain size of the developed emulsion. In the projection method an intermediate magnification is sometimes obtained by using a film that permits an optical enlargement to be made by a factor of 5–10 or more. In this case a film of higher resolving power than that used for conventional projection microradiography is required.

13.4 COMPARISON BETWEEN THE CONTACT AND PROJECTION METHOD

With both these methods the best resolution of the optical and ultraviolet microscope of the order of 0.2–1 microns, can be obtained. The chief advantage of contact-microradiography is that it can be carried out with almost any X-ray tube having a suitable window and a focal spot of the order of 0.3–1 mm diameter. *The optical microscope* to be used for observing or photographing the microradiographic image on a very fine grained photographic emulsion *limits the resolution* slightly if compared with the projection method. Alternatively, electron microscope methods can be applied for examining stripped ultrafine grained material, but this is a very delicate method. The contact method in general requires longer exposure times than its rival method chiefly because of the low sensitivity of the fine grained emulsion.

The projection method requires a special X-ray apparatus, but offers the advantage that the specimen is not in contact with the photographic material, so that the latter can easily be exchanged without disturbing the sample. Furthermore any chemical interference due to specimen-emulsion contact is avoided. Using the projection method the field of view is very small, whereas it is practically unlimited in the contact method.

The fact that emulsion and sample are separated also offers an advantage in micro-stereoradiographic techniques where the specimen has to be tilted between the two exposures.

13.5   REFLECTION X-RAY MICRORADIOGRAPHY

A further method of X-ray microradiography is based on total reflection of X-rays at glancing incidence from polished concave mirrors. Only a very brief outline will be given, as this method has not been used in routine applications.

Since the refractive index for X-rays is of the order of unity, focusing by refraction, although physically feasible, would practically be impossible because of the extraordinarily great focal lengths (many meters) involved. Total reflection of X-rays from polished surfaces or mirrors however, appears to be a more useful approach. Total reflection is based on the fact that the refractive index for X-rays is a little less than one, hence a critical angle exists for X-rays incident from a less dense to a dense medium. The critical angle is near 90° and total reflection occurs within glancing angles of about 10–30 minutes. Assuming a wavelength of 1 Å, the critical angle for aluminum is about 0.1° and that for gold about 0.5°. The intensity of the reflected beam is strongly wavelength dependent. A point image can be achieved by the use of a second mirror which is placed perpendicularly by to the first in order to correct for astigmatism (see Figure 13.10).

The construction of apparatus for reflection microradiography is extremely delicate. Nevertheless, some of the difficulties have been overcome and resolution better than 1 micron has been achieved in instruments developed purely for research purposes. (P. Kirkpatrick and A.

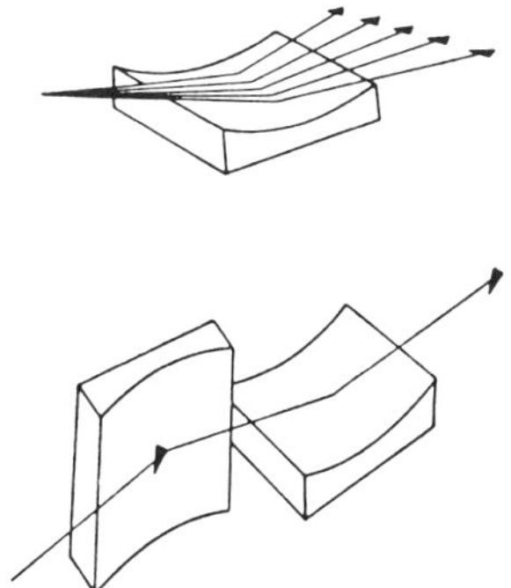

Figure 13.10. Reflection of X-rays at grazing incidence. Upper drawing: a single mirror leads to a line focus. Lower drawing: two mirrors in planes at right angle leads to a point focus. (*After* P. Kirkpatrick.)

V. Baez, 1948, A. V. Baez 1957; W. Ehrenberg, 1949; J. F. McGee, 1957, 1959; H. H. Pattee, 1957.)

13.6 ABSORPTION AND EMISSION MICRO-ANALYSIS

The contact and projection methods described in the previous sections permit a 'quantitative and qualitative analysis' of the specimens under examination. This determination may be based on absorption or on emission analysis. The latter is achieved either by electron or by X-ray excitation (Fluorescence analysis). It is intended in this section to point out some of the principles of the method.

13.6a Absorption Micro-analysis

This method has been developed particularly by A. Engström (1962) and his school for the investigation of biological materials. Using the contact method Engström achieved such a high sensitivity that quantities of an element as small as $10^{-12} - 10^{-14}$ grams could be detected in a volume of 1–10 cubic microns. The quantitative analysis of elements and compounds is facilitated by the rapid change of the mass-absorption coefficients with the atomic number of elements. The differentiation between substances can be further improved in some cases by utilizing the presence of absorption edges at particular wavelengths.

The physical concept of quantitative absorption-analysis is based on the arguments presented on page 574 and on the equations (13.2) and (13.8). In order to determine the mass thickness (i.e. density $\times$ thickness) of a given substance, one makes use of the simple equation

$$I = I_0 \exp - \left(\frac{\mu}{\rho}\right) \rho x \qquad (1.18)$$

where I_0 is the incident, I the transmitted intensity, μ/ρ the mass absorption coefficient, ρ the density, and x the thickness of the material. ρx is the mass thickness of the substance. For the evaluation of ρx we can determine I and I_0 from a separate experimental calibration of the photographic emulsion in use, by plotting X-ray intensity against optical density. Alternatively I can be found from a reference step wedge sample made of the same composition as the sample under examination. The sample and the reference step wedge are exposed side by side simultaneously and the mass thickness ρx of the substance can be assessed from a graph showing the relation between thickness of the stepwedge and optical density. Since both sample and wedge are exposed simultaneously any changes in the response of the photographic material are eliminated. I_0 is the same for both the sample stepwedge and the specimen and can be determined separately. This method was used by A. Engström mainly with a continuous X-ray spectrum for measuring the dry weight of biological material.

In a quantitative analysis of various elements embedded in a matrix two X-ray wavelengths at both sides of an absorption edge can be used. The absorption coefficients must be known and the intensities of the radiations incident and transmitted can be evaluated by the method indicated before.

Information on mass absorption coefficients of many elements of low atomic number within the soft region of the X-ray spectrum is now available* and this facilitates these investigations. If compounds instead of elements are to be examined the respective mass absorption coefficients can be found from equation (4.25) or from

$$\left(\frac{\mu}{\rho}\right)_{\text{total}} = \frac{k_1}{100}\frac{\mu_1}{\rho_1} + \frac{k_2}{100}\frac{\mu_2}{\rho_2} + \frac{k_3}{100}\frac{\mu_3}{\rho_3} + \ldots$$

where $k_1 k_2$ and k_3 are the weight percentages of the elements present in the molecular compounds or mixtures. The sensitivity of the method depends on the height of the absorption edge which, as mentioned before, increases sharply with decreasing atomic number. The height of the absorption edge for a given element increases in the order of K, L and M discontinuities. A. Engström (1956, 1962), R. Glocker and W. Frohnmayer (1925).

Reviews on the subject will be found in V. E. Cosslett and W. C. Nixon (1960), B. Lindström (1960, 1963), and V. E. Cosslett (1965).

* See Appendix p. 605.

13.6b Emission Micro-analysis (Scanning Methods)

In this method the electron spot (or probe) of about 1 micron in diameter or less excites X-rays on the surface of the specimen whose chemical composition is to be examined. The small volume which is irradiated by the electron beam emits an X-ray spectrum containing the characteristic radiations of the various elements present in the small irradiated volume of about 1 cubic micron. The spectral emission is detected by a vacuum spectrometer consisting basically of an analysing crystal and a proportional counter. The signal from the counter is amplified and led to a cathode ray display tube. By scanning the specimen, the brightness of the screen of the cathode ray tube changes with the variations of local intensity of emission of the specimen. Concentration microanalysis is then carried out by a comparison between the intensity of a given characteristic line with the intensity of the same characteristic line emitted by a reference standard sample of the same composition, using the same conditions of electron bombardment. This now very popular method has the advantage of providing an image of the specimen together with a quantitative analysis. (R. Castaing, 1951, 1955, 1960, 1963; P.Duncumb, 1962; H. H. Pattee, V. E. Cosslett, and A. Engström, 1963; V. E. Cosslett, 1965.) Bibliography on Electron-Probe Micro-analysis by K. F. J. Heinrich (1964).

13.7 APPLICATIONS OF MICRORADIOGRAPHY

Microradiography is chiefly applied in the fields of medicine, biology, metallurgy, mineralogy, and also in industry. A few of these applications will be reviewed in this section, so that the reader may appreciate the potentialities of the methods. The illustrations Figures 13.11–13.18 show examples of microradiographs in the fields of medicine, biology and metallurgy.

13.7a Medicine and Biology

Microradiographic methods have been successfully used by various authors in studies of peripheral blood circulation, for instance of the ears of rabbits. (R. L. de C. H. Saunders, J. Lawrence, and D. A. Maciver, 1957; R. L. de C. H. Saunders, 1961, 1962; R. L. de C. H. Saunders and V. R. Carvalho, 1962.) Contrast media were used in the blood vessels and the animals were anaesthetized for the exposure. (A. E. Barclay, 1951.) Intramuscular vascular patterns of humans revealed coarse and fine interconnected vascular networks which apparently have never been observed in such completeness. Kidney

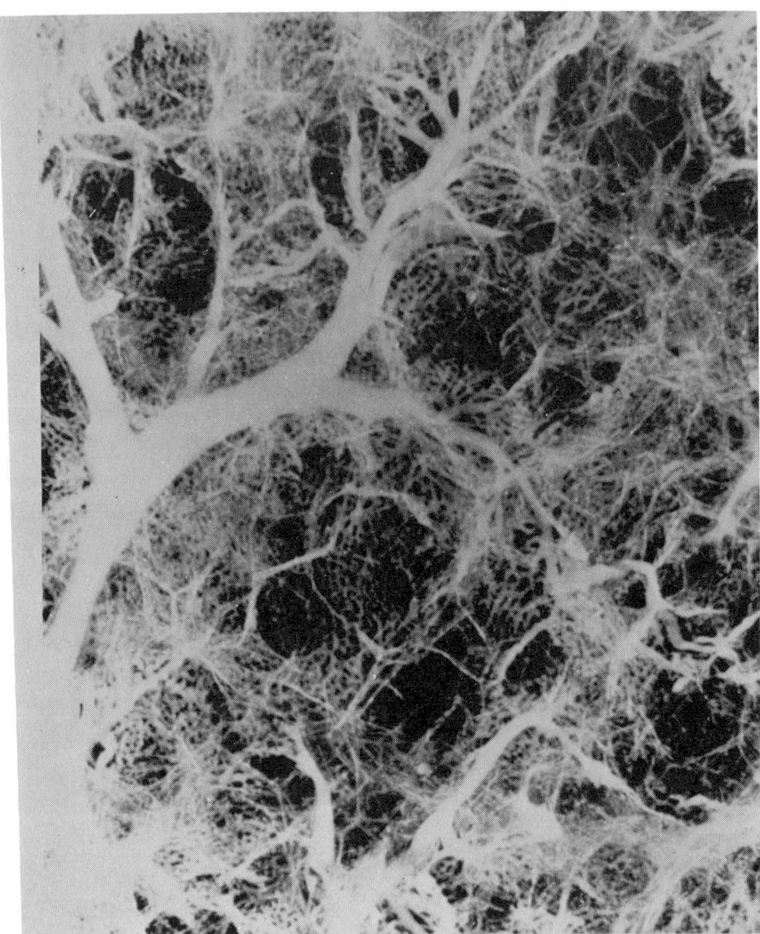

Figure 13.11. Microangiogram of the human lung showing the terminal branches of the pulmonary artery breaking up into the capillary network within the walls of the air sacs or alveoli. Projection microradiograph ×150. (*By courtesy of* Professor R. L. de C. H. Saunders, X-ray Optics and Microanalysis, Academic Press, New York, 1963.)

circulation has been studied in the rabbit by means of direct and indirect cineradiography at low magnification. (A. E. Barclay, 1951.) The alveoli and alveolar ducts in lung structures in 100–300 micron sections have been examined in uninjected inflated dry lungs. (C. P. Oderr, 1960.)

The distribution of hydroxyapatite in bone tissues has been

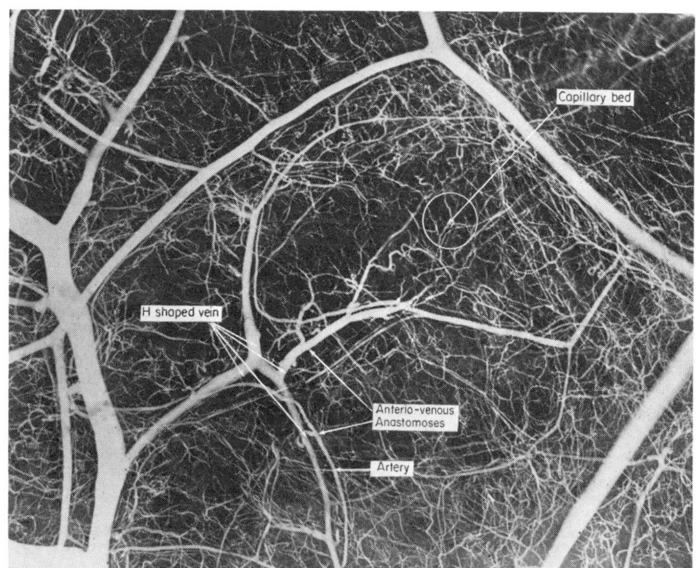

Figure 13.12. Microangiogram showing the microcirculatory pattern within the rabbit ear. Small arteries and veins, as well as the capillary bed are clearly visualized. Two arterio-venous anastomoses can be seen connecting two small arteries to a large H. shaped vein. Projection Microradiograph ×34. (*By courtesy of* Professor R. L. de C. H. Saunders. *J. Roy. Microscop. Soc.* **83,** 55 (1964).)

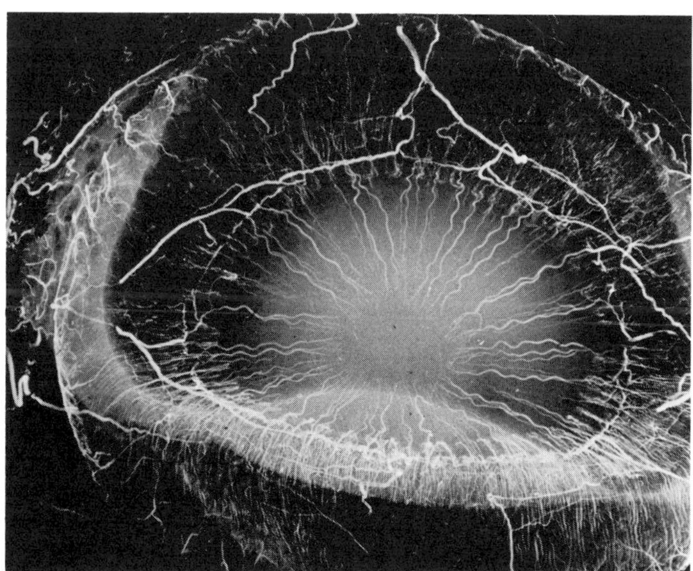

Figure 13.13. Contact microradiograph of a human eye taken after injection of the vessels with a radiopaque (25 per cent Micropaque) solution. The pupil of the eye can be identified and surrounding it the vessels of the iris, ciliary body, and choroid ×13. (*By courtesy of* Professor R. L. de C. H. Saunders, X-ray Microscopy and X-ray Microanalysis, Elsevier, Amsterdam, 1960.)

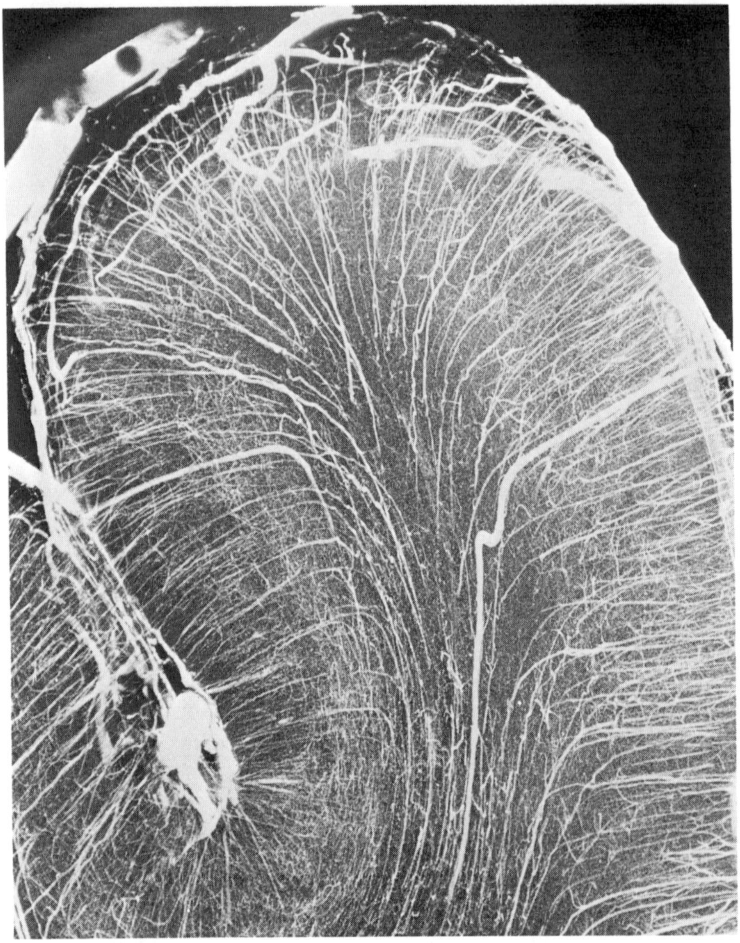

Figure 13.14. Cross section of a gyrus of the human brain, showing the short cortical arteries and capillary bed which supply the grey matter. Long transcerebral vessels can be seen passing through the grey matter into the subjacent white matter. Projection micrograph ×19. (*By courtesy of* Professor R. L. de C. H. Saunders. Anatomy Department, Dalhousie University, Halifax, Canada.)

investigated by means of microradiography, and also a quantitative study of bone renewal and other features of bone structure has been made on human beings of different ages (H. A. Sissons, J. Jowsey, and L. Stewart, 1960.) A great deal of work has been done on the microradiography of teeth sections with the aim of finding the distribution of mineral salts. (A. Engström and others, 1958.)

A considerable amount of study has been devoted to insects, for

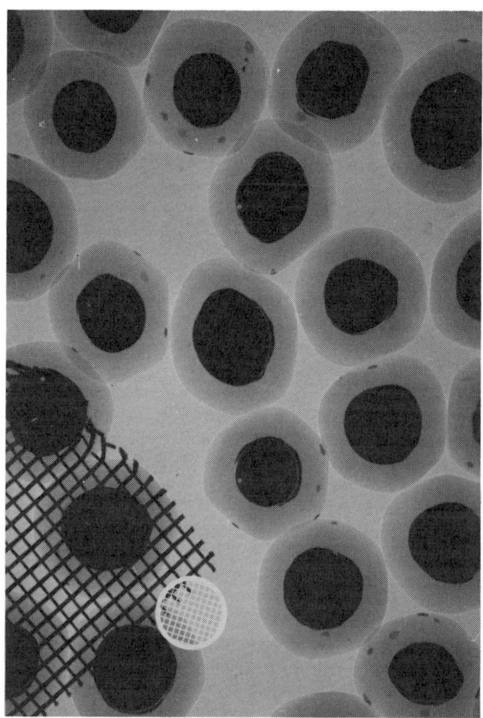

Figure 13.15. Microradiograph of graphite coated uranium carbide microspheres, showing uranium contamination spots on the surface of the spheres and also diffusion effects at the interface between the graphite and the uranium carbide. Twelve kV, using a copper transmission target. Dark mesh is 600 t per inch placed in the plane of the microspheres during the initial exposure and white mesh is 100 t per inch electron microscope grid placed on the negative when the secondary enlargement was made. (*By courtesy of* R. S. Sharpe, The Non-Destructive Testing Centre, A.E.R.E., Harwell, England.)

instance the study of the digestive channels after ingestion of food and contrast media; the tracheal system of living anaesthetized beetles has also been examined. In plant tissues crystalline structures often occur, e.g. rows of crystals of calcium oxalate in epidermal cells of the bulb of Allium and Hyacinthus. The distribution of these crystals and of others and their occurrence under various conditions has been studied by microradiography. (J. Salmon, 1957.)

13.7b Metallurgy

In metallurgy the applications may be classified according to R. S. Sharpe (1961) into

1. The determination of the spatial distribution of structural defects.

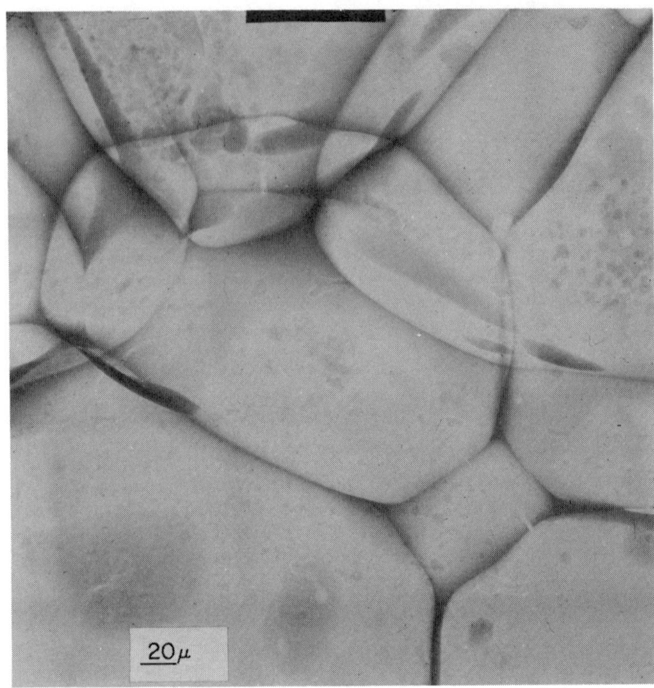

Figure 13.16. Microradiograph of aluminum 5 per cent tin alloy, 0.5 mm thick (Projection method) Tin appears dark at the grain boundaries. ×400, 20 kV, 5 min. exposure (*After* W. C. Nixon. Some New Applications of the X-ray Microscope. Proc. Int. Conference Electron Microscop. London 307. 1954. *J. Royal Microscop. Soc.* London).

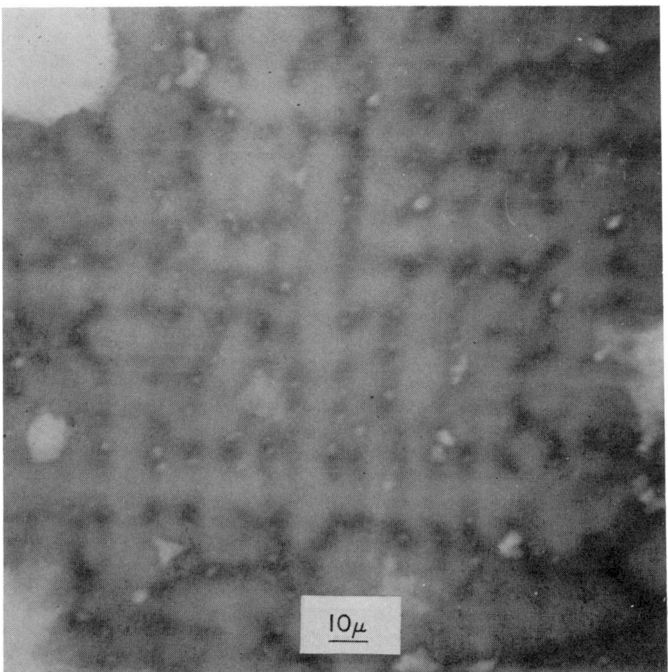

Figure 13.17. Microradiograph of porous bronze (95 per cent copper, 5 per cent tin). The tin appears in dentritic form and small white circles show the position of voids within the darker tin. (Projection method.) × 1000, 20 kV, 10 min exposure. (*By courtesy of* W. C. Nixon, in X-ray Microscopy by V. E. Cosslett and W. C. Nixon, Cambridge University Press, Cambridge (1960), and *Brit. J. Radiol.* **28.** 532. 1955.)

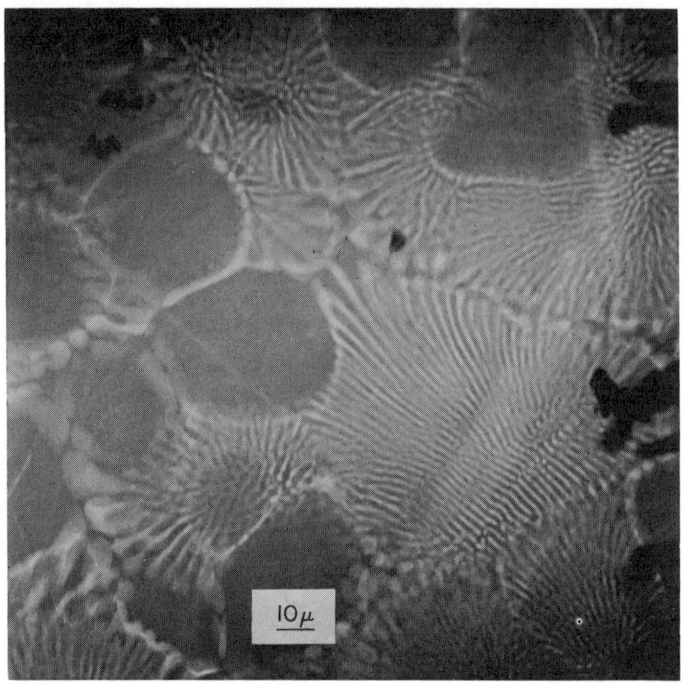

Figure 13.18. Microradiograph of zinc, 5 per cent aluminum alloy, dark areas represent zinc and the striped regions show an eutectoid mixture of the two. (Projection Method.) ×1000, 20 kV, 4 min. Exposure. (*After* W. C. Nixon, Some new applications of the X-ray Microscope. Proc. Int. Conference Electron Microscopy. London 307, (1954). *J. Royal Microscop. Soc.* London.)

2. The qualitative detection and distribution of specific elements.
3. The quantitative assessment of element content.

Practical examples of microradiography in metals entail for instance the study of inclusions of manganese sulfide and tungsten carbide in steel, using Co-radiation, and microporosity in an aluminum alloy, using Mn radiation. (R. S. Sharpe, 1957, 1960.) Segregation studies in cast iron have been made by J. Cohen and others (1955). Diffusion of metals, defects in castings, spot welding, and phase transformation have also been subjected to microradiographic investigations.

In industrial applications the study of the structure of paper, fibers, and fabrics has given useful information.

This review of applications of microradiography covers only a small section of papers published in this field.

BIBLIOGRAPHY

Cosslett, V. E., and Nixon, W. C. (1960). *X-Ray Microscopy*. Cambridge Monographs on Physics. The University Press, Cambridge.

Cosslett, V. E., Engström, A., and Pattee, H. H. (1957). *X-Ray Microscopy and Microradiography. 1st. Internat. Symposium*. Academic Press, New York.

Engström, A., Cosslett, V. E., and Pattee, H. H. (1960). *X-Ray Microscopy and Microanalysis. 2nd. Internat. Conference*. Elsevier, New York & Amsterdam.

Ely, R. V. (1913–1963). *Micro-X-ray Radiography and Analysis*. Bibliography. Guildford, England.

Pattee, H. H., Cosslett, V. E., and Engström, A. (1963). *X-Ray Optics and X-Ray Microanalysis. 3rd Internat. Conference*. Academic Press, New York.

Heinrich, K. F. J. (1964). Bibliography on Electron-Probe Microanalysis. *Proceedings of the Electron Microprobe Symposium. Washington*, Wiley, New York.

REFERENCES

Asunmaa, S. K. (1960). X-ray sensitive recording materials for electron-optical contrast. In *X-Ray Microscopy and Microanalysis*. Elsevier, Amsterdam. p. 66.

Asunmaa, S. K. (1963). Electronmicroscope enlargements of X-ray absorption micrographs. In *X-ray Optics and X-ray Microanalysis*, Academic Press, New York. p. 33.

Baez, A. V. (1957). Design of X-ray microscopic mirrors using an electronic computer. In *X-ray Microscopy and Microradiography*, Academic Press, New York.

Barclay, A. E. (1951, 1947, 1948). *Micro-Arteriography*. Blackwell, Oxford also *Brit. J. Radiol.*, **20**. 394, and *Am. J. Roentgenol.*, **60**. 1.

Castaing, R. (1951). *Applications of Electron-Probes to Local Chemical and Crystallographic Analysis*. Ph.D. Thesis, Paris.

Castaing, R. (1955). The electron beam microanalyser. *Rev. Mét. (Paris)* **50**. 625.

Castaing, R. (1960). The fundamentals of quantitative electron-probe microanalysis. *Proc. 9th Ann. Conf. X-ray Analysis*, University of Denver. Plenum Press. p. 351.

Castaing, R. (1963). X-ray microprobe techniques. In *X-Ray Optics and X-Ray Microanalysis*. Academic Press, New York. p. 263.

Cohen, J., Hall, E., Leonard, L., and Ogilvie, R. (1955). Investigation of segregation in cast iron by radiographic techniques. *Non-destructive Testing*. **13**. 33.

Combée, B., and Recourt, A. (1957, 1958). A simple apparatus for contact microradiography between 1.5–5 kV. *Philips Techn. Rev.* **19.** 221.

Cosslett, V. E., and Nixon, W. C. (1951, 1952). X-ray shadow microscope. *Nature,* **168.** 24. *Nature,* **170.** 436.

Cosslett, V. E., and Nixon, W. C. (1952). An experimental X-ray shadow microscope. *Proc. Roy. Soc.* B. **140.** 422.

Cosslett, V. E. (1965). X-ray microscopy. *Reports on Progress in Physics.* *Inst. Physics and Phys. Soc., London.* **28.** 381.

Duncumb, P. (1962). The design of the electron-probe microanalyser. *J. Inst. Metals.* **90.** 154.

Ehrenberg, W. (1949). X-ray optics. The production of converging beams by total reflection. *J. Opt. Soc. Am.* **39.** 741.

Ehrenberg, W. (1949). Imperfections of optical flats and their effect on the reflection of X-rays. *J. Opt. Soc. Am.* **39.** 746.

Ehrenberg, W., and White, M. (1957). Some physical aspects of contact microradiography. In *X-Ray Microscopy and Microradiography.* Academic Press, New York.

Ehrenberg, W., and White, M. (1959). Resolution in contact microradiography. *Brit. J. Radiol.* **32.** 380. 558.

Ehrenberg, W. (1947). X-ray Optics. *Nature, London.* 160. 330.

Engström, A. (1956). Historadiography. In G. Oster and A. W. Pollister (Ed.), *Physical Techniques in Biological Research,* Vol. III. Cells and Tissues. Academic Press, New York.

Engström, A. (1962). *X-ray Microanalysis in Biology and Medicine.* Elsevier, Amsterdam.

Engström, A., Björnerstedt, R., Clemedson, C. J., and Nelson, A. (1958). Microscopic distribution of mineral salts studied by X-ray microscopy. In *Bone and Radiostrontium,* Ch. 2B. Almguist and Wiksell. Stockholm. pp. 29–61.

Engström, A., and Lindström, B. (1951). The properties of fine-grained photographic emulsions used for microradiography. *Acta Radiol.* Vol. 35. 33.

Glocker, R., and Frohnmayer, W. (1925). X-ray spectroscopic determination of the proportion by weight of an element in mixtures and compounds. *Ann. Phys.* 76. 369.

Granboulan, P. (1963). Resolving power and sensitivity of a new emulsion in electron microscope autoradiography. *J. Roy. Microscop. Soc.* 81. 165.

Henke, B. L. (1963). Production, detection, and applications of ultra-soft X-rays. In *X-ray Optics and X-Ray Microanalysis.* Academic Press, New York. p. 157.

Kirkpatrick, P., and Baez, A. V. (1948). Formation of optical images by X-rays. *J. Opt. Soc. Am.* 38. 766.

Lindström, B. (1960). An evaluation of quantitative microradiographic procedures applied to biological and medical research. In *X-Ray Microscopy and Microanalysis.* Elsevier, Amsterdam.

Lindström, B. (1963). X-ray absorption micro-analysis. In *X-Ray Optics and Microanalysis*. Academic Press, New York.

McGee, J. F. (1957). A long wavelength X-ray reflection microscope. In *X-Ray Microscopy and Microradiography*. Academic Press, New York.

McGee, J. F. (1959). An introduction to total reflection X-ray microscopy. *Proc. 8th Ann. Conf. on X-Ray Analysis*. University of Denver. Plenum Press. p. 213.

Oderr, C. P. (1960). Lung structure studied by microradiography. In *X-ray Microscopy and Microanalysis*. Elsevier, Amsterdam. p. 290.

Pattee, H. H. (1957). The compound four-mirror reflection X-ray microscope. In *X-Ray Microscopy and Microradiography*. Academic Press, New York.

Pattee, H. H., and Warnes, R. (1957). Discussion remarks in high resolution X-ray microscopy. By W. A. Ladd and M. W. Ladd. In *X-Ray Microscopy and Microradiography*. Academic Press, New York. p. 387.

Recourt, A. (1957). Contact microradiography below 1 kV. In *X-Ray Microscopy and Microradiography*. Academic Press, New York. p. 234.

Salmon, J. (1957). Microradiographic techniques and plant histology. In *X-Ray Microscopy and Microradiography*. Academic Press, New York. p. 465.

Saunders, R. L. de C. H. (1961). X-ray projection microscopy of the skin. *Blood Vessels and Circulation*. Pergamon Press, London. pp. 38–56.

Saunders, R. L. de C. H. (1962). Medical applications of X-ray microscopy. *Canad. J. Surg.* 5. 299.

Saunders, R. L. de C. H., and Carvalho, V. R. (1963). X-ray microscopy of the microvascular system of the human lung. In *X-ray Optics and Microanalysis*. Academic Press, New York. p. 109.

Saunders, R. L. de C. H., Lawrence, J., and Maciver, D. A. (1957). Microradiographic studies of the vascular patterns in muscle and skin. In *X-ray Microscopy and Microradiography*. Academic Press, New York.

Sharpe, R. S. (1957). The microradiography of metals. Its advantages and limitations. In *X-ray Microscopy and Microradiography*. Academic Press, New York. p. 590.

Sharpe, R. S. (1960). Some applications of microradiography to fuel element inspection. *Fuel Element Sympos.* 2. Vienna, Academic Press, New York. p. 45.

Sharpe, R. S. (1961). X-ray microscopy. *Progress Non-destruct. Testing* Vol. III. Heywood, London.

Sissons, H. A., Jowsey, J., and Stewart, L. (1960). Quantitative microradiography of bone tissue. In *X-ray Microscopy and Microanalysis*. Elsevier, Amsterdam. p. 199.

Sissons, H. A., Jowsey, J., and Stewart, L. (1960). The microradiographic appearance of normal bone tissue at various ages. In *X-ray Microscopy and Microanalysis*. Elsevier, Amsterdam. p. 206.

14

Appendixes

14.1 APPENDIX TO CHAPTER 1 AND OTHERS

TABLE 14.1. Physical Constants

c	Velocity of light	2.99793×10^{10} cm/sec
h	Planck's constant	6.6252×10^{-27} erg. sec
N_0	Avogadro's number	6.02472×10^{23} mole^{-1}
V_0	Standard volume of a perfect gas	22420.7 cm^3/mole
R_0	Gas constant	8.31662×10^7 erg/deg./mole
m_e	Electron restmass	0.510985 MeV $= 9.1085 \times 10^{-28}$ gram
m_n	Neutron restmass	939.526 MeV $= 1.67470 \times 10^{-24}$ gram
m_p	Proton restmass	938.232 MeV $= 1.67243 \times 10^{-24}$ gram
e	Electronic charge	4.8028×10^{-10} esu
eV	Electron volt	1.6021×10^{-12} erg
Ci	Curie	3.7×10^{10} disint./sec
b	Barn	10^{-24} cm^2

TABLE 14.2. Alphabetical List of Elements

Element	Symbol	Atomic number Z
Actinium	Ac	89
Aluminum	Al	13
Americium	Am	95
Antimony	Sb	51
Argon	A	18
Arsenic	As	33
Astatine	At	85
Barium	Ba	56

TABLE 14.2. *Contd.*

Element	Symbol	Atomic number Z
Berkelium	Bk	97
Beryllium	Be	4
Bismuth	Bi	83
Boron	B	5
Bromine	Br	35
Cadmium	Cd	48
Calcium	Ca	20
Californium	Cf	98
Carbon	C	6
Cerium	Ce	58
Cesium	Cs	55
Chlorine	Cl	17
Chromium	Cr	24
Cobalt	Co	27
Copper	Cu	29
Curium	Cm	96
Dysprosium	Dy	66
Einsteineum	Es	99
Erbium	Er	68
Europium	Eu	63
Fermium	Fm	100
Fluorine	F	9
Francium	Fr	87
Gadolinium	Gd	64
Gallium	Ga	31
Germanium	Ge	32
Gold	Au	79
Hafnium	Hf	72
Helium	He	2
Holmium	Ho	67
Hydrogen	H	1
Indium	In	49
Iodine	I	53
Iridium	Ir	77
Iron	Fe	26
Krypton	Kr	36
Lanthanum	La	57
Lawrencium	Lw	103
Lead	Pb	82
Lithium	Li	3
Lutecium	Lu	71
Magnesium	Mg	12
Manganese	Mn	25
Mendelevium	Md	101
Mercury	Hg	80
Molybdenum	Mo	42
Neodynium	Nd	60
Neon	Ne	10

TABLE 14.2. *Contd.*

Element	Symbol	Atomic Number Z
Neptunium	Np	93
Nickel	Ni	28
Niobium	Nb	41
Nitrogen	N	7
Nobelium	No	102
Osmium	Os	76
Oxygen	O	8
Palladium	Pd	46
Phosphorus	P	15
Platinum	Pt	78
Plutonium	Pu	94
Polonium	Po	84
Potassium	K	19
Praseodynium	Pr	59
Promethium	Pm	61
Protractinium	Pa	91
Radium	Ra	88
Radon	Rn	86
Rhenium	Re	75
Rhodium	Rh	45
Rubidium	Rb	37
Ruthenium	Ru	44
Samarium	Sm	62
Scandium	Sc	21
Selenium	Se	34
Silicon	Si	14
Silver	Ag	47
Sodium	Na	11
Strontium	Sr	38
Sulfur	S	16
Tantalum	Ta	73
Technetium	Tc	43
Tellurium	Te	52
Terbium	Tb	65
Thallium	Tl	81
Thorium	Th	90
Thulium	Tm	69
Tin	Sn	50
Titanium	Ti	22
Tungsten	W	74
Uranium	U	92
Vanadium	V	23
Xenon	Xe	54
Ytterbium	Yb	70
Yttrium	Y	39
Zinc	Zn	30
Zirconium	Zr	40

Range-Energy Relationship for electrons, α-particles and protons.

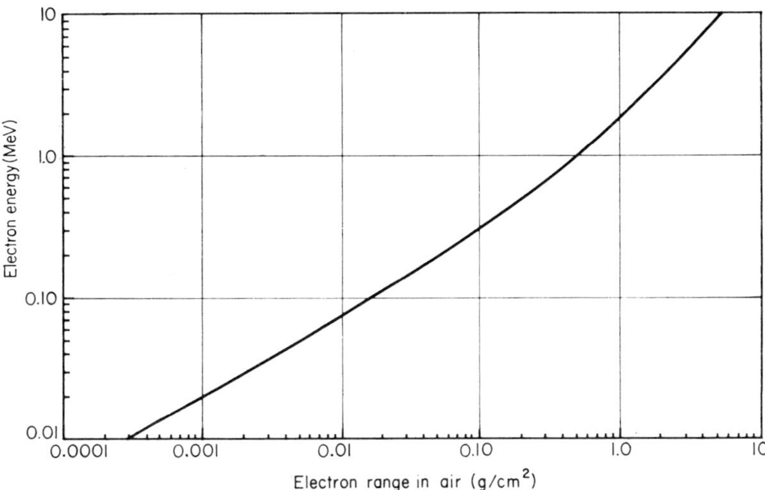

Figure 14.1. Range of electrons in air (gm/cm²) as a function of electron energy in MeV. (*After* I.C.R.U. Handbook No. 62. 1956. National Bureau of Standards, Washington D.C.)

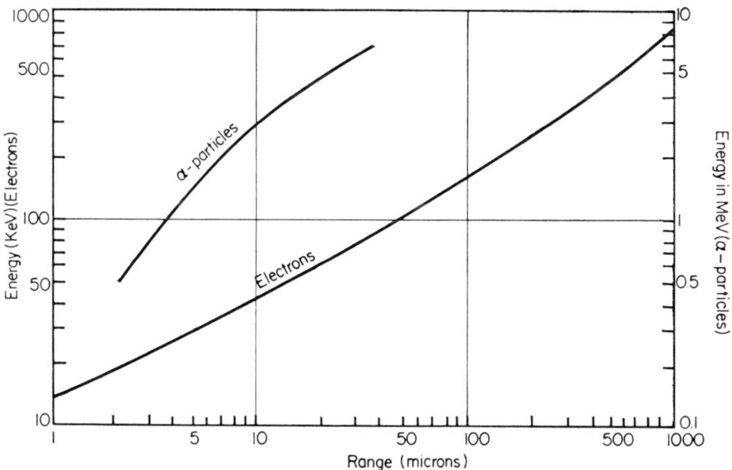

Figure 14.2. Range of electrons and α-particles in nuclear track type emulsions (of physical density 3.9 grams per cm³) as a function of particle energy in KeV (electrons) and in MeV (α-particles).

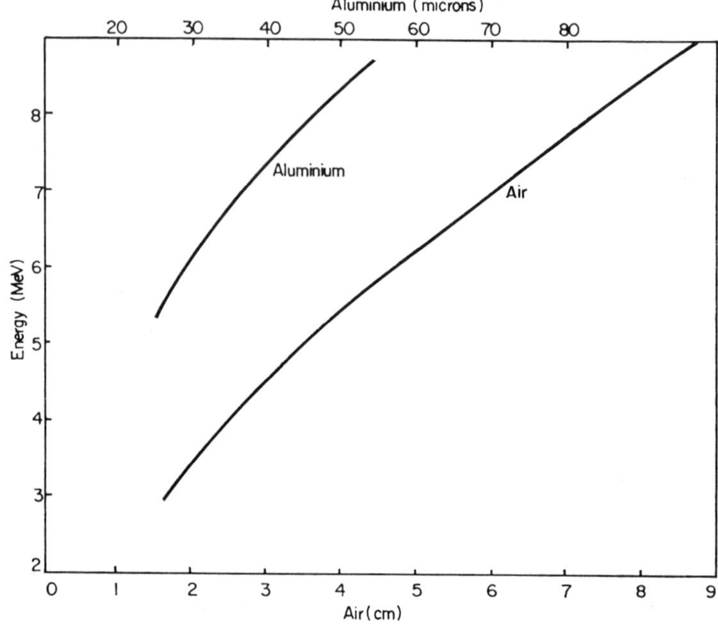

Figure 14.3. Range of α-particles in air and in aluminum as a function of energy (in MeV). The range in aluminum is expressed in microns (1 micron $= 0.001$ millimeters). The range in air is expressed in cm.

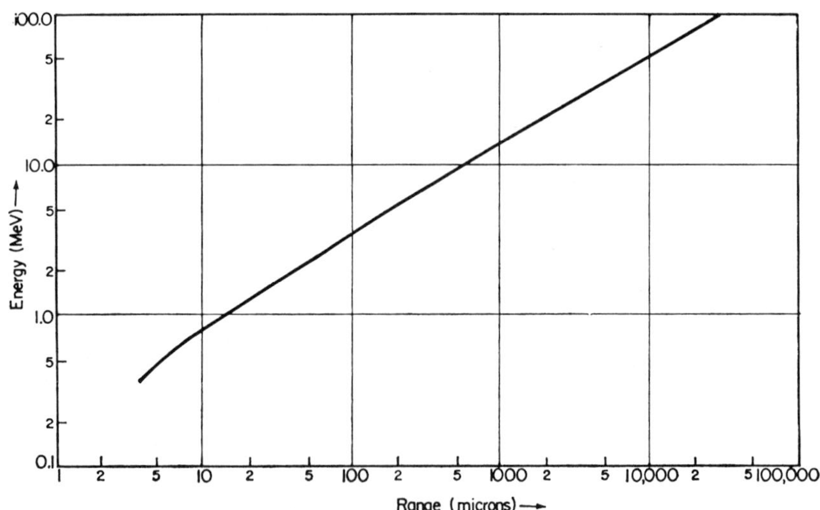

Figure 14.4. Range of protons in microns in nuclear track type emulsion as a function of proton energy. (Average physical density of emulsion 3.9 grams/cm³.)

TABLE 14.3.

MASS ABSORPTION COEFFICIENTS FOR X AND γ RAYS

Radiation traversing a layer of substance is reduced in intensity by a constant fraction μ per centimeter. After penetrating to a depth x the intensity is $I = I_0 e^{-\mu x}$ where I_0 is the intensity at the surface. μ/ρ is the mass absorption coefficient where ρ is the density of the material.

Values of μ/ρ for λ = .005 Å to λ = 44.6 Å. Where two values of μ/ρ for one value of λ occur they represent the maximum and minimum values at an absorption discontinuity.

Compiled by S. J. M. Allen

λ = 44.6 − 2.74 Å   λ = 2.50 − .900 Å

From Handbook of Chemistry and Physics. 48th Edition. pp. E 121–122.
By courtesy of "The Chemical Rubber Company", Cleveland, Ohio, U.S.A.

λ, Å	He	C	N	O	Ne	Al	S	Cl	A
44.6	3600	2170	3850	5765	13100	....	....	....	45700
11.88	....	....	....	....	6850	850	....	....	....
9.87	....	1063	1796	2540	4310	500	1320	1570	1860
8.32	....	656	1109	1585	2750	330	794	962	1160
7.94	....	....	....	....	....	280 3700	....	....	....
6.97	....	390	645	976	1727	2800	500	610	748
5.39	....	185	312	476	865	1450	249	310	360
5.17	....	160	273	413	763	1350	221	277	324
5.01	....	....	....	....	....	....	210 2260	....	....
4.38	....	....	....	....	....	....	....	178 1830	....
4.36	....	97.8	166	258	478	815	1570	1800	202
4.15	....	84.6	144	222	416	720	1350	1476	174
3.93	....	71.0	121	189	356	635	1175	1256	153
3.87	....	....	....	....	....	....	....	....	148 1460
3.69	....	....	....	....	....	....	....	....	....
3.59	....	55.2	96	150	279	500	928	966	1215
3.51	....	....	....	....	....	....	....	....	....
3.38	....	46.0	79.5	117	231	425	795	880	1025
3.35	....	43.0	....	....	....	417	780	870	1015
3.24	....	....	....	....	....	....	....	....	....
3.03	....	35.0	....	84.0	175	323	595	670	760
2.74	....	25.0	....	60.0	135	250	454	520	600

λ, Å	Mg	Al	S	Cl	A	Ca	Fe	Ni	Cu
2.50	161	193	355	400	475	620	147	180	197
2.29	....	150	285	315	355	480	115	137	153
1.93	77.2	93.5	173	198	235	306	71.2	89.5	96.2
1.74	....	83.0	....	....	....	....	54 465	....	....
1.656	....	60.7	110	126	143	195	410	59.2	63.5
1.539	40.8	49.0	91	103	114	163	325	48.0	50.9
1.484	....	....	....	....	....	....	....	40.5 338	....
1.432	....	40.0	75	85	93	130	285	325	42
1.389	31.5	36.8	68.5	76.7	85.7	125	252	275	38.5
1.377	....	....	....	....	....	....	....	....	37.0 307
1.293	....	29.8	55.3	60	72	102	212	233	260
1.280	....	28.8	....	....	....	....	....	225	252
1.235	21.4	26.3	49.5	55.5	62.5	90	181	208	230
1.104	....	18.6	38.0	44	50	67	135	155	175
1.071	....	....	....	....	....	....	....	....	....
1.038	....	....	....	....	....	....	....	....	....
1.000	11.8	14.12	26.7	29.7	34.5	49	100	121	130
.980	....	....	....	....	....	....	....	....	....
.949	....	12.0	22.0	24.5	....	42	86	99	114
.932	....	....	....	....	....	....	....	....	....
.900	....	10.4	....	....	....	....	74.5	86.5	98.5

λ = 44.6 − 2.74 Å

λ, Å	Fe	Ni	Cu	Zn	Kr	Ag	Sn	Xe	Pt	Au
44.6					31800			6740		12500
11.88		6900	7550							
9.87		4540	5030			2700			2440	
8.32		3140	3450			1800			1560	
7.94										
6.97		2000	2130			1300			1190	
5.39		1250	1290			845			1645	
5.17		1150	1190			790				
5.01										
4.38										
4.36	610	715	760	910		535	640			
4.15	540	630	690	820		461	550		1290	
3.93	470	555	610	715		408	490			
3.87										
3.69						354				
						1410				
3.59	375	450	495	575		1360			1370	
3.51						1300				
						1510				
3.38	320	380		495						
3.35	312	375	404	480		1310			1120	
3.24						1230				
						1440				
3.03	245	290	315	375		1290			939	
2.74	185	239	262	283		925			756	

λ = 2.50 − .900 Å

λ, Å	Zn	Br	Mo	Ag	Sn	I	W	Pt	Au	Pb
2.50	228			710	850			596		
2.29	180			550	670			480		
1.93	110			405	470		300	358	385	428
1.74										
1.656	72.5			285				228		
1.539	58.6	89		217	247	290	176	202	213	230
1.484										
1.432	49.3			192	220		130	172	179	202
1.389	45.2			174	209			155	166	185
1.377										
1.293	39			146	176			132	138	154
1.280	36			127	146					
	287									
1.235	250			125	140		95	115	122	137
1.104	208			96.5	115			99	107	120
1.071								77.5		
								198		
1.038									76.5	
									194	
1.000	145		52	73.0	86.0			165	174	75
.980								155	168	73
.949	129			63.0	75.5			146	156	68
										168
.932								136	148	159
								184		
.900	112	150		54.2	65.0			168	134	145
									182	

λ = 2.50 − .900 Å

λ, Å	H	Li	Be	B	C	N	O	Ne	Na
2.50	.52	4.0	6.1	9.1	17.8		44.5	100	128
2.29					15.0		36.4	75.5	
1.93	.50	2.10	3.05	4.7	8.75	14.0	21.7	49.0	61.3
1.74									
1.656									
1.539	.48	1.10	1.60	2.45	4.52	7.45	11.1	24.0	32.1
1.484									
1.432									
1.389	.47	.86	1.25	1.87	3.35	5.50	8.1	17.0	23.4
1.377									
1.293									
1.280									
1.235	.46	.67	.95	1.35	2.42	3.95	5.7	12.4	17.1
1.104									
1.071									
1.038									
1.000	.45	.43	.55	.76	1.36	2.10	3.13	6.5	8.8
.980									
.949					1.20				
.932									

λ = .892 − .184 Å

λ, Å	H	Li	Be	B	C	N	O	Ne	Na	Mg	Al
.892											9.75
.880	.440	.350	.425	.580	.990	1.50	2.20	4.55	6.10	8.34	
.862					.907						8.85
.850					.814						7,85
.814					.750						6.86
.780					.598						
.710	.435	.260	.315	.365	.550	.870	1.22	2.50	3.30	4.30	5.22
.680					.467						4.52
.631	.435	.225	.255	.305	.370	.610	.900	1.80	2.30	3.0	3.73
.618											
.560											2.60
.497	.435	.198	.210	.220	.315	.400	.520	.930	1.18	1.52	1.90
.485					.308						1.77
.476	.430			.215	.304		.485				1.74
.424											1.23
.417	.390	.180	.185	.198	.256	.310	.372	.580	.750	.940	1.170
.380					.230						.950
.331											
.260	.385	.156	.166	.175	.185	.200	.210	.270	.305	.343	.402
.220					.178						.300
.200	.375	.151	.160	.165	.175	.180	.183	.210	.225	.250	.270

MASS ABSORPTION COEFFICIENTS FOR X AND γ RAYS (Continued)

λ = .892 — .184 A

λ, Å	S	Cl	A	Ca	Fe	Ni	Cu	Zn	Br	Sr	Mo
.892											
.880	18.2	20.7	24.0	34.8	69.5	82	91.2	103			36.0
.862											
.850					63.5	74	84.5	96.5			
.814					57	66	75.7	86			28
.780					50.5	59.5	67.5	77			
.710	9.90	11.6	13.0	18.6	38.5	48.1	51.0	59.0	80	106.	19.9
.680					32.7	41	45.3	52.7			
.631	6.90	8.40	9.80	13.3	27.0	34	36.2	41.0	56.8	72.5	15.0
.618											12.5
											88.0
.560					18.2	24	25.5	30.7			
.497	3.50	4.20	5.0	6.60	13.9	17.9	18.4	21.0	32.0	40.5	50.2
.485					12.4	15.4	16.9	19.5			
.476							16.6				42
.424											
.417	2.10	2.47	2.95	3.97	8.45	10.5	11.45	12.3	19.0	24.0	30.0
.380					6.32	7.70	8.42	9.95			22
.331											
.260	.650	.750	.850	1.10	2.28	2.89	3.16	3.58	5.30	6.50	8.20
.220					1.42	1.80	2.00	2.32			
.200	.400	.445	.500	.630	1.10	1.45	1.55	1.78	2.4	3.32	4.30
.184						1.24					

λ = .178 — .005 A

λ, Å	H	Li	Be	B	C	N	O	Ne	Na	Mg	Al
.178					.164						.235
.175	.360	.144	.150	.155	.163	.166	.169	.185	.195	.205	.228
.158					.160						.208
.155											
.146	.340				.155		.162		.170	.176	.195
.142	.330				.153						.191
.130	.320	.132		.149	.152		.157		.160	.168	.186
.120					.150		.154			.163	.172
.113	.310				.147		.153		.155	.160	.166
.107											
.098	.280	.125		.138	.142		.144		.150	.152	.156
.080	.255				.137						.146
.072	.250	.118		.132	.136		.137		.139	.140	.143
.064	.245	.110		.126	.130		.130		.130	.130	.130
.050						120					.115
.040	.205				.110						.106
.030	.180				.095						.093
.024	.165				.080						.079
.010	.117				.059						.058
.005	.078				.0385						.0380

$\lambda = \cdot892 - .184$ Å

λ, Å	Ag	Sn	I	Ba	Ta	W	Pt	Au	Pb	Bi	U
.892	....	....	....	....	....	....	165 201	178	142	....	....
.880	50	60	....	....	....	....	195	170	135	....	....
.862	....	....	....	....	....	....	185	163 193	130	....	....
.850	46	56	....	....	....	....	179	186	124	....	....
.814	41	49.5	....	....	....	....	160	167	111 150	....	....
.780	36	44.5	....	....	....	....	144	150	136 166	....	....
.710	27.5	34.0	38.5	42.0	100	104	115	120	136	....	....
.680	23.5	28.4	....	....	....	....	102	108	120	....	....
.631	19.6	23.0	26.4	31.1	72	75	84.5	87	98	....	....
.618	....	....	....	....	....	....	....	....	....	....	....
.560	13.3	16.2	....	....	....	....	62	66	75	....	....
.497	10.5	11.8	15.6	17.8	36	38	47	48.5	52.8	....	....
.485	9.8 62.5	11.1	....	....	....	....	....	....	....	....	....
.476	60	....	....	....	....	....	42	....	47.5	....	....
.424	43.5	8.0 46.6	....	....	....	....	....	....	....	....	....
.417	41	45	9.2	10.5	21.5	22.5	27.4	28.4	32.0	....	....
.380	31.2	34	....	....	....	17.3	21.1	22	26.4	27.8	....
.331	21.7	24.5	....	5.4 28.0	....	....	....	....	18.1	19.5	....
.260	11.4	12.8	14.2	16.1	6.7	6.85	8.0	8.3	10.0	11.0	....
.220	7.05	7.80	....	....	....	4.25	5.25	5.50	5.92	6.4	....
.200	5.48	6.20	7.0	8.0	3.4	3.50	4.25	4.40	4.90	5.1	5.40
.184	4.45	....	....	....	2.8 11.8	....	3.45	3.60	4.05	4.2	....

$\lambda = .178 - .005$ Å

λ, Å	S	Cl	A	Ca	Fe	Ni	Cu	Zn	Br	Sr	Mo
.178	....	....	....	....	....	....	1.15	....	....	....	....
.175	.335	.341	.400	.460	.800	1.05	1.12	1.26	1.90	2.24	2.95
.158	....	....	....	....	.640	.815	.862	.990	....	....	....
.155	....	....	....	....	....	....	....	....	....	....	....
.146	.249	.280	....	.345	.520	....	.680	....	....	....	....
.142	....	....	....	....	.515	.630	.670	.780	....	....	1.55
.130	.220	.230	....	.290	.424	....	.551	....	....	....	....
.120	.200	....	....	....	.368	.430	.455	.537	....	....	....
.113	.189	.195	....	.230	.337	....	.422	....	....	....	....
.107	....	....	....	....	....	....	....	....	....	....	....
.098	.166	.176	....	.200	.265	....	.325	....	....	....	.790
.080	....	.164	....	....	.235	.264	.268	.308	....	....	....
.072	.150	.158	....	.180	.202	....	.232	....	....	....	....
.064	.139	.142	....	.155	.178	....	.198	....	....	....	.413
.050	....	....	....	....	.140	....	.155	....	....	....	....
.040	....	....	....	....	.118	....	.126	....	....	....	....
.030	....	....	....	....	.095	....	.100	....	....	....	....
.024	....	....	....	....	.080	....	.081	....	....	....	....
.010	....	....	....	....	.058	....	.057	....	....	....	....
.005	....	....	....	....	....	....	.0380	....	....	....	....

λ, Å	Ag	Sn	I	Ba	Ta	W	Pt	Au	Pb	U
.178	....	....	....	....	....	2.7 11.3	3.16	3.30	3.55	....
.175	3.96	4.50	5.10	5.70	10.0	10.5	2.97	3.13	3.48	3.95
.158	3.00	3.40	....	....	....	8.6	2.45 9.40	2.43	2.60	....
.155	....	....	....	....	....	....	....	2.30 8.80	....	....
.146	2.48	2.66	....	....	6.75	....	7.60	7.85	2.35	2.70
.142	2.31	2.64	....	....	....	6.75	7.20	7.33	2.10 7.75	....
.130	1.97	2.12	....	....	5.10	....	6.30	6.40	6.55	2.20
.120	1.61	1.77	....	2.20	....	4.60	4.92	4.98	5.20	1.90
.113	1.47	1.60	....	....	3.80	....	4.40	4.50	4.75	....
.107	....	....	....	....	....	....	....	....	....	1.62 4.65
.098	1.05	1.17	....	....	2.80	....	3.15	3.21	3.50	3.90
.080	.73	.79	....	....	....	2.30	2.40	2.42	2.50	2.70
.072	.584	.614	....	....	1.75	....	2.00	2.05	2.10	2.25
.064	.465	.490	....	....	1.35	....	1.52	1.55	1.64	1.80
.050	....	.320	....	....	....	....	.86	.88	1.00	....
.040	....	.21	....	....	....	....	....	....	.62	....
.030	....	.13	....	....	....	....	....	....	.38	....
.024	....	.10	....	....	....	....	....	....	.21	....
.010	....	.060	....	....	....	....	....	....	.071	.082
.005	....	.0385	....	....	....	....	....	....	.0425	.044

TABLE 14.4. Some Slow Neutron Resonance Cross-Sections

Natural element	Resonance energy (eV)	Cross-sections in barns
Cadmium	0.176	7,800
Gadolinium	0.028	46,000
Gold	4.8	15,000
Indium	1.4	28,000
Rhodium	1.28	44,000
Silver	5.1	7,600
Samarium	0.096	15,500

14.2 APPENDIX TO CHAPTER 6. PHOTOGRAPHIC DOSIMETRY (SEE p. 208)

Filtration Required to Obtain Narrow Energy Bands of Radiation

When calibrating photographic material using X-rays of various qualities, it is useful to know the type and thickness of filters to be employed in order to obtain narrow energy bands of radiation from the heterogeneous X-ray beams generated by X-ray apparatus.

In the table below approximate data are quoted as a guide only, as the values change with the type of X-ray unit, its calibration and inherent filtration in use. The following values were obtained from ionization measurements using increasing filter thicknesses as described on p. 28.

The evaluation of these data is based on the use of a fully rectified X-ray unit. From the approximate half value layer thicknesses in mm Cu thus evaluated, the equivalent kilovoltages (KeV) were derived from

TABLE 14.5. Filtration Required at Various kVp to Obtain Narrow X-ray Energy Bands

kVp	Filter in mm Cu	Filter in mm Al	HVL mm Cu	Narrow energy bands (approx. keV)
50	0.3	0.5	0.08	33
60	0.35	0.5	0.17	42
80	2.0	—	0.50	62
110	3.0	—	0.85	75
165	4.0	—	1.28	89
200	7.0	—	2.75	140

Figure 14.5. This figure shows the relationship between the half value layer in mm Cu and KeV. Figure 14.5 is based on published values of the mass-absorption coefficients of copper for various wavelengths in Å. (see Table 14.3.) Using the μ/ρ values, the linear absorption coefficients were determined and the Half Value Layers in mm Cu were derived from HVL $= 0.693/\mu$, see equation 1.20 p. 27.

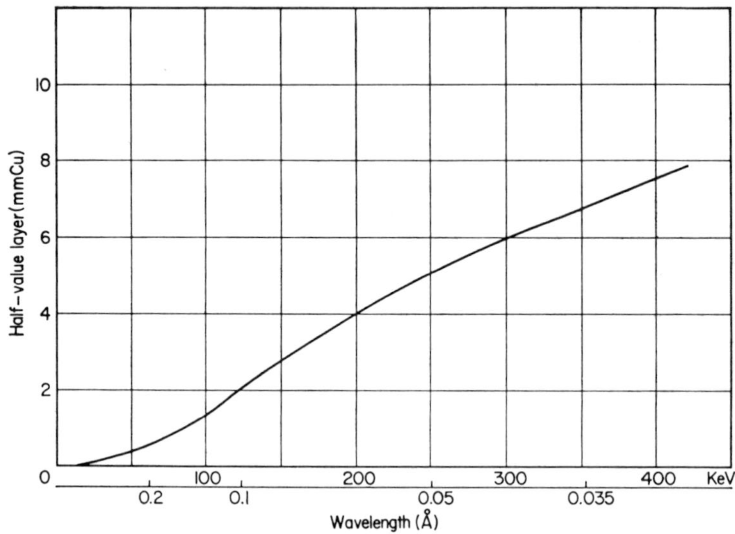

Figure 14.5. Relationship between half value layer in mm Cu and equivalent wavelength in Å and KeV.

14.3 APPENDIX TO CHAPTER 8
MEDICAL RADIOGRAPHY (p. 321)

Radiographic Exposure Chart

For those readers who are unfamiliar with medical radiography, but would like to know some of the average exposure conditions used in this field, a typical exposure chart covering some of the important regions of the human body is given on the next page.

The conditions set out in this table should be taken as a guide only, as they vary not only amongst various hospitals, but are also dependent on apparatus circuit, type of film and screens in use, processing procedures etc. The data quoted in the Table 14.6 refer to a four valve X-ray unit. The no-screen X-ray films refer to the fast type medical no-screen film and the screen films to a medium variety in conjunction with medium fast intensifying screens. The meaning of the abbreviations used in the column under 'projection', in Table 14.6 is given below.

D.P.	Dorsi-Plantar
P.A.	Postero-anterior
A.P.	Antero-posterior
Lat.	Lateral
F.F.D.	Focus-film distance

TABLE 14.6. Exposure Chart (Medical Radiography)

Region	Projection	kVp	mAs.	F.F.D. inch	F.F.D. cm	Film	Movable Grid
Hand	P.A.	55	5	30	75	No-scr.	No
Hand	Lat.	60	10	30	75	No-scr.	No
Wrist	P.A.	60	8	30	75	No-scr.	No
Elbow	A.P. and Lat.	60	18	30	75	No-scr.	No
Shoulder	A.P.	60	20	36	90	Scr. Film	No
Foot	D.P.	70	8	36	90	No-scr.	No
Knee	A.P.	60	8	36	90	Scr. Film	No
Pelvis	A.P.	65	80	36	90	Scr. Film	Yes
Spine (Dorsal)	A.P.	75	75	36	90	Scr. Film	Yes
Skull	Lat.	60	100	36	90	Scr. Film	Yes
Lung	P.A.	60	12	60	150	Scr. Film	No
Stomach	P.A.	80	35	30	75	Scr. Film	Yes
Colon	P.A.	80	50	30	75	Scr. Film	Yes
Gall Bladder	P.A.	70	50	36	90	Scr. Film	Yes
Pregnancy	P.A.	80	50	36	90	Scr. Film	Yes

14.4 APPENDIX TO CHAPTER 9
THE PROCESSING OF RADIOGRAPHS

Processing formulae as recommended by Kodak Ltd., England

(1) *D19 developer (for X-ray films)*

2.0 grams	Elon
90.0 grams	Sodium sulfite (anhydrous)
8.0 grams	Hydroquinone
45.0 grams	Sodium carbonate (anhydrous)
5.0 grams	Potassium bromide
	Water to make up to 1000 c.c.

The chemicals should be dissolved in the order given above. Use without dilution for X-ray films. Recommended minimum development time 5 min at 20°C (68°F).

(2) *D19 R Replenisher (for X-ray films)*

4.5 grams	Elon
90.0 grams	Sodium sulfite (anhydrous)
17.5 grams	Hydroquinone
45.0 grams	Sodium carbonate (anhydrous)
7.5 grams	Sodium hydroxide
	Water to make up to 1000 c.c.

Dissolve the chemicals in the order given above. The D19 developer should be maintained at a constant level in the tank by frequent additions of the above replenisher solution. The total volume of replenisher used should not be greater than the original volume of the developer.

(3) *Stop Bath*

1000 c.c. of water
17 c.c. glacial acetic acid

(4) *Acid Hardening Fixing Bath*

240 grams	Sodium thiosulfate (hypo crystals)
15 grams	Sodium sulfite (anhydrous)
17 c.c.	Acetic acid (glacial)
7.5 grams	Boric acid (cryst.)
15.0 grams	Potassium alum
	Water to make up to 1000 c.c.

The chemicals should be dissolved in the order given above.

(5) *Hardener Stock Solution*

600 c.c. Water at about 52°C (125°F)
75 grams	Sodium sulfite (anhydrous)
88 c.c.	Acetic acid (glacial)
37.5 grams	Boric acid (cryst.)
75.0 grams	Potassium alum
	Water to make up to 1000 c.c.

The chemicals should be dissolved in the order given above.

(6) *Tropical Hardening Bath*

(Kodak Ltd. Formula SB4)

1000 c.c.	Water
30 grams	Potassium chrome alum
140 grams	Sodium sulfate (cryst.)

Agitate films for 30–45 seconds and bathe for 3 minutes. Replace hardening bath after 20 (8 × 10 inch) films per gallon have been treated, otherwise scum markings are likely to occur.

14.5 APPENDIX TO CHAPTER 10
INDUSTRIAL RADIOGRAPHY

Derivation of K-factor (See p. 432)

As pointed out in Chapter 10 page 432 the K-factor is defined as the rate of exposure in roentgens per hour at 1 cm from a γ-ray point source of 1 millicurie activity. This factor may be obtained experimentally, but it may also be derived theoretically using the following arguments.

We assume first for simplicity that the decay scheme of the isotope in question involves only one photon per disintegration. We know that the number of disintegrations per millicurie per hour for any radioisotope is $3.7 \times 10^7 \times 3600 = 13.32 \times 10^{10}$.

If we consider a point source of 1 millicurie, then the photon flux per cm^2 hour at 1 cm distance is

$$\frac{13.32 \times 10^{10}}{4\pi} = 1.06 \times 10^{10} \qquad (14.1)$$

From equation 4.24 page 149, we find that the photon flux per cm^2 roentgen is

$$\frac{N}{R} = \frac{54.31 \times 10^6}{(\mu_{en}/\rho)_{(air)} \times W} \qquad (14.2)$$

where $(\mu_{en}/\rho)_{(air)}$ is the mass energy absorption coefficient for air and W the energy in units of MeV. By dividing the photon flux per cm^2 hour, i.e. 1.06×10^{10} by the photon flux per cm^2 roentgen, we obtain

$$K = \frac{1.06 \times 10^{10} (\mu_{en}/\rho)_{(air)} \times W}{54.31 \times 10^6}$$

$$= 195.3 (\mu_{en}/\rho)_{(air)} W \text{ R/hr per mCi at 1 cm} \qquad (14.3)$$

The K-factor as a function of photon energy is plotted in Figure 14.6.

It is seen that the K-factor decreases initially with increasing photon energy from 0.01 to 0.065 MeV and then rises again steeply with photon energy. The reason for the drop is caused by the decrease in

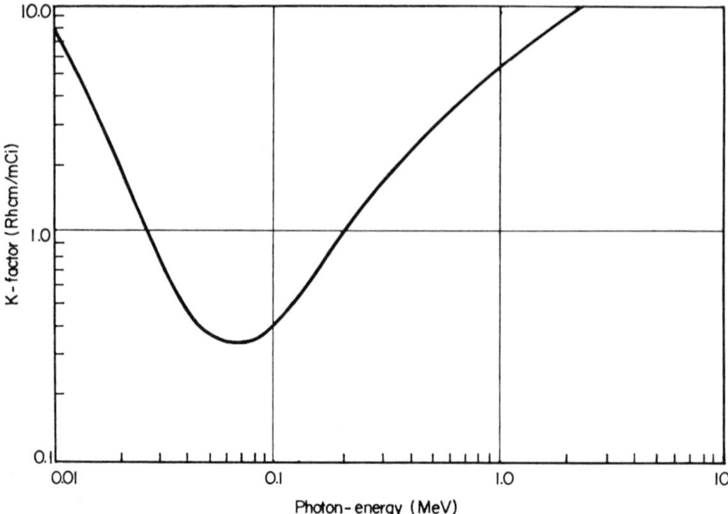

Figure 14.6. K-factor as a function of photon energy in (MeV) (calculated).

energy absorption coefficient for air from 0.01–0.08 MeV (see Figure 14.7). The rise of the K-factor is essentially caused by the increasing photon energy, as $(\mu_{en}/\rho)_{(air)}$ then remains fairly constant.

As pointed out before, the derivation refers to 1 photon per disintegration. If several photons per disintegration, are involved, as for instance in the case of ^{60}Co and many other isotopes, the total K factor is the sum of the individual K-factors, as is shown in the following example for ^{60}Co.

The two γ-ray photons emitted by ^{60}Co have energies of 1.17 and 1.33 MeV and using Figure 14.7 and equation (14.3) we find that the number of roentgens per mCi hour at 1 cm is

$$\text{at } 1.17 \text{ MeV: } 195.3 \, (\mu_{en}/\rho)_{(air)} \times W = 195.3 \times 0.0275 \times 1.17 = 6.28$$
$$\text{at } 1.33 \text{ MeV: } 195.3 \, (\mu_{en}/\rho)_{(air)} \times W = 195.3 \times 0.0270 \times 1.33 = \underline{7.01}$$
$$\overline{13.29}$$

Thus the total rate of exposure in roentgens per mCi/hr at 1 cm distance is 13.29. The $(\mu_{en}/\rho)_{(air)}$ values which have been introduced in the above calculations are taken from the diagram Figure 14.7 which is plotted from data of the National Bureau of Standards, Handbook 85 (1964).

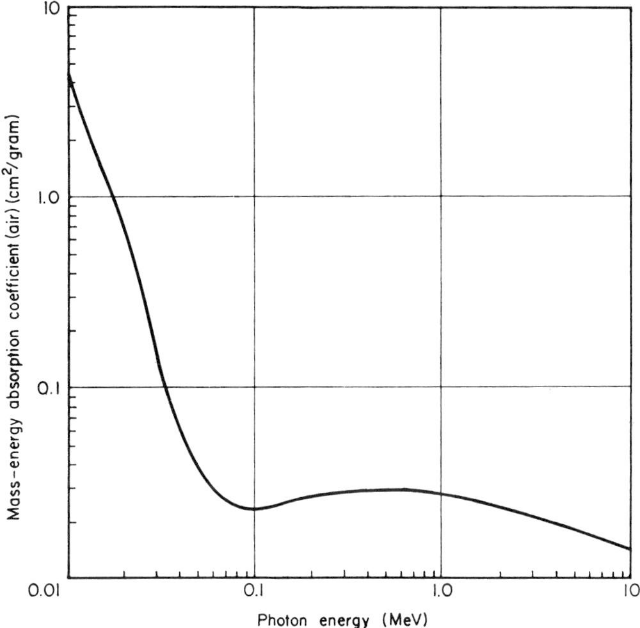

Figure 14.7. Mass-energy absorptions coefficients (cm^2/ft) for air as a function of photon energy (MeV). (This graph is plotted from tables of the National Bureau of Standards, Handbook 85, 1964.)

14.6 APPENDIX TO CHAPTER 12
AUTORADIOGRAPHY

Subbing of Glass Slides for Wet-Adhesion of Stripping Emulsions

As mentioned on page 531 the glass slides used for stripping emulsions in autoradiography should preferably be 'subbed' in order to achieve wet adhesion of the emulsion during the processing procedure. The following method which is given in Wall's 'Photographic Emulsions' Chapman and Hall (1929) is found suitable for this purpose. The glass slides should be cleaned by soaking in

100 grams	Potassium bichromate
100 c.c.	Sulfuric acid (conc.)
	Water to make up to 1000 c.c.

Dissolve the bichromate in the water, then add the sulfuric acid slowly to the cold solution with constant stirring. The slides should

then be washed and be dipped in the following solution at about
70°F (21°C)

5 grams	Gelatin
0.5 grams	Chrome alum
	Water to make up to 1000 c.c.

The slides should be drained in a rack and dried. The specimens
should then be mounted directly on these 'subbed' slides.

Subject Index